QUARTÄRPALÄONTOLOGIE

Abhandlungen und Berichte des Instituts für Quartärpaläontologie Weimar

Herausgeber: H. D. Kahlke, Weimar

Band 6 · 1986

AKADEMIE-VERLAG BERLIN

Quartärpaläontologie · Berlin 6 (1986) · S. 1—243 · 66 Abbildungen · 21 Tafeln · 47 Tabellen

Gedenkmünze auf der Titelseite

Anläßlich des Todes von WILHELM ERNST TENTZEL (1659—1707, königlicher und chursächsischer Rat und Histograph zu Gotha) ausgegebene Silbermedaille. Der Medailleur war CHRISTIAN WERMUTH (1661—1739).

WILHELM ERNST TENTZEL war der Verfasser der später so berühmt gewordenen Schrift: „Epistola de Sceleto elephantino Tonnae nuper effosso ad Virum toto orbe celeberrimum ANTONIUM MAGLIABECHIUM" (Gotha 1696, Jena 1696) als Antwort auf die „Kurtze doch ausführliche Beschreibung des Unicornu fossilis oder gegrabenen Einhorns / welches in der Herrschaft Tonna gefunden worden / Verfertiget von dem Collegio Medico in Gotha / den 14. Febr. 1696".

Mit dieser Schrift von 1696 begründete WILHELM ERNST TENTZEL — obgleich er sich damals nicht gegen die Ansichten des Collegio Medico zu Gotha durchzusetzen vermochte — die moderne Quartärpaläontologie in Thüringen.

Gesamtredaktion: Dipl.-Geol. ERIKA BIELEFELDT, Akademie-Verlag Berlin

Die russischen Texte wurden redigiert von:

Dr. A. SHER, Akademie der Wissenschaften der UdSSR, Moskau, Institut für Evolutionäre Morphologie und Ökologie der Tiere

Die englischen Texte wurden redigiert von:

Dr. A. CURRANT, British Museum (Natural History), London, Großbritannien
Dr. A. W. GENTRY, British Museum (Natural History), London, Großbritannien
Dr. A. HALL, University of York, Großbritannien
Dr. A. LISTER, University of Cambridge, Großbritannien
Dr. T. STUART, University of Cambridge, Großbritannien
Dr. A. SUTCLIFFE, British Museum (Natural History), London, Großbritannien
Dr. CH. TURNER, University of Cambridge, Großbritannien

Die französischen Texte wurden redigiert von:

Prof. Dr. E. HEINTZ, C.N.R.S., Paris, Frankreich
zur Zeit C.N.R.S. — Bureau de Bonn, BRD

ISBN 3-05-500087-0
ISSN 0138—3116

Erschienen im Akademie-Verlag Berlin, DDR-1086 Berlin, Leipziger Straße 3—4

Lizenznummer: 202 · 100/450/86
Printed in the German Democratic Republic
Gesamtherstellung: VEB Druckerei „Thomas Müntzer", 5820 Bad Langensalza
LSV 1475
Bestellnummer: 763 559 9 (2164/6)
09500

Neue Forschungsergebnisse zur Paläontologie des Tertiärs und Quartärs im europäisch-asiatischen Raume

Inhalt

Praeovibos priscus STAUDINGER
(Ölgemälde, Prof. Dr. C. C. FLEROV, 1978)

Ovibos moschatus (ZIMMERMANN)
(Ölgemälde, Prof. Dr. C. C. FLEROV, 1978)

Zum Geleit

Der vorliegende Band umfaßt eine Reihe von Beiträgen zur Paläontologie des Tertiärs und Quartärs des europäisch-asiatischen Raums, die von Schülern und Freunden dem Altmeister der sowjetischen Paläontologie, Prof. Dr. C. C. FLEROV, *Moskau, gewidmet sind.*

Nur wenige Tage nach einem seiner zahlreichen Arbeitsbesuche in unserem Institut in Weimar verstarb unser ausländischer Mitarbeiter und Freund am 26. Juli 1980. Sein umfassendes Wissen, seine künstlerische Phantasie, gepaart mit großem Können, und sein steter Humor werden uns unvergessen bleiben.

In seinen wissenschaftlichen Arbeiten und in seinen Gemälden — Rekonstruktionen des Lebens der Vorzeit — lebt er weiter.

Weimar, im Dezember 1984

HANS DIETRICH KAHLKE

CONSTANTIN CONSTANTINOVICH FLEROV

4. Februar 1904 bis 26. Juli 1980

Ein prominenter Zoologe, Paläontologe und Tiermaler, Professor Dr. CONSTANTIN CONSTANTINOVICH FLEROV, verstarb am 26. Juli 1980 in Moskau.

C. C. FLEROV wurde am 4. Februar 1904 als Sohn von C. F. FLEROV, Professor für Infektionskrankheiten, in Moskau geboren. C. C. FLEROV schloß 1925 sein Studium an der Universität Moskau, Abt. für Naturgeschichte, Mathematisch-Naturwissenschaftliche Fakultät, ab. Hervorragende Wissenschaftler wie die Akademiemitglieder M. A. MENZBIR, A. N. SEVERTSEV und M. V. PAVLOVA zählten zu seinen Lehrern. Bereits als Student zeigte C. C. FLEROV großes Interesse für Säugetiere und für wissenschaftlich-künstlerische Darstellungen, wobei er erfolgreich seine Studien und Arbeiten als wissenschaftlicher Sammler der Zoologischen Universität im Rahmen der Staatlichen Universität Moskau und als Leitungsmitglied der Staatlichen Arbeitsstelle für Anschauungsunterricht, National-Komitee für Ausbildung, unter Leitung von V. A. VATAGIN, einem bekannten Tiermaler, verband.

Nach Abschluß seines Studiums arbeitete er seit 1926 am Zoologischen Institut der Akademie der Wissenschaften der UdSSR, zuletzt als Senior-Wissenschaftler bis 1937. Danach erhielt er durch die Akademiemitglieder A. A. BORISIAK und Y. A. ORLOV einen Ruf als Senior-Wissenschaftler an das Paläontologische Institut der Akademie der Wissenschaften der UdSSR, und von 1947 bis 1973 leitete er das Paläontologische Museum der Akademie der Wissenschaften der UdSSR. Von 1937 bis 1946 verband er seine Arbeiten mit einer weiteren Aufgabe am Staatlichen Darwin-Museum als Senior-Wissenschaftler und von 1974 bis zu seinem letzten Tage war er Wissenschaftlicher Konsultant am Institut für Evolutionäre Morphologie und Ökologie der Tiere, Akademie der Wissenschaften der UdSSR.

FLEROVS erste wissenschaftliche Arbeit „Beobachtungen zum Leben von *Arvicola terrestris*" wurde 1925 publiziert, zu einer Zeit, da er noch Student war.

1935 schloß er mit dem Biologie-Diplom sein wissenschaftliches Studium ab, wobei ihm dieses ohne Verteidigung seiner Thesen zugesprochen wurde. 1946 verteidigte er erfolgreich seine Dissertation „Moschustiere und Hirsche" für den wissenschaftlichen Grad eines Doktors der Biologie. 1949 erhielt er den Professoren-Titel.

C. C. FLEROV ist der Autor von 110 wissenschaftlichen Arbeiten, die zoologischen und paläontologischen Fragen der Wirbeltiere, insbesondere solchen der Säugetiere, gewidmet sind. Er nahm an 36 hauptsächlich zoologischen Expeditionen teil, und er war ebenfalls Teilnehmer an einer Vielzahl internationaler wissenschaftlicher Kongresse.

Weiterhin publizierte er seine Arbeiten über den Europäischen Mink und über verschiedene Ungulaten (*Bison*, *Capreolus*, *Cervus* und andere Cerviden) und er lieferte große Abschnitte zu den Kollektiv-Monographien unter der Leitung unserer großen Zoologen E. N. PAVLOVSKY und B. S. VINOGRADOV, nämlich „Arktische Säugetiere" und „Säugetiere Tadžikistans" (1935). Diese Arbeiten sind als Ergebnis seiner zahlreichen Expeditionen, durch den Ural, den Kaukasus, durch Mittelasien und den Fernen Osten, zu betrachten und mit diesen Arbeiten wurde FLEROVS Name auf zoologischen Gebiet sehr berühmt.

Sein wissenschaftliches Interesse war in seiner Breite und Vollständigkeit bemerkenswert. Er war ein brillanter Vertreter eines allgemein biologischen Trends in Zoologie und Paläontologie. Dabei waren seine Forschungsarbeiten durch eine Komplex-Kombination von morphologischen, ökologischen und entwicklungsgeschichtlichen Gedankengängen ausgezeichnet. C. C. FLEROV liebte keine rein deskriptiven Arbeiten und schätzte solche Forscher wenig, deren Interessen-Sphäre auf die Beschreibung neuer Taxa begrenzt blieb.

In seiner bemerkenswerten Arbeit „Morphologie und Ökologie der Cervoidea im Verlauf ihrer Evolution" (1950) und ebenfalls in seiner großen Monographie „Moschustiere und Hirsche" (1952) gab er eine Zusammenfassung der Entwicklungsgeschichte und eine detaillierte systematische Beschreibung aller dieser Formen. Darüber hinaus diskutierte er geographische Variabilität, morpho-funktionelle Unterschiede und ökologische Anpassungen, Fragen, die ihn ein Leben lang interessierten.

Aufeinander folgend publizierte C. C. FLEROV eine Serie paläontologischer Monographien und Artikel, die sich hauptsächlich mit den „exotischen" Dinocerata und Pantodonta befaßten, deren fossile Reste aus Paläogen-Ablagerungen der Mongolei stammten. In der Monographie „Dinocerata der Mongolei" (1957) sind die originellsten Kapitel jene über morpho-funktionelle Anpassungen und über das Leben dieser Tiere. Fragen des sexuellen Dimorphismus, der Veränderungen im Alter und taphonomische Besonderheiten werden ebenfalls in der Arbeit diskutiert.

C. C. FLEROV war auch der Inspirator und der Hauptbearbeiter der Kollektiv-Monographie „Zoogeographie des asiatischen Paläogen" (1974). Das umfassende Ziel dieser Arbeit war es, die Gesetzmäßigkeiten der Entwicklung und der Verbreitung der wichtigsten Vertebraten-Gruppen des großen asiatischen Raums auf dem Hintergrund der Umweltänderungen im Paläogen darzustellen.

CONSTANTIN CONSTANTINOVICH war stets an den Problemen der organischen Welt der Quartär-Periode interessiert, als sich die modernen Bedingungen der Erde herausbildeten. C. C. FLEROV sah in der Quartär-Periode die Zeit des „Großen Gerichts". Auf dem Gebiet des Studiums quartärer Säugetiere war CONSTANTIN CONSTANTINOVICH eine Autorität. Mehr als zehn Jahre lang arbeitete er eng mit dem Institut für Quartärpaläontologie Weimar (DDR) zusammen und war ein enger Freund des Institutsdirektors.

In den letzten Jahren publizierte er zahlreiche Arbeiten über Morphologie, Systematik und Evolutionsgeschichte der Gattung *Bison*, und er war auch der führende Autor der Kollektiv-Monographie „Der Europäische Bison" (1979), in der er die Kapitel über die Systematik und Geschichte der Gattung *Bison* übernahm.

Als perfekter Kenner des Museumswesens und als talentierter Tiermaler hat C. C. FLEROV auch viel für den künstlerischen Wert der Ausstellungen des Paläontologischen Museums der Akademie der Wissenschaften der UdSSR getan, dessen Direktor er 26 Jahre lang war. Er arbeitete an den Skelett-Gruppen neuer fossiler Reptilien einschließlich der Dinosaurier wie auch über die Säugetiere der Mongolei.

Ausgezeichnete Ölbilder von der Hand C. C. FLEROVS, Rekonstruktionen fossiler Tiere in ihrer Umwelt, sind Stolz des Alten und des Neuen Paläontologischen Museums der Akademie der Wissenschaften der UdSSR. Ebenso sind seine Bilder im Zoologischen Institut in Leningrad zu bewundern, im Staatlichen Darwin-Museum, im Institut für Quartärpaläontologie Weimar (DDR) und im Staatlichen Museum der Mongolischen Volksrepublik in Ulan-Bator.

CONSTANTIN CONSTANTINOVICH FLEROV popularisierte in großem Umfange wissenschaftliche Erkenntnisse, er hielt Vorlesungen und leitete Diskussionen, er sprach über Radio und Fernsehen, er schrieb allgemeinverständliche Arbeiten über die Entwicklung des Lebens auf der Erde sowie über die verschiedensten Probleme der Zoologie und Paläontologie, die in Magazinen und Zeitungen erschienen.

Als Teilnehmer einer großen Zahl internationaler Symposien und Kongresse demonstrierte C. C. FLEROV im Ausland sowjetische Zoologie und Paläontologie.

Als ständiger Vorsitzender der Kommission Säugetiere des Wissenschaftlichen Rates „Wege und Gesetzmäßigkeiten der historischen Entwicklung der Pflanzen- und Tier-Organismen" unternahm C. C. FLEROV große Anstrengungen, die sowjetischen Paläontologen zu vereinigen. Später wurde er Mitglied des Präsidiums der Theriologischen Gesellschaft der UdSSR sowie Mitglied zahlreicher anderer wissenschaftlicher Gesellschaften der UdSSR, ebenfalls ausländisches Mitglied der Gesellschaft für Wirbeltier-Paläontologie der USA und der Argentinischen Paläontologischen Assoziation.

Die Arbeit von C. C. FLEROV und seine publizistischen Aktivitäten wurden durch die Sowjetische Regierung durch solche hohe Auszeichnungen wie den „Orden der Arbeit Rotes Banner" und die Medaillen „Für die Verteidigung der Stadt Moskau", „Für tapfere Arbeit während des Großen Vaterländischen Krieges", „Zur Erinnerung des 800. Jahrestages von Moskau" anerkannt.

CONSTANTIN CONSTANTINOVICH FLEROV besaß ein großes Talent, mit Leuten umzugehen. Oft konnte man ihn von Freunden und Bewunderern umgeben antreffen, vielfach auch von jungen Menschen, denen er Episoden aus der Geschichte der Wissenschaft oder aber die eindrucksvollsten und wichtigsten Ereignisse seines eigenen Lebens schilderte.

C. C. FLEROV zeigte dazu immer Interesse für Fortgang und Richtung der Arbeiten seiner jüngeren Kollegen, und die Diskussionen mit jungen Wissenschaftlern trugen viel dazu bei, Richtung und Umfang der wissenschaftlichen Interessensphären vieler Zoologen und Paläontologen zu bestimmen.

Sein Tod ist ein unwiederbringlicher Verlust für die Paläontologie und Zoologie und für alle diejenigen, die ihn näher kannten.

L. P. TATARINOV B. A. TROFIMOV

CONSTANTIN CONSTANTINOVICH FLEROV

4. February 1904 to 26. July 1980

A prominent zoologist and palaeontologist, and an animal-painter, Professor Dr. CONSTANTIN CONSTANTINOVICH FLEROV, died in Moscow on 26. July 1980.

C. C. FLEROV was born on 4. February 1904 in Moscow, the son of C. F. FLEROV, a professor of infectious diseases. C. C. FLEROV graduated in 1925 from the Department of Natural History, Physical-Mathematical Faculty, University of Moscow. Prominent scientists like the Academicians M. A. MENZBIR, A. N. SEVERTSEV, and M. V. PAVLOVA were among his teachers. As a student, C. C. FLEROV already was showing a great interest in mammals and in scientific-artistic representation, successfully combining his studies and work as a scientific collector of the Zoological University within the Moscow State University and as a member of the staff of the State Workshop of Visual Aids, National Education Committee under leadership of V. A. VATAGIN, a well-known animal-painter. After graduating from the University, in 1926, he entered the Zoological Institute of the USSR Academy of Sciences where he worked finally as a senior scientist, up to 1937. Since then, having been invited by the Academicians A. A. BORISIAK and Y. A. ORLOV, C. C. FLEROV worked as a senior scientist at the Palaeontological Institute of the USSR Academy of Sciences, and from 1947—1973 he was in charge of the Palaeontological Museum of the USSR Academy of Sciences. 1937—1946 he combined his work with another task at the State Darwin Museum as a senior scientist and from 1974 up to his last day he was a scientific consultant at Academician A. N. Severtsev's Institute of Evolutionary Morphology and Ecology of Animals, the USSR Academy of Sciences.

FLEROV's first scientific work "Observations on the way of life of *Arvicola terrestris*" was published in 1925, when he was still a student. In 1935 C. C. FLEROV awarded the Master's Degree of Biology without defending his Master's thesis, and in 1946 he successfully defended his dissertation on "Musk deer and Deer" for the degree of Doctor of Biology. In 1949 he was awarded the title of Professor.

C. C. FLEROV is the author of 110 scientific works, the main subjects of which are zoology and palaeontology of vertebrates, mainly mammals. He took part in 36 mainly zoological expeditions, and he also was a participant of a large number of international scientific congresses.

Further he published his works on the European mink and on different ungulates (*Bison*, *Capreolus*, *Cervus* and other deer) and he contributed large sections to the collective monographs published under leadership of our great zoologists E. N. PAVLOVSKY and B. S. VINOGRADOV, namely "Arctic Mammals" and "Mammals of Tadjikistan" (1935). These works have been the result of his frequent expeditions through the Urals, the Caucasus, Middle Asia, and the Far East and due to these works FLEROV's name became very famous in the field of zoology.

His scientific interests were remarkable for their breadth and completeness. He was a brilliant representative of a general biological trend in zoology and palaeontology. His researches are distinguished by a complex combination of morphological, ecological, and evolutionary approaches. C. C. FLEROV did not like any pure descriptive works and did not value highly those researchers whose sphere of interest was limited to the description of new taxa.

In his remarkable work "Morphology and ecology of Cervoidea in the process of their evolution" (1950) and also in his great monograph "Musk deer and Deer" (1952) he gave a summary of the evolution and a detailed systematical description of all these forms. Moreover, he discussed geographical variability and variation in morpho-functional and ecological adaptations, that is, the questions in which he was always interested.

Subsequently C. C. FLEROV published a series of palaeontological monographs and articles, dedicated mainly to the "exotic" Dinocerata and Pantodonta, the fossil remains of which are found in the Palaeogene deposits of Mongolia. In the monograph "Dinocerata of Mongolia" (1957) the most original are the chapters on morpho-functional adaptations and on the way of life of these animals. The questions of sexual dimorphism, age alternations, and taphonomic pecularities are also discussed.

C. C. FLEROV was the inspirer and the main performer of the collective monograph "Zoogeography of the Asian Palaeogene" (1974). The bold aim of this work was to ascertain the laws of development and of distribution of the basic vertebrate groups within the vast territory of Asia against the background of Palaeogene environmental changes.

CONSTANTIN CONSTANTINOVICH was always interested in the problems of the organic world in the Quaternary period, when modern conditions were formed on earth. C. C. FLEROV was seeing in the Quaternary period the time of the "Great Trial". In the field of studying Quaternary mammals, CONSTANTIN CONSTANTINOVICH was an authority. Over more than ten years he closely cooperated with the Institute of Quaternary Palaeontology in Weimar (GDR), and he was a close friend of the director of the Institute.

In the last years FLEROV published many papers on morphology, systematics, and evolutionary history of the genus *Bison*, and he was also the leading author of the collective monograph "European *Bison*" (1979), in which he wrote the chapters on systematics and history of *Bison*.

As he was a perfect connoisseur of Museum activities and a talented animal-painter, C. C. FLEROV had done much for extending and increasing the art value of Museum exhibitions in the Palaeontological Museum of the USSR Academy of Sciences, where he was a director for 26 years. He was working on the skeleton-assemblages of new fossil reptiles, including dinosaurs, as well as on mammals from the territory of Mongolia.

Wonderful pictures were painted by C. C. FLEROV: reconstructions of fossil animals in their environment are the adornment and pride of the old and new Palaeontological Museums of the USSR Academy of Sciences. His pictures are also on exhibition in the Zoological Institute in Leningrad, in the Darwin State Museum, in the Institute of Quaternary Palaeontology in Weimar (GDR), and in the State Museum of the MPR in Ulan-Bator.

CONSTANTIN-CONSTANTINOVICH FLEROV widely popularized scientific knowledge, giving lectures and hold discussions, spoke on radio and television, he wrote general papers on the development of life on the earth as well as on various problems of zoology and palaeontology, published in magazines and newspapers.

C. C. FLEROV demonstrated abroad Soviet zoology and palaeontology when he attented a large number of international symposia and congresses.

Having been a permanent chairman of the Commission on Mammals of the Scientific Council "Ways and laws of the historical development of plant and animal organisms", FLEROV made great efforts to unite the Soviet palaeontologists. Later he was a member of the praesidium of the Theriological Society of the USSR and a member of many other scientific societies in the USSR, also a foreign member of the USA Society of Vertebrate Palaeontology, and a member of Argentine Palaeontological Association.

FLEROV's work and public activities were honoured by Soviet Government awards such as the "Order of Labour Red Banner", and the medals "For Defence of the City of Moscow", "For Valiant Labour during the Great Patriotic War", "In the Memory of the 800th Anniversary of Moscow".

CONSTANTIN CONSTANTINOVICH possessed a great talent for dealing with people. He always could be seen surrounded by friends and admirers, often by rather young people, with whom he shared both the episodes from the history of science and the brightest and most interesting events of his own life.

C. C. FLEROV was constantly interested in the trends of the work of his junior colleagues, and his discussions with those young scientists did much for determination and enlargement of the sphere of scientific interest of many zoologist and palaeontologist.

His death is an irretrievable loss for palaeontology and zoology, and for all those people who know him closely.

L. P. TATARINOV B. A. TROFIMOV

Quartärpaläontologie 6 13—21 Berlin 1986

New Data on the Evolution of Bovine of the Tribe Bovini

By Irina Aleksandrovna Dubrovo, Moscow, and Nikolai Iosipovich Burchak-Abramovich, Tbilisi

With 2 Figures, 2 Tables and 2 Plates

Contents

Abstractum

Results of detailed investigations of the skull of a Late-Pliocene bovine from Transcaucasia — *Adjiderebos cantabilis* Dubrovo & Burchak-Abramovich — are given.

A comparative analysis of craniological features of *Adjiderebos* and other known genera of Bovini has brought to light an affinity of *Adjiderebos*, *Bison*, and *Leptobos* and has permitted the authors to express some considerations on phylogenetic relationships in the tribe Bovini.

Die Ergebnisse eingehender Untersuchungen an einem spätpliozänen Bovinen-Schädel aus Transkaukasien — *Adjiderebos cantabilis* Dubrovo & Burchak-Abramovich — werden vorgelegt.

Eine vergleichende Analyse kraniologischer Eigenheiten von *Adjiderebos* und anderen bekannten Bovini-Genera ergab Beziehungen zwischen *Adjiderebos*, *Bison* und *Leptobos*. Die Resultate ermöglichten es den Autoren, phylogenetische Verwandtschaftsverhältnisse im Tribus Bovini aufzuzeigen.

Приведены результаты детального изучения черепа позднеплиоценового быка из Закавказья — *Adjiderebos cantabilis* Dubrovo & Burchak-Abramovich.

Сравнительный анализ краниологических признаков *Adjiderebos* и других известных родов быков показал родство *Adjiderebos*, бизонов и лептобосов и позволил высказать ряд соображений о филогенетических связях в трибе Bovini.

1. Introduction

A morphological similarity between the skulls of *Leptobos* and primitive *Bison* was discussed by Flerov (1972) more than ten years ago. He spoke of the possibility of including *Leptobos* and *Bison* in the subtribe Bisontina (Flerov, 1977), but without giving reasons for such an association, apparently because of the absence of material that would enable the phylogenetic relationship between the genera in this subtribe to be established.

The study of the nearly complete skull of the Late Pliocene bovine *Adjiderebos cantabilis* Dubrovo & Burchak-Abramovich from Western Azerbaijan makes it possible to elucidate the phylogenetic relationship of this genus, *Leptobos* and *Bison*, thus providing confirmation for Flerov's opinion.

Late Neogene Bovidae were fairly widespread in the Soviet Union's southern territories. Finds of their remains are known from Moldavia (Alexeeva 1977), the Ukraine (Dubrovo and Kapelist 1979), the Caucasus (Baryshnikov 1981), and Transcaucasia (Gadzhiev and Akhundov 1976).

Relatively complete materials are available from the Kvabebi Locality in Georgia and were used to distinguish the genus and species *Ioribos aceros* Vekua (Vekua 1972). The rests of the Late Neogene Bovini in the USSR territory are represented by cranial fragments or separate bones of the postcranial skeleton. Thus inadequacy of material precluded the compilation of a complete diagnosis of the Mio-Pliocene *Urmiabos azerbaidzanicus* Burchak-Abramovich (Burchak-Abramovich 1950), Pliocene *Protobison kuschkunensis* Bur.-Abr. Gad. & Vek. (Burchak-Abramovich et al. 1980) and *Bison tamanensis* Vereshchagin (Vereshchagin 1957) and the identification of the species of *Leptobos* found in the USSR.

This accounts for the particular interest attributed to the nearly complete, uniquely preserved skull of *Adjiderebos cantabilis* (Table 1).

The skull was excavated in 1973 from Late Pliocene sandstones outcropping in the upper part of the Adzhidere ravine, Kazakhskij district, Azerbaijan SSR, during expedition work conducted by a party from the Palaeontological Institute, USSR Academy of Sciences, (party chief I. A. Dubrovo, members P. E. Vulykh, V. A. Presnyakov, L. P. Galukhina).

The study of the skull revealed it to be different from those of all known genera of Bovidae. The significantly shortened posthorn part of the skull, a broad basicranium, elaborate ethmoidal fissures, choana undivided by the vomer, and oval in cross-section and the homonymously bent horns all indicate that the described genus belongs to the subtribe Bovina (Sokolov 1953).

The basis for substantiating the genus *Adjiderebos*, apart from the great number of craniological differences, is its special position in the evolutionary series of Bovina.

The authors suggest distinguishing several groups within the subtribe Bovina for the convenience of further description:

I. Taurina with the genera *Bos* Linnaeus, *Poephagus* Gray, *Urmiabos* Burchak-Abramovich, *Bibos* Hodgson and *Novibos* Coolidge.

II. Bisonina with the genera *Bison* Smith, *Probison* Sahni & Khan and *Protobison* Burchak-Abramovich & Vekua.

III. Leptobovina with the genera *Leptobos* Rütimeyer, *Proleptobos* Pilgrim, *Ioribos* Vekua and *Epileptobos* Hooijer.

IV. Adjiderebovina with the genus *Adjiderebos* Dubrovo & Burchak-Abramovich.

The study of *Adjiderebos* reveals that it has some similar traits, on one hand, with Bisonina and, on the other, with Leptobovina and is, in fact, a representative of a special group.

All the aforementioned groups are morphological distinct and represent separate phylogenetic lines.

The last three groups are indubitably significantly different from the first group. However, in our opinion, these differences are not so great as to reach the rank of differences between subtribes; consequently we use in the present work the classification of the Bovidae proposed by SOKOLOV (1953).

Names and temporal attributions of geological units in the present paper corresponds to the stratigraphic scheme officially recognized in the USSR.

2. Systematics

Family **Bovidae** GRAY, 1821

Tribe **Bovini** SIMPSON, 1945

Subtribe **Bovina** SOKOLOV, 1953

2.1. Genus Adjiderebos DUBROVO & BURCHAK-ABRAMOVICH, 1984

Fig. 1; Table 1, 2; Plate I, II

Type species: *Adjiderebos cantabilis*, Upper Pliocene, Azerbaijan, USSR.

Diagnosis: Males have well-developed horn cores. The cranium is wide. The posthorn part of the cranium shorter than the frontal one and slightly shorter than the postorbital part of the front. It is clearly divided into an anterior-parietal, and a posterior-supraoccipital area lying at an obtuse angle to one another. The parietal area is greatly convex in the median portion. Bregmal fossa is practically non-existent. The dorsal surface of the posthorn cranial part lies at an angle of about 170° to the frontal surface. The rear parts of the temporal fossae are of average width and deep. The distance between them is appreciably less than that between the external margins of condyli occipitalae. Parietal crests are double. The occiput is wide and semiround. Dorsal protuberance of the zygomatic process of the temporal bone steeply rises superiorly, significantly constricting temporal fossa here. The glenoid surface for the mandible is sagittally narrow, the postglenoid processus is very high. The forehead is broad and relatively short. The interhorn elevation is undeveloped. The forehead at the level of horn cores is strongly concave. Orbits are non-tubular, with a thin lower wall. Supraorbital foramina lie behind the level of posterior orbital margins. Praeorbital fossae are non-developed. Ethmoidal fissures are conspicuous. Intermaxillary bones contact the nasal bones. The postorbital cranium width is slightly less than cranium width in zygomatic arches, much less than the width across supraear tubers and more than the width in tuber malare. The inferior surface of the main occipital bone is smooth and weakly convex transversely. The palate in front of the tooth row is strongly concave in the median portion. The constriction of interalveolar crests in front of the alveoli P^2 is abrupt. The horn cores are of average length and are in the first half compressed superiorinferiorly. They insert on the cranium at a moderate distance from the orbits, which is approximately equal to the horizontal diameter of the orbit. The lower margin of the horn cores lies slightly below the level of the upper margin of the occipital crest. The horn cores are rather significantly bent and at the base are directed laterally and slightly posteriorly, without going backward beyond the occipital crest level. Approximately half way along their length the horn cores, while continuing to go laterally, rise upwards and slightly bend anteriorly. The teeth are hypsodont. The posterolateral/vertical crest of M^3 tends backwards and outwards.

Table 1. Measurements of the skulls of bovines of the tribe Bovini

Measurements [in mm]	*Adjiderebos cantabilis*	*Bison priscus*			*Leptobos stenometopon*	*Leptobos falconeri*	*Epileptobos groeneveldtii*	*Ioribos aceros*
		Yakutia			QSV 982	N 39567, N 48026	N 2766	Georgian Academy of Sciences
	Palaeontological Institute, USSR Academy of Sciences				(cast)	HOOIJER (1958)		K 1
	3724-1	2125/1	2368	2125/2				
1. Length from inferior margin of foramen magnum to anterior margin of alveole P^2	278	408	415?		318			291
2. Length of frontal bones (from nasion to bregma)	130	210	223	200	172	142; 140		171
3. Length of frontal zone (from anterior orbital margin to median level of horn core bases)	150	203	195		146			
4. Length of posthorn zone (from the median level of horn core bases to occipital crest)	80	144	155	132	113			
5. Distance from posterior orbital margin to anterior margin of horn core base	75	97	106	107	88	98; 100	125	
6. Longitudinal orbital diameter	71	77	84		65			57
7. Maximum occiput width	220	320	300	298	158	197	235	134
8. Skull width across zygomatic arches	206	265	270?					145
9. Maximum frontal width (across external orbital margins)	230	352	378	380	238	218; 222	240	155
10. Minimum postorbital frontal width	210	306	305	299	202	205; 213	215	
11. Skull width across tubers malare	146	210	213		145			103
12. Minimum skull width in the diastema region	87	116?	125		60			

Table 1 (continue)

Measurements [in mm]	*Adjiderebos cantabilis*	*Bison priscus*			*Leptobos stenometopon*	*Leptobos falconeri*	*Epileptobos groeneveldtii*	*Ioribos aceros*
		Yakutia			QSV 982	N 39567 N 48026	N 2766	Georgian Academy of Sciences
	Palaeontological Institute, USSR Academy of Sciences				(cast)			
	3724-1	2125/1	2368	2125/2		Hooijer (1958)		K 1
13. Minimum distance between posterior margins of temporal fossae	65	213	211	191	55	60; 70	95	33
14. Distance between external margins of occipital condyles	106	147	147	140	101	106; 105	110	80
15. Dimensions of condyles	49 × 34	70 × 41	78 × 48	63 × 38	× 33			52 × 27
16. Height of occiput (from occipital crest to inferior margin of foramen magnum)	102	183	185?	158	110		156	80
17. Length of horn core along maximum curvature	280	420	575	415			650	
18. Length along straight line	173	320	414	312?				
19. Frontal diameter of horn core	72	110	123	115	83	67; 70	94	
20. Diameter perpendicular to the above	55	120	107	107	59	64; 59	89	
21. Horn-core base circumference	200	355	362	360	223			
22. Length of tooth row (at alveolar level)	138	157	163		136			137
23. Length of the molars row	84	103	99		83			85

Table 2. Indexes of the skulls of bovines of the tribe Bovini

Indexes [%]	*Adjiderebos cantabilis*	*Bison priscus*			*Leptobos stenometopon*	*Leptobos falconeri*	*Epileptobos groeneveldtii*	*Ioribos aceros*
		Yakutia			QSV 982	N 39567 N 48026	N 2766	Georgian Academy of Sciences
	Palaeontological Institute, USSR Academy of Sciences				(cast)	Hooijer (1958)		K 1
	3724-1	2125/1	2368	2125/2				
1. Ratio of the length of frontal zone to that of the inferior margin of foramen magnum to anterior margin of alveola P^2	53.9	49.7	46.9?		49.5			
2. Ratio of the length of frontal zone to that of frontal bones	115.4	96.7	87.4		85.9			
3. Ratio of posthorn to frontal zone length	47.3	70.9	79.5		77.3			
4. Ratio of posthorn zone length to occiput height	98.0	78.7	83.7?	83.5	102.7			
5. Ratio of maximum occiput width to occiput height	215.7	174.8	162.2?	188.6	143.6		150.6	167.5
6. Ratio of minimum postorbital frontal width to maximum frontal width	91.3	86.9	80.7	88.8	84.8	94.0; 95.9	89.6	
7. Ratio of skull width across zygomatic arches to the length from inferior margin of foramen magnum to the anterior margin of alveola P^2	74.1	64.9	65.0					49.8
8. Ratio of skull width in zygomatic arches to the length of frontal bones	158.5	126.2	121.1?					84.8
9. Ratio of minimum distance between posterior margins of temporal fossae to that between external margins of occipital condyles	61.3	144.9	143.5	136.4	54.4	56.6; 66.7	86.4	41.2
10. Ratio of distance from orbital margin to horn-core margin to orbital diameter	105.6	125.4	126.2		135.4			

Comparison: Compared to all known genera of the subtribe Bovina the genus in question differs by the thin lower orbital wall and, presumably, by the steep elevation of processus zygomaticus of ossis temporalis i.e. by abrupt constriction at this area of temporal fossae; from all genera, except the Leptobovina group, it differs by a significant concavity of the frontals at the level of horn cores.

From the genera of the group Taurina (BURCHAK-ABRAMOVICH 1950; GROMOVA 1931; SOKOLOV 1953) the genus *Adjiderebos* clearly differs by the greater development of the posthorn cranial part, its division into parietal and supraoccipital areas, a double parietal crest.

From the genera *Probison* and, presumably, *Protobison*, the genus in question differs by the longer posthorn cranial part and its different morphology, a short distance between the rear ends of temporal fossae which is much less than that between the external ends of occipital condyles, and a slightly greater width of the rear ends of temporal fossae.

From the genera *Proleptobos*, *Ioribos* and *Perimia* the genus *Adjiderebos* differs by the shorter posthorn cranial part and a greater relative width of the cranium.

In addition, there are some other important differences from each of the above-listened genera. Thus, for instance, from *Proleptobos* the genus *Adjiderebos* differs by the absence of praeorbital fossae, a more posterior position of supraorbital foramina, a smaller width of the frontal part of the cranium. From *Ioribos* (VEKUA 1972) by a wider cranium, a different shape of the occiput, narrower and deeper temporal fossae, more protruding orbits, more hypsodont teeth, etc.

From *Epileptobos* (HOOIJER 1958) the genus *Adjiderebos* differs by a longer posthorn cranial part, the absence of sagittal crest, a different structure of the posthorn cranial part, shorter and less massive horn cores whose forward bend is less pronounced. From the genus *Perima* (PILGRIM 1939), the placing of which in the subtribe Bovina is not substantiated, it differs by a larger angle between the frontal plane and the posthorn cranial part, the presence of parietal crests, lateral protrusion of the orbits, a significant distance between the orbit and the base of the horn core, a less pronounced rearward departure of horn cores and their elevation above the frontal level, not a straight but a curved shape and not a round but an oval cross-section of the base of the horn core. From *Probison* (SAHNI & KHAN, 1968), *Adjiderebos* differs by the quite different form and course of horn cores: flattened not anteroposteriorly but vertically and directed not inferioanteriorly but slightly posterolatero-superiorly, a double but not ordinary (as is the case with *Probison*) parietal crest, a greater convergence of the rear ends of temporal fossae. The generic rank of *Probison* is by no means universally recognized, but not substantiation has been given for placing *P. dehmi*, its only species, in the subgenus *Eobison* of the genus *Bison* (BURCHAK-ABRAMOVICH et al. 1980); therefore, we deem it necessary to point out the differences between the genus concerned and *Probison*.

Comparison with the genera *Bison* and *Leptobos*, with which the genus in question has maximum similarity, is dealt in much greater detail (see Tables 1, 2).

From the genus *Bison* (GROMOVA 1935; FLEROV 1972)[1]) *Adjiderebos* differs by a shorter posthorn cranial part clearly divided into the parietal and supraoccipital parts (Fig. 1) by a double rather than ordinary parietal crest. The distance between the rear ends of temporal fossae in *Adjiderebos* is significantly less, while in *Bison* it is larger than the distance between the outward margins of occipital condyles. The difference between these two genera usually lies in a different relationship of postorbital and zygomatic cranium width; in *Adjiderebos* the former is slightly inferior to the latter, while in the case of *Bison* it is much greater and only in the subgenus *Eobison* are they equal. Another difference of *Adjiderebos* from *Bison* lies in a different structure of the parietal region: in *Adjiderebos* both the parietal bones converge towards their apexes, displacing rearward the interparietal bone, while in *Bison* the interparietal bone divides parietal bones at the front and reaches as far as bregmatic fossa. Orbits in *Adjiderebos* protrude sideways scarcely at all. In the case of *Bison* they protrude quite significantly. *Adjiderebos* differs, moreover, by the presence of contact between nasal and maxillary bones, a sharp convergence of interalveolar crests in front of P^2, and the presence of a deep fossa in this part of the palate which are less significant in *Bison*. Horn cores in *Adjiderebos* are directed more significantly posteriorly at the base than in *Bison* and their lower margin does not descent below the level of the upper part of occipital crest — in the case of *Bison* even the upper edge of the horn core base mostly descends below this level.

From *Leptobos* (RÜTIMEYER 1877—1878; MERLA 1949; VIRET 1954; et alii apart from literary data, use was made for comparison of a skull cast of an adult male *L. stenometopon* RÜTIM., QSV 982) the genus *Adjiderebos* differs by a greater relative skull width, the shorter posthorn and parietal-cranial parts, a less close approach of posterior ends of temporal fossae, their lesser width here, a lower and broader occiput. The occipital crest in *Adjiderebos* does not protrude posteriorly beyond the occipitale plane, whereas in *Leptobos* it does so. The angle between the frontal plane and parietal area in *Adjiderebos* is much larger — about 170° — as compared to about 135° in *Leptobos*, the parietal crest in *Adjiderebos* being broader. Horn cores in *Adjiderebos* are relatively short, without pronounced ridges, they bend slightly rearward (not proceeding beyond the occiput plane) the lower surface of horn cores lying slightly below the level of the upper part of occipital crest. In *Leptobos* the horn cores have clear-cut ridges, are strongly bent rearward (proceeding beyond the occiput plane), their lower surface being above the level of the upper part of occipital crest. Supraorbital grooves in *Leptobos* are nearly parallel to one another, while in *Adjiderebos* they converge towards their front ends. The nasion in *Adjiderebos* lies beyond the level of anterior orbit margins, while in *Leptobos* it lies on the same level.

As revealed by the above comparison, the differences of the described form from all of the genera included by different authors in the subtribe Bovina are indisputably of generic value.

---

[1]) Comparison with *Bison* were made by using an abundant collection of skulls of this genus at the Palaeontological Institute of the USSR Academy of Sciences.

2.1.2. Adjiderebos cantabilis DUBROVO & BURCHAK-ABRAMOVICH, 1984

Fig. 1; Plate I, II

Holotype: Coll. No. 3723-1, Palaeontological Institute, Academy of Sciences of the USSR. A practically intact skull of an adult animal, presumably, male; locality Kushkuna, Adjidere ravine, Kazakhskii district, Azerbaijan SSR; Upper Pliocene, Apsheronian.

Description: An animal not of large dimensions (see Table 1). The skull is relatively broad, with well-developed horn cores. The posthorn part enters the dorsal cranium surface and lies at a very obtuse angle to the frontal zone.

Dorsal view of the cranium: The forehead (Fig. 1a; Plate I) is broad and transversely concave. This concavity is particularly pronounced on the level of horn cores, due to the fact that frontal bones at the areas adjoining the horn core rise conspicuously. The anterior and interior slopes of these tubers are gently inclined, while the rear one is steep. The maximum depth of frontal concavity at this level (along the median line) is about 2 cm. No interhorn ridge is present. The postorbital narrowing of the frontals is not sharp. The orbits slightly protrude laterally and are oval in shape, with a large longitudinal diameter. The orbital walls are thin. Supraorbital foramina, two on each side, lie one above the other at the level of postorbital constriction in a deep, narrow fossa at the rear end of a sulcus supraorbitalis. On the orbital level the sulci supraorbitalis become very gentle-sloping. From them towards the median line the frontal surface rises slightly and on the level of anterior orbital edges the forehead in its middle part is slightly convex. This convexity extends, moreover, to the region of nasal bones. Nasal bones are fairly broad and strongly convex transversely. Frontal and nasal bones lie on the same level, lacrimal bone is slightly concave. There are clear-cut fissura ethmoidalis. Intermaxillary bones with their nasal processus contact on the nasal bones. On the os maxillari there is a well expressed tuber malare lying on the level of M^1. Foramen infraorbitale is oval and lies very far forwards-below, at the root of P^2.

The postorbital part of the dorsal surface is represented by the parietal platform formed by the parietal and interparietal bones and the supraoccipital platform, i.e. the upper part of squama occipitalis. Generally, the posthorn part of the cranium is convex both longitudinally and transversely, but in details its relief is more complicated. Thus, parietal bones in the median part of parietal area give rise to a fairly low relief in the form of a gently elevated tuber. Interparietal bone directly adjoins at the back on the two parietal bones, its forward part contributing to the formation of the tuber described. Laterally on either side of the tuber there lie horizontal platforms progressively narrowing towards the bases of the horn cores.

The angle between frontal bone and the upper part of the parietal area is obtuse, about 170°. Bregmatic fossa is practically nonexistent. The posterior part of the parietal area is formed by interparietal bone, the anterior portion of which has the outlines of a semicircular segment. An arcuate boundary separates the anterior and posterior parts of interparietal bone, which lie at about 115° angle to one another. The surface of the posterior part of the interparietal bone is fine-tubercular and rugose, while that of the anterior part is smooth. Behind the interparietal bone the supraoccipital area is separated by the medial rib of double parietal crest. The supraoccipital area is flat and fine-tubercular, its surface being steeply inclined posteiooinferiorly but somewhat more gently than the posterior part of interparietal bone. The supraoccipital area is restricted from below by the occipital crest.

Laterally the parietal zone is limited by a broad parietal crest having an external and internal rib. The external rib separates the parietal zone from the temporal fossa. The internal rib runs laterally parallel to the external one over the parietal bone and thereby restricts interparietal bone from below.

Occipital view: Occiput (Fig. 1b; Plate I) is relatively wide and low, its upper boundary being almost regularly arcuate. The overall trend of occipital surface is vertical. Crista occipitalis protrudes backward only insignificantly and does not reach the level of posterior margins of occipital condyles. External occipital protuberance has the form of an equilateral triangle (side length 24 mm). Its surface is concave vertically and nearly planar transversely. Beneath occipital crest the occiput surface is concave, underneath the end of external occipital protuberance it becomes flatter and in the region of pars lateralis transversely convex. Occipital condyle is wide at the top and narrows progressively towards the lower end. Its longitudinal axis lies at an angle of about 30° to the vertical cranial axis. At the upper-external angle of the condyle there is a sharpened projection — an unciform torus. The dorsal edge of the condyle lies somewhat above the level of the summit of incisura condylojugularis. In the depth of the subcondylar fossa there is the foramen hypoglossum. Processus jugularis has an anteroposteriorly extended base. The end of the processus descends below the level of the ventral end of the condyle.

Ventral view: The palate (Plate II) is wide, contracting anteriorly and flat. However, at the anterior margin, within the margo interalveolaris, from the level of anterior border of P^2 and forward, in the middle portion of palatine there lies a deep valley (maximum depth of about 60 mm, in the diastema area about 12 mm). Maxillary margines interalveolares are quite conspicuous and significantly converge anteriorly, the distance between them at the P^2 level being 18 mm. The front boundary between maxillary and palatine bones lies at the level of M^1. Foramen palatinum aborale lies directly behind the anterior margin of palatal notch in the narrow recess of palatine sinus. Foramen maxillare lies slightly posteriorly. Foramen sphenopalatinum has the outlines of an elongated oval (28 mm × 12 mm) with a longitudinal major axis. The choanas are not separated by the vomer. The anterior margin of choana staphylion lies slightly posteriorly with respect to the anterior margin of palatine notch, which is nearly at the level of the posterior margin of the alveolus of M^3.

The teeth are hypsodont and of the typical bovine structure. Additional columns are well developed, the styles are conspicuous and sharpened. The rear crest of third molar is bent posteroexternally. The enamel is rugose and dull. The tooth row is relatively long. The row of premolars is shorter compared to that of molars.

Os basisphenoidale has a conspicuously transversely convex inferior surface across which an eakly conspicuous longitudinal ridge passes medially. The posterior end of os basisphenoidale carries the endings of the anterior parts of tubercula muscularis which are principally located on os basioccipitale.

Os basioccipitale has a more or less longitudinally flat and conspicuously transversely convex inferior surface. A sagittal ridge runs longitudinally in the anterior half of this surface. Tubercula pharyngea are elongated-oval with a sagittal major axis.

Facies articularis mandibularis is slightly convex sagittaly and rather more convex transversely, the transversal convexity increasing medially. Tuberculum articulare has the form of a rounded oval with a sagittaly minor axis. Fossa mandibularis at the postero-external part of facies articularis is inconspicuous. Processus postglenoidalis is relatively high.

Lateral view: Fossa temporalis is conspicuously expanded in the posterior half (Fig. 1c; Plate II), is constricted anteriorly and its wall, inclined at about 45°, becomes nearly vertical at the front. Fossa temporalis is posteriorly restricted by occipital crest and mediosuperiorly by the external edge of the parietal crest. Its lower boundary is formed by the crista temporalis. The crest rises conspicuously upwards in the area of the zygomatic processus, significantly constricting the temporal fossa. The distance from the top of the crest to the underside of the horn pedicle is significantly less here than the width of foramen occipitale magnum. The orbital fossa is relatively deep and wide.

Horn cores: Horn cores (Fig. 1; Plates I, II) are relatively short, homonymously curved and are located behind the orbitals at a distance approximately equal to the longitudinal orbital diameter. The surface of the horn cores is rugose, whithout any more or less conspicuous ridges, with two longitudinal, very short grooves at the anterior and posterior sides in the middle of the horn core. The maximum sweep of the horn cores is 545 mm; the distance between their apexes is 517 mm. Departing from the skull, the horn cores tend laterally and slightly posteriorly, so that they do not go beyond the occipital crest plane. Their inferior surface is slightly below the upper part of occipital crest. The superior surface is above the frontal level. The tops of horn cores are significantly elevated above the forehead, i.e. by 90 mm. Starting approximately at one-third along their length the horn cores tend laterally upwards and then in the last third upwards and slightly forwards. The horn cores are flattened at the base superoinferiorly; this flattening fairly soon decreases and in the distal third of their length the horn cores have a circular cross-section. The horn cores thin gradually from base to apex. Corona cornus is inconspicuous. The horn core pedicle is short, i.e. shorter than the longitudinal orbital diameter.

Material: Holotype — a nearly complete skull of an adult but not old animal. Collection of the Palaeontological Institute, the USSR Academy of Sciences, No. 3724-1. Sutures between the bones are seen either quite clearly or less clearly or are quite inconspicuous. Particularly conspicuous are sutures between occipital squama and lateral occipital bones, the median sagittal suture between frontal bones, the longitudinal suture between horizontal branchings of maxillary bones and palatine bones. Quite conspicuous are all sutures in temporal fossa and in the facial part of the skull. Sutures in the parietal zone have overgrown and can be identified only with the aid of the magnifying glass.

The left half of the skull is better preserved than the right one, being practically free from damage or deformation. On the right side, part of the occiput with occipital condyle and jugular process is broken off, the zygomatic arch is absent and the surfaces of the frontal bone and orbital are deformed. The anterior part of the muzzle and tooth crowns are broken off, only sinistral M^2, M^3 and dextral M^3 have been preserved.

Geological age and geographical distribution: Late Pliocene (Apsheronian), Transcaucasia.

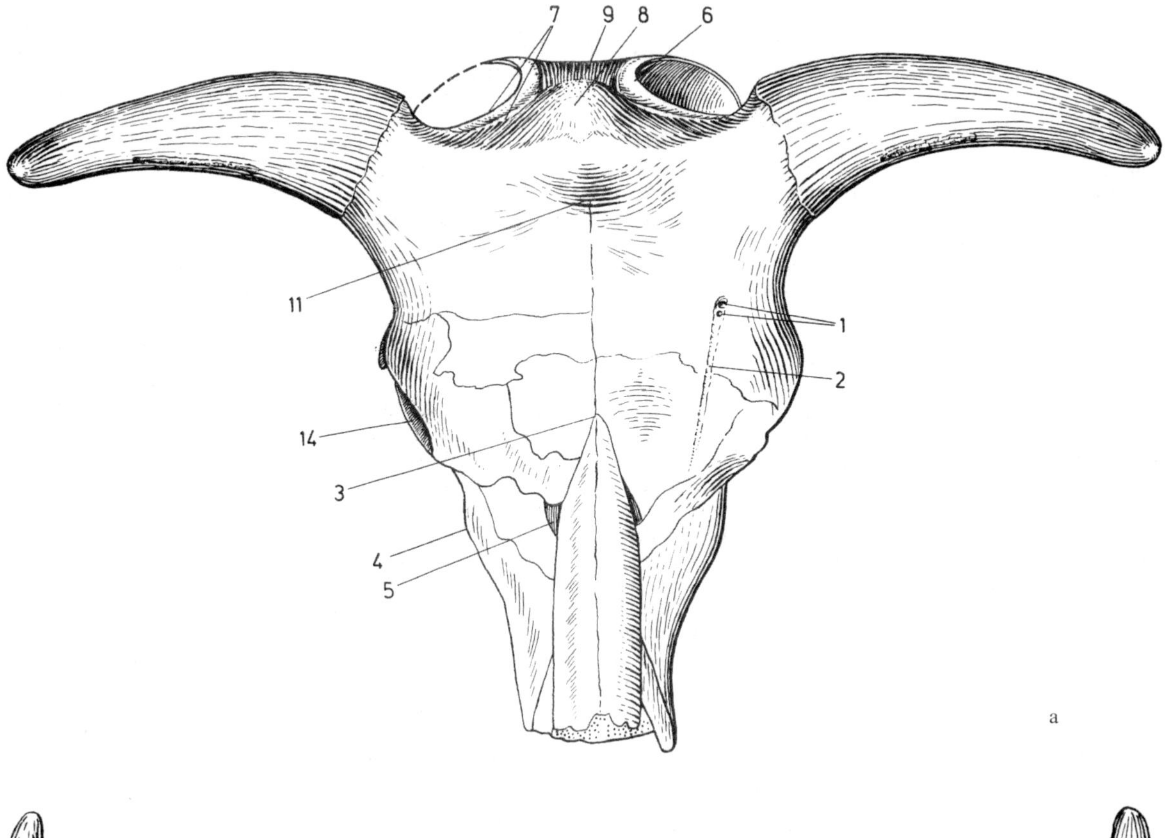

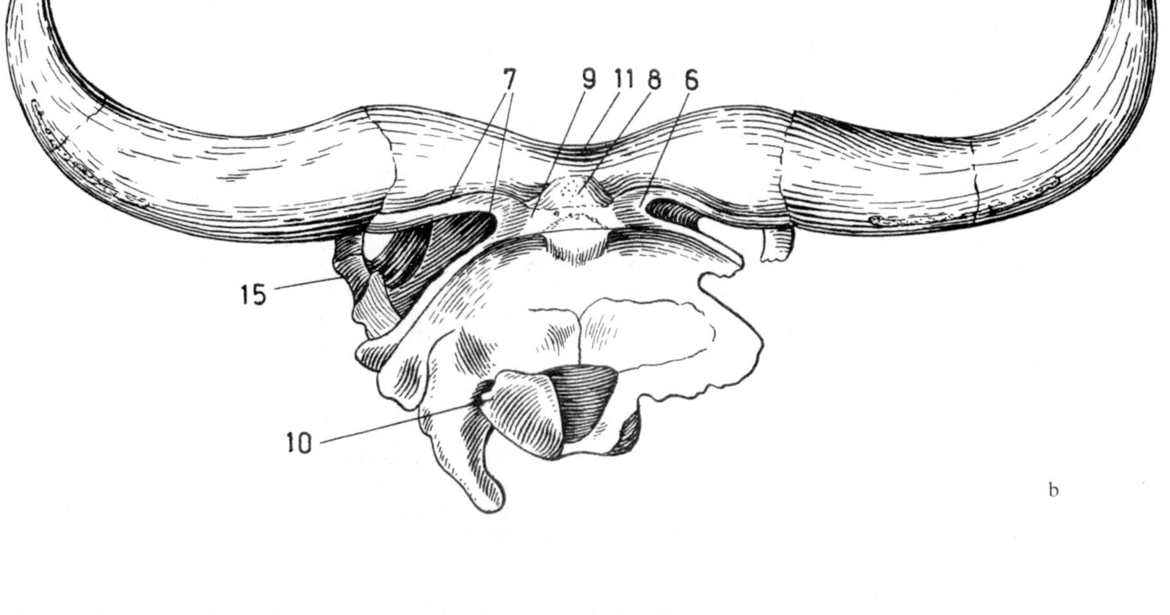

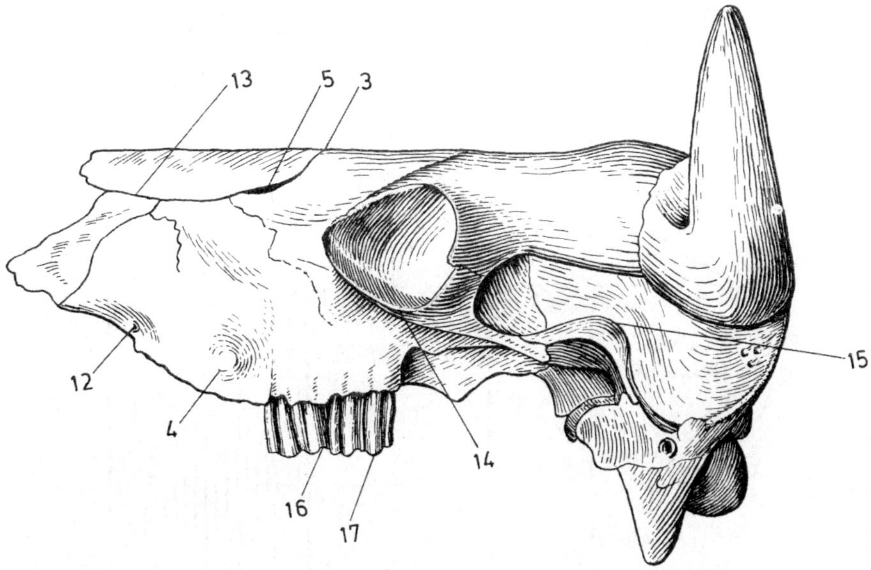

Fig. 1. *Adjiderebos cantabilis* DUBROVO & BURCHAK-ABRAMOVICH Kushkuna, Kazakhskii district, Azerbaijan SSR; Late Pliocene (Apsheronian). Skull, about 1/4 nat. size

a — view from above; b — rear view; c — lateral view
1 — supraorbital foramina; 2 — supraorbital groove; 3 — nasion; 4 — tuber malare; 5 — ethmoidal fissure; 6 — posterior margin of temporal fosa; 7 — parietal crest; 8 — parietal area; 9 — supraoccipital area; 10 — unciform rim of occipital condyle; 11 — conspicuous frontal concavity; 12 — infraorbital foramen; 13 — maxillary and nasal bones contact; 14 — thin inferior margin of the orbital; 15 — conspicuous elevation of zygomatic arch; 16 — M^2; 17 — M^3

3. Discussion

In the course of evolution the skull of Bovini grew wider, its posthorn part shortening and praeorbital fossae disappearing. Temporal fossae became deeper and narrower and the hypsodonty of teeth increased.

Thus, in the Mio-Pliocene *Proleptobos* there are still present praeorbital fossae which are absent in the later genera. In the Middle Akchagylian *Ioribos* the skull is long, with a narrow occiput, with an elongated parietal zone, with non-deep temporal fossae, wide at their posterior ends and conspicuously converging here. In the Pleistocene *Epileptobos* the skull is significantly broader, with deeper temporal fossae.

On the basis of these craniological features which are changing in the process of evolution we may deduce the significant progressivness of *Adjiderebos*: its skull is broad, with a relatively shortened posthorn area, temporal fossae are deep, with converging posterior ends and teeth are hypsodont. Most of these features indicate the closeness of *Adjiderebos* to the progressive group Bisonina. However, certain no less important features of Adjiderebovina are typical for Leptobovina. It is interesting to note that with respect to certain features common for Leptobovina and Adjiderebovina the Late Pliocene *Adjiderebos* is more progressive than the Pleistocene *Epileptobos*. For example the index of convergence of posterior ends of temporal fossae, expressed as the ratio of minimum width between posterior edges of temporal fossae to the distance between external margins of occipital condyles (Table 2).

The craniological features of *Adjiderebos* undoubtedly point to its affinity with the genera in the groups Bisonina and Leptobovina.

In the cranial morphology of *Adjiderebos* there are some definite features pointing to its affinity with *Bison*, particularly with the most primitive species in the subgenus *Eobison*: *Bison (Eobison) sivalensis* (LYDEKKER, 1878) and *B. (Eobison) palaeosinensis* (TEILHARD DE CHARDIN & PIVETEAU, 1930). Similarities between *Adjiderebos* and *Bison* relate to overall proportions of the skull which is relatively short and wide, with a low broad occiput, the superior part of which forms in both the genera a nearly regular semicircle. The parietal zone in both genera is part of the dorsal cranial surface, the posterior ends of temporal fossae being visible when looking at the skull from above. On the occipital condyles in either genus there are present hook-shaped rims. Horn cores in these genera show a relatively insignificant posterior trend, a laterally upward tendency, their inferior surface not lying above the upper part of the occipital crest. In either genus the nasal bones terminate behind the anterior margin of orbits and the supraorbital grooves slightly converge in their front margins.

Numerous similar features of the cranial structure are observed in *Adjiderebos* and *Leptobos* (MERLA 1949; PETROCCHI 1956). Thus, in both genera the posthorn part of the skull is much shorter than the frontal zone, its dorsal surface is formed by the parietal and supraoccipital areas, parietal crests are double, temporal fossae considerably converge, the frontal surface between the horn core bases is concave, supraorbital foramina lie behind posterior orbital margins, orbits protrude little laterally, nasal bones contact on intermaxillary ones, a clear-cut recess is present in the middle of the anterior part of the palate.

Thus, *Adjiderebos*, being endowed with features typical of it alone, comprises certain traits common with the groups

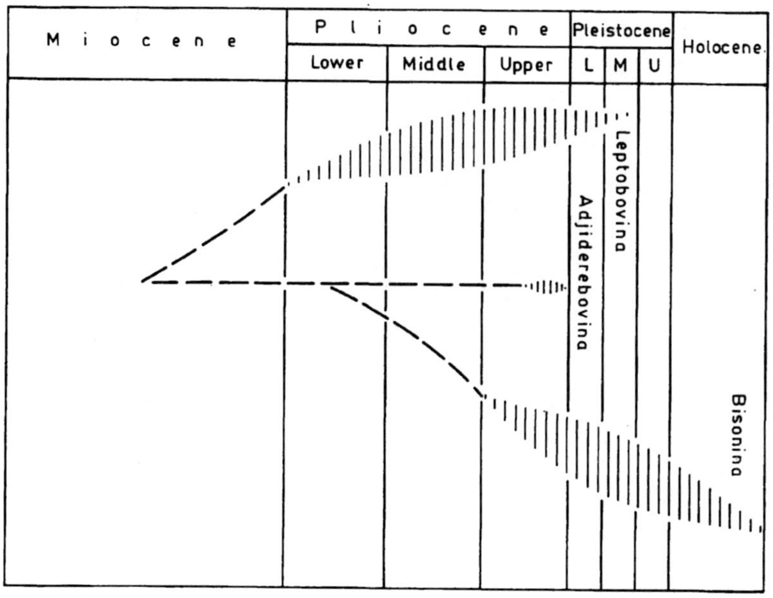

Fig. 2. Scheme of presumed phylogenetic relationships of the groups Leptobovina, Adjiderebovina and Bisonina and their stratigraphic distribution

Leptobovina and Bisonina, thus enabling us, so to speak, to place *Adjiderebos* at the basis of the branches Bisonina and Leptobovina. However, *Adjiderebos* cannot be a direct ancestor of the group Bisonina: the remains of representatives of this group have been found not only in deposits dated similarly as the Apsheronian ones from which *Adjiderebos* skull originates but also in geologically earlier deposits. Thus, *Protobison* and *Bison* (*Eobison*) *sivalensis* were found in the Tatrot horizon, *Probison* in the layers of the Middle Akchagylian, *Bison* (*Eobison*) *palaeosinensis* — of the Sanmenian, *Bison tamanensis* — of the Apsheronian. The group Leptobovina had appeared even earlier — in the Mio-Pliocene (genus *Proleptobos*).

On the basis of what has been said above we may easily presume that Adjiderebovina and Leptobovina originated from some even older form unknown to us, which must have possessed features of the *Leptobos-Bison* type in the cranial structure. From such an ancestral form there must have developed later (Fig. 2) two lines: that of leptoboses having no bison features and the line whose representatives combined the features of leptoboses and bisons. The first of the branches — Leptobovina with the genera *Leptobos*, *Proleptobos*, *Ioribos*, *Epileptobos* — existed until the Pleistocene (Java). The second one was Adjiderebovina terminating in the genus *Adjiderebos* in the Apsheronian of Transcaucasia. Presumably from Adjiderebovina at a certain time period, in the Early or Middle Pliocene, there had separated the branch Bisonina with the predominance of bison features, which has survived through various genera until today as the genus *Bison*.

New data on the Late Pliocene bovines, on the genus *Adjiderebos*, confirm the close relation between leptoboses and bisons. Adjiderebovina link up the evolutionary branches Leptobovina and Bisonina into a single generic stem.

Summary

In the course of work carried out by the Palaeontological Institute in Transcaucasian party a nearly complete skull of a fossil bovine was found. Lithological and stratigraphic data have enabled N. V. PASHALY and I. A. DUBROVO to age the sandstones outcropping in the summit of the Adjidere revine, from which the described skull originated, as corresponding to the Late Pliocene — Apsheronian.

As a result of detailed comparative-morphological study of the skull concerned and of the skulls of the known genera of the subtribe Bovini it was possible to identify a new genus and species — *Adjiderebos cantabilis* DUBROVO & BURCHAK-ABRAMOVICH, 1984.

The authors propose that four groups should be distinguished in the composition of the subtribe Bovini: Taurina, Bisoñina, Leptobovina and Adjiderebovina which are different morphologically and represent separate phylogenetic branches.

Craniological features of *Adjiderebos* point to a close affinity of Adjiderebovina both with Leptobovina and Bisonina. The authors consider it to be most likely that the first two groups originated from a common ancestral form which was endowed with the features of both, leptoboses and bisons. From this hypothetic, presumably Miocene form, there developed two branches: Adjiderebovina preserving mixed traits and the branch Leptobovina which was already free from bisonone features.

Later, presumably in the Early or at the beginning of the Middle Pliocene the branch Bisonina separated from Adjiderebovina.

Thus, the new material on the Late Pliocene bovine — on *Adjiderebos cantabilis* — give every reason to come to the conclusion of the relationship of leptoboses and bisons.

Leptobovina, Bisonina and Adjiderebovina related to a common phylogenetic stem.

Zusammenfassung

Im Verlaufe von Expeditionsarbeiten des Paläontologischen Instituts, Akademie der Wissenschaften der UdSSR, im Transkaukasus konnte ein nahezu vollständiger Bovinen-Schädel entdeckt werden. Die lithologischen und stratigraphischen Untersuchungen ermöglichten es N. V. PASHHALY und I. A. DUBROVO die Sandsteine, die am Hang der Adjidere-Schlucht zu Tage treten und aus denen der Schädel stammt, in das späte Pliozän (Apsheron) zu stellen.

Als Ergebnis detaillierter vergleichend-morphologischer Untersuchungen an diesem Schädel sowie an Schädeln der bekannten Gattungen des Subtribus Bovini war es möglich, eine neue Gattung und Art: *Adjiderebos cantabilis* DUBROVO & BURCHAK-ABRAMOVICH, 1984, aufzustellen.

Die Autoren schlagen vor, innerhalb des Subtribus Bovini vier Gruppen zu unterscheiden: Taurina, Bisonina, Leptobovina und Adjiderebovina, die sich morphologisch unterscheiden und verschiedenen phylogenetischen Linien angehören.

Die craniologischen Merkmale von *Adjiderebos* zeigen eine nahe Verwandtschaft der Adjiderebovina mit den Leptobovina und Bisonina auf. Die Autoren halten es für am wahrscheinlichsten, daß zuerst zwei Gruppen aus einer gemeinsamen Ausgangsform entstanden, die Merkmale beider, der *Leptobos*- und der *Bison*-Gruppe, noch einschloß. Von dieser hypothetischen, wahrscheinlich miozänen Ausgangsform entwickelten sich zwei Linien: die Adjiderebovina, die noch gemischte Merkmale bewahrten, und der Zweig der Leptobovina, der bereits frei von „bisoninen" Merkmalen war.

Später, wahrscheinlich im frühen oder aber zu Beginn des mittleren Pliozäns, schied sich der Zweig der Bisonina von den Adjiderebovina. So ermöglicht der neue Fund des spätliozänen *Adjiderebos cantabilis* das Verwandtschaftsverhältnis der Gruppe *Leptobos* und der Gruppe *Bison* aufzuzeigen.

Leptobovina, Bisonina und Adjiderebovina gehören einem gemeinsamen phylogenetischen Stamm an.

Резюме

Во время работ Закавказского отряда Палеонтологического института в Казахском районе Аз. ССР был найден почти полный череп ископаемого быка. Литологические и стратиграфические данные позволили Н. В. Пашалы и И. А. Дуброво датировать песчаники, вскрывающиеся в верховьях оврага Аджидере, из которых происходит череп быка, поздним плиоценом — апшероном.

В результате детального сравнительно-морфологического изучения найденного черепа и черепов известных родов подтрибы Bovini были установлены новые род и вид — *Adjiderebos cantabilis* Dubrovo & Burchak-Abramovich, 1984.

Авторы предлагают выделить в составе подтрибы Bovini четыре группы: Taurina, Bisonina, Leptobovina и Adjiderebovina, отличающиеся морфологически и представляющие отдельные филогенетические ветви.

Краниологические признаки *Adjiderebos* указывают на близкое родство Adjiderebovina и с Leptobovina и с Bisonina. Авторы считают наиболее вероятным, что первые две группы происходят от общей предковой формы, имевшей признаки как лептобосов так и бизонов. От этой гипотетической, вероятно, еще миоценовой формы, развились две ветви: Adjiderebovina, сохранявшая смешанные признаки, и ветвь Leptobovina — уже без бизоньих признаков.

Позже, видимо, в раннем или начале среднего плиоцена от Adjiderebovina отходит ветвь Bisonina.

Таким образом новые материалы по позднеплиоценовым быкам — Adjiderebovina позволяют обоснованно говорить о близком родстве лептобосов и бизонов, относящихся вместе с *Adjiderebos* к одному филогенетическому стволу.

References

Alekseeva, L. I., 1977: Teriofauna rannego antropogena Vostočnoj Evropy (Theriofauna of the Early Anthropogene of Eastern Europe). — Moskva: Nauka. — 214 pp.

Baryshnikov, G. F., 1981: Otrad Artiodactyla Owen 1848 — parnopalye. — In: Katalog mlekopytajuščich SSSR (Order Artiodactyla Owen, 1848. — In: Catalogue of the mammals of the USSR). — Moskva: Nauka. — P. 343—408

Burchak-Abramovich, N. I., 1950: Novye predstaviteli iskopaemych bykov *Taurina* v gipparionovoj faune Maragi, *Urmiabos azerbaidzanicus* Bur.-Abram. gen. et sp. nov. (A new representative of fossil bovid *Taurina* in the *Hipparion* fauna of Maraga, *Urmiabos azerbaidzanicus* Bur.-Abram. gen. et sp. nov.). — In: Dokl. Akad. Nauk SSSR. — Moskva **70** (5). — P. 875—878

Burchak-Abramovich, N. I.; Gadzhiev, D. V.; Vekua, A. K., 1980: Paleobiologičeskaja istorija pozdnekajnozojskich bykov Kavkaza (Palaeobiological history of the Late-Cenozoic Bovidae of the Caucasus). — Tbilisi: Micniereba. — 104 pp.

Dubrovo, I. A.; Burchak-Abramovich, N. I., 1984: Novyj rod pliocenovogo byka (*Adjiderebos* gen. nov.) iz Zakavkaja (A new genus of Pliocene bovine — *Adjiderebos* gen. nov. — from Transcaucasia). — In: Dokl. Akad. Nauk SSSR. — Moskva **276** (3). — P. 717—720

Dubrovo, A. I.; Kapelist, K. V., 1979: Katalog mestonachoždenij tretičnych pozvonočnych Ukr. SSR (Catalog of localities of Tertiary vertebrates of the Ukraine). — Moskva: Nauka. — 155 pp.

Flerov, C. C., 1972: Drevnejšie predstaviteli i istorija roda *Bison* (The oldest representatives and the history of the genus *Bison*). — In: Teriologija. — Novosibirsk **1**. — P. 81—86

—, 1979: Sistematika i evoljucija. In: Zubr. (Systematics and Evolution. In: Bison). — Moskva: Nauka. — P. 9—127

Gadzhiev, D. V.; Akhundov, F. M., 1976: Kratkij obzor mestonachoždenij pozvonočnoj fauny v otloženijach apšeronskogo jarusa Azerbajdžana. In: Voprosy paleontologii i stratigrafii Azerbajdžana (A brief review of localities of the vertebrate fauna in deposits of the apsheron stage of Azerbaijan, in: Problems of the palaeontology and stratigraphy of Azerbaijan). — Baku. — P. 173—186

Gromova, V. I., 1931: Pervobytnyj byk ili tur (*Bos primigenius* Boj.) v SSSR (Primary aurochs [*Bos primigenius* Boj.] in the USSR). — In: Ežegodn. zool. Muz. Akad. Nauk SSSR. — Leningrad **32** (3). — P. 293—364

—, 1935: Pervobytnyj zubr. (*Bison priscus* Boj.) v SSSR (Primary bison — *Bison priscus* Boj. — in the USSR). — In: Tr. zool. Inst. Akad. Nauk SSSR. — Moskva, Leningrad 2. — P. 77—202

Hooijer, D. A., 1958: Fossil Bovidae from the Malay Archipelago and the Punjab. — In: Zool. Verh. — Leiden **38**. — 112 pp.

Lydekker, R., 1878: Crania of Ruminants from the Indian Tertiaries. — In: Palaeont. India. — Calcutta **10** (1). — P. 88—181.

Merla, G., 1949: *Leptobos* Rütim. — In: Palaeontogr. Ital. — Pisa **46**. — P. 41—155.

Petrocchi, C., 1956: *Leptobos* di Sahabi. — In: Boll. Soc. Geol. Ital. — Roma **75**. — P. 1—35.

Pilgrim, G. I., 1939: The fossil Bovidae of India. — In: Palaeont. India. — Calcutta **24**. — 356 pp.

Rütimeyer, L., 1877—1878: Die Rinder der Tertiär-Epoche nebst Vorstudien zu einer natürlichen Geschichte der Antilopen. — In: Abh. schweiz. paläont. Gesellsch. — Basel 4. — P. 1—72; **5**. — P. 73—208.

Sahni, M. R.; Khan, E. 1968: *Probison dehmi* n. gen., n. sp. — a recent find of an Upper Siwalik Bovid. — In: Mitt. bayer. Staatssamml. Paläont. u. hist. Geol. — München **8**. — P. 247 bis 251.

Sokolov, I. I., 1953: Opyt estestvennoj klassifikacii polorogich (Bovidae) (An attempt of natural classification of cavicornians — Bovidae) — In: Tr. zool. Inst. Akad. Nauk SSSR. — Moskva, Leningrad 14. — 295 pp.

Teilhard de Chardin, P.; Piveteau, J., 1930: Les mammifères fossiles de Nihowan (Chine). — In: Ann. Paleont. — Paris **29**. — 134 pp.

Vekua, A. K., 1972: Kvabebskaja fauna akčagylskich pozvonočnych (The Kvabebi fauna of Akchagyl vertebrates). — Moskva: Nauka. — 258 pp.

Vereshchagin, N. K., 1957: Ostatki mlekopitajuščich iz nižnečetvertičnych otloženij Tamanskogo poluostrava (The remains of mammals from the Lower Quaternary deposits of the Taman peninsula). — In: Tr. zool. Inst. Akad. Nauk SSSR. — Moskva, Leningrad 22. — P. 9—74

Viret, M. J., 1954: Le loess a bancs durcis de Saint-Vallier (Drome) et sa faune de mammifères villafranchiens. — In: Nouv. Arch. Mus. Hist. Natur. Lyon. — Lyon **4**. — 200 pp.

PLATE I
Adjiderebos cantabilis DUBROVO & BURCHAK-ABRAMOVICH
Kazakhskii district, Azerbaijan SSR, Late Pliocene (Apsheronian), about 2/7 nat. size

PLATE II

Adjiderebos cantabilis DURBROVO & BURCHAK-ABRAMOVICH

Kazakhskii district, Azerbaijan SSR. Late Pliocene (Apsheronian), about 1/3 nat. size

Quartärpaläontologie 6 25—28 Berlin 1986

The Late Cenozoic Faunistic Complexes of Transbaikalia with Special Reference to the Micromammalia

By MARGARITA ALEKSANDROVNA ERBAEVA, Ulan-Ude

With one Figure

Contents

Abstractum

The Villafranchian small-mammal faunal sequence of Transbaikalia has many elements in common with contemporaneous European faunas. Within the Pleistocene the fauna of Transbaikalia has been reduced to a certain extent due to environmental changes specific to this area.

Die villafrankische Kleinsäuger-Faunenfolge Transbaikaliens hat zahlreiche gemeinsame Elemente mit zeitgleichen Faunen Europas. Während des Pleistozäns wurde die Fauna Transbaikaliens in dem Umfange reduziert, der den spezifischen Umweltänderungen dieses Raums entsprach.

В плиоценовой фауне Западного Забайкалья встречено много общих с европейскими фаунами видов мелких млекопитающих. В плейстоцене их число значительно сократилось, что было связано со специфическими особенностями природных условий Забайкалья.

1. Introduction

Fossil remains of small mammals are abundent in all types of Cenozoic continental deposits of Transbaikalia, and their great importance for biostratigraphy as well as for reconstructing the palaeogeographical environment is often stressed.

Fossil mammal remains have been used successfully for interregional correlations of Cenozoic deposits in all parts of the USSR. This has become possible due to the investigations of late Cenozoic-Quaternary mammals from Moldavia and Belorussia in the West as far as to the North-East regions of the USSR. By now the fossil mammalian faunas of Transbaikalia, West Siberia, Kazakhstan, Georgia, Ukraine, etc. as well as the main stages of their development have been well studied. The analogues of the European faunistic complexes as distinguished by GROMOV (1948) are found in each region mentioned above including Transbaikalia.

2. The Chikoi Faunistic Complex

In Western Transbaikalia Pliocene associations are known as the Chikoi faunistic complex and correlated to the Pre-Chaprovian faunas of Eastern Europe (VANGENGEJM and ZAZHIGIN 1982). The mammal fauna includes the following taxa (ERBAEVA 1974; VANGENGEJM et al. 1975) (Fig. 1/6):

Beremendia fissidens (PETENYI)
Petenyia hungarica KORMOS
Sorex sp.
Pliolagus sp.
Ochotonoides complicidens BOULÉ & TEILHARD DE CHARDIN
Ochotona minor ERBAEVA
Ochotona intermedia ERBAEVA
Sinocastor cf. *zdanski* YOUNG
Sicista pliocaenica ERBAEVA
Orientalomys sibiricus ERBAEVA
Cricetinus cf. *varians* ZDANSKY
Cricetulus sp.
Villanyia eleonorae ERBAEVA
Mimomys minor FEJFAR
Mimomys cf. *reidi* HINTON
Mimomys pseudointermedius ERBAEVA
Prosiphneus ex gr. *praetingi* TEILHARD DE CHARDIN
Canis cf. *chihliensis* var. minor TEILHARD DE CHARDIN & PIVETEAU
Nyctereutes cf. *sinensis* (SCHLOSSER)
Euryboas cf. *lunensis* DEL CAMPANA
Felis (*Lynx*) *shansius* TEILHARD DE CHARDIN
Acinonyx sp.
Felis (*Puma*) sp.
Hipparion ex gr. *houfenense* TEILHARD DE CHARDIN & YOUNG
Hipparion tchikoicum IVÅNIEV
Dicerorhinus sp.
Palaeotragus sp.
Gazella cf. *sinensis* TEILHARD DE CHARDIN & PIVETEAU
Antilospira sp.

By the end of the Middle and the beginning of the Late Pliocene (NIKIFOROVA et al. 1980) the climate of Western Transbaikalia was rather arid and warm. This period was characterized by the absence of rootless toothed voles, by the abundance of Lagomorphs and by the predominance of endemic, Central Asian mammals. But there are at least some elements in common with the contemporaneous European faunas: *Beremendia fissidens*, *Petenyia hungarica*, *Mimomys minor*, *Mimomys* cf. *reidi*, and *Pliolagus*, known as

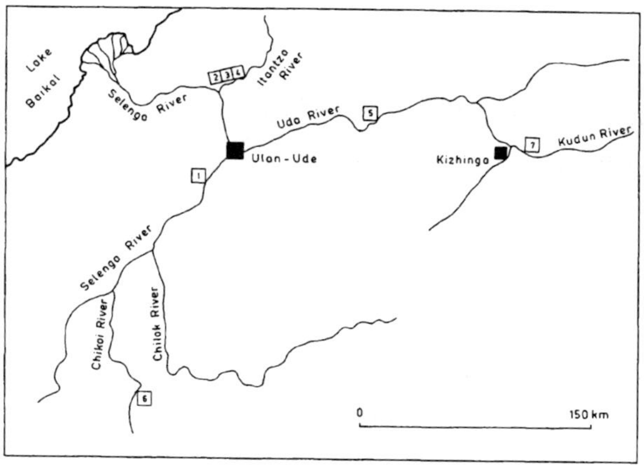

Fig. 1. Late Cenozoic fossil localities east of Lake Baikal

elements of the continental facies of steppe and forest-steppe environment of Europe. It is interesting, however, to note that the corresponding European faunas also include such Central Asian elements as Ochotonoides, widely distributed from Greece and Hungary as far as Northern China.

3. The Itantza Faunistic Complex

The Itantza faunistic complex is the following faunistic level in the West Transbaikal sequence. The corresponding level of Eastern Europe is perhaps the Chaprov complex. The Itantza complex includes the following taxa (VANGENGEJM 1977; AGADJANIAN and ERBAEVA 1983) (Fig. 1/2, 3, 4):

Ochotona sp.
Citellus (Citellus) cf. *tologoicus* ERBAEVA & POKATILOV
Citellus (Urocitellus) itancinicus ZAZHIGIN
Sinocastor sp.
Cricetulus (Tscherskia) triton varians (ZDANSKY)
Clethrionomys sp.
Mimomys cf. *newtoni* MAJOR
Villanyia ex gr. *laguriformes* ERBAEVA
Prosiphneus (ex gr. youngi-*pseudarmandi*) sp.
Equus ex gr. *sanmeniensis* TEILHARD DE CHARDIN & PIVETEAU
Itanzatherium angustirostre BELYAEVA
Cervidae gen. et sp. indet.
Alces latifrons (JOHNSON)
Gazella cf. *sinensis* TEILHARD DE CHARDIN & PIVETEAU
Ovibovini gen. et sp. indet.

Towards the end of this time interval the climate of Transbaikalia became still more arid and semideserts and dry steppes came to predominate as is confirmed by palynological data too. This fauna is also characterized by the predominance of Central Asian mammals, but the number of Lagomorphs is much smaller and species of both subgenera *Citellus* and *Urocitellus* appeared. *Villanyia* and *Mimomys* are not numerous. In the Itantza faunistic complex *Villanyia*, *Mimomys*, and *Prosiphneus* are represented by more progressive types than in the Chikoi complex. *Prosiphneus* was developing hypsodont teeth with poorly developed roots.

4. The Dodogol Micromammal Fauna

The next stage of development of the micromammal fauna is the stage of Dodogol (VANGENGEJM et al. 1975). The fauna is known from the reddish clays of the lower part of the Dodogol locality and from Zasuchino (the second faunistic horizon). The Dodogol fauna includes the following taxa: (Dodogol Loc. by AGADJANIAN and ERBAEVA 1983) (Fig. 1/1):

Ochotona cf. *hyperborea* PALLAS
Citellus (Urocitellus) sp.
Villanyia laguriformes ERBAEVA
Mimomys sp.
Allophaiomys sp.
Prosiphneus (ex gr. *youngi-pseudarmandi*) sp.

(Zasuchino Loc. by AGADJANIAN and ERBAEVA 1983; VANGENGEJM and SOTNIKOVA 1981)

Ochotonoides cf. *complicidens* BOULÉ & TEILHARD DE CHARDIN
Ochotona sp.
Citellus sp.
Villanyia cf. *laguriformes* ERBAEVA
Mimomys sp.
Allophaiomys cf. *pliocaenicus* KORMOS
Prosiphneus sp.
Rhinocerotidae gen. et sp. indet.
Coelodonta sp.
Equus ex gr. *sanmeniensis* TEILHARD DE CHARDIN & PIVETEAU
Cervidae gen. et sp. indet.
Bison sp.
Spirocerus wongi TEILHARD DE CHARDIN & PIVETEAU

It is possible to consider the Dodogol fauna as chronological equivalent of the Odessa complex of East Europe with *Mimomys* and *Villanyia*. *Villanyia* is represented in the Dodogol fauna by a more progressive species: *Villanyia laguriformes*. This species combines characters of both *Villanyia* and rootless *Lagurodon* by developing hypsodont teeth.

The Dodogol fauna is characterized by the appearance of *Allophaiomys*. *Allophaiomys*, *Mimomys*, and *Villanyia* are represented in both areas, in Transbaikalia as well as in Europe.

5. The Zasuchino Mammal Fauna

The post-Villafranchian stage of the Transbaikalian sequence is represented by the Zasuchino mammal fauna. It is known from the Zasuchino locality (brown sandy layer, the third faunistic horizon by VANGENGEJM and SOTNIKOVA 1981 AGADJANIAN and ERBAEVA 1983). The Zasuchino fauna includes the following taxa (Fig. 1/5):

Ochotona sp.
Marmota cf. *sibirica* RADDE
Citellus (Urocitellus) undulatus PALLAS
Clethrionomys sp.
Alticola sp.
Cricetulus sp.
Allophaiomys pliocaenicus KORMOS
Lasiopodomys cf. *brandti* RADDE
Pitymys ex gr. *hintoni-gregaloides* sp.
Myopus cf. *schisticolor* LILLJEBORG
Microtus ex gr. *oeconomus* PALLAS
Microtus ex gr. *mongolicus* RADDE
Castor sp.
Nyctereutes sp.
Canis variabilis PEI
Xenocyon sp.
Ursus sp.
Gulo sp.
Hyaena brevirostris cf. *sinensis* ZDANSKY
Panthera sp.
Felis sp. cf. *macrotis* MILNE-EDWARDS

Elephantinae gen. et sp. indet.
Equus ex gr. *sanmeniensis* TEILHARD DE CHARDIN & PIVETEAU
Coelodonta cf. *tologojensis* BELYAEVA
Rhinocerotidae gen. et sp. indet.
Cervus sp.
Alces latifrons (JOHNSON)
Megacerini gen. et sp. indet.
Bison sp.
Spirocerus cf. *wongi* TEILHARD DE CHARDIN & PIVETEAU
Ovibovini gen. et sp. indet.
Bovidae gen. et sp. indet.

The Zasuchino fauna is characterized by the appearance of *Pitymys*, *Lasiopodomys*, and some species of *Microtus*, *Allophaiomys* — a more progressive type than in the Dodogol fauna — *Pitymys*, *Microtus* (*Stenocranius*) are European elements in the Transbaikal fauna.

6. The Tologoi Faunistic Complex and the Further Faunistic Development in Transbaikalia

The following Tologoi complex of Transbaikalia is correlated with the Tiraspol complex of Eastern Europe and characterized by following taxa:

Ochotona daurica gureevi ERBAEVA
Citellus undulatus gromovi ERBAEVA
Allactaga saltator transbaikalieus ERBAEVA
Ellobius tancrei BLASIUS
Meriones unguiculatus MILNE-EDWARDS
Myospalax aspalax wongi YOUNG
Eolagurus simplicidens YOUNG
Microtus (Stenocranius) gregalis PALLAS
Microtus mongolicus RADDE
Lasiopodomys brandti RADDE
Archidiskodon sp.
Equus ex gr. *sanmeniensis* TEILHARD DE CHARDIN & PIVETEAU
Coelodonta tologoijensis BELYAEVA
Cervus ex gr. *elaphus* LINNAEUS
Spirocerus cf. *peii* YOUNG
Bison sp.

With this faunistic complex the climate of Siberia has changed to more or less periglacial environments. Probably, climate and vegetation resembled the recent conditions of Northern Mongolia. The main character of the Tologoi faunistic complex is the absence of voles with rooted teeth and the predominance of Central Asian elements: *Eolagurus*, *Meriones*, *Allactaga*, *Ellobius*, *Lasiopodomys*, etc. Of these elements *Eolagurus* and *Microtus* (*Stenocranius*) were widespread in Eurasia, as far as the Ukraine in the West.

Summary

The earliest Transbaikalian fauna complex to be discussed here is the Chikoi complex. It is very interesting to note that this complex includes a certain number of elements of the contemporaneous European faunas: *Beremendia fissidens*, *Petenyia hungarica*, *Mimomys minor*, *Mimomys* cf. *reidi*, *Pliolagus*, and *Ochotonoides*.

The following Itantza fauna complex has the following taxa in common with the European faunas of the same horizon: *Mimomys*, *Villanyia*, *Citellus*, *Allophaiomys*.

A later stage of Transbaikalian faunal evolution is represented by a fossil fauna recently discovered near Zasuchino, having the following taxa in common with East Europe: *Allophaiomys*, *Pitymys*, and *Microtus*.

The Tologoi fauna complex of Lower Pleistocene age (= lower Middle Pleistocene of European sequence) shows only *Eolagurus* and *Microtus (Stenocranius)* which are widespread at this period in Europe too.

During the Late Pleistocene *Microtus (Stenocranius) gregalis*, *Lagurus lagurus*, and *Microtus oeconomus* are known in both areas.

Zusammenfassung

Der früheste Faunenkomplex Transbaikaliens, der hier zur Diskussion steht, ist der Chikoi-Komplex. Es ist interessant aufzuzeigen, daß dieser Komplex eine gewisse Anzahl zeitgleicher europäischer Elemente umfaßt: *Beremendia fissidens*, *Petenyia hungarica*, *Mimomys minor*, *Mimomys* cf. *reidi*, *Pliolagus*, und *Ochotonoides*.

Der folgende Itantza-Faunenkomplex Transbaikaliens hat folgende Taxa gemeinsam mit den entsprechenden europiäschen Faunen: *Mimomys*, *Villanyia*, *Citellus*, *Allophaiomys*.

Die hangende Stufe der Faunenentwicklung Transbaikaliens wird durch die kürzlich entdeckte fossile Fauna von Zasuchino repräsentiert, die folgende Taxa mit den entsprechenden Faunen Osteuropas gemeinsam hat: *Allophaiomys*, *Pitymys* und *Microtus*.

Der Tologoi-Faunenkomplex unterpleistozäner Zeitstellung (= unteres Mittelpleistozän der europiäschen Folge) zeigt nur noch *Eolagurus* und *Microtus (Stenocranius)* als gemeinsame Formen, die zu dieser Zeit noch eine weite Verbreitung auch in Europa besaßen.

Aus dem Spätpleistozän beider Gebiete sind lediglich *Microtus (Stenocranius) gregalis*, *Lagurus lagurus* und *Microtus oeconomus* als gemeinsame Elemente zu nennen.

Резюме

В Западном Забайкалье наиболее древняя фауна млекопитающих известна как фауна чикойского комплекса. Она содержит много общих с европейскими фаунами видов мелких млекопитающих. Это — *Beremendia fissidens*, *Petenyia hungarica*, *Mimomys minor*, *Mimomys* cf. *reidi*, *Ochotonoides* и другие. Следующая во времени фауна — это фауна итанцинского комплекса. Общими с европейскими фаунами формами мелких млекопитающих являются *Villanyia*, *Mimomys*, *Citellus*.

За итанцинской фауной следует додогольскоая фауна, в составе которой известны прогрессивные формы родов *Mimomys* и *Villanyia* и появляются представители рода *Allophaiomys*.

Фауна, следующая за додогольской известна из местонахождения Засухино. Общими с восточноевропейскими фаунами элементами ювляются *Allophaiomys*, *Pitymys* и *Microtus*.

Фауна раннего плейстоцена — фауна тологойского комплекса — содержит только остатки широкораспространённых в Евразии представителей родов *Eolagurus* и *Microtus*. В среднем плейстоцене ареал *Lagurus transiens* и *Microtus (Stenocranius)* охватывал большую часть Евразии, однако, остатки *Lagurus transiens* в Забайкалье немногочисленны.

В позднем плеуйстоцене широко распространёнными формами, общими с европейскими фаунами, являются *Lagurus lagurus*, *Microtus (Stenocranius) gregalis*, *Microtus oeconomus*.

References

AGADJANIAN, A. K.; ERBAEVA, M. A., 1983: Late Cenozoic Rodents and Lagomorphs of the USSR. — Moskva. — 189 pp., in Russian.

ERBAEVA, M. A., 1974: Villafranchian fauna of small mammals of Western Transbaikalia. — In: Vth Congr. Néogène Méditerranèen. — In: Bur. Rech. Geol. et Minieres, Mém., — Paris **78**. — 137—139.

GROMOV, V. I., 1948: Palaeontological and archaeological arguments on the stratigraphy a of continental Quaternary deposits of the territory of the USSR. — Akad. Nauk SSSR, Tr. geol. Inst. — Moskva 64. — 520 pp., in Russian.

Nikiforova, K. V., 1982: Problems of the geology and history of the Quaternary (Antropogene). — In: Quaternary. — Moskva. — P. 6—35, in Russian.

Vangengejm, E. A., 1977: Palaeontological argument on the Anthropogene stratigraphy of Northern Asia. — Moskva. — 172 pp., in Russian.

Vangengejm, E. A.; Erbaeva, M. A.; Zhegallo, V. I.; Sotnikova, M. V., 1975: Palaeontological argument on the stratigraphy of the Antropogene of Western Transbaikalia. — Tr. Akad. Nauk SSSR, Geol. Ser. — Moskva (10). — P. 87—95, in Russian.

Vangengejm, E. A.; Sotnikova, M. V., 1981: Geology and mammal fauna of Zasuchino, Western Transbaikalia. — In: Bjull. Kommiss. izuč., četvertič. per. — Moskva **51**. — P. 106—117, in Russian.

Vangengejm, E. A.; Zazhigin, V. S., 1982: Faunistic complexes and fauna review of the USSR. — In: Stratigraphy of the USSR, Quaternary system. — Moskva. — P. 267—279, in Russian.

Quartärpaläontologie 6 29—42 Berlin 1986

Elephants of the Genus Archidiskodon from the Late Pliocene and Pleistocene of the Kuznetsk Basin (southeastern West Siberia)

By IRINA VLADIMIROVNA FORONOVA, Novosibirsk

With 4 Figures, 5 Tables and 2 Plates

Contents

Abstractum

Elephant teeth referred to four species of *Archidiskodon* are described: *Archidiskodon* cf. *gromovi* GARUTT & ALEXEEVA, *Archidiskodon* cf. *meridionalis* (NESTI), *Archidiskodon trogontherii* (POHLIG), and *Archidiskodon* (?) sp. They come from the Kuznetsk Basin in the south-eastern part of West Siberia and were found in the late Pliocene and Quaternary deposits (according to the scheme of the Quaternary adopted by the USSR Interdepartmental Stratigraphic Committee, MSK). Features of dental morphology of the Kuznetsk Basin archidiskodonts as well as their stratigraphic position are discussed.

In der vorliegenden Arbeit werden fossile Elefantenzähne der Gattung *Archidiskodon* beschrieben, die vier verschiedenen Spezies zugewiesen werden: *Archidiskodon* cf. *gromovi* GARUTT & ALEXEEVA, *Archidiskodon* cf. *meridionalis* (NESTI), *Archidiskodon trogontherii* (POHLIG) und *Archidiskodon* (?) sp. Die untersuchten Materialien stammen aus dem Kuznetsker Becken, südöstlicher Teil Westsibiriens und wurden in Ablagerungen pliozänen und quartären Alters (entsprechend dem Quartär-Schema des UdSSR Stratigraphischen Komitees, MSK) gefunden. Weiter werden Zahn-Morphologie und stratigraphische Stellung der Archidiskodonten aus dem Kuznetsk Becken diskutiert.

В статье описываются зубы слонов рода *Archidiskodon*, относящиеся к четырем видам: *Archidiskodon* cf. *gromovi* GARUTT & ALEXEEVA, *Archidiskodon* cf. *meridionalis* (NESTI), *Archidiskodon trogontherii* (POHLIG) и *Archidiskodon* (?) sp. Находки сделаны в верхнеплиоценовых и четвертичных отложениях (по схеме МСК СССР) Кузнецкой котловины, расположенной на юго-востоке Западной Сибири. Рассматриваются особенности строения зубов архидискодонтных слонов этой территории, а также их стратиграфическое положение.

1. Introduction

Fossils of *Archidiskodon* are rather rare in the Pliocene and Pleistocene of the Asiatic part of the USSR, and within the Kuznetsk Basin the author has identified them for the first time. Late Pleistocene *Mammuthus primigenius* (BLUMENBACH) and *Bison priscus* (BOJANUS) were the earliest records of fossil mammals in this region, mentioned by B. K. POLENOV (1897), A. A. INOSTRANTSEV (1897), I. P. TOLMACHOV (1909), V. I. YAVORSKY and P. I. BUTOV (1927), P. M. RYZHKOV (1927). Later elements of the Pliocene/early Pleistocene fauna were reported from the region (ALEXEEVA 1970, 1980), but there were no proboscideans belonging to *Archidiskodon* among them.

In recent years some new fossil mammal localities have appeared in the Kuznetsk Basin. They are associated with open-cast coal mining activities in which the Neogene and Pleistocene cap deposits are washed away with hydromonitors. During the period 1974 to 1981 the author collected more than 5000 fossils from these deposits referred to 40 species of mammals, including carnivores, proboscideans, and ungulates.

Only the molar teeth of archidiskodonts are considered in the present discussion, being the most reliable material to solve the problems of proboscidean taxonomy and phylogeny. System of measurements worked out by V. E. GARUTT and the author is followed in this study (GARUTT and FORONOVA 1976). Osteological collections of the Zoological and Geological Institutes (USSR Academy of Sciences) were studied for comparison. Dr. V. E. GARUTT has kindly given to me the opportunity to use measurements and occlusal surface prints of elephant teeth from collections elsewhere in the Soviet Union and other countries (G.D.R., Roumania, ČSSR).

The author is sincerely grateful to her teachers — Drs. S. A. ARKHIPOV and V. E. GARUTT — for their valuable consultations and help during the work.

2. Stratigraphy and Faunal Localities

The Kuznetsk Basin lies in the south-eastern part of West Siberia surrounded by the Kuznetsk Alatau Mountains in the east, Gornaya Shoriya in the south, Salair Ridge in the west, and Kolyvan-Tom Upland in the north-west (Fig. 1). The Basin was a broad fluvio-lacustrine plain during the late Cenozoic. The sequence of the Neogene and Quaternary deposits overlying Palaeozoic bedrock is more than 100 m thick.

The collection of fossil bones along with geological and other evidence allowed the author, with Dr. ARKHIPOV's consultation, to subdivide this sequence into 10 biostratigraphical units of late Pliocene/Pleistocene age (GALKINA and FORONOVA 1980; FORONOVA 1982a). With some modifications of range and faunal characteristics these units are ranked at

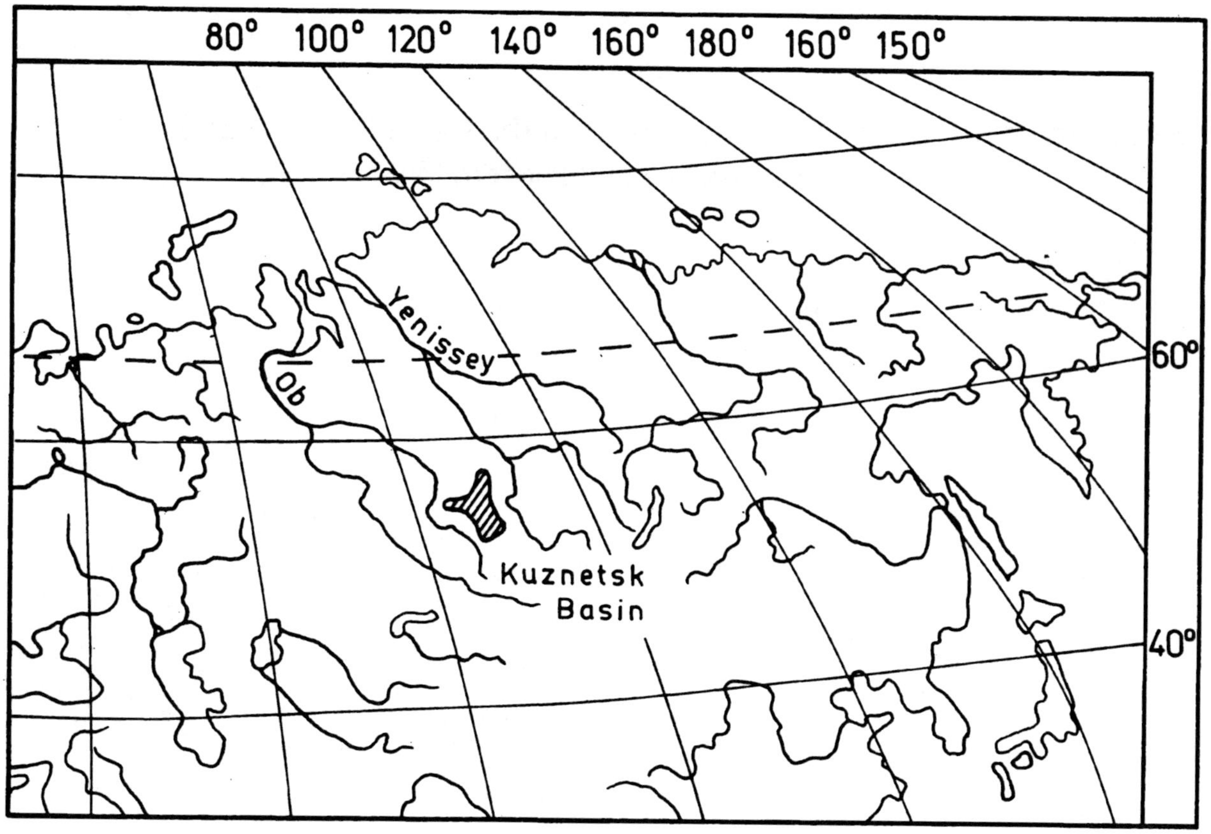

Fig. 1. Location of the Kuznetsk Basin (shaded)

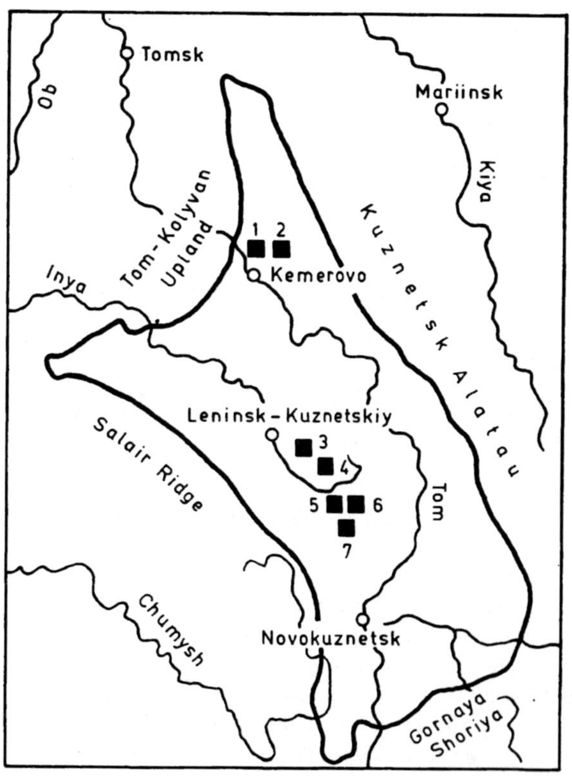

Fig. 2. Scheme of the Kuznetsk Basin, showing location of fossil mammal localities

1 — Kedrovsk; 2 — Chernigovsk; 3 — Mokhovsk; 4 — Gramoteinsk; 5 — Bachatsk; 6 — Novosergeyevsk; 7 — Krasnobrodsk

present as suites within the new stratigraphic scheme for the Altai-Sayan region (ZUDIN et al. 1982). The Neogene/Quaternary boundary in this scheme is traditionally drawn at the base of the Siberian equivalents of the Bakinian Stage, or approximately at the datum of 0.7—0.8 mill. yrs. Correlation of the mentioned units and suites is shown in the Table 1.

The lowermost part of the Neogene/Quaternary succession comprises variegated clays overlying Palaeozoic bedrock and is interpreted as the worked products of an ancient residual soil. These clays are distinguished as Unit I. This unit contains no fossils but it can be compared lithologically with the Miocene clays of West Siberia and is called the Meret Suite (ZUDIN et al. 1982). The overlying Unit II is presumed to be of Middle(?)-Late Pliocene age and comprises red-brown compact, unstratified clays. This unit corresponds to the Mokhovsk Suite. Of the larger mammals only rare an mainly indeterminate bones, deformed and highly mineralized (or enclosed in concretions) are known from this unit. L. I. GALKINA reports the occurrence of *Villanyia hungaricus* KORM., *Mimomys intermedius* NEWT.[1]), *Allophaiomys pliocaenicus* KORM., *Prolagurus praepannonicus* TOP., *Eolagurus argyropuloi* GROM. & PARF., *Prosiphneus* sp. and some other rodents in the upper part of these deposits (ZUDIN et al. 1982).

The fluvio-lacustrine sediments which make up Unit III fill erosional cuts in the red-brown clays of Unit II or even in Palaeozoic rocks. These very compact grey clays and loams with sand and gravel beds and lenses are distinguished as the Sagarlyk Suite. Here L. I. GALKINA has found fossils of *Eola-*

[1]) = *Cromeromys intermedius* (NEWT.) ZAZHIGIN, 1980

Table 1. Stratigraphy of the Neogene and Quaternary deposits of the Kuznetsk Basin

Standard scale		Suites (ZUDIN et al. 1982)	Units (GALKINA and FORONOVA 1980)
Pleistocene	Upper	Elovka	X
	Upper	Krasnobrodsk	IX
	Upper / Middle	Bachatsk	VIII
	Middle	Kedrovsk	VII
	Lower		VI
	Lower		V
	Lower	Sergeyevsk	IV
Pliocene		Sagarlyk	III
		Mokhovsk	II
Miocene		Meret	I

gurus argyropuloi GROM. & PARF., *Prolagurus pannonicus* (KORM.) of *posterius*-type, *Sicista* sp. and other rodents. Unit IV, comprising pale yellow-brown compact unstratified loams with buried soils, overlies Unit III. Fossils of a very large, archaic horse have been found in situ in the upper member of Unit IV as exposed in the Bachatsk Quarry. This horse is comparable in size with *Equus* (*Plesippus*) *verae* SHER from Olyor Suite of the Kolyma Lowland and is sometimes even larger. Some bones of large *Bison* sp. as well as of *Spirocerus* sp. and Cervidae have also been found here. To judge by the presence of an archaic horse which corresponds to pre-caballoid horses of Eurasia and North America in its evolutionary level, the pale yellow-brown loams of Unit IV are suggested to be of Late Pliocene age. This unit is distinguished as the Sergeyevsk Suite (Table 1).

One tooth of *Archidiskodon* cf. *gromovi* GARUTT & ALEXEEVA has been found in these loams in the course of their being washed out with hydromonitors in the Novosergeevsk Quarry (Fig. 3-A; Plate I, Fig. 1). Interlamellar spaces and cracks on the surface of the tooth are filled with a compact oxidized sand and gravel consisting of fragments of Palaeozoic rocks and coal crumbs, which suggests that it may be possible to relate the tooth to a gravel bed of similar composition in the basal part of Unit IV, but at the same time, as the tooth is rounded, it is possible that it may have been derived from still earlier sediments of Unit III, or even Unit II. One tooth of *Archidiskodon* cf. *meridionalis* (NESTI) has been found in the talus of the same loams in the Mokhovsk Quarry (Fig. 3-B, Plate I, Fig. 2). Assignment of this tooth to Unit IV is considered justified as in this particular section this unit is underlain by Palaeozoic rocks and overlain by the late Pleistocene loams.

Later deposits distinguished as Units V and VII correspond partially to the Kedrovsk Suite in the new stratigraphic scheme of the Kuznetsk Basin (ZUDIN et al. 1982) but this suite does not include our Unit VI composed of loess-like loams with buried soils (Table 1).

These loams have only been preserved as local patches probably on some undenuded interfluves. Corresponding to the Kedrovsk Suite are Units V and VII. They occupy erosional features and are commonly separated by a gravel bed. Both units are of fluvial and lacustrine origin and are composed of peculiar bluish-grey compact stratified silts and clays. They include wood remains, lenses of plant detritus, freshwater molluscs and ostracods. Subdividing these lithologically similar deposits into two units, V and VII, has been made on the basis of their contained mammal fauna (FORONOVA 1982a, b). The lower part of the clays (Unit V) contains numerous fossils indicating an early Pleistocene age (according to the MSK scheme, Middle Pleistocene according to central European stratigraphy) including *Archidiskodon trogontherii* (POHL.), *Equus* cf. *mosbachensis* REICH., *Coelodonta* sp., *Cervus* cf. *elaphus* L., *Bison priscus* (BOJ.). Also a tooth of *Archidiskodon* (?) sp. has been found in the hydromonitor face in the Bachatsk Quarry where the basal part of the clays and upper part of the underlying pale yellow-brown loams of Unit IV are exposed (Fig. 3-C).

The upper part of the bluish-grey clays (Unit VII) and the lower part of overlying light-brown loams with buried soils are correlated with the Bachatsk Suite (Unit VIII) and have yielded fossils of *Spelaearctos rossicus* BORISS., *Mammuthus chosaricus* DUBROVO, *Coelodonta antiquitatis* (BLUM.), *Megaloceros giganteus* (BLUM.), *Cervus elaphus* L., *Rangifer tarandus* (L.), *Bison priscus* (BOJ.) and of a large broad-footed caballine horse (Fig. 3-C). The author believes that the upper part of the Kedrovsk Suite and the lower part of the Bachatsk Suite correspond to the Middle Pleistocene of the MSK scheme (FORONOVA 1982b).

Two upper units of loess-like loams (IX and X) correlated respectively with the Krasnobrodsk and Elovka Suites contain an abundance of bone remains of Late Pleistocene mammals. Among them *Mammuthus primigenius* (BLUM.), *Equus caballus* L., and *Bison priscus* (BOJ.) (both now small in size), *Coelodonta antiquitatis* (BLUM.) and *Panthera spelaea* (GOLDF.) are dominant (Fig. 3-C).

3. Systematic Description

Order Proboscidea ILLIGER, 1881

Family **Elephantidae** GRAY, 1821

There are many opinions regarding the systematic position of the animals included in this family. Most students classify its members into three subfamilies: Mammuthinae, Loxodontinae, and Elephantinae.

Subfamily **Mammuthinae** SIMPSON, 1945

Genus **Archidiskodon** POHLIG, 1885

The scope of this genus is variously understood by palaeontologists. The author follows V. E. GARUTT and E. A. VANGENGEJM in considering it to include three species: *Archidiskodon gromovi* GARUTT & ALEXEEVA, *A. meridionalis* (NESTI), and *A. trogontherii* (POHL.) (= *Elephas wüsti* M. PAVL.) (GARUTT 1981; GARUTT and VANGENGEJM 1982).

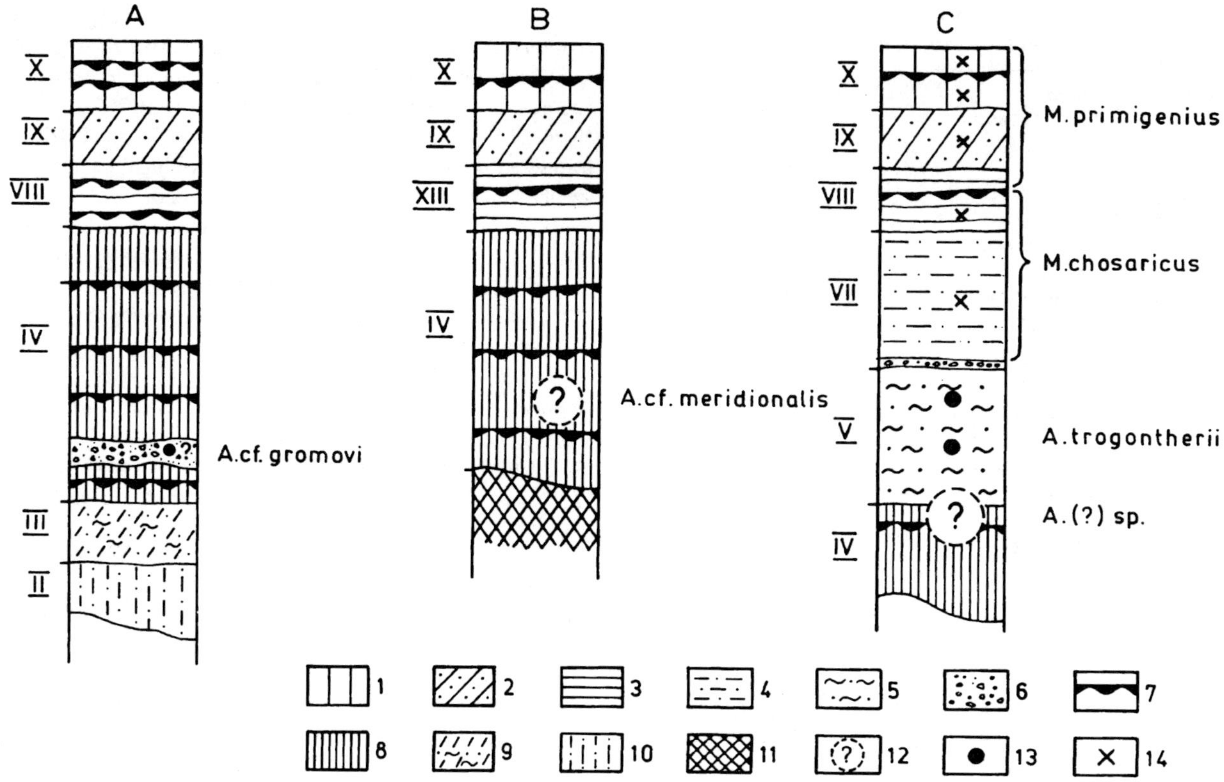

Fig. 3. Sections of the Pliocene-Quaternary deposits in the Kuznetsk Basin containing fossil Proboscidea

Sections: A — Novosergeevsk; B — Mokhovsk; C — Bachatsk I—X — units according to GALKINA and FORONOVA (1980) (see also Table 1 for comparison) 1 — loess-like loams; 2 — silty stratified loams; 3 — light-brown loess-like loams; 4 — bluish-grey stratified silts and clays of fluvio-lacustrine origin; 5 — compact fluvio-lacustrine bluish-grey stratified clays; 6 — sand-gravel bed; 7 — buried soils; 8 — pale-brown compact unstratified loams; 9 — compact grey sandy clays and loams; 10 — compact red-brown unstratified clays; 11 — Palaeozoic rocks; 12 — supposed place of occurrence of the fossil in the section; 13 — fossils of archidiskodont elephants; 14 — fossils of elephants of the genus *Mammuthus*

3.1. **Archidiskodon** cf. **gromovi** GARUTT & ALEXEEVA, 1965
Plate I, Fig. 1

Material and locality: Anterior fragment of the right upper M^3, heavily worn and rounded (IGG[2]) No. 3580), Novosergeevsk Quarry near Krasnobrodsky settlement. Lower part of the Sergeyevsk Suite? (probably derived from earlier sediments, see discussion above).

Geological age: Late Pliocene?

Description: The M^3 fragment consists of three very thick plates. The height of the remaining fragment is 65 mm, the length 79 mm and the width 97 mm. The tooth appears to consist of plates only, interlamellar spaces being virtually absent. This seems to be an effect of heavy wear. Average anteroposterior length of a plate is 20 mm. Though deeply worn the tooth retains a pronounced attrition figure due to deep plate dissection consisting of two rounded lateral ovals and a broad median sinus (— o —). The enamel is thick, roughly folded and strongly projecting on the occlasal surface. Its thickness averages 3.5 mm.

Comparison: The tooth in question differs from *A. meridionalis* (NESTI) and *A. trogontherii* (POHL.) in the larger average plate length and enamel thickness as well as in the peculiar attrition figures. The distinct figure can be traced almost down to the base of the crown, each plate being deeply dissected with vertical slits into three roughly equal parts. Special importance is attached to this feature which is considered to be more pronounced in the geologically most ancient forms. In enamel thickness and the nature of subdivision of the plates and their frequency, the tooth does not differ from the *A. gromovi* teeth reported from Liventsovka and Khapry in the European part of the USSR (ALEXEEVA and GARUTT 1965; ALEXEEVA 1977; GARUTT and BAJGUSHEVA 1981) and from Podpusk in southern West Siberia (VISLOBOKOVA 1974). Measurements in Table 2 indicate this tooth to fall inside the limits of variation of *A. gromovi* teeth. However, the fragmentary nature of this fossil allows it to be classified only as resembling GROMOV's elephant, i.e. as *A.* cf. *gromovi* GARUTT & ALEXEEVA.

Remarks: *Archidiskodon gromovi* is the earliest species of archidiskodonts so far known in the USSSR. It was defined as a distinct species by GARUTT and ALEXEEVA (1965) based on specimens from the sand quarries of Liventsovka and Khapry in Rostov on Don region. The earliest *A. gromovi* remains have been recorded in the south of the European part of the USSR, being members of local faunas related to the Moldavian complex (Kosyakino, Kotlovina), i.e. faunas of Rousillon-type dated from the earliest part of the late Pliocene (GARUTT and BAJGUSHEVA 1981). In West Europe, according to AZZAROLI (1977), *A. gromovi* is characteristic of early and middle Villafranchian faunas but some students (AGUIRRE 1969; MAGLIO 1973) rank the distinction between *A. gromovi* and *A. meridionalis* as of subspecies level only.

[2]) Institute of Geology and Geophysics, Siberian Branch of the USSR Academy of Sciences, Novosibirsk.

Table 2. Measurements of *Archidiskodon gromovi* teeth (in mm)

	Kuznetsk Basin *A.* cf. *gromovi* IGG, No. 3580 M^3	Azov Sea Region *A. gromovi*, M^3 (GARUTT and BAJGUSHEVA 1981)	Podpusk *A. gromovi*, M^3 IGG, No. 288/314, (VISLOBOKOVA 1974)
Crown length	—	203—(240)—295*)	232.3
Crown width	97	85—(110)—120	87.2
Crown height	—	105—(134—139)—147	104.0
Total number of plates	—	11—(13)—15	11
Average anteroposterior length of a plate	20	15.8—(20.2)—25.6	—
Plate frequency (number of plates in 100 mm)	5	3.3—(4.5)—5.5	4.5—5.0
Enamel thickness	3.5	2.6—(3.2—3.5)—4.4	3.6—3.8

*) In brackets — optimum numerical characteristics most frequently noted in the particular set of specimens.

In the southern part of West Siberia *A. gromovi* remains have been found near Podpusk village in deposits correlated with the Kochkov Suite of the Pavlodar Priirtyshye. The species is characteristic of the Podpusk-Lebyazhean faunal complex (VISLOBOKOVA 1974).

3.2. **Archidiskodon** cf. **meridionalis** (NESTI, 1885)
Plate I, Fig. 2

Material and locality: Left upper M^3, IGG No. 4096, about five anterior plates missing. Mokhovsk Quarry near Mokhovsky settlement, Sergeyevsk Suite (?).

Geological age: Final part of the Late Pliocene?

Description: Size large (width 105 mm, height of unworn plate 175 mm). Of the nine preserved plates four are worn. Average anteroposterior length of a plate is 19 mm.

Enamel thick (averages 3.2 mm) and roughly folded. The plates are clearly separated into three ovals. Dissection reaches the middle of the crown, being clearly distinctive in transverse section. A second section at 4 cm from the crown base reveals the plates loosing the three section structure and the ovals merging into one, slightly broadened in the middle.

The very dense arrangement of plates is characteristic of this tooth. The section through the middle of the crown shows interlamellar spaces to be present between the ends of plates only, while in basal section the spaces are virtually absent, the plates being tightly pressed one to another. This peculiar feature is common both to the tooth under discussion and to the above *A.* cf. *gromovi* tooth.

Comparison: The tooth examined differs from *A. gromovi* teeth in having shallower plate dissection into three parts, in its lower average plate anteroposterior length and enamel thickness.

It shows some resemblance to *A. trogontherii* in the pattern of plate dissection and in the mean anteroposterior length of a plate. The latter value is taken as the anteroposterior length of a plate itself plus half of the interplate space on the anterior and posterior sides of the plate. As noted above, this particular tooth has a mean plate length virtually equal to the length of the plate itself. The plates of *A. trogontherii* teeth are considerably thinner while the interplate space much broader. Hence, the two forms compared have metrical similarity with respect to the mentioned value but not mor-

Table 3. Measurements of *Archidiskodon meridionalis* molars [in mm]

	Kuznetsk Basin	South part of the European part of the USSR (ALEXEEVA 1977)		Sinyaya Balka (DUBROVO and BAJGUSHEVA 1964 cited in ALEXEEVA 1977)
	A. cf. *meridionalis* IGG No. 4096	*A. m. meridionalis*		*A. m. tamanensis*
	M^3	M^3	M^3	M^3_3
Crown length	—	300	265?	**252—317** 159—328
Crown width	105	88—117	92—104	**95—115** 82—122
Crown height	175	115?—123	95?—105	**127—185** 115—120
Total number of plates	15	15—16	—	**12— 17** 12— 17
Plate frequency (100 mm)	6	$4^{4/5}$—$5^{1/2}$	$4^{1/3}$—$5^{1/2}$	**4.5—6.5** 6 —6
Enamel thickness	3.2	3.0—3.5	3.2—3.5	**2.5—3.0** 2.5—4.0

Table 4. Measurement of *Archidiskodon trogontherii* molars [in mm]

	Kuznetsk Basin		East Siberia (VANGENGEJM 1961)		Ural (GARUTT et al. 1977)	
	A. trogontherii		*E.* cf. *wüsti*	*M.* cf. *trog.*	*A. trogontherii wüsti*	
	M^3	M_3	M_3	M_3	M^3	M_3
Crown length	—	336	—	—	225—305	223—293
Crown width	177—122	97— 98	70	70	83—112	78—95
Crown height	150—180	150—155	135	—	139—195	120—217
Total number of plates	—	19—20	—	—	19—23	19—21
Plate frequency (100 mm)	4.3—5.3	4.3—6.8	5.5—6	6—6.5	6.5—9.0	5.5—7.5
Enamel thickness	2.6—2.8	2.5—2.8	2.2	2—2.5	1.6—2.5	1.7—2.3

phological similarity (Tables 3, 4). Moreover, the two forms differ in enamel thickness.

The tooth examined resembles most closely *A. meridionalis* (NESTI) in crown proportions, enamel thickness and plate number and frequency. At the same time it differs inhaving essentially narrower interlamellar spaces and in the absence of a median sinus and a *meridionalis*-like attrition figure (—0—) at the base of the crown. A similar absence of the mentioned features has also been reported for teeth resembling *A. meridionalis* from Ust-Talovka (Altai foothills) by VISLOBOKOVA (1973). The tooth described also differs from the Altai specimen in having shorter interlamellar spaces. In crown height and plate frequency the tooth from the Kuznetsk Basin bears the closest resemblance to *A. meridionalis tamanensis* DUBROVO, but taking account of all the tooth features as well as the fragmentary nature of the specimen we are inclined to identify it as *A.* cf. *meridionalis* (NESTI). It may correspond to the late, i.e. Tamanian, subspecies.

Remarks: Being the lineal descendant of *A. gromovi A. meridionalis* was widely distributed in Europe and Asia in the Late Pliocene and Early Pleistocene. Having reached the southern regions of the European part of the USSR (where one of their migrational routes might lie) these elephants spread in two directions, to south-western Europe and to Siberia via southern Ural and further to the North-East (OSBORN 1942; GARUTT 1965; GARUTT and BAJGUSHEVA 1981).

Recent studies (ALEXEEVA 1977) recognize seven subspecies of *A. meridionalis*, only three of which are known from the USSR: *A. m. meridionalis*, *A. m. taribanensis* and *A. m. tamanensis*[3]). Geologically younger subspecies, such as *A. m. cromerensis*, *A. m. voigtstedtensis*, *A. m. jockgrimensis* and *A. m. vestinus*, are characteristic of the Günz/Mindel and early Mindel of the Alpine scheme and are described on the basis of West European specimens[4]).

Distinction of some subspecies may quite probably be accounted for by different evaluation of their characters. Some students think *A. m. taribanensis* GABUNIA & VEKUA from Taribana in Georgia should merit separation as a separate species (ALEXEEVA 1977), while others hold to the opinion that it does not differ essentially in skull structure from the nominate subspecies *A. m. meridionalis* (AZZAROLI 1977). The latter author recognizes three advanced subspecies of *A. meridionalis*: *A. m. cromerensis*, *A. m. tamanensis* and *A. m. vestinus* besides the nominate one.

In the south of West Siberia a form similar to *A. meridionalis* existed during the final part of the Pliocene (upper Eopleistocene, according to the scheme of GROMOV et al. 1969). It was a member of the Kizikhan and Razdolean faunal complexes (VISLOBOKOVA 1973). Later on some subspecies may be recognized here corresponding to *A. m. meridionalis* and *A. m. tamanensis* in evolutionary level (VISLOBOKOVA 1973; FORONOVA 1982c).

3.3. **Archidiskodon trogontherii** (POHLIG, 1889)
Plate II, Fig. 1

Material and locality: a) Fragment of left upper M^3, IGG No. 8, b) Fragment of left upper M^3, IGG No. 5, c) Fragment of left lower M_3, IGG No. 686, Novosergeevsk Quarry, Krasnobrodsky settlement. d) Right lower M_3, IGG No. 607, e) Left lower M_3, IGG No. 605, Bachatsk Quarry, Starobochaty settlement. f) Fragment of left lower M_3, IGG No. 23, Kedrovsk Quarry, Latyshy village.

All the teeth are from the bluish-grey clays and silts of the Kedrovsk Suite.

Geological age: Early Pleistocene.

Description: a) M^3 No. 8 has 3 1/2 plates preserved. Average anterioposterior length of a plate is 21 mm. The interpalate spaces are broad (up to 15 mm) and the crown is rather wide (117 mm). The height of the least worn plate is 150 mm. Attritional figures in the initial stage of wear consist of three ovals (———). In the middle part of the crown (in cut section) the side and median ovals fuse to form a complete loop broadened in the middle. Average enamel thickness is 2.8 mm.

b) M^3 No. 5 has 15 plates preserved, about four plates are broken away in the anterior part of the crown. Size large; the fragment is 256 mm long. The crown is considerably widened in the middle, narrowing uniformly posteriorly. The least unworn plate is 180 mm high. Average anteroposterior plate length 17.6 mm. Seven plates are affected by attrition. In the early stage of wear they show a pronounced pattern of three equal ovals (———). On the anterior plates which are worn down to form complete loops, three oval patterns can still be observed on account of two waists separating the plate into three almost equal parts. Average enamel thickness is 2.6 mm. In a section cut through the middle of the crown the mean length of a plate is 19 mm; almost all plates show median broadening but at this level vertical dissection into three parts is almost unpronounced. Enamel thickness in section is 2.7—2.8 mm.

[3]) GARUTT (1981) is inclined to consider *A. m. tamanensis* DUBROVO as a separate species *A. tamanensis* DUBROVO.

[4]) French scientists have recently described one more subspecies of Southern elephant, *Mammuthus meridionalis depereti* (COPPENS & BEDEN, 1982).

Table 4 (continue)

Tiraspol (DUBROVO 1971)		Süssenborn (DUBROVO 1971)		Tiraspol (ALEXEEVA 1977)	
M. trogontherii trogontherii		*M. trogontherii trogontherii*		*Archidiskodon wüsti*	
M^3	M_3	M^3	M_3	M^3	M_3
303—370	300—410	246—400	225—383	300	310—350
86—115	79—115	73—120	71—114	95—109	80—132
160—211	130—173	129—212	127—167	165—210	125—200
18—23	16—21	16—23	16—21	22	19—$20^{1/2}$
5.0—8.5	4—7	4.0—8.25	4—8.25	4—$7^{1/5}$	$5^{1/2}$—7
2—3	2—3	2—3	2—3.3	1—3	2—3

c) M_3 fragment No. 686 includes nine plates. Crown measurements are not listed because of heavy damage. Average anteroposterior length of a plate is 17.5 mm. Attrition figures are distinctive (———), the median broadening being well-pronounced. Average enamel thickness 2.6 mm.

d) and e) Two M_3's Nos. 607 and 605 are very similar and may belong to the same individual. The teeth are large and well preserved. The right one is complete, the posterior part of the left is broken. Crown lengths are 336 and 266 mm respectively, crown widths 98 and 97 mm, heights of unworn plates 150 and 155 mm. The total number of plates is 19, the left tooth lacking about 3—4 plates. Average anteroposterior lengths of plates 15.4 and 16.2 mm respectively. In the initial stages of wear the attrition figure of the plates looks like three ovals, later becoming band-like with pronounced median broadening. Enamel is roughly folded, average thickness ranging 2.5—2.7 mm.

f) M_3 No. 23 includes 7 1/2 plates. The measurements are omitted, the crown being partially damaged and broken posteriorly. The plates are large and widely separated. Average anteroposterior length of a plate is 20.6 mm. Each plate shows a complete loop with pronounced median broadening. Enamel thickness averages 2.8 mm.

Comparison: A trogonthere elephant holding an intermediate position between *Archidiskodon* and *Mammuthus*. The teeth under discussion are here compared with both genera. Kedrovsk Suite teeth differ from *A. gromovi* and *A. meridionalis meridionalis* in larger crown size, in attrition figure, in larger number of plates and less thick enamel. They have some resemblance to the latter species in lamellar frequency.

The teeth from the Kedrovsk Suite resemble an advanced subspecies of the southern elephant, *A. meridionalis tamanensis*, in almost all features with the exception of the total number of plates, being 19—20 in the former, while maximum 17 in the latter. They differ from *Mammuthus primigenius* having fewer plates which are thick and widely spaced, in their regular attrition figures, deep plate dissection and thick enamel. Some of them resemble teeth of *Mammuthus chosaricus* DUBROVO in the crown size, while being distinguished by their plate dissection pattern, thicker plates and greater enamel thickness. In the Khozar elephant, as in *M. primigenius*, the plates are not dissected deeply; the vertical slits are arranged unsystematically, whereby the attrition figure varies and is usually indefinite. The median expansions are absent as a rule.

In the teeth under study deep dissections of plates into three parts can be observed, forming peculiar attrition figure characteristic mainly of *A. trogontherii* (———). The vertical slits do not penetrate to the very base of the crown and in well worn plates the ovals fuse into complete loops with median expansions.

Thus, the number and frequency of plates, enamel thickness and distinct attrition figure make it possible to refer the teeth examined to *Archidiskodon trogontherii* (POHLIG).

The extreme values of measurable features[5]) of the Kuznetsk Basin *A. trogontherii* teeth have been compared with the same of *A. meridionalis* from the south of the European part of the USSR (ALEXEEVA 1977) of *Elephas* (?) cf. *wüsti* and *Mammuthus* cf. *trogontherii* from East Siberia (VANGENGEJM 1961), *A. trogontherii wüsti* from Ural (GARUTT et al. 1977) and *Mammuthus trogontherii trogontherii* from Süssenborn and Tiraspol localities (DUBROVO 1971). *Archidiskodon trogontherii* teeth from the Kuznetsk Basin have been found to be more similar in morphology to the European ones, as well as to the teeth of late representative of southern elephant (*A. m. tamanensis*). The teeth described differ from those of the East Siberian and Uralian *A. trogontherii* in having somewhat lower plate frequencies and in having thicker enamel (Table 4) (FORONOVA 1982c).

Remarks: *Archidiskodon trogontherii* (POHLIG) (= *Elephas wüsti* M. PAVL.) the last representative of the archidiskodonts, was widely distributed in Eurasia during the Early Pleistocene. In the USSR territory it was a member of the Tiraspolian (East Europe), Koshkurganian (Kazakhstan) and Vyatkinian (West Siberia) faunal complexes.

Elephas trogontherii was first described by POHLIG (1885, 1889), who distinguished two varieties. One, an archaic form, close to the southern elephant in tooth morphology, was named *E. meridionalis trogontherii*, the other, an advanced form approaching the woolly mammoth, was named *E. primigenius trogontherii*. According to POHLIG, this species occupies an intermediate position between the southern elephant and woolly mammoth both in morphological and geological respects.

Repeated attempts have since been made to define the taxonomic and stratigraphic limits of this species more precisely. PAVLOVA (1910) distinguished a new species, *Elephas wüsti*, based on elephant teeth from the Kolkotov gully near Tiraspol and on teeth from Süssenborn which had been previously referred by H. POHLIG and E. WÜST

[5]) Results of comparison are more reliable when optimum numerical characteristics of the features are used (GARUTT and FORONOVA 1976). In the present case comparison of optimum values was not possible.

to the earlier variety of *E. trogontherii*, thus narrowing the scope of the POHLIG's species to include only the advanced variety close to the woolly mammoth. However, H. POHLIG clearly indicated that the regarded the Süssenborn specimens as being most typical of his species (GARUTT 1971).

Up to now the taxonomic position of *A. trogontherii* has been variously interpreted. Soviet students have also been unable to agree on the point. DUBROVO (1963) places *E. wüsti* in synonomy with *E. trogontherii* POHLIG. She puts *E. trogontherii* into the genus *Mammuthus* and recognizes two subspecies. The earlier one is her *M. trogontheri trogontherii* corresponding to POHLIG's variety *E. meridionalis trogontherii* from the lower Pleistocene (Mindel I—II). The later one, more advanced, has been named by her as *M. trogontherii chosaricus*. The latter corresponds to *E. primigenius trogontherii* POHLIG from the Middle Pleistocene (Mindel/Riß, Riß) (DUBROVO 1966).

YATSKO (1948) and ALEXEEVA (1977) hold to the opinion that *E. wüsti* is a separate species of *Archidiskodon* — *A. wüsti* M. PAVL., 1910 —, while *E. trogontherii* is an early representative of the mammoth group (*Mammuthus*).

The present author shares another view according to which *E. wüsti* is identified as an early variety of *E. trogontherii* referred to the genus *Archidiskodon* based on diagnostic features of generic value (GARUTT 1971, 1972, 1981; GARUTT and VANGENGEJM 1982)[6]). According to this view it should be named *Archidiskodon trogontherii* (POHLIG) (= *Elephas wüsti* M. PAVL.).

3.4. **Archidiskodon** (?) sp.
Plate II, Fig. 2

Material and locality: Left lower M_2, IGG No. 15. About two plates broken off in the anterior part of the crown. Bachatsk Quarry, Starobochaty settlement. Upper member of the Sergeyevsk Suite or basal part of the Kedrovsk Suite.

Geological age: Early Pleistocene?

Description: A tooth of large size comprising 7 1/2 plates preserved (the original number of plates was apparently no more than nine. The sides of the tooth are covered with a thick deposit of cement[7]). Crown length is 157 mm, width 95 mm, incomplete height is 100 mm (all the plates are partially worn down). A deep depression in the posterior part of the crown represents a contact facet with M_3. Average anteroposterior length of a plate is 20 mm, plate frequency being 5. The enamel is thick (ranging 2.4—3.4 mm, average 3.0) and folded, especially on the anterior plates.

Different stages of development of enamel figures can be seen on the occlusal surface due to the oblique wear of the tooth (non-parallel to the crown base) which is typical of elephants. The two most worn anterior plates have rhombic median expansions — "loxodont sinuses" — and fuse at these expansions (Plate II, Fig. 2). The less worn third and fourth plates show median expansions in the form of angular projections on the posterior enamel wall. The fifth, sixth and seventh plates consists of several styles, worn but not fused into the complete loop. Median styles appear to be shifted posteriorly. One-sided expansion (like on the third and fourth plates) seem to be formed when these median shifted-back styles fuse with the standing-in a-line ones.

Comparison and taxonomic remarks: The tooth under discussion has no common features with the teeth of elephants referred to the genus *Mammuthus*. It differs markedly from *Palaeoloxodon* in its proportions and the considerable width of its crown, as well as in having thicker enamel. However, it resembles *Palaeoloxodon* in having loxodont sinuses on the anterior, most worn plates.

In most features the present specimen resembles *Archidiskodon*, especially the earliest representatives of the genus. Apart from enamel pattern, it differs from *A. trogontherii* in having thick enamel and a more archaic appearance. Metrical values of crown size, plate number and frequency and enamel thickness of this tooth fall within the limits of variation of the same features in *A. gromovi* and *A. meridionalis*. At the same time, it differs essentially from them in the rather peculiar pattern of the worn plates.

The tooth from the Kuznetsk Basin shows the closest resemblance to teeth described by TEILHARD DE CHARDIN and TRASSAERT (1937) as *Archidiskodon tokunagai* MATSUMOTO from a Plio-Pleistocene formation in south-eastern Shansi (China). Considerable similarity in overall appearance, notably in the plate attrition figures (Fig. 4) is supported by the agreement between metrical values for the main characters (Table 5).

Similar teeth from the Lower Pleistocene (?) of China (Shanghai, Loc. 50) had been described previously by HOP-

Table 5. Comparison of measurements of *Archidiskodon* (?) sp., *Palaeoloxodon tokunagai* and *Archidiskondon tokunagai* teeth (in mm)

	Archidiskodon (?) sp. IGG No. 15 Kuznetsk Basin	*Palaeoloxodon tokunagai* MATS., Shanghai, Loc. 50 (HOPWOOD 1935)		*Archidiskodon tokunagai* (MATS.), China, Locs. 6, 15, 2, and 39 (TEILHARD DE CHARDIN and TRASSAERT 1937)	
	M_2	M_2	M_2	M_2	M^2
Crown length	157?	256	242	143, 148, 197	122
Crown width	95	75	78.5	92, 86, 99	93
Crown height	100?	—	—	73, 99, 115	—
Total number of plates	7.5 ×	× 10 ×	× 9 ×	× 6 ×, × 8 ×	× 6 ×
Plate frequency (number of plates in 100 mm)	5	4.5	4.5—5.0	4.6; 4.3	4.8
Enamel thickness	3	2	2	3, 3, 3	3

[6]) These authors place the late form of *E. trogontherii* POHLIG into the genus *Mammuthus* as a separate species *M. chosaricus* DUBROVO (= *M. trogontherii chosaricus* DUBROVO, 1966).

[7]) One more small plate (talon) may be present in the posterior part of the crown under the cement. Talons are taken into account in plate number estimation as separate plates if they are included in the grinding surface and could be involved in chewing (GARUTT and FORONOVA 1976).

1

2

3

Fig. 4. Molar teeth of *Archidiskodon* (?) sp. and of *Archidiskodon tokunagai* MATSUMOTO

1 — *Archidiskodon* (?) sp., M_2 sin., IGG No. 15, Bachatsk Quarry; 2 — *Archidiskodon tokunagai* MATSUMOTO, M^3 sin., No. 10459, Shanxi, China (TEILHARD DE CHARDIN and TRASSAERT 1937); 3 — *Archidiskodon tokunagai* MATSUMOTO, M_3 dext., No. 10462, Shanxi, China (TEILHARD DE CHARDIN and TRASSERT 1937) 1/2 nat. size

WOOD (1935) as *Palaeoloxodon tokunagai* MATSUMOTO. HOPWOOD chose as the type specimen a left lower M_3 from Japan previously described by MATSUMOTO (1924, 1929) as *Loxodonta (Palaeoloxodon) tokunagai*. In the diagnosis HOPWOOD indicates that the Chinese teeth had lower plate frequencies ($4^1/_2$—5) than the Japanese specimens and had loxodont enamel figures on heavily worn plates. The same features are characteristic of the tooth from the Kuznetsk Basin. Moreover, he noted the teeth from China to be more archaic than the Japanese ones, and *P. tokunagai* to be the most primitive species of the genus *Palaeoloxodon*.

In describing the teeth from the Plio-Pleistocene formation in south-eastern Shanxi, China (the lowermost subzone of the Zone III?, Locs. 2, 6, 15 and 39) TEILHARD DE CHARDIN and TRASSAERT (1937) indicated that the form they were studying was identical with *Palaeoloxodon tokunagai* MATSUMOTO, previously described by HOPWOOD. They retained the species name *tokunagai* but preferred not to include it in "dubious" (as they put it) *Palaeoloxodon*. In their opinion, this species was more closely related to the primitive archidiskodont type, resembling the most archaic European *A. meridionalis* specimens in some advanced characters (plate frequency and enamel thickness), while being unlike it in having a more primitive pattern of plate wear. Discussing the genus *Palaeoloxodon* MATSUMOTO in the same paper the authors pointed out that MATSUMOTO had included into his new genus the teeth of two different types: one with low and wide crowns and a small number of plates resembling the *planifrons-meridionalis*-group (*P. tokunagai* teeth fall within this type) and the other with high and narrow crowns and a larger number of plates resembling *P. antiquus*. TEILHARD DE CHARDIN and TRASSAERT suggested giving the name *Palaeoloxodon* to neither of these two types, naming the first, primitive, type as *Archidiskodon tokunagai* MATSUMOTO.

As mentioned above the species name *tokunagai* was given by MATSUMOTO to certain Japanese elephants. Recently DUBROVO (1981) has made a revision of this material. She recognizes only one elephant species of this group to have definitely existed on the Japanese Islands, *Palaeoloxodon naumanni* (MAKIYAMA). The name *P. tokunagai* MATSUMOTO is in the synonymy of *P. naumanni*. Thus, the species name *tokunagai* is not applicable either to the Chinese teeth, or to the Siberian one.

Close morphological resemblance between the tooth No. 15 from the Kuznetsk Basin and the teeth described by TEILHARD DE CHARDIN and TRASSAERT from Shansi, their similar geological age and relative nearness of the two localities (south West Siberia and China) make it possible to refer them to the same species, but our fossil cannot be named as *Archidiskodon tokunagai* MATSUMOTO because of particular morphological features as well as through the complicated and rather confused history of the naming of the Chinese species. It seems unlikely that its taxonomic position can be definitely established until the skull of this elephant is known. Prevalence of archidiskodont features in the teeth under discussion makes it possible to identify our fossil only as *Archidiskodon* (?) sp.[8]).

4. Conclusion

During the Pliocene and Early Pleistocene the Kuznetsk Basin environment was apparently favourable for archidiskodont elephants. Different species of the *Archidiskodon-Mammuthus* phyletic lineage inhabited this region, one giving place to another through time.

Cranial and dental morphology of archidiskodonts as well as their general construction (relatively short body with long extremities) are known to evidence their adaptation to browsing, mainly leaves and twigs of trees (ALEXEEVA and GARUTT 1965). However, in GROMOVA's opinion (1965), they had not yet obtained a marked specialization in this direction.

Early archidiskodonts represented in our collection by *A.* cf. *gromovi* and *A.* cf. *meridionalis* may have existed in the Kuznetsk Basin, like *A. trogontherii*, in steppe and forest-steppe environments. Thus a greater number of dental plates in the Kuznetsk elephants against the typical *A. gromovi* and *A. meridionalis* may indicate adaption to coarser vegetable diet. *Archidiskodon* cf. *gromovi* teeth are the most primitive in morphology. Evolution of *A.* cf. *meridionalis* dentition is reflected in a closer setting of the enamel plates accompanied by decrease of their anteroposterior length and enamel thickness.

Climatic cooling and progressive continentalization are known to have taken place in Eurasia around the Plio-Pleistocene boundary. Fossils of the last representative of the archidiskodonts, *A. trogontherii* (POHLIG), are recorded in sediments dating from the Early Pleistocene (= Middle Pleistocene, Elster/Mindel of European sequence) glaciation. According to POHLIG (1885), this species holds an intermediate position between *A. meridionalis* and *M. primigenius* both in taxonomic and stratigraphic respects, being the immediate ancestor of the latter. *Archidiskodon trogontherii* teeth are characterized by their large total number of plates (up to 22) shortened anteroposteriorly when compared with ancestral species and by decreased enamal thickness. However, *A. trogontherii* teeth from the Kuznetsk Basin show some peculiar features of morphology. They differ from the East-Siberian and Uralian *A. trogontherii* in having somewhat lower plate frequencies and thicker enamel. In spite of these distinctions the general character of dental evolution of the genus *Archidiskodon* as reflected in the increasing of the total number of plates and simultaneous decrease of plate length and enamel thickness can be clearly traced on the material from the Kuznetsk Basin.

Summary

Fossils of the species of the *Archidiskodon-Mammuthus* phyletic lineage have been discovered in the upper Pliocene (according to the USSR MSK Scheme) and Quaternary deposits of the Kuznetsk Basin (south-eastern part of West Siberia). Teeth of *Archidiskodon* cf. *gromovi*, *A.* cf. *meridionalis*, *A. trogontherii*, and *A.* (?) sp., recognized in this region for the first time, are described.

Early archidiskodonts of the Kuznetsk Basin do not differ from typical forms of *A. gromovi* and *A. meridionalis* in dental morphology

[8]) According to H. D. KAHLKE (pers. comm.) it is possible that this form will be included in a new genus of elephants distributed over southern Asia during the Late Pliocene — Early Pleistocene (?).

(with the exception of plate frequency). *Archidiskodon trogontherii* teeth examined are peculiar in having lower plate frequencies and greater enamel thickness as compared with the same species from East Siberia and the Ural. In these features they resemble most closely the European representatives of the species. At the same time, the general character of dental evolution of the genus *Archidiskodon* reflected in the increasing of the total number of plates and decreasing of plate anteroposterior length and enamel thickness can be clearly traced on the Kuznetsk Basin collection.

Zusammenfassung

Molaren verschiedener Arten der *Archidiskodon-Mammuthus* Entwicklungslinie konnten in Ablagerungen der oberen Pliozäns (dem UdSSR MSK-Schema entsprechend) und des Quartärs im Kusnezkbecken geborgen werden. Zähne von *Archidiskodon* cf. *gromovi*, *A.* cf. *meridionalis*, *A. trogontherii* und *A.* (?) sp. wurden zum ersten Mal aus dieser Region bekannt und beschrieben.

Die frühen Archidiskodonten des Kusnezkbeckens unterscheiden sich in der Morphologie der Zähne — mit Ausnahme der Lamellenfrequenz — nicht von den typischen Formen des *A. gromovi* und des *A. meridionalis*. Dagegen unterscheiden sich die vorliegenden Molaren von *A. trogontherii* durch eine geringere Lamellenfrequenz und größere Schmelzstärke von denen Ostsibiriens und solchen des Ural-Gebiets. In diesen Merkmalen stimmen sie weitgehend mit den entsprechenden Formen Europas überein.

Gleichzeitig ist der allgemeine Entwicklungstrend der Gattung *Archidiskodon* in der Zunahme der Lamellenzahl, in der Abnahme der anteroposterioren Länge der Lamellen sowie der Schmelzstärke am Material aus dem Kusnezker Becken zu verfolgen.

Резюме

На территории Кузнецкой котловины (юго-восток Западной Сибири) в верхнеплиоценовых и четвертичных отложениях (по схеме МСК) обнаружены остатки слонов всех видов филогенетической линии *Archidiskodon-Mammuthus*. В статье описаны зубы представителей рода *Archidiskodon*: *A.* cf. *gromovi*, *A.* cf. *meridionalis*, *A. trogontherii* и *A.* (?) sp., установленные здесь впервые.

Древние архидискодонты котловины по строению зубов (за исключением частоты пластин) не отличаются от типичных *A. gromovi* и *A. meridionalis*. Зубы *A. trogontherii*, по сравнению с таковыми из Восточной Сибири и Урала, характеризуются меньшей частотой пластин и большей толщиной эмали и наиболее сходны по этим признакам с европейскими представителями вида. Вместе с тем, общий характер изменения зубной системы слонов рода *Archidiskodon* в филогенезе проявившийся в увеличении количества пластин при одновременном уменьшении их длины и толщины эмали, также отчетливо прослеживается и на материале из Кузнецкой котловины.

References

AGUIRRE, E., 1969: Revision sistematica de los Elephantidae por su morfologia y morfometria dentaria (Caputilo lx). — In: Estud. geol. — Madrid **25**. — P. 123—177.

ALEXEEVA, E. V., 1970: Plejstocenovaja teriofauna Kuzbassa. — In: Priroda i prirodnye Altaja i Kuzbassa. C. 2 (Pleistocene mammal fauna of the Kuznetsk Basin). — Novosibirsk: P. 118 to 120.

—, 1980: Mlekopitajuščie plejstocena jugo-vostoka Zapadnoj Sibiri (Chiščnye, chobotnye, kopytnye) (Pleistocene mammals of the southeastern part of west Sibiria (carnivores, proboscideans, ungulates)). — Moskva: P. 188.

ALEXEEVA, L. I., 1977: Teriofauna rannego antropogena Vostočnoj Evropy (Early Anthropogene mammal fauna of Eastern Europe). — Moskva: P. 214.

ALEXEEVA, L. I.; GARUTT, V. E., 1965: Novye dannye ob ėvoljucii slonov roda *Archidiskodon*). — In: Bjull. Komm. izuč. četvert. per. — Moskva **30**, p. 161—166.

AZZAROLI, A., 1977: Evolutionary patterns of Villafranchian elephants in Central Italy. — In: Mem. Acad. naz. Lincei, Ser. VIII. — **14** (2, 4). — P. 149—168.

COPPENS, Y.; BEDEN, M., 1982: *Mammuthus meridionalis depereti* n. subsp. (Mammalia, Elephantidae), nouveau Mammonth du Pleistocene inférieur d'Europe occidentale. — In: C. R. Acad. Sci. — Paris 294, Ser. II, 4. — P. 291—294.

DUBROVO, I. A., 1960: Drevnie slony SSSR (Ancient elephants of the USSR). — Moskva: P. 78.

—, 1963: O sistematičeskom položenii *Elephas wüsti* (On the taxonomic position of *Elephas wüsti*). — In: Paleont. Ž. — Moskva 4. — P. 95—102.

—, 1966: Sistematičeskoe položenie slona chozarskogo faunističeskogo kompleksa (Taxonomic position of the elephant of the Khazarian faunal komplex). — In: Bjull. Komm. izuč. četvert. perioda. — Moskva **32**. — P. 63—74.

—, 1971: Otrjad Proboscidea. Sem. Elephantidae (Order Proboscidea. Fam. Elephantidae). — In: Plejstocen Tiraspolja. — Kišinev. — P. 92—108.

—, 1981: Die fossilen Elefanten Japans. — In: Quartärpaläontologie. — Berlin **4**. — P. 49—84.

FORONOVA, I. V., 1982: Novye nachodki mlekopitajuščich v plejstocennovych otloženijach Kuzneckoj kotloviny (New finds of mammals in the Pleistocene deposits of the Kuznetsk Basin). — In: Mamontovaja fauna Aziatskoj časti SSSR. — In: Tr. zool. Inst. Akad. Nauk SSSR. — Leningrad 3. — P. 50—57, (1982a).

—, 1982: Plejstocenovye mlekopitajuščie Kuzneckoj kotloviny i ich stratigrafičeskoe značenie (Pleistocene mammals of the Kuznetsk Basin and their stratigraphic significance). — In: Tez. dokl. k XI Kongr. INKVA, T. I. — Moskva. — P. 274 to 275, (1982b).

—, 1982: Archidiskodontnye slony Kuzneckoj kotloviny (Archidiskodont elephants of the Kuznetsk Basin). — In: Problemy stratigrafii i paleogeografii plejstocena Sibiri, k XI Kongr. INKVA. — Novosibirsk: P. 72—82, (1982c).

GALKINA, L. I.; FORONOVA, I. V., 1980: Antropogenovaja teriofauna Batčaskogo kar'era Kuzneckoj kotloviny (Kuzbass) [Anthropogene mammal fauna of the Bachat Quarry in the Kuznetsk Basin (Kuzbass)]. — In: Fauna i ekologija pozvonočnych Sibiri. — Novosibirsk: P. 176—188.

GARUTT, V. E., 1965: Iskopaemye slony Sibiri (Fossil Elephants of Siberia). — In: Antropogenovyj period v Arktike i Subarktike. — Moskva: P. 106—130.

—, 1971: K istorii vyjavlenija promežutočnych zven'ev v filogenetičeskoj linii slonov *Archidiskodon-Mammuthus* (On the history of ascertainment of intermediate links in the elephant phyletic lineage *Archidiskodon-Mammuthus*). — In: Chronologija lednikovogo veka. — Moskva: P. 78—88.

—, 1972: Sovremennyj slon Afriki i vopros o proischoždenii podsemejstva Elephantinae (Recent African elephant and the problem of the subfamily Elephantinae origin). — In: Tez. dokl. otčetn. sess. poitogam rabot ZIN Akad. Nauk SSSR. — Moskva: P. 9—10.

—, 1981: Otrjad Chobotnye — Proboscidea (Order Proboscidea).— In: Katalog mlekopitajuščich SSSR/ed. by I. M. GROMOV; G. I. BARANOVA. — Leningrad: P. 304—318.

GARUTT, V. E.; BAJGUSHEVA, V. S., 1981: *Archidiskodon gromovi* GARUTT & ALEKSEEVA — Der älteste Elephant der Mammutlinie in Eurasien. — In: Quartärpaläontologie. — Berlin **4**. — P. 7—18.

GARUTT, V. E.; FORONOVA, I. V., 1976: Issledovanie zubov vymeršich slonov (metodičeskie rekomendacii) [Investigation of fossil elephants teeth (recommendations of the procedure)]. — Novosibirsk: P. 35.

GARUTT, V. E.; SHOKUROV, A. B.; YACHIMOVICH, V. L., 1977: Novye mestonachoždenija tiraspols'skoj fauny na granice Baškirii i Tatarii (New localities of the Tiraspol fauna at the border of Bashkiriya and Tatariya). — In: Itogi biostratigrafičeskich, litologičeskich i fizičeskich issledovanij pliocena i plejstocena Volgo-Ural'skoj oblasti. — Ufa: P. 91—96.

GARUTT, V. E.; VANGENGEJM, E. A., 1982: Sistematičeskij obzor osnovnych otrjadov mlekopitajuščich. Otrjad Proboscidea Chobotnye (Taxonomic review of the main orders of mammals. Order Proboscidea). — In: Stratigrafija SSSR. Četvertičnaja sistema. Polutom I. — Moskva: P. 311—317.

GROMOV, V. I.; KRASNOV, I. I.; NIKIFOROVA, K. V.; SHANTSER, E. V., 1959: Schema podrazdelenij antropogena (Scheme of subdivision of the Anthropogene). — In: Bjull. Kom. izuč. četv. per. — Moskva **36**.

HOPWOOD, A. T., 1935: Fossil Proboscidea from China. — In: Palaeont. Sinica, Ser. C. — Peking **9** (3).

INOSTRANTSEV, A. A., 1898: Geologičeskoe opisanie Tomskoj gubernii (Geological description of the Tomsk Gubernia). — In: Tr. geol. časti, Kab. II. — St. Petersburg. — P. 106.

MAGLIO, M., 1973: Origin and evolution of the Elephantidae. — In: Trans. Amer. phil. Soc., N.S. — Philadelphia **63** (3). — P. 1 to 149.

MATSUMOTO, H., 1924: Preliminary Note on Fossil Elephants in Japan. — In: J. geol. Soc. — Tokyo **31**. — P. 267.

—, 1929: On *Loxodonta (Palaeoloxodon) tokunagai* MATS., with remarks on the Descent of Loxodontine Elephants. — In: Sci. Rep. Imp. Univ., (2), Geol. — Tohoku **13** (1). — P. 7—11.

OSBORN, H. F., 1942: Proboscidea. A monograph of the discovery, evolution, migration and extinction of the mastodonts and elephants of the world. II. Stegodontoidea, Elephantidea. — New York: Amer. Mus. Press. — P. 803—1676.

PAVLOV, M., 1910: Les éléphants fossiles de la Russie. — In: Nouv. Mem. Spec. Imp. Nat. — Moscou **17**. — P. 1—56.

POHLIG, H., 1885: Über eine Hipparionen-Fauna von Maragha in Nordpersien. — In: Z. dt. geol. Gesellsch. — Berlin **37**. — P. 1022—1027.

—, 1889: Dentition und Kraniologie des *Elephas antiquus* FALC., mit Beiträgen über *Elephas primigenius* BLUM. und *Elephas meridionalis* NESTI. — In: Nov. Acta Leopold. — Halle **53**.

POLENOV, B. K., 1897: Geologičeskoe opisanie Tomskoj gubernii (Geological description of the Tomsk Gubernia). — In: Tr. geol. časti, Kab. II, vyp. 2. — St. Petersburg 1. — P. 159.

RYZHKOV, P. M., 1927: *Elephas primigenius* BLUM. i *Bison priscus* H. v. MEYER iz rajona Kemerovskogo rudnika na r. Tomi (*Elephas primigenius* BLUM. and *Bison priscus* H. v. MEYER from the district of Kemerovo mine on the Tomsk River). — Tomsk: Sib. technol. Inst., **47** (I) (3). — P. 54—70.

TEILHARD DE CHARDIN, P.; TRASSAERT, M., 1937: The Proboscidians of South-Eastern Shansi. — In: Palaeont. Sinica, Ser. C. — Nanking **13** (1). — P. 1—58.

TOLMACHOV, I. P., 1909: Geologičeskoe opisanie Tomskoj gubernii (Geological description of the Tomsk Gubernia). — In: Tr. geol. časti, Kab. II, T. VIII. — St. Petersburg. — P. 150.

VANGENGEJM, E. A., 1961: Paleontologičeskoe obosnovanie stratigrafii antropogenovych otloženij severa Vostočnoj Sibiri (Paleontological basis for the stratigraphy of fhe Anthropogene deposits of the northern part of East Siberia. — In: Tr. geol. Inst. Akad. Nauk SSSR. — Moskva 48. — P. 182.

VISLOBOKOVA, I. A., 1973: Paleontologičeskoe obosnovanie stratigrafii eoplejstocenovych otloženij juga Zapadnoj Sibiri (Palaeontological basis for the stratigraphy of the Eopleistocene deposits of the southern part of West Siberia). — Novosibirsk, Avtoref. kand. dis., 24 P.

—, 1974: O novoj *Archidiskodon gromovi* GARUTT et ALEKSEEVA v eoplejstocenovych otloženijach juga Zapadnoj Sibiri (On a new find of *Archidiskodon gromovi* GARUTT et ALEKSEEVA from the Eopleistocene deposits of the south West Siberia). — In: Bjull. Kom. izuc. četv. perioda. — Moscow **42**. — P. 171—176.

YAVORSKY, V. I.; BUTOV, I. I., 1927: Kuzneckij kamennougol'nyj bassejn (The Kuznetsk Coal Basin). — In: Tr. geol. Komit., N.S., — 177. — 244 Pp.

YATSKO, I. YA., 1948: Skelet *Elephas wüsti* M. PAVL. iz terrasovych otloženij Chadžubejskogo limana vblizi g. Odessa (Skeleton of *Elephas wüsti* M. PAVL. from the terrace deposits of the Khadzhibeysk Liman near Odessa). — In: Tr. Odess. Univ. — **2** (2) (54).

ZUDIN, A. N.; et al., 1982: Obosnovanie stratigrafičeskoj schemy neogenovych i cetvertyčnych otloženij Kuzneckoj kotloviny (The basis for the stratigrafic scheme of the Neogene and Quaternary deposits of the Kuznetsk Basin). — In: Problemy stratigrafii i paleogeografii plejstocena Sibiri. — Novosibirsk: P. 133—149.

1

2

PLATE I
Molar teeth of *Archidiskodon* cf. *gromovi* GARUTT & ALEXEEVA and *Archidiskodon* cf. *meridionalis* (NESTI)
Fig. 1. *Archidiskodon* cf. *gromovi*, M^3 dext., IGG No. 3580, Novo-sergeyevsk Quarry

Fig. 2. *Archidiskodon* cf. *meridionalis*, M^3 sin., IGG No. 4096, Mokhovsk Quarry

Nat. size

1

2

PLATE II

Molar teeth of *Archidiskodon trogontherii* (POHLIG) and *Archidiskodon* (?) sp.

Fig. 1. *Archidiskodon trogontherii*, M^3 sin., IGG No. 5, Novosergeyevsk Quarry

Fig. 2. *Archidiskodon* (?) sp., M_2 sin., IGG No. 15, Bachatsk Quarry

Nat. size

Quartärpaläontologie 6 43—52 Berlin 1986

A Review of the Süssenborn Horses and the Origin of Equus hydruntinus REGALIA

By ANN FORSTEN, Helsinki

With 6 Figures

Contents

Abstractum

The fossil horse material of Süssenborn is reviewed and compared with the record of other fossil localities.

Das fossile Pferde-Material von Süßenborn wird revidiert und mit Fossilfunden anderer Fundstellen verglichen.

Приводится обзор материала по ископаемым лошадям из Зюссенборна и сравнение его с ископаемыми остатками из других местонахождений.

1. Introduction

The mammalian fossil locality Süssenborn in Thuringia (GDR) is one of the classical lower middle Pleistocene ones. The Süssenborn fossil equids were discussed by WÜST (1900), v. REICHENAU (1915), and MUSIL (1969). MUSIL recognized three equid species: *Equus süssenbornensis*, *E. marxi*, and *E. altidens*. *Equus marxi* is not well differentiated, wether in size or morphology, from the larger *E. süssenbornensis*, nor from the smaller *E. altidens* (see also GROMOVA and DUBROVO 1975, p. 122). The chances are that that *E. marxi* is based on material referable to either of the two other forms at Süssenborn.

2. Analysis of the Süssenborn Horse Sample

I measured and analysed the measurements on the equid teeth and limb bones from Süssenborn kept in the Institut für Quartärpaläontologie, Weimar (GDR), and the Staatliches Museum für Naturkunde, Stuttgart-Ludwigsburg (FRG). Comparative material from some other localities was studied in the same collections and in the Institute of Palaeobiology, Tbilisi (Grusinian SSR). My sincere thanks are due the keepers of these collections.

Most variables in the Süssenborn sample are bimodally, instead of, as expected, trimodally, distributed. The limb bones from Süssenborn, when plotted in scattergrams (Figs. 3—5), in all cases fall into well separated size groups. Among the MC III, one specimen IQW 1964/1040 (Süss. 1139), falls outside either of the two subsamples, being shorter than the bones in both (Fig. 3), but does not differ in preservation or colour from the rest of the bones. Specimen IQW 1964/1042 (Süss. 9244) which MUSIL (1969, p. 637—638) believed to be small for *E. süssenbornensis*, falls in among the large MC III. On the basis of dental size and morphology, I separated two subsamples of the cheek teeth from Süssenborn. The subsamples overlap slightly for most size variables (Figs. 1a—e), but each is ± unimodal and the coefficients of variation are normal. Comparison with related local samples of limb bones and cheek teeth (Figs. 1—5) also indicates that only two size groups are represented at Süssenborn.

On morphological ground the question as to how many species of horse are present at Süssenborn is not quite as simple. In addition there is probably some mixing of fossils of different ages, possibly even of different provenances. The enamel pattern of the lower cheek teeth from Süssenborn is stenoid (= zebroid). Three lower cheek teeth IQW 1965/2208 (Süss. 7296), IQW 1964/888 (Süss. 3430), and IQW 1964/1127 (Süss. 5254) differ from the rest in having a caballoid double knot (Fig. 6). Of these, IQW 1965/2208 (Süss. 7296) also differs in preservation and colour, and may even derive from a different locality (R.-D. KAHLKE, pers. comm.). A right P^2, P^4—M^3, and left P^4—M^3 (Süss. 9180-1) of one individual differs from the rest of the uppers in morphology and yellowish colour; this specimen may be of a younger age than the rest of the material.

Because there is no clearly intermediate size group, either for the dental or limb bone measurements, I believe that the Süssenborn equid sample belongs to two species of stenoid horse, about equally frequent in the sample. Two lowers belong to a caballoid horse, probably contemporaneous with the stenoid ones, judging by the colour and preservation of the fossils. In addition the sample comprises rare specimens of a probably younger age.

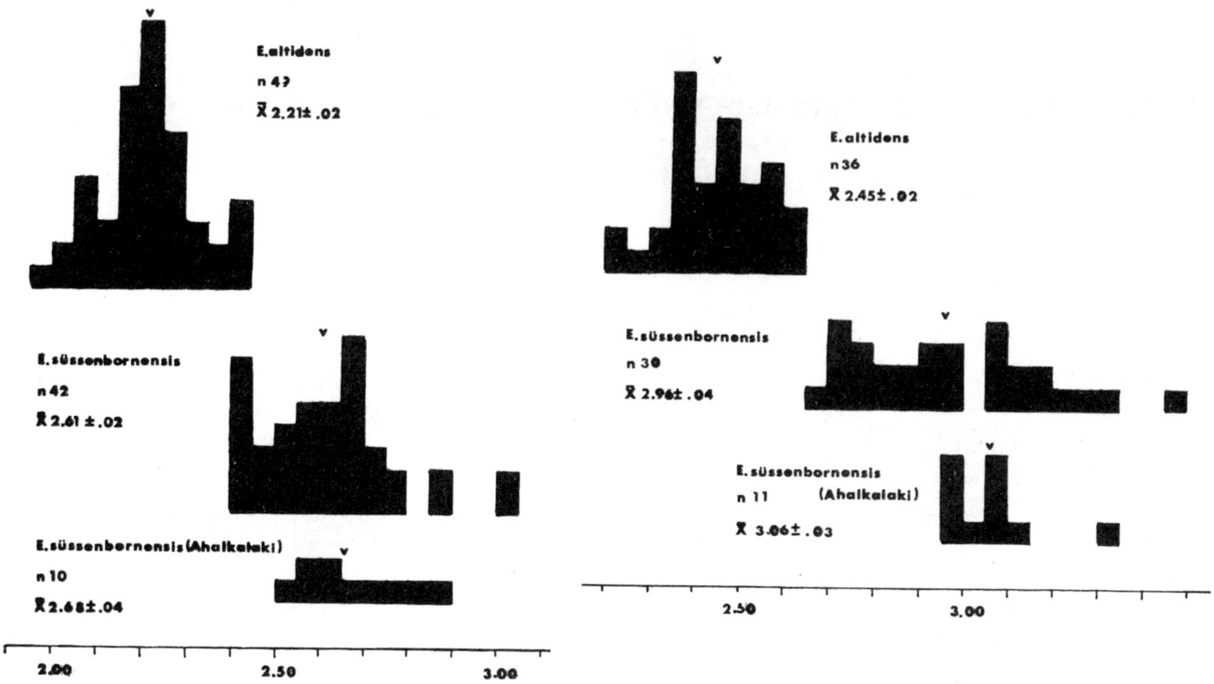

Fig. 1a. Length at the tooth base of M^1—M^2, resp. P^3—P^4, in *E. altidens* and *E. süssenbornensis* from Süssenborn, and *E. süssenbornensis* from Ahalkalaki

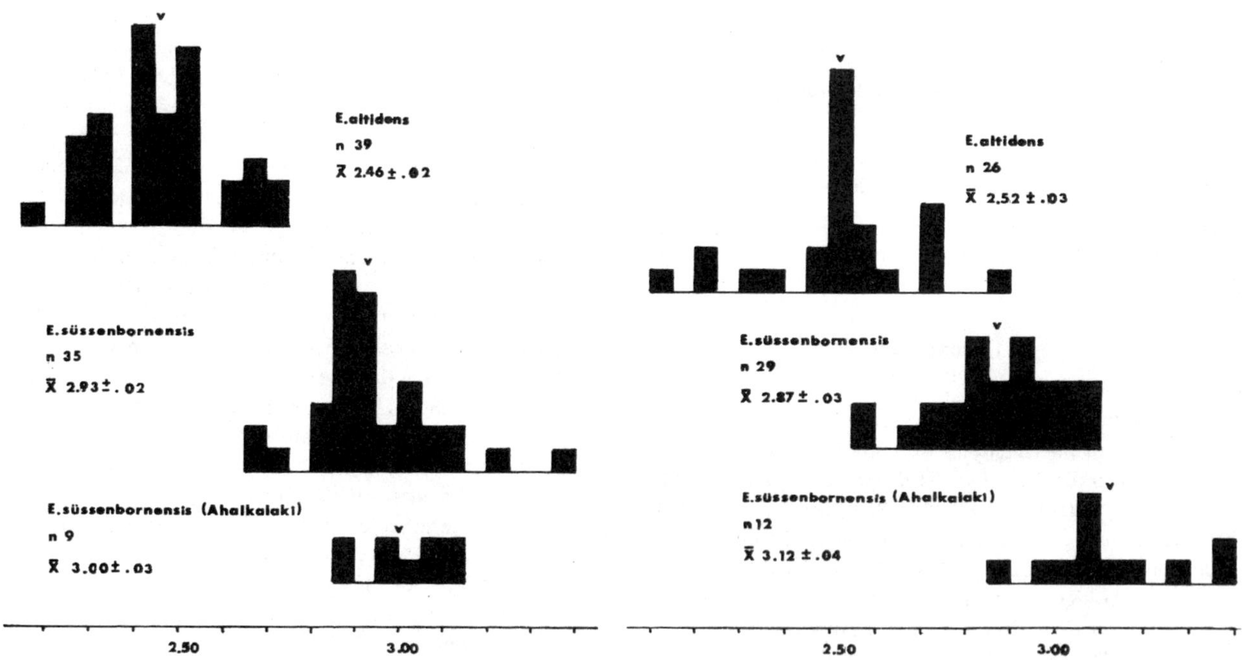

Fig. 1b. Breadth at the tooth base of M^1—M^2, resp. P^3—P^4

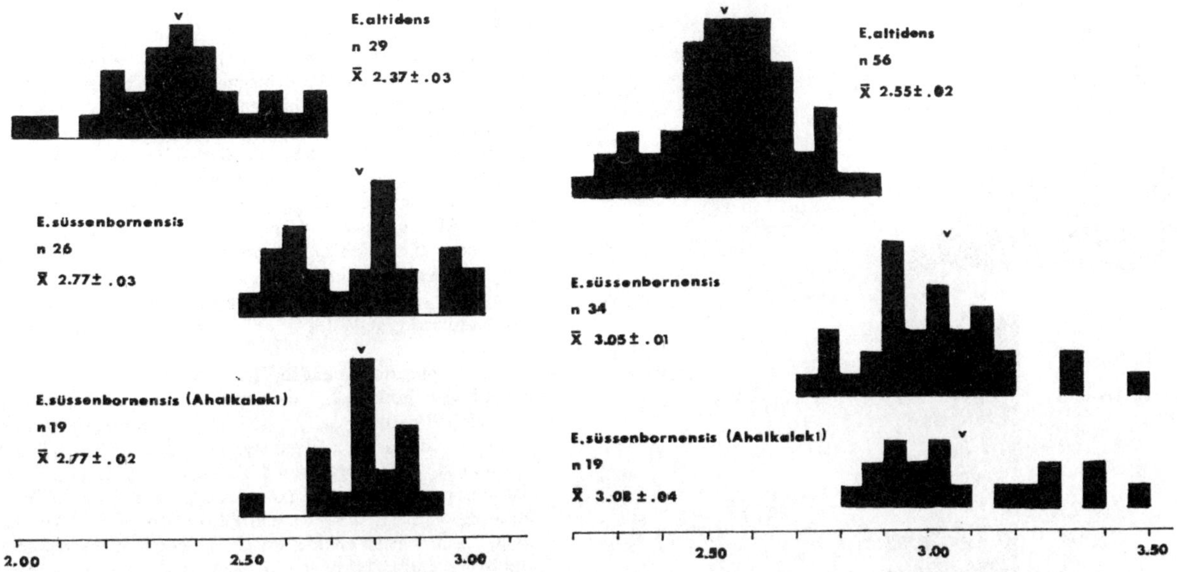

Fig. 1c. Length at the tooth base of M_1-M_2, resp. P_3-P_4

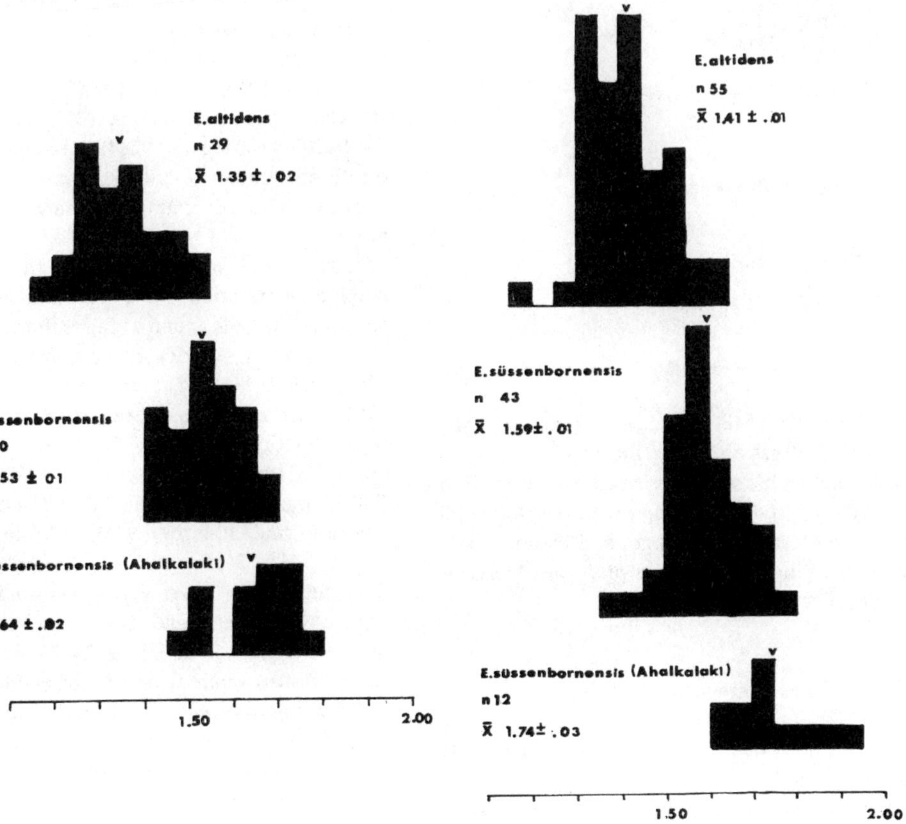

Fig. 1d. Breadth at the tooth base of M_1-M_2, resp. P_3-P_4

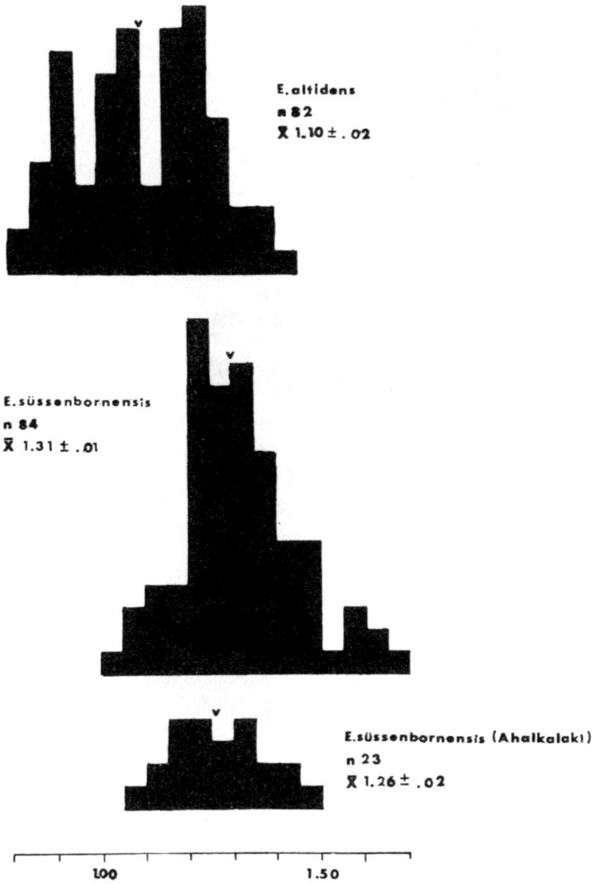

Fig. 1e. Protoconal length

2.1. Taxonomy

2.1.1. **Equus** sp.

Two lowers, IQW 1964/888 (Süss. 3430) (left M_1—M_2) and IQW 1964/1127 (Süss. 5254) (right P_3—P_4), have a caballoid double knot with a wide, U-shaped, valley between the metaconid and metastylid, an incipient or developed pli caballinid, but no protostylid. For length and breadth these teeth fall within the range of variation of *E. mosbachensis* v. REICHENAU (see EISENMANN 1981, Table 22).

2.1.2. **Equus süssenbornensis** WÜST

Lectotype: Left P^2—M^3 IQW 1964/1177 (Süss. 514) (MUSIL 1969, p. 645, Fig. 11:1).

Paralectotype: Left P_2—M_3 IQW 1965/2592 (Süss. 430—435) (MUSIL 1969, p. 648, Fig. 11:3); Right P_2—M_2 IQW 1964/1302 (Süss. 9279) (MUSIL 1969, Fig. 11:2); Left P_3—M_1 IQW 1964/1300 (Süss. 637) (MUSIL 1969, Fig. 11:4).

Equus süssenbornensis was recently discussed by MUSIL (1969). Some features of the teeth should be additionally stressed:

— In the upper cheek teeth the protocone is usually heeled and as a rule quite long (MUSIL 1969, Fig. 11, 1). GROMOVA and DUBROVO (1975, p. 121) criticized MUSIL's choice of IQW 1964/1177 (Süss. 514) as lectotype of *E. süssenbornensis*. They believe that the specimen shows dental characters too derived for a stenoid horse, i.e. a long protocone with concave lingual border and, in the premolars, a bifid mesostyle. They conclude that the specimen, in fact, belongs to *E. mosbachensis*. However, the protocone in *E. süssenbornensis* is almost always concave or shows a lingual indentation, except at the base of the crown where the lingual border of the protocone may be straight or convex. Although not as deeply bifid as in the caballoid horses in general, the styles of the premolars in *E. süssenbornensis* are grooved at least at some point along the crown, those of the molars are faintly grooved or simple. I believe that MUSIL is right in referring IQW 1964/1177 (Süss. 514) to *E. süssenbornensis*.

— The mean plication count in 71 P^3—M^2 is 12.8 ± 0.7.

— In the lower cheek teeth cingular structures occur: a protostylid is developed in almost all P_3—M_3 as a seam and/or as a basal plication, never as an isolated stylid, however. About 1/2 of the P_2 show a weak "protostylid". SAMSON (1975, p. 224—225) also observed a parastylid (= protostylid) plication in P_4 and M_1—M_2 of *E.* cf. *süssenbornensis* from Roumania. An "ectostylid" occurs as an enamel thickening between the protoconid and hypoconid in 1/2 of the M_1—M_3 and in 1/5 of the P_3—P_4.

— In M_1—M_3 the ectoflexid is consistently deep, often broad and lingually flattened, reaching into the metaconid-metastylid commissure. The pli caballinid is situated at the mouth of the ectoflexid, instead of within it.

To this species belong left and right P_2—M_3 IQW 1964/1303—1304 (Süss. 9284), which MUSIL (1969, Fig. 8:1—2) referred to *E. marxi*, as well as the paratype of that species (left P_3, P_4 and M_1) (WÜST 1900, Taf. VI, Figs. 5—7; v. REICHENAU 1915, p. 21—24, Taf. VI, Figs. 1, 2, 5, 6, 9, 10); see also SCHWARZ 1928, p. 437). The premolars of IQW 1964/1303—1304 (Süss. 9284) are abberant in having a deep ectoflexid, similar to that in the molars.

The limb bones are among the largest there are of any fossil horse.

MUSIL (1969, p. 651) discussed the occurrence and geographic distribution of *E. süssenbornensis*. *Equus süssenbornensis* has been found at Süssenborn, Voigtstedt, and Bad Frankenhausen (GDR), at Stranska Skala (ČSSR) (MUSIL 1965, 1969, 1971), and at Ahalkalaki (Grusinian SSR) (VEKUA 1962). The species has been cited from Solilhac (France) (PRAT 1968, Fig. 36; but Fig. 36:3, a P_3—P_4, is caballoid), Sinyaya Balka (Taman Peninsula, USSR), and Kolkotova Balka/Tiraspol (Moldavian SSR) (GROMOVA 1949, p. 105; GROMOVA and DUBROVO 1975), and from Roumania (SAMSON 1975, p. 224—225, Fig. 9: 7—9). These finds span a time range from the Taman lower Pleistocene to the Cromerian lower middle Pleistocene and cover most of Europe. Recently SHER (1971, p. 145—186, Figs. 24, 25) described a new species, *E.* (*Plesippus*) *verae*, from the lower Pleistocene of Kolyma (NE Siberia), morphologically and metrically almost identical to *E. Süssenbornensis* (Figs. 2—5). The morphological traits believed to characterize *E. verae* and to differentiate *E. verae* from *E. süssenbornensis* (SHER 1971, p. 185), are common to the two forms. I believe that *E. verae* is a synonym of *E. süssenbornensis*. MUSIL (1969, p. 651) may be right in stressing that the centre of *E. süssenbornensis* distribution lies to the East. The species does not seem to occur in France, in spite of PRAT (1968).

The possible relationship of the so-called *E. robustus* POMEL from the upper Pliocene and lower Pleistocene of Senèze (France), Tegelen (The Netherlands), Oasele (Roumania), Gombasek (ČSSR), and Kisláng (Hung. P. R.) (not corres-

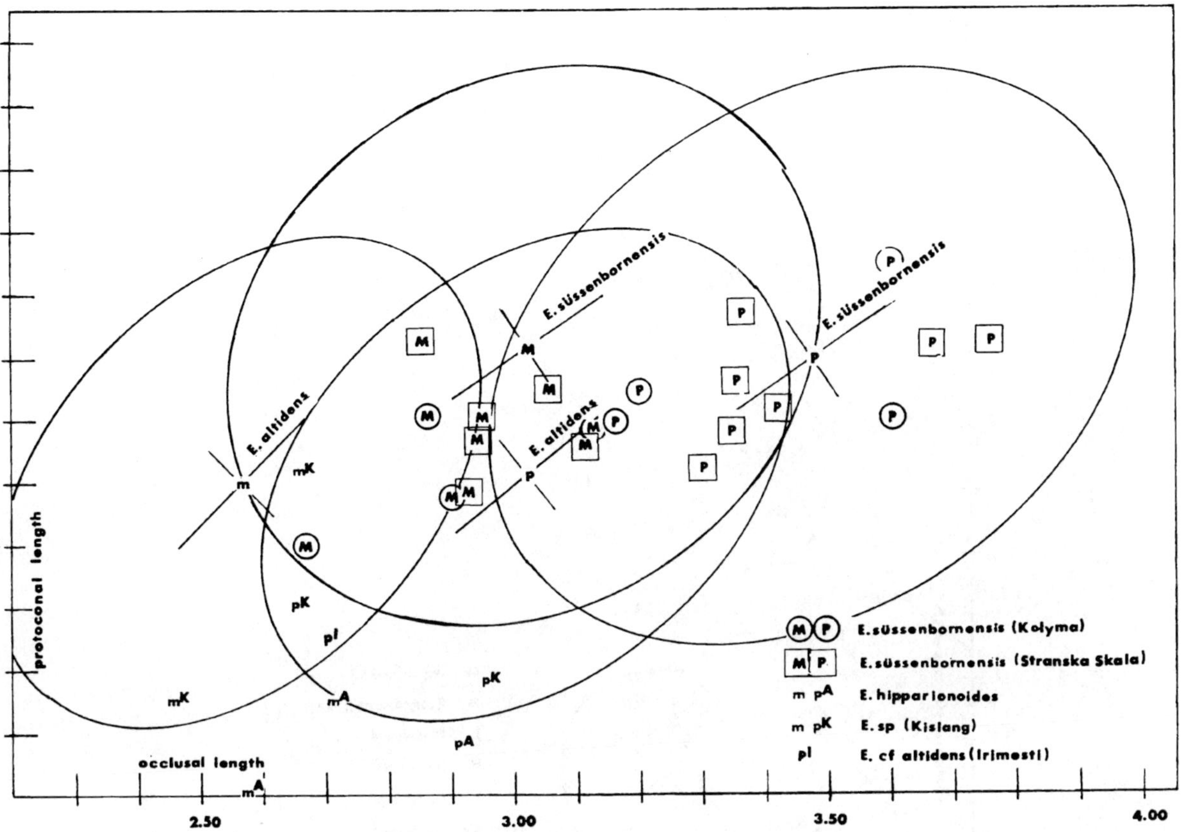

Fig. 2. Protoconal length plotted to occlusal length of P^3—P^4 (P and p) and M^1—M^2 (M and m) of some species of *Equus*; metric data. Ellipses constructed on the data of *E. altidens* (p and m) and *E. süssenbornensis* (P and M) from Süssenborn

Data on *E. verae* = *E. süssenbornensis* from Kolyma according to SHER (1971, Table 12); from Stránská Skála according to MUSIL (1971, Table 1); on *E. hipparionoides* according to VEKUA (1960, Table 1); on *E.* sp. from Kislang according to NOBIS (1971, Table 32), and on *E.* cf. *altidens* from Irimeşti according to SAMSON (1975, p. 220)

ponding to *E. robustus* from Peyrolles, France; see SAMSON 1975, p. 201—202) and *E. süssenbornensis*, should be investigated. The former are as large as *E. süssenbornensis*, but in the teeth of *E. robustus* from Kislang the protocone is said to be short (KRETZOI 1954, P. 252). In this character *E. robustus* from Kislang resembles the large stenoid horse (*E. stenonis* "*major*" BOULE = *E.* cf. *bressanus* VIRET = *E. livenzovensis* BAJGUSHEVA) from the lower Villafranchian of Chapry, Valovaya Balka, and Liventsovka (USSR); the contemporaneous form from Morskaya (USSR) has a relatively long protocone, however (Coll. 301, Geological Institute, Academy of Sciences of USSR, Moscow). SAMSON (1975, S. 200—207) referred to these early, large stenonids as subspecies of *Plesippus* (*Allohippus*) *major* BOULE.

At Kislang and Liventsovka there also occurs a smaller horse, referred to *E. stenonis* COCCHI, but resembling *E. altidens* (KRETZOI 1954, p. 252; NOBIS 1971, Plates CIX, CX, CXI, CXII; BAJGUSHEVA 1971, p. 19—20).

2.1.3. **Equus altidens** v. REICHENAU

Type: On the hypodigm described by v. REICHENAU (1915) see MUSIL (1969, p. 658), SCHWARZ (1928, p. 437) chose as the lectotype the right P_2 figured by v. REICHENAU (1915, Plate VI, Fig. 17).

The second stenoid equid species in the Süssenborn samples is smaller than *E. süssenbornensis*.

The tooth crowns are tall and slender. In the uppers (MUSIL 1969, Fig. 7) the protocone is variable, flattened or indented lingually, occasionally lacking a heel. The postprotoconal valley tends to be deep and the pli caballin often becomes obliterated even in moderately worn teeth, especially molars. MUSIL's (1969; p. 659) statement that in the uppers of *E. altidens* the mesostyle is never bifid, is disproved by his Fig. 7. Almost all P^2 and P^3—P^4 in *E. altidens* have a bifid mesostyle. In the molars the styles are usually simple. The mean plication count in 80 P^3—M^2 is 7.4 ± 0.4.

VON REICHENAU (1915, p. 25—28, Plate VII, Figs. 1—5), without designating a holotype, described *E. marxi* on the basis of three upper cheek teeth (right P^3—P^4, M^1—M^2, left M^1—M^2, all lacking number), probably of one individual (see MUSIL 1969, p. 656—657, Fig. 15:1). These little worn teeth, referred originally to *E.* cf. *germanicus* NEHRING by WÜST (1900, Plate VI, Fig. 17), are tall and slender. On the basis of their length and breadth proportions, they fall in among the measurements on the teeth of *E. altidens*, but the protocones are long. As in *E. altidens*, these teeth have a deep postprotoconal groove and lack a pli caballin. I believe that these teeth belong to *E. altidens*. This makes *E. marxi* a synonym of *E. altidens*, which has page priority.

The lower cheek teeth (MUSIL 1969, Fig. 10) have cingular stylids: almost all P_3—M_3 have a protostylid seam, often

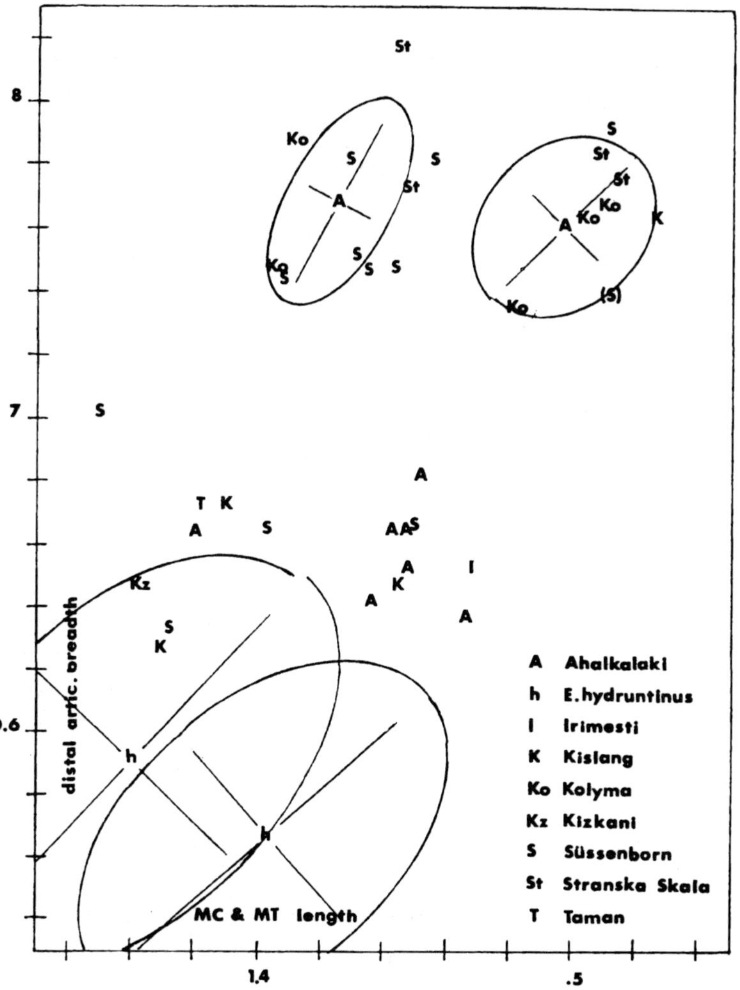

Fig. 3. MC III and MT III distal articular breadth plotted to total length; log data

Upper ellipses constructed on the data of *E. süssenbornensis* from Ahalkalaki; lower ellipses on data of *E. hydruntinus*, various localities and sources. Data on Kitzkani according to DAVID and SUSHPANOV (1972, p. 15—16); on Kislang and Taman according to NOBIS (1971, Plates C, CIX, CXI); Kolyma according to SHER (1971, Plates 16, 20); on Irimeşti according to SAMSON (1975, Plate 15)

forming a basal plication; in 1/3 of P_2 a "protostylid" is indicated. An ectostylid enamel thickening occurs in 1/5 of P_2—P_4 and M_1—M_3. The ectoflexid in M_1—M_3 is deep, penetraiting into the double knot; in a few molars the ectoflexid is shallow, barely reaching the metaconid-metastylid stalk.

The limb bones are slender; MUSIL (1969, p. 639) compared them to those of *E. hydruntinus* and Recent hemiones and asses. On their size and proportions, the two proximal phalanges from Voigtstedt IQW 1966/5851 (Voi. 822) and IQW 1966/7350 (Voi. 3584) belong to *E. altidens* (Fig. 5).

The occurrence and distribution of *E. altidens* is more poorly known than that of *E. süssenbornensis*. The two species are found together at Voigtstedt and Süssenborn (MUSIL 1965, 1969). SAMSON (1975, p. 220—221, Figs. 9:6, 24) referred "*Allohippus*" cf. *altidens* on a M^1—M^2 (possibly rather P^3—P^4) from the lower Günz of Rotbav-Silvestru and "*A.*" cf. *marxi* on a MT III from the Cromer-Mindel of Irimeşti (both Roumania). The MT III is very slender, resembling those of *E. altidens* (Fig. 3). At Ahalkalaki, *E. süssenbornensis* occurs together with a smaller horse, *E. hipparionoides* VEKUA (VEKUA 1960); *E. hipparionoides* closely resembles *E. altidens* (Figs. 3—5), but the protocones of the type teeth, left P^4—M^3 (VEKUA 1960, Fig. 1, Plate 1), is very short (Fig. 2).

Slender-built, stenonid horses from the lower and lower middle Pleistocene have often been referred to subspecies of *E. stenonis*. *Equus altidens* differs in length and slenderness of the metapodials and phalanges from all forms of *stenonis*, e.g. *E. stenonis senezensis* PRAT, as well as from the small *E. stehlini* AZZAROLI. In the Cromerian fauna of Solilhac, PRAT (1968, p. 144) identified *E.* cf. *süssenbornensis* and a smaller horse *E. stenonis* var., which SAMSON (1975, p. 221) compares with *E. altidens*. Of the bones referred to *E. stenonis* var., the MT III (PRAT 1968, Plate 96) is too short and the proximal phalanx (PRAT 1968, Plates 63, 100) too massive, for *E. altidens*. Of the bones of *E. stenonis* var. from the Cromerian of Sainzelles, the MT III (PRAT 1968, Plate 96) resembles those of *E. altidens*, as does one of the two proximal phalanges (Plate 100), whereas the other is very short for

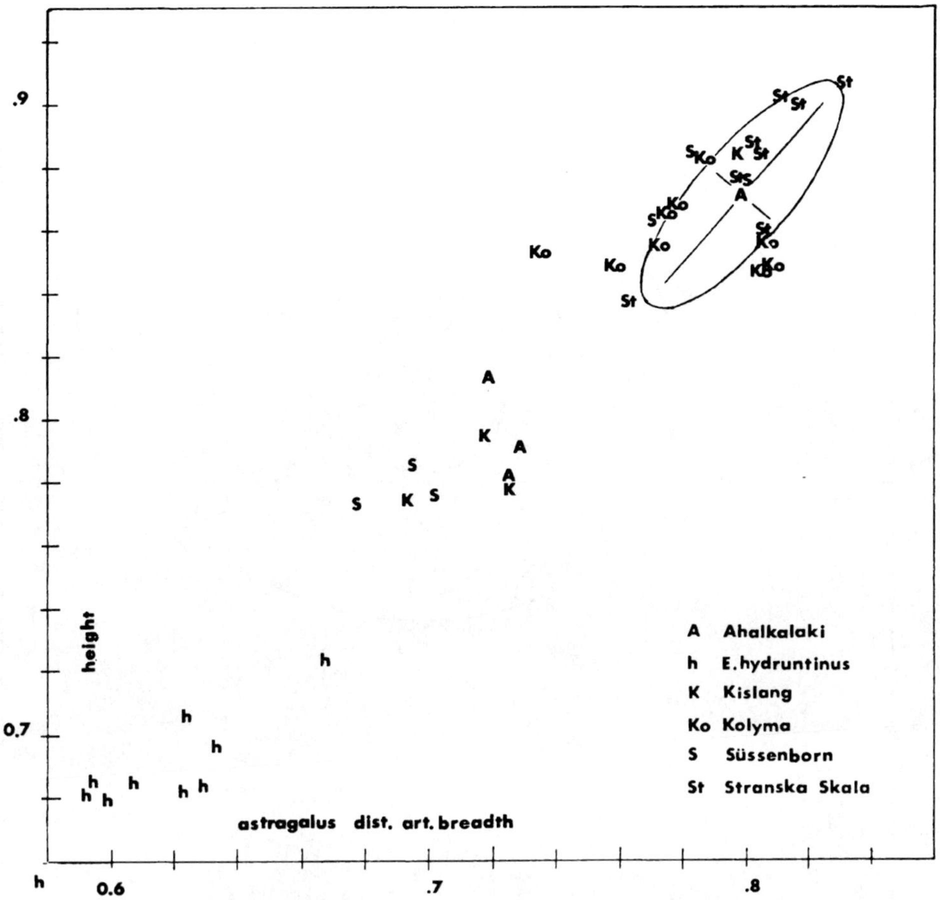

Fig. 4. Astragalus height plotted to distal articular breadth; log data

Ellipse constructed on the data of *E. süssenbornensis* from Ahalkalaki. Data on *E. hydruntinus* from various localities and sources; on Kislang according to Nobis (1971, Table CX); on Kolyma according to Sher (1971, Table 18)

that species. In the fauna of Kislang, although considerably older (Calabrian according to Kretzoi 1954) than that of Süssenborn, occur two horses (the larger *E. robustus* and smaller *E. stenonis*; Kretzoi 1954, p. 252—253), which resemble *E. süssenbornensis* and *E. altidens*. The small metapodials and phalanges from Kislang (Nobis 1971, Plates CIX, CXI, CXII) are similar to those of *E. altidens* (Fig. 3). From the Taman Peninsula (USSR), in addition to *E.* cf. *süssenbornensis*, there has been found a smaller equid, represented by a MC III (Nobis 1971, Plate C), which resembles those of *E. altidens* (Fig. 3), as do the MC III from Kitzkani (Moldavian SSR) (*Allohippus* sp. according to David and Sushpanov 1972, p. 15—16, see Samson 1975, p. 221, Fig. 23) (Fig. 3), the MT III from Solenoe Zajmyshe (USSR) (*E. hydruntinus* according to Gromova 1949, 1965, p. 73; but see Samson 1975, p. 224, Fig. 24), and the MC III from Abbeville (France) (*E. stenonis* var. according to Prat 1968, p. 141; see Samson 1975, p. 221). Mottl (1942, p. 21—22, 25) cited *Allohippus stenonis* race *minor* from Varberg (Hungary), together with *E. mosbachensis*. The measurements on, and morphology of the teeth of *A. stenonis minor* resemble those of *E. altidens* and *E. hydruntinus*. According to Azzaroli (1979, p. 39) slender metapodials also occur in the Cromer Forest Bed (Great Britain). Azzaroli does not give measurements, however.

From Tiraspol (Moldavian SSR) Gromova and Dubrovo (1975, p. 121) reported *E.* cf. *süssenbornensis* and a smaller equid, *E.* (*Allohippus*) sp., earlier referred by Gromova (1949) to *E. hydruntinus*. Gromova and Dubrovo believe that this horse may represent *E. altidens* (see also Musil 1969, p. 660). Although single specimens of the metapodials from Tiraspol, on the basis of Gromova and Dubrovo's measurements, fall in among those of *E. altidens*, the total range of the Tiraspol metapodials spans the gap between the large and the small size group at Süssenborn, the bones as a mean being shorter and more massive than those of *E. altidens*.

3. The Origin of Equus hydruntinus

Equus altidens from the Cromer-Mindel is morphologically, and stratigraphically a connecting link between Villafranchian middle-sized stenoid horses with slender limbs, e.g. from Kislang and Kitzkani, on the one hand, and the European "ass" *E. hydruntinus* on the other (Musil 1969, p. 660, 1975, Fig. 8; Samson 1975, p. 221). The latter was widely distributed in Europe in the upper middle and upper Pleistocene, occurring as late as in the Neolithic (Bökönyi 1954). Dentally primitive, *E. hydruntinus* is derived in small size

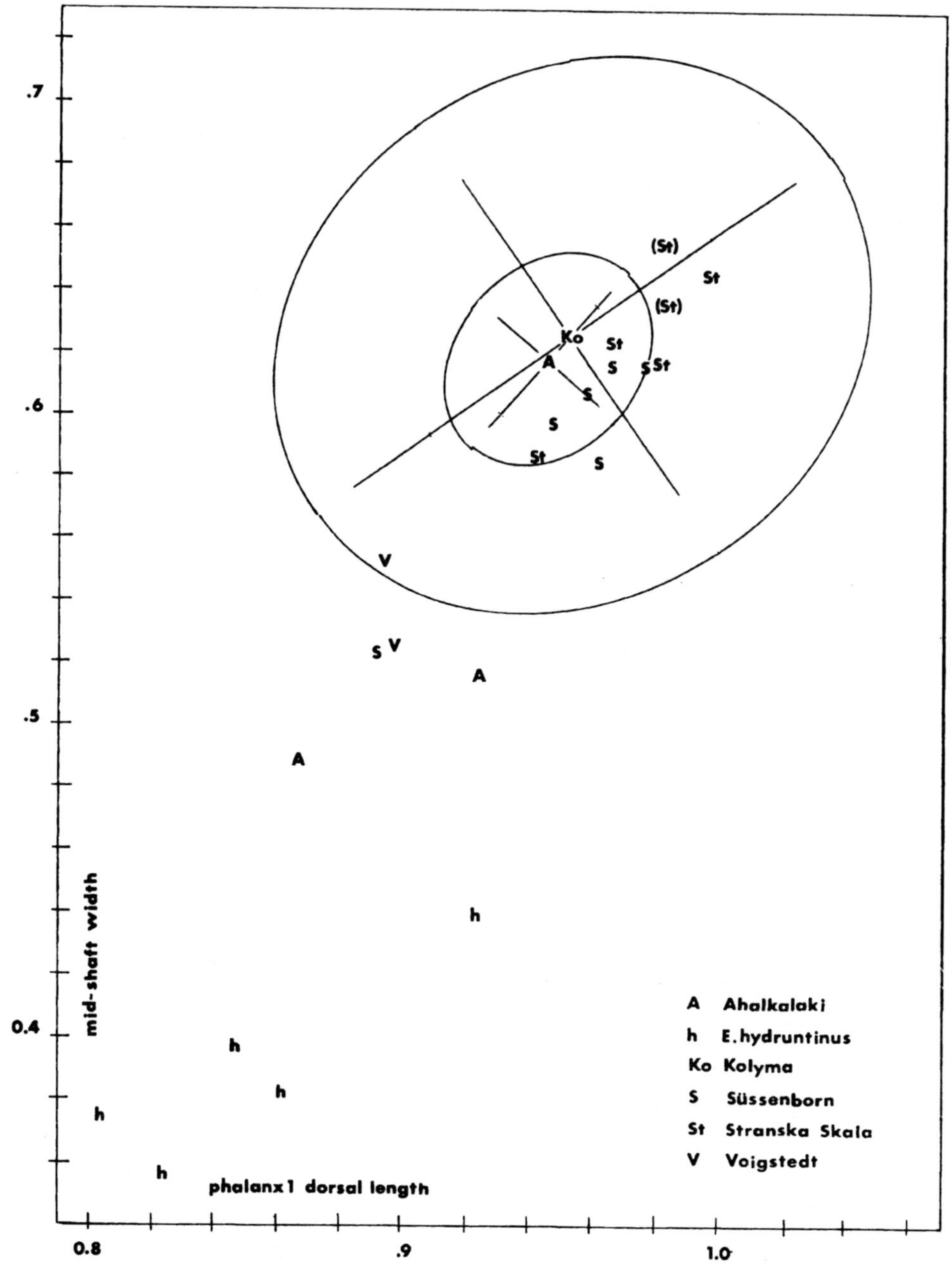

Fig. 5. Phalanx 1 mid-shaft width plotted to dorsal length; log data

Small ellipse constructed on the data of *E. süssenbornensis* from Ahalkalaki; large ellipse on the data of *E. verae* = *E. süssenbornensis* from Kolyma (data SHER 1971, Table 21). Data on *E. hydruntinus* from various localities and sources

and slender limb proportions. In its size and slender proportions, *E. hydruntinus* resembles the Recent ass and hemiones, as well as *E. altidens* and some still earlier forms similar to that species, but *E. hydruntinus* differs from the ass and hemiones in dental morphology, particularly in the deep ectoflexid of M_1—M_3 and stenonid double knot. *Equus hydruntinus* appears to have evolved from a horse similar to *E. altidens* through decrease in size, but with little change in proportions (Figs. 3—5). *Equus hydruntinus* is not an ass, but a stenonid (= zebroid) horse, the last representative in Eurasia of a once flourishing lineage (ALIMEN 1946; GROMOVA 1949, p. 213; PRAT 1968, p. 561; AZZAROLI 1979, p. 37—38; EISENMANN and PATOU 1980, p. 646; DAVIS 1980, p. 299; but see GROMOVA 1965, p. 71—75).

4. Conclusion

Equus süssenbornensis and *E. altidens* appear to be the descendants of very large and middle-sized stenoid horses respectively, characterized by exceptionally slender limb proportions, extant in Europe since the Villafranchian. *Equus süssenbornensis* evidently became extinct with the

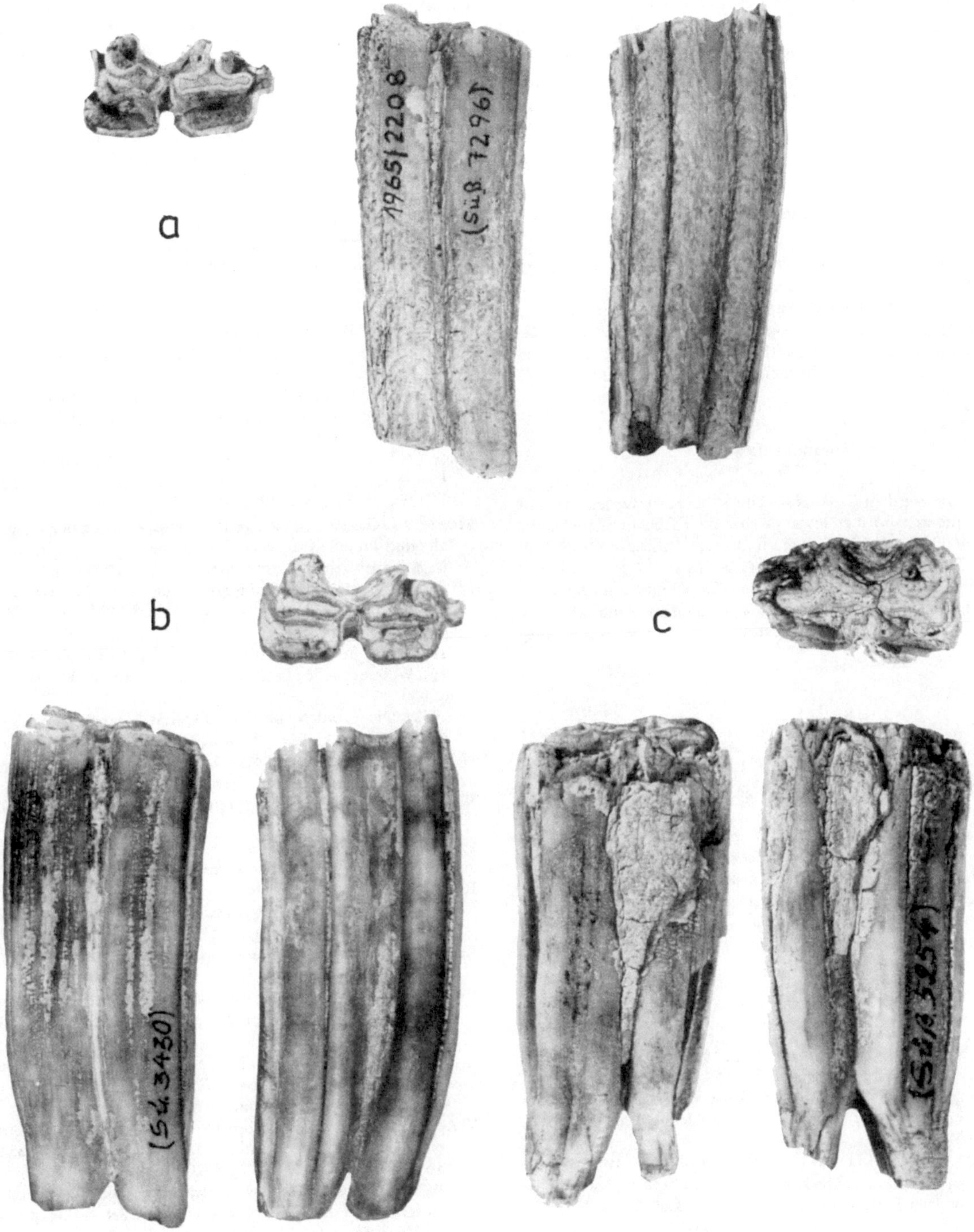

Fig. 6. Lower cheek teeth from Süssenborn with a caballoid double knot

a — IQW 1965/2 208 (Süss. 7 296); b — IQW 1964/888 (Süss. 3430); c — IQW 1964/1 127 (Süss. 5 254)

Cromerian forms and left no descendants (see also MUSIL 1969, p. 651). This contradicts PRAT's (1968, p. 516) opinion that there is a phyletic connection between *E. bressanus — E. süssenbornensis* and *E. mosbachensis*. The lineage represented at Süssenborn by *E. altidens* may have continued on into the Neolithic with *E. hydruntinus*.

Summary

The horses from the Cromerian of Süssenborn are reviewed and compared with local forms of similar and earlier age. The two common species of horse in the Süssenborn fauna, *E. süssenbornensis* WÜST and *E. altidens* v. REICHENAU, represent evolutionary lineages evidently extant in Europe since the Villafranchian. *Equus altidens* may be the ancestor of the European "ass", *E. hydruntinus*.

Zusammenfassung

Die Pferde aus dem Cromerium von Süßenborn werden revidiert und mit anderen Formen etwa gleicher oder früherer Zeitstellung verglichen. Die beiden häufigsten Pferdearten der Fauna von Süßenborn, *E. süssenbornensis* WÜST und *E. altidens* v. REICHENAU, sind als Evolutionslinien zu betrachten, die offensichtlich seit dem Villafranchium in Europa auftreten. *Equus altidens* kann als Ausgangsform des europäischen „Esels" *E. hydruntinus* angesehen werden.

Резюме

Приводится обзор лошадей из кромерия Зюссенборна и их сравнение с местными формами сходного и более раннего возраста. Два обычных для зюссенборнской фауны вида лошадей, *E. süssenbornensis* WÜST и *E. altidens* v. REICHENAU, представляют эволюционные линии, несомненно существовавшие в Европе с виллафранка. *E. altidens*, возможно, является предком европейского „осла", *E. hydruntinus*.

References

ALIMEN, H., 1946: Remarques sur *Equus hydruntinus* REGALIA. — In: Soc. géol. Fr., — Paris **7**. — P. 585—595.

AZZAROLI, A., 1979: On a late Pleistocene ass from Tuscany, with notes on the history of asses. — In: Palaeont. Ital. — Pisa **71** (n.s. 41). — 47 pp.

BAJGUSHEVA, V. S., 1971: Iskomaemaja teriofauna Livenzovkovo kariera (severo-vostočnoe Priazove). — In: Materiali po faunam Antropogena SSR. — In: Tr. zool. Inst. Akad. Nauk SSSR. — Leningrad 49. — P. 5—25 (Engl. summary).

BÖKÖNYI, S., 1954: Eine Pleistozän-Eselsart im Neolithikum der Ungarischen Tiefebene. — In: Acta Archaeol. — Budapest **4**. — P. 9—24.

DAVID, A. I.; SUSHPANOV, K. I., 1972: Ostatki mlekopitajuščich iz neogenovich otloženii Moldavii. — In: Posvonočnie neogena i plejstocena Moldavii. — Kižinev: Stiinza. — P. 3—24.

DAVIS, S. J., 1980: Late Pleistocene and Holocene equid remains from Israel. — In: Zool. J. Linn. Soc. — London **70**. — P. 289 to 312.

EISENMANN, V., 1981: Étude des dents jugales inférieurs des *Equus* (Mammalia, Perissodactyla) actuels et fossiles. — In: Palaeovertebrata. — Montpellier **10** (3—4). — P. 127—226.

EISENMANN, V.; PATOU, M., 1980: La faune de la Grotte de Felines-Termenes (Aude). Résultats préliminaires et étude détaillee des restes d'*Equus hydruntinus* (Mammalia, Perissodactyla). — In: L'Anthropologie. — Paris **84** (4). — P. 633—649.

GROMOVA, V. I., 1949: Istoria lošadei (Roda *Equus*) v Starom Svete, I—II. — In: Tr. paleont. Inst. Akad. Nauk SSSR. — Moskva **17** (1—2). — 374 pp.; 162 pp.

—, 1965: Kratkij obzor četvertičnych mlekopitajuščich Evropi. — In: Bull. Komm. izuč. četv. perioda. — Moskva. — 142 pp.

GROMOVA, V. I.; DUBROVO, I. A., 1975: Fossile Pferde aus dem Faunenkomplex von Tiraspol. — In: Quartärpaläontologie. — Berlin **1**. — P. 107—124.

KRETZOI, M., 1954: Bericht über die Calabrische (Villafranchische) Fauna von Kislang, Kom. Féjer. — In: Magyar Állami Földtani Intezet. — Budapest 1953 (I). — P. 213—265.

MOTTL, M. G., 1942: Beiträge zur Säugetierfauna der ungarischen alt- und jungpleistozänen Flußterrassen. — In: Mitt. u. Jahrb. ungar. geol. Anst. — Budapest **36** (2). — P. 1—70.

MUSIL, R., 1965: Die Equidenreste aus dem Altpleistozän von Voigtstedt in Thüringen. — In: Paläont. Abh., A. — Berlin **2** (2/3). — P. 435—449.

—, 1969: Die Equiden-Reste aus dem Pleistozän von Süßenborn und Weimar. — In: Paläont. Abh., A. — Berlin **3** (3/4). — P. 617 to 666.

—, 1971: Die Pferdefunde der Lokalität Stránská Skála. — In: Anthropos, N. S. 2. — Brno **20**. — 204 pp.

—, 1975: Die Equiden aus dem Travertin von Ehringsdorf. — In: Abh. zentr. geol. Inst. — Berlin 23. — P. 265—335.

NOBIS, G., 1971: Vom Wildpferd zum Hauspferd. — Köln u. Wien: Springer-Verl. — 96 pp., (Fundamenta; B, **6**).

PRAT, F., 1968: Recherches sur les Equides pléistocenès en France. — Bordeaux, Univ., These roneo, No. 226, 670 pp.

REICHENAU, W. v., 1915: Beiträge zur näheren Kenntnis fossiler Pferde aus deutschem Pleistozän. — In: Abh. hess. geol. Landesanst. — Darmstadt **7** (1). — 115 pp.

SAMSON, P., 1975: Les equides fossiles de Roumanie. — In: Geol. Romania. — Roma **14**. — P. 165—352.

SCHWARZ, E., 1928: Über diluviale Pferde der *Equus caballus*-Gruppe. — In: Jahrb. preuß. geol. Landesanst. — Berlin **48**. — P. 429—476.

SHER, A. V., 1971: Mlekopitajuščie i stratigrafija plejstocena krainevo severo-vostoka SSSR i Severnoi Ameriki. — In: Bull. Komm. izuč. četv. perioda. — Moskva. — 310 pp.

VEKUA, A. K., 1960: Svoevobrasnaja iskopaemaja lošad iz plejstocena Ahalkalaki (jug Grusii). — In: Dokl. Akad. Nauk SSSR. — Moskva **132** (6). — P. 1417—1420.

—, 1962: Ahalkalakskaja nižneplejstocenovaja fauna mlekopitajuščich. — Tbilisi: Akad. Nauk Grus. SSR. — 208 pp. (Georgian; Russian and English summaries).

WÜST, E., 1900: Untersuchungen über das Pliozän und das älteste Pleistozän Thüringens. — In: Abh. naturforsch. Gesellsch. Halle. — Halle **23**. — 352 pp.

Quartärpaläontologie 6 53—65 Berlin 1986

Funde von Archidiskodon meridionalis und von Trogontherium cuvieri aus den interglazialen Tegelen-Schichten

Von EKKE WOLFGANG GUENTHER, Ehrenkirchen bei Freiburg im Breisgau

Mit 3 Abbildungen, einer Tabelle und 2 Tafeln

Inhalt

Abstractum

Die altpleistozänen Tone und Sande aus der Umgebung von Tegelen in den Niederlanden haben eine reiche Fauna geliefert. In erster Linie Reste des Altbibers *Trogontherium*, aber auch einige Backenzähne von Elefanten.

Diese werden untersucht und es wird überprüft, ob sie etwas zur Altersdatierung der Tegelen-Schichten aussagen.

Weiter wird bei den Tierarten die Frage der Sterbegemeinschaften diskutiert. Besondere Merkmale an den Backenzähnen aus den Tegelen-Schichten und zum Vergleich die einer weit reicheren Fauna aus den Villafranchiano-Sedimenten des Valdarno südlich von Florenz werden herausgestellt.

From the Lower Pleistocene clays and sands of the area around Tegelen in The Netherlands a rich fossil fauna is known, mainly remains of the beaver *Trogontherium* and a few elephant molars which are studied in the present paper in connection with biostratigraphical correlations of the Tegelen deposits. Questions of the taphonomy of both fossil taxa are also examined.

Specieal features of the elephant molars of the Tegelen deposits are discussed in comparison with rich fossil elephant material from Villafranchian deposits of the Valdarno near Florence.

В нижнеплейстоценовых глинах и песках в районе Тегелена в Нидерландах известна богатая ископаемая фауна, состоящая главным образом из остатков трогонтериевых бобров и месkольких моляров слонов. В данной статье приводятся результаты изучения этих остатков в связи с биостратиграфической корреляцией тегеленских отложений. Рассмотрены также вопросы тафономии для обоих ископаемых таксонов.

Обсуждаются характерные особенности моляров слона из тегеленских отложений в сравнении с обширными материалами по ископаемым слонам из виллафранкских отложений в Вальдарно около Флоренции.

1. Die Ablagerungen des Tiglium

Die altpleistozänen Tone, Schluffe, Sande und Kiese des Tegelen werden in einer Anzahl von Gruben abgebaut, die auf beiden Seiten der Grenze BRD—Niederlande liegen, südlich von Venlo und westlich von Mönchengladbach und Krefeld. Sie entstanden als Ablagerungen eines ehemaligen Maaslaufs, zu erheblichen Teilen in Altwassern. In ihrer Verbreitung, Schichtenfolge und deren Mächtigkeit variieren sie stark, so daß nahezu jede Grube ihre eigene Sedimentfolge aufweist und innerhalb von größeren Gruben ändert sich sogar mitunter das geologische Profil.

Fauna und Flora bestätigen, daß es sich um die Ablagerungen einer warmen Zeitphase handelt, die dem ältesten Pleistozän zuzuordnen ist. Die Tegelen-Schichten sind von den darunter liegenden pliozänen Reuver-Schichten durch die Ablagerungen einer Prätegelen-Kaltzeit, die man als Brüggen-Kaltphase bezeichnet hat, getrennt. Dem warmzeitlichen Tegelen folgt dann wiederum eine Kaltphase, die Eburon-Kaltzeit.

Die Tegelen-Schichten sind gleichaltrig mit den Villafranchiano-Ablagerungen aus dem Valdarno (die jedoch einen größeren Zeitraum umfassen) und mit Sedimenten der Britischen Norfolkküste wie Bacton und West Runton. Andere Sedimente aus Norfolk, wie East Runton und Trimingham scheinen nach ihrer Proboscidier-Fauna jünger zu sein.

Einige Beobachtungen sprechen dafür, daß nicht alle Tegelen-Ablagerungen im gleichen Zeitraum entstanden sind. Während die meisten Tegelen-Schichten durch eine *Archidiskodon*-Fauna gekennzeichnet sind, wurden in der Grube Maalbeck bei Belfeld außer diesen Faunenelementen auch *Anancus arvernensis* (CROIZET & JOBERT) sowie *Tapirus arvernensis* (CROIZET & JOBERT) nachgewiesen, was nach KARTENBOUT VAN DER SLUIJS (1960) vielleicht an ein weiter zurückliegendes Alter denken lassen könnte. Diese Annahme ist jedoch keineswegs restlos gesichert. *Anancus arvernensis* lebte bereits im Pliozän (Asti), doch fand man seine Reste ebenso in Schichten des Villafranchiano, zunächst vor dem Erscheinen von *Archidiskodon meridionalis* (NESTI), später (nach KURTÉN 1968) auch gleichzeitig mit diesem. Auch *Tapirus arvernensis*, eine kleine Tapirart, lebte bereits zur Zeit des Asti und wurde ebenfalls in den pleistozänen Ablagerungen des Villafranchiano nachgewiesen.

FRENZEL (1967) gibt ein Diagramm der Vegetationsentwicklung vom Unterlauf des Rheins seit dem Ende des Tertiärs. Es basiert auf Pollenanalysen. Hierin stellt er mit anderen Autoren das Tegelen zwischen eine Prätegelen-Kaltzeit, die dem pliozänen Reuver folgt und eine Eburon-Kaltzeit, welche die Tegelen-Warmzeit ablöst. Nach dem Diagramm besteht das Tegelen aus zwei aufeinander folgenden warmen Phasen, deren ältere sich durch ein Dominieren thermophiler Laubhölzer auszeichnet, während die jüngere Phase vor allem Holzarten feuchter bis nasser Standorte enthält.

Ein 1977 gut aufgeschlossenes Profil konnte in der auf deutscher Seite liegenden Ziegeleigrube von Eyck bei dem Dorfe Bracht aufgenommen werden. Hier ließen sich folgende Lagen voneinander trennen, wobei die Schichtenmächtigkeiten nur geschätzt sind, was bei ihrem starken Wechsel jedoch zu tolerieren ist.

1 m	Podsol (quartäre Bodenbildung)
4 m	grober Kies, bis 2 cm ∅ (im unteren Abschnitt kryoturbates, kaltzeitliches Lagerungsgefüge)
1 m	graues Tegelen, tonig-schluffig (warmzeitlich)
1 m	schwarzes Tegelen. Korngröße 2—10 µm (Auskunft im Ziegelwerk), warmzeitlich
3 m	Sande, Grobsande und Kiese. Tongerölle vor allem an der Basis, kaltzeitlich
0,70 m	hellgraue Schluffe und Tone
3 m	Mittelsande mit einzelnen Grobsandlagen
0,30 m	schwarzer Torf mit Stubben, warmzeitlich
3 m	Reuverton, mit einem Tongehalt höher als bei den Tegelen-Schichten, warmzeitliches Pliozän.

In den Sanden zwischen dem Tegelen und den pliozänen Reuvertonen liegt also eine etwa 0,70 m mächtige, tonigschluffige Schicht, deren Klimacharakter noch nicht bekannt ist.

2. Die Säugerfunde

Die Tegelen-Schichten sind insbesondere durch ihre altpleistozäne Säugetierfauna bekannt geworden. Diese lagert heute in verschiedenen, vor allem niederländischen Museen.

Zu dieser Untersuchung wurden folgende Museen aufgesucht:[1])

Goltziusmuseum der Stadt Venlo
Rijksmuseum van Nature Historie zu Leiden
Naturhistorisches Museum zu Maastricht
Missionsmuseum zu Steyl bei Tegelen
Teyler-Museum zu Haarlem[2])
Institut für Taxonomische Zoologie der Universität Amsterdam.

Die Knochen und Zähne der Säugetiere stammen, wie ihr Erhaltungszustand zeigt, nicht alle aus den gleichen Sedimenten, vorherrschend jedoch, wie z. B. die meisten Reste von *Trogontherium*, aus schluffig-tonigen Lagen, den Absätzen von Stillwassern. In mehreren Fällen haftet den Fossilien aber auch sandiges Material an. Einige der Funde zeigen Manganflecken, ähnlich wie die meisten Reste aus Mosbach bei Wiesbaden, die vorwiegend aus Sanden und Kiesen stammen.

Die Knochen und Zähne aus den Tegelen-Schichten lagen zumeist nicht gleichmäßig im Sediment verteilt, sondern sie waren oft in Nestern angereichert. Diese befinden sich vorwiegend dicht am Rande der Hochterrasse. In den heutigen Gruben stößt man nur selten auf Fossilien (LÜTTSCHWAGER und VAN BEMMEL 1962). Einige fossile Reste, darunter mehrere Kieferfragmente von *Trogontherium*, waren in Konkretionen eingeschlossen. Bei diesen ist klastisches Sediment durch Eisenverbindungen (Limonit) verkittet worden. Ihre Farbe variiert zwischen der eines grauen Tons und der von Rost. Das klastische Material schwankt in seiner Korngröße zwischen tonig, schluffig und feinsandig.

In den Konkretionen können Mandibeln, bisweilen aber auch ganze Schädel von *Trogontherium* stecken (MAI 1977).

Im Museum von Venlo befindet sich eine weitgehend hohle Konkretion, in welcher der Abdruck der Kaufläche der unteren Molaren und des Mandibelkörpers eines *Trogontherium* noch zu erkennen ist. Die Präparation der in den Konkretionen erhaltenen fossilen Reste ist schwierig, da das umhüllende Material härter ist als Zähne und Knochen. Auch sind letztere zumeist zerbrochen. Funde von Fossilien in Konkretionen liegen im Goltziusmuseum der Stadt Venlo, dem Naturhistorischen Museum der Stadt Maastricht, in Leiden und in Amsterdam (MAI 1977).

Die meisten Knochen sind in kleine Stücke zertrümmert. Es gibt kein vollständiges Hirschgeweih, nur ganz wenige Schädel von *Trogotherium* und fast nur Fragmente von Unterkiefern derselben.

Einige Elefantenzähne sind bereits als Fragmente in das Sediment eingebettet worden. Ein Teil der Brüche entstand dann später bei der Fossilisation. So fand man einen Elefantenzahn, der im Sediment in mehr als hundert Einzelstücke zerfallen war. Ein weiterer Teil der Beschädigungen entstand beim Abbau der tonig-schluffigen Sedimente und bei der Bergung der Fossilien.

Die Farbe der fossilen Reste wechselt häufig, was zum großen Teil die Folge einer unterschiedlichen Einbettung sein mag.

Es bleibt also festzustellen, daß das umfangreiche Fossilmaterial, das die Tegelen-Schichten geliefert haben, im ganzen nicht sehr gut erhalten ist. Es fällt jedoch auf, daß selbst kleine Bruchstücke nur in seltenen Fällen Spuren einer Abrollung zeigen. Sie sind also nicht in einem stärker strömenden Flußlauf abgelagert worden, sondern eher in Stillwassern.

Unter den Resten der Mammalier dominieren bei weitem diejenigen, die von *Trogontherium* und von zwei Hirscharten stammen. Ein Vergleich der Funde, die das Naturhistorische Museum in Maastricht besitzt, ergab folgende prozentuale Verteilung: *Trogontherium cuvieri* war mit etwa 60% vertreten. *Eucladoceras tegulensis* (DUBOIS), nach KURTÉN (1968) gleich *Euctenoceras ctenoides* (NESTI), eine größere Hirschart mit mittellangen bis langen Geweihstangen, folgt mit ungefähr 20% und *Cervus rhenanus* (DUBOIS) — nach KURTÉN identisch mit *Cervus etuerarium* (CROIZET & JOBERT) —, eine kleine Hirschart mit einem dreisprossigen Geweih, liegt bei

---

[1]) Den jeweiligen Wissenschaftlern dieser Institutionen bin ich für ihre wertvolle Hilfe zu besonderem Dank verpflichtet.

[2]) Die Tegelen-Fossilien der Kollektion P. WEINGÄRTNER zu Vechta in Oldenburg i. O. (Dominikan Kloster) befinden sich jetzt in Haarlem.

etwa 10%. In nur geringen Mengen (von etwa drei Tieren) fanden sich Reste des Bibers (*Castor fiber* LINNAEUS), des etruskischen Nashorns [*Dicerorhinus etruscus* (FALC.)] mit etwa sechs Tieren und des Südelefanten *Archidiskodon meridionalis* (NESTI) mit zwei Zähnen und Knochenresten. Selten, zum Teil nur durch ein Exemplar belegt, gab es Knochen und Zähne eines Affen (*Macaca florentina* COCCHI), eines nahen Verwandten des heute noch lebenden Gibraltar-Affen, ferner einer großen Katze (*Panthera* cf. *schreuderi* VON KOENIGSWALD), der im Villafranchiano häufigen *Hyaena perrieri* CROIZET & JOBERT, von der Knochenbruchstücke, Gebißteile und Koprolithe erhalten sind. Nachgewiesen ist weiter der für das Villafranchiano typische *Ursus etruscus* CUVIER. Hinzu kommen noch einige Überreste von *Sus strozzi* MENEGHINI, von *Equus robustus* POMEL (nach KURTÉN synonym mit *Equus bressanus* VIRET) und von *Leptobos* cf. *elatus* POMEL. Gefunden wurden Reste von *Hystrix etruscus* BOSCO, *Desmana tegulensis* SCHREUDER, von *Talpa* sp. und von *Hypolagus*. KURTÉN (1968) nennt von Tegelen *Mimomys newtoni* (MAJOR), eine kleine Wühlmausform, die vom späten Villafranchiano bis zur Cromer-Warmzeit nachgewiesen ist. Ferner werden *Emys orbicularis* LINNAEUS angeführt und Eierbruchstücke, die von Vögeln stammen.

Die Fauna bestätigt damit den warmzeitlichen Charakter der Tegelen-Ablagerungen sowie deren Einordnung in die Zeit des Villafranchiano. Einiges spricht dafür, daß die Hauptmenge der Sedimente in eine mittlere oder höhere Phase dieses Zeitabschnitts einzuordnen ist und damit wohl dem höheren Valdarno entspricht.

3. Die Pflanzenwelt

Im pliozänen Reuver finden sich zahlreiche wärmeliebende Pflanzenarten. Sie verschwinden in den Tegelen-Schichten bis auf geringe Reste. Die Flora des Tegelen wurde vor allem von C. und E. M. REID (1915) und von ZAGWIJN in verschiedenen Publikationen (1957, 1959, 1960, 1963) diskutiert. FRENZEL (1967) veröffentlichte ein Schema, das die Vegetationsentwicklung vom Unterlauf des Rheins seit dem Ende des Tertiärs bis zum Postglazial zusammenstellt.

Die tertiären Baumarten nehmen nach C. und E. M. REID im Reuver noch etwa 50%, nach FRENZEL etwa 20—35% ein. In den Schichten des Tegelen geht der Anteil der tertiären Baumarten auf ganz geringe Reste zurück, nach C. und E. M. REID auf etwa 15%. In der Cromer-Warmzeit fehlen sie.

Die Überlieferung an Hölzern, Früchten, Samen und Pollen bezeugt für die Tegelen-Schichten eine echte Warmzeit. Aus dem Diagramm von ZAGWIJN und noch deutlicher aus dem bei FRENZEL erkennt man bei den Tegelen-Schichten zwei Gipfel thermophiler Laubholzarten trockener und feuchter Standorte. Bei dem älteren Maximum erreichen diese bei FRENZEL einen Wert von 50% und bei dem oberen Höchststand einen solchen von noch 30% der Gesamtflora.

In erster Linie haben die ausgezeichneten Aufschlüsse der Ville, ostwärts von Köln, eine detaillierte Gliederung der ältest- und altpleistozänen Ablagerungen des Rheins ermöglicht. Vor allem in den Gruben Fortuna und Frechen hat URBAN (1978a, b, c, 1979, 1980) an Hand von Pollenanalysen eine Gliederung erarbeitet. Sie vergleicht diese mit Gliederungen, wie sie sich aus Arbeiten von DOPPERT und ZONNEFELD (1955) sowie ZAGWIJN (1957, 1960, 1963, 1974) der Tegelen-Schichten ergeben.

Abbildung 1 zeigt Ergebnisse der Arbeiten von URBAN. Danach läßt sich das Tiglium in drei warme Phasen unterteilen, die durch Kältevorstöße voneinander getrennt sind. Driftblöcke, die durch Flußeis transportiert wurden, fanden sich in Grobkiesablagerungen aus der Frechen-Kaltzeit in der Grube Frechen. Tongerölle, wie sie sich in der Grube van Eik nachweisen ließen und die nur in gefrorenem Zustand im Fluß transportfähig sind, wurden nicht beobachtet.

4. Backenzähne des Südelefanten aus den Tegelen-Schichten

In den verschiedenen Museen der Niederlande konnten insgesamt 11 Backenzähne von Elefanten aus den Tegelen-Schichten untersucht werden. Es handelt sich ohne Ausnahme um Fragmente, die zumeist aus wenigen Lamellen, in einem Falle sogar nur aus einer Lamelle bestehen. Diese geringen Bruchstücke sind nicht immer mit ausreichender Sicherheit den einzelnen Kieferteilen, aus denen sie stammen, zuzuordnen, doch ist die Bestimmung der Dentitionsstufen, denen sie angehören, fast immer möglich, so daß das ungefähre Alter der betreffenden Elefanten sich ermitteln läßt.

Man kann davon ausgehen, daß die Tiere in dem Gebiet gelebt haben, in dem ihre Reste gefunden wurden, daß also nicht ihre Kadaver von weither angeschwemmt wurden. Hierfür sprechen allein die Auswurfstücke.

Es überrascht, daß die Backenzähne vorwiegend von jüngeren Tieren stammen. Nur zweimal sind Zähne von älteren Individuen (M^2 oder M^3) überliefert. Sechs Molaren stehen fünf Milchmolaren gegenüber. Große Backenzähne werden in den Tongruben weniger leicht übersehen als kleine Zähne. Es ist daher unwahrscheinlich, daß die Alterszusammensetzung die Folge einer Auslese bei der Fossilbergung ist. Daß kein erster Milchmolar vorliegt ist verständlich, da dieser kleine Zahn besonders oft unbeachtet bleibt. Nichts spricht dafür, daß die Auslese des Fundguts die Folge menschlicher Jagd ist, man muß andere Todesursachen in Erwägung ziehen. Es wäre zum Beispiel denkbar, daß Raubtiere jüngere Elefanten bevorzugt erbeuteten oder aber die jüngeren Tiere kamen durch unvorsichtiges Benehmen in Sümpfen oder im Wasser zu Tode.

Die Fundangaben, die den Elefantenzähnen und den Resten von *Trogontherium* beigegeben sind, erlauben keine Eingliederung in die Schichtprofile und nur bei einem Teil der Fossilien ist die Grube genannt, aus der sie geborgen wurden. Es ist also durchaus denkbar, daß ein Teil der Funde aus unterschiedlich alten Schichten stammt, ohne daß es heute noch möglich wäre, dieses zu überprüfen.

Die wichtigsten Maße, Rechenwerte und Beobachtungen an den Elefantenzähnen sind in Abb. 2 zusammengestellt. In den Einzelbeschreibungen finden sich weitere Mitteilungen.

T1. *Archidiskodon meridionalis* (NESTI). Zweiter oder dritter Molar des rechten Oberkiefers

Von dem Zahn sind lediglich die beiden letzten Lamellen und der hintere Talon erhalten, wodurch die Bestimmung, ob es sich um einen zweiten oder dritten Molaren handelt, sehr erschwert wird. Die Breite des Zahns, die mit ihrem maximalen Wert vielleicht 100 mm erreicht hatte, könnte für einen dritten Molaren sprechen. Doch liegen zweite Molaren (z. B.

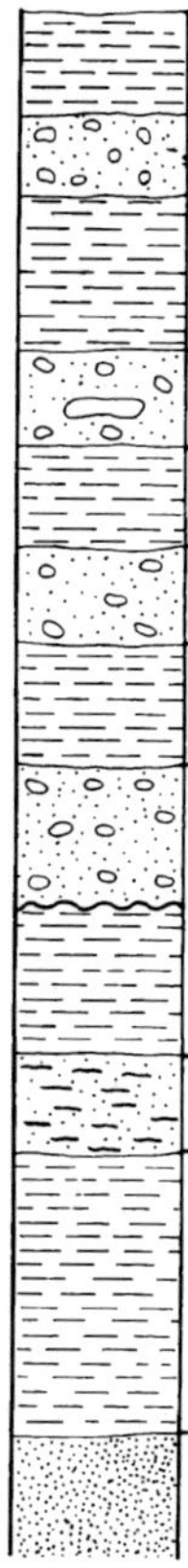

z.T. Kowalczyk 1969	Ville nach Urban 1978		Zagwijn 1957, 1960, 1963, 1974 Doppert & Zonneveld 1955	
Ton - E	Tagebau Frechen Pinus, Juniperus	Kühles Klima	? Eburon od. Waal ?	Post - Tegelen
Schotter - e	Hauptterrassen - Folge	Kalte Phase	? Eburon ?	
Ton - D	Tagebau Frechen, schluffig - tonige Linsen, Phellodendron, Pterocarya, Fagus, Carpinus	Fechen - Intergl. III von Eyck - Intergl.	Tiglium C 5-6 (? Waal ?)	Tegelen - Sedimente
Schotter - d	Grobkies mit Driftblöcken	Frechen - Kaltzeit III	Tiglium C4c	
Ton - C	Tagebau Frechen, Schluff, Mittelsd., Pinus, Picea, Ulmus, Pterocarya	Frechen - Intergl. II	Tiglium C 1-4b	
Schotter - c	Kiesschüttung	Frechen - Kaltzeit II	Tiglium - B	
Ton - B2	Tagebau Frechen, Fagus, Tsuga, Carpinus, 7 Pollenzonen	Frechen - Intergl. I Belfeld - Interglaz.	Tiglium - A	
Schotter - b2	Schotter u. Sande	Frechen - Kaltzeit Ib	Kaltes Praetiglium	Prae - Tegelen
Erosionsdiskord.				
Ton - B1	Tagebau Fortuna Koniferen	Fortuna - Oscillation Interstadial	warmgemäßigtes Praetiglium	
Schotter b1	sandig - schluffig - tonig	erster Kälteeinbruch Ia	kaltes Praetiglium	
Ton A2	Tagebau Fortuna, nach oben Abnahme der warmen tertiären Florenbestandteile	warmes Tertiär	A_2 Reuver	
Ton A1	Reine Tertiär - Floren		und A_1	
Sande				

Abb. 1. Gliederung altpleistozäner Schichtenfolgen im Gebiet des Niederrheins und der Maas

Nr.	Museum Nr.	Museum Ort	Fundstelle	Dentit.-Reihe	Lam. Form.	Länge	Breite	Höhe	Schmelz-stärke	Abrasion Lamel.	Abrasion %	Erhaltung	Alter	LLQ.	Bemerkungen
T1	-	Maastricht	Tegelen Ton	M^{2} d.	-2x	-	-92 (100)	>100	3,5	alle	80	kleines Fragment	35 (69)	~20,5	vielleicht M^{3}
T2	615	Maastricht	Tegelen Ton	M^{2} d.	-½ 1x	-50	91 o.	-105	3,8	alle nicht prox. x	50	prox. Fragment	30 (60)	20,0	vielleicht M^{3}
T3	53067	Rijksmus. Leiden	Tegelen Gr. Russel	M^{2} d.	∞4 -	-111 -	91 1/2 Dz.	-55	3,0	alle	90	mehrere Bruchst.	40	19,5	Resorption u. Impression
T4	-	Steyl bei Tegelen	Jammertal Gr. Canoy	M_{2} d.	-2x	~50	83	-22	3,0	alle	80	prox. Fragment	38	>20,0	Impression
T5	53066	Rijksmus. Leiden	Grube Denissen	M_{2} d.	-3x	-74	85 o.	-56	4,0	alle	90	prox. Fragment	40	~21,0	x als Suffix
T6	St. 94062	Rijksmus. Leiden	Grube St. Ophet Veld	M^{1} d.	∞2-	-61 -	75 o.	-41 -	3,2	alle	90	Auswurf-Rest	25	16,1	Resorption u. Impression
T7	-	Univers. Amsterd.	Tegelen 1935	mm^{3} d	-4x	-90	~64 w. DZ.	-57	3,0	alle	50	prox. Fragment	8	~18,6	vielleicht M^{1} d.
T8	St. 32266	Rijksmus. Leiden	teilweise Kies	mm^{3}	-1½-	-41 -	56 o.	-33,5	3,2	alle	85	Fragm.	12	17,0	–
T9	St. 53066	Rijksmus. Leiden	Grube Russel	mm 2	-1 -	-18 -	29 o.	-32	3,0	alle	-	Einzel-Lamelle	5	19,0	–
T10	St. 35254	Rijksmus. Leiden	Tegelen Kies	? mm 3	-2 - x2 -	~24 - ~30 -	~48 ~56	-	~4,0 ~3,0	alle	-	2 Lam.-Bruchstücke	8	17,0 18,0	–

Abb. 2. Wesentliche Daten zu Backenzähnen von *Archidiskodon meridionalis* aus den Tegelen-Schichten

T3) vor, die gleiche Breitenwerte aufweisen. Das Fragment T1 hat immer noch eine Breite von 92 mm. Die Höhe der letzten bereits angekauten Lamellen von T1 liegt bei 95 mm, in nicht angekautem Zustande dürfte das Maß etwa 100 mm betragen haben. Die Dicke des Schmelzes auf der Kaufläche beträgt 3—3,5 mm. Die Spalten sind tief eingesenkt, die Mamillen sind besonders kräftig ausgebildet und an der letzten Lamelle noch in der beachtlichen Länge von 40 mm erhalten. Die Furchen, die sich von den Mamillen ausgehend in Richtung zur Wurzel fortsetzen, durchziehen das Dentin in seiner vollen Länge. Die ursprüngliche Anzahl der Mamillen dürfte bei sieben gelegen haben, am Talon sind noch vier Mamillen erhalten. An der vorletzten Lamelle ist die basale Schmelzbrücke zu erkennen, durch welche diese mit der nächsten Lamelle verbunden war.

Der Zahn ist bereits von BERNSEN (1930) beschrieben und abgebildet (Fig. 2 und 3) worden. Auch er ist sich nicht ganz sicher, ob es sich um einen zweiten oder dritten Molaren handelt. Wenn die S-förmige Biegung der Lamellen die Folge einer Pression sein sollte, meint er, könne es nur ein zweiter Molar sein. Derartige Biegungen kann man jedoch ebenfalls an dritten Molaren, allerdings vorwiegend des Unterkiefers, beobachten.

T2. *Archidiskodon meridionalis* (NESTI). Zweiter oder dritter Molar des rechten Oberkiefers

Von dem Zahn sind nur eine Lamelle ganz und die vorausgehende etwa zur Hälfte sowie der Talon des proximalen Zahnabschnitts erhalten. Bei einer Länge von 50 mm heißt die Lamellenformel somit —1/2 1x. Zwischenzement und Deckzement fehlen. Die Breite der letzten Lamelle liegt bei 91 mm. Mit Deckzement muß die Gesamtbreite des Zahns weiter nach der distalen Seite (die nicht mehr erhalten ist) gegen 100 mm betragen haben.

Die Schmelzstärke auf der Kaufläche beträgt 3,8—4,0 mm. Die letzte Lamelle ist nur wenig angekaut. Der Talon war wohl noch nicht in den Kauprozeß einbezogen, doch läßt sich dies nicht mit Sicherheit sagen, da die Spitzen seiner Mamillen abgebrochen sind. Die letzte Lamelle endet in fünf langen und kräftig gebauten sowie zwei schwächeren, an der Seite liegenden Mamillen. Der Schmelz auf der Kaufläche ist, wohl infolge der geringen Abrasion, kaum gefältelt. Der Längen-Lamellen-Quotient läßt sich bei dem kleinen Zahnfragment nur abschätzen, er liegt zwischen 18 und 22, im mittleren Wert etwa bei 20.

T3. *Archidiskodon meridionalis* (NESTI). Zweiter Molar des rechten Oberkiefers

Der Zahn wurde in der Grube Russel, Tiglia-Egypt bei Tegelen gefunden. Er ist stark niedergekaut und die Wurzel zeigt deutliche Spuren von Resorption. Der Molar war — anders als dies beim Mammut der Fall ist — mit seiner distalen Seite während des Nachschubs nicht sehr stark gegen den Zahn des Unterkiefers gekippt worden, so daß wenigstens die erhaltenen Lamellen verhältnismäßig gleich stark abradiert wurden. An der Vorderseite der Kaufläche ist ein 24 mm langes Dentinplanum ausgebildet, hier fehlen also die letzten Reste der Lamellen.

Das Fundstück war zerbrochen und wurde aus zwei großen und sieben kleinen Teilen zusammengesetzt. Etwa 85—95% des Molaren sind bereits abgekaut. Die Kaufläche ist infolge der kräftigen Abrasion nahezu glatt, Dentin, Zement und Schmelz bilden eine Ebene. Die für die Zerkleinerung der Nahrung so wichtige Raspel ist also nicht mehr vorhanden.

Die Breite der letzten Lamelle liegt bei 91 mm, wobei nur an einer Seite noch Deckzement erhalten ist. Die maximale Breite in nicht abgekautem Zustande mag bei 96—98 mm gelegen haben. Auf der proximalen Zahnseite erkennt man eine kräftige Impression, die so stark ist, daß hier bereits der Schmelz fehlt und das Dentin der letzten Lamelle freigelegt ist. Die Impression liefert den Beweis, daß es sich nur um einen zweiten Molaren handeln kann.

Die Lamellen sind, schon infolge der starken Abrasion, ganz besonders dick, das Zwischenzement ist nur als schmaler Streifen erhalten. Sämtliche Lamellen bestehen aus einem rechten und einem linken Teil, sind also zweigeteilt; das Zwischenzement hat zwei Inseln gebildet. Der Längen-Lamellen-Quotient liegt zwischen 19,0 und 19,6, im mittleren Wert also bei 19,3.

T4. *Archidiskodon meridionalis* (NESTI). Zweiter Molar des rechten Unterkiefers

Im Missions-Museum zu Steyl bei Tegelen befindet sich ein Zahnfragment, das bereits von BERNSEN (1930) eingehend beschrieben worden ist, wobei vor allem die Frage diskutiert wurde, zu welcher Elefantenart das Fossil gehört. Das Fragment stammt aus der Grube Canoy von Herfkens & Co., dem Jammertal bei Venlo. Es besteht aus den zwei letzten Lamellen und dem proximalen Talon. Von dem sehr stark niedergekauten Molaren sind bereits etwa 80% abradiert. Wurzeln sind nicht mehr vorhanden. Das Zement hat eine hellbraune Farbe, das Dentin ist dunkelbraun und der Schmelz ist bläulich-grau.

Der Molar hat an der proximalen Seite rechts eine sehr deutliche Pressionsmarke als Folge des Drucks, den der nachfolgende Molar — es ist sicher der dritte Molar — ausgeübt hat. Dementsprechend kann es sich nur um einen zweiten Molaren handeln. Die maximale Breite des Reststücks liegt bei 82 mm.

T5. *Archidiskodon meridionalis* (NESTI). Zweiter Molar des rechten Unterkiefers

Erhalten sind die drei letzten Lamellen und der proximale Talon. Dieser ist klein ausgebildet und haftet als Suffix der letzten Lamelle an. Das Fragment ist aus drei Stücken zusammengesetzt. Der Molar ist sehr stark niedergekaut, so daß nur noch etwa 10% des Gesamtzahns vorliegen.

RICHARZ (1921) und BERNSEN (1930) veröffentlichten bereits Beschreibungen des Molaren und BERNSEN gab eine Abbildung. Der Molar zeigt an seinem proximalen Ende die Eintiefung des Pressionsdrucks des nachfolgenden Zahns. Unter Berücksichtigung der Zahnbreite, die ohne Deckzement — das nicht mehr erhalten ist — bei den hintersten Lamellen noch bei 85 mm liegt, kann es sich wohl nur um einen zweiten Molaren handeln.

Der Schmelz hat eine große Dicke, die zwischen 3 und 4 mm liegt. Infolge der groben, plizidenten Fältelung ist dieses Maß nur abzuschätzen. Die beachtliche Schmelzstärke ist zum Teil durch die starke Niederkauung bedingt.

T6. *Archidiskodon meridionalis* (NESTI). Erster Molar, wahrscheinlich des rechten Oberkiefers

Der Zahn stammt aus der Sammlung Prof. VAN DER VLERK und KORTENBOUT VAN DER SLUIJS. Das Etikett nennt das Datum vom 18. 12. 1951, an welchem Tage der Molar wahrscheinlich gefunden wurde.

Es handelt sich um den letzten Rest eines Backenzahns, der so stark niedergekaut ist, daß der Elefant ihn wohl ausgeworfen hat. Er gibt also keinen Hinweis auf das Sterbealter des Tiers, doch bezeugt er, daß die Elefanten in dem Fundgebiet gelebt haben und nicht als Kadaver im Wasser aus einem anderen Gebiet angetrieben sind.

Infolge der starken Abrasion ist die Kaufläche bereits weitgehend geglättet und die Wurzelseite ist schon kräftig resorbiert. Der Abkauungsrest besteht aus zwei Lamellen und dem hinteren Talon. In diesem zeigt sich eine sehr tiefe Impression des nachfolgenden Zahns. Der Schmelz ist schwarz, das Dentin braun gefärbt. Die Schmelzstärke liegt zwischen 2,5 und 3,2 mm. Der Schmelz ist gekörnt. Die Dicke der Lamellen beträgt im mittleren Wert 16 mm. Das Zwischenzement zwischen den Lamellen ist nur schmal ausgebildet. Der Längen-Lamellen-Quotient liegt zwischen 15 und 17, im mittleren Wert bei 16,15.

Die Bestimmung als erster Molar wurde an Hand der Breite und des Längen-Lamellen-Quotienten getroffen. Sie ist am wahrscheinlichsten, jedoch nicht ganz sicher.

T7. *Archidiskodon meridionalis* (NESTI). Dritter Milchmolar oder erster Molar des rechten Oberkiefers

Das Fragment wurde bereits bei SCHREUDER (1946) genannt. Es ist aus zwei Stücken zusammengesetzt. Erhalten sind die vier letzten Lamellen und der hintere Talon. Die Lamellenformel des voll erhaltenen Zahns könnte etwa ×6× gewesen sein. Die volle Höhe könnte bei 68 mm gelegen haben und die volle Breite bei etwa 66 mm. Die am Fragment meßbaren Werte sind auf Tabelle 2 angeführt. Sämtliche Lamellen (außer dem hinteren Talon) sind angekaut. Die Abrasion hat etwa 50% des Zahns abgetragen. Der Längen-Lamellen-Quotient liegt zwischen 17,8 und 19,5, im mittleren Wert bei 18,6. Die Schmelzstärke ist für einen dritten Milchmolaren oder einen ersten Molaren mit etwa 3 mm recht beachtlich.

T8. *Archidiskodon meridionalis* (NESTI). Wahrscheinlich dritter Milchmolar des Oberkiefers

Erhalten sind 1 1/2 Lamellen aus dem mittleren Zahnteil. Von einer weiteren Lamelle sind noch geringe Reste vorhanden. Da der dritte Milchmolar und der erste Molar häufig einander sehr ähnlich sind, ist die Eingliederung des Zahns in der Dentitionsreihe nicht mit letzter Sicherheit zu treffen, zumal lediglich ein sehr geringes Fragment überliefert ist.

SCHREUDER (1946), die den Zahn ebenfalls anführt, hält ihn für einen D_4. Bei einem so kleinen Bruchstück ist oftmals die Lage im Kiefer, ob rechts oder links, oben oder unten, nicht mit letzter Sicherheit festzustellen. Die Abkauung ist so weit fortgeschritten, daß man mit einer Abrasion von wenigstens 85% zu rechnen hat.

Die Zwischenräume zwischen den Lamellen, gesehen auf der Kaufläche, sind klein, das Zwischenzement also nur schmal ausgebildet. Deutlich zu erkennen ist, daß die Lamellendicke in der Mitte des Zahns etwa ebenso groß ist wie auf den Seiten, so daß sich auf der Kaufläche ein etwa viereckiger Querschnitt ergibt (tetragonal, mit einer sehr langen und einer kurzen Achse). Der Schmelz ist grob gekörnt. Der Längen-Lamellen-Quotient liegt zwischen 16,5 und 17,5, im mittleren Wert also bei 17. Die Schmelzstärke ist mit ungefähr 3,2 mm beachtlich groß.

Als Fundschicht ist angegeben: Gedeelt Kies, also teilweise Kies.

T9. *Archidiskodon meridionalis* (NESTI). Wahrscheinlich zweiter Milchmolar

Der Zahnrest besteht aus nur einer Lamelle, die jedoch einige wichtige Merkmale erkennen läßt. Er stammt aus der Sammlung von KORTENBOUT VAN DER SLUIJS (Datum: 28. Okt. 1955). Die Breite mißt sich ohne Deckzement mit 29 mm, dürfte also wohl mehr als 32 mm betragen haben, was den Schluß zuläßt, daß es sich um einen zweiten Milchmolaren handelt, weniger wahrscheinlich um einen dritten Milchmolaren. Die Schmelzstärke ist allerdings mit über 3 mm für einen zweiten Milchmolaren sehr groß. Die Schmelzfältelung ist kräftig. Es zeigt sich also, daß bereits die Milchmolaren dieses für *A. meridionalis* kennzeichnende Merkmal besitzen.

Die Stellung des Zahns im Kiefer ist bei einem so geringen Fragment nicht bestimmbar.

T10. *Archidiskodon meridionalis* (NESTI). Zwei Fragmente eines dritten Milchmolaren oder ersten Molaren

Im Rijksmuseum von Leiden befinden sich einige Einzelteile von Elefantenbackenzähnen, die aus Kiesen von Tegelen stammen sollen (Kollektion Prof. VAN DER VLERK).

Es sind dies sechs Mamillen eines sehr großen Backenzahns und zwei kleinere Lamellenstücke aus der Mitte von zwei Backenzähnen oder desselben Zahns. Die Längen-Lamellen-Quotienten dieser beiden Bruchstücke lassen Werte von 17 und 18 errechnen, doch sind dies nur Näherungswerte. Die Schmelzstärke ist mit 3 und 4 mm auffallend groß. Der Schmelz ist kräftig gekörnt. Die Breitenwerte betragen etwa 48 und 56 mm, die Fragmente könnten daher von einem dritten Milchmolaren oder einem ersten Molaren stammen. Auch hier besagen Stärke und Ausbildung des Schmelzes, daß es sich um Fragmente von *Archidiskodon meridionalis* handelt.

5. Merkmale der Backenzähne von Archidiskodon meridionalis nach Funden aus den Tegelen-Schichten und aus dem Valdarno in Italien

5.1. Merkmale der Backenzähne von Archidiskodon meridionalis (NESTI)

Der Südelefant hat niedrige, ovale oder auch tetragonale Molaren von besonderer Breite, bei den dritten Molaren zumeist über 100 mm, die aus verhältnismäßig wenigen, dafür aber kräftigen Schmelzlamellen zusammengesetzt sind. Der dritte Molar von *Archidiskodon meridionalis* hat im Höchstfalle 16 Lamellen, die Zähne aus dem Valdarno erreichen 12—14 Lamellen. Der Schmelz ist besonders dick, auf der Kaufläche fast immer über 3 mm, selbst bei den Zähnen einer niedrigeren Dentitionsstufe. Er zeigt nach einiger Abrasion grobe Zacken, die oft unregelmäßig verzerrt erscheinen und deren Aussehen man als plizident, geperlt oder gekörnt bezeichnen kann. Die Schmelzfiguren sind breit, oft weniger rhombisch sondern mehr tetragonal ausgebildet, das heißt, daß die Querschnitte der beiden Seiten ähnlich groß sind wie diejenigen der Mitte. Das Zwischenzement zwischen den einzelnen Lamellen ist häufig auffallend schmal. Vielfach sind auch bei früheren Dentitionsstufen acht Mamillen vorhanden, die nicht selten besonders kräftig sind. Die Mamillen sind zumeist besonders lang, was zur Folge hat, daß bei fortschreitender Abrasion noch acht Lamellen Mamillen zeigen können.

Die Zähne von *Mammuthus primigenius* erfahren bei dem Nachschub von proximal nach distal eine Kippung derart, daß der vordere Zahnteil verstärkt gegen den im Kiefer gegenüberliegenden Zahn des Ober- oder Unterkiefers verschoben wird, was zur Folge hat, daß der distale Zahnabschnitt bereits abgekaut ist, während der proximale Zahnteil noch nicht oder nur wenig in Abrasion ist. Bei *Archidiskodon meridionalis* ist diese Kippung weit geringer, wodurch ein größerer Abschnitt der Lamellen gleichzeitig in Funktion ist. Es können alle Lamellen abradiert werden, ohne daß der distale Zahnabschnitt bereits völlig abgekaut und dann der Rest mit dem Wurzelteil, soweit dieser nicht resorbiert ist, abgebrochen ist, was dem Normalverhalten der Molaren von *Mammuthus primigenius* entspricht.

Der Längen-Lamellen-Quotient hat bei den dritten Molaren des Südelefanten Werte um 20; diese liegen also weit höher als bei *Parelephas trogontherii*. Lediglich Bullen mit besonders großen Molaren können beim Steppenelefanten in Ausnahmefällen ähnlich hohe Längen-Lamellen-Quotienten haben. Bei den Zähnen einer geringeren Dentitionsstufe von *Archidiskodon meridionalis* geht der LLQ auf ±15 zurück.

5.2. Vergleich der Archidiskodon meridionalis Zähne aus dem Gebiet um Tegelen mit italienischen Funden aus dem Valdarno

Im Valdarno wurde aus Schichten des Altpleistozäns verschiedener Fundstellen eine umfangreiche Fauna mit zahlreichen Elefanten-Backenzähnen geborgen. Diese liegen vor allem in den Museen der Universitäten Florenz und Pisa sowie dem Naturhistorischen Museum in Basel. Hiervon standen mir 51 Zähne verschiedener Dentitionsstufen zum Vergleich zur Verfügung. Ihr Erhaltungszustand war zufriedenstellend. Sie gaben ein wertvolles Vergleichsmaterial zu den wenigen, erheblich schlechter erhaltenen Elefanten-Backenzähnen aus den Tegelen-Schichten.

Bei dem jungpleistozänen, hochentwickelten Elefanten der *Mammuthus*-Stammreihe, dem *Mammuthus primigenius* (BLUMENBACH) der Würm-Kaltzeit, ist der zweite Molar an Hand von Form sowie Anzahl und Bau der Lamellen stets leicht vom dritten Molaren zu unterscheiden (GUENTHER 1981). Bei den altpleistozänen Backenzähnen des *Archidiskodon meridionalis* ist die Trennung der Zähne der beiden Dentitionsstufen zumeist schwieriger, vor allem dann, wenn nur ein Teil des Zahns erhalten ist, die Gesamtzahl der Lamellen nicht mehr festzustellen und oft auch der Längen-Lamellen-Quotient nicht mehr exakt zu errechnen ist. Allerdings ließen sich bei den Zähnen aus den Tegelen-Schichten mehrfach an deren Proximalseite die Impressionen des nachfolgenden Zahns, der nur der dritte Molar sein konnte, erkennen, womit die Bestimmung als zweiter Molar gesichert war. Bei einem Teil der Tegelen-Zähne ist jedoch die Bestimmung der Dentitionsstufe nicht eindeutig.

Die Werte, die man am besten selbst bei Fragmenten ermitteln kann, sind Zahnbreite und Längen-Lamellen-Quotient. Die Abb. 3 zeigt daher diese Werte der zweiten Molaren aus den Tegelen-Schichten und aus dem Valdarno. Dabei sind die niederländischen Zähne mit einem Dreieck und die italienischen Funde mit einem Kreis gekennzeichnet. Die Signatur der Oberkiefermolaren ist ausgefüllt und zudem ist das Dreieck der Oberkieferzähne der Tegelen-Schichten mit der Spitze nach oben dargestellt.

Es zeigt sich, daß die Felder der beiden aus verschiedenen Teilen Europas stammenden Zahngruppen zwar ineinander liegen, doch sind die Tegelen-Zähne im mittleren Wert etwas breiter und zeigen einen höheren Längen-Lamellen-Quotienten. Diese Merkmale charakterisieren phylogenetisch

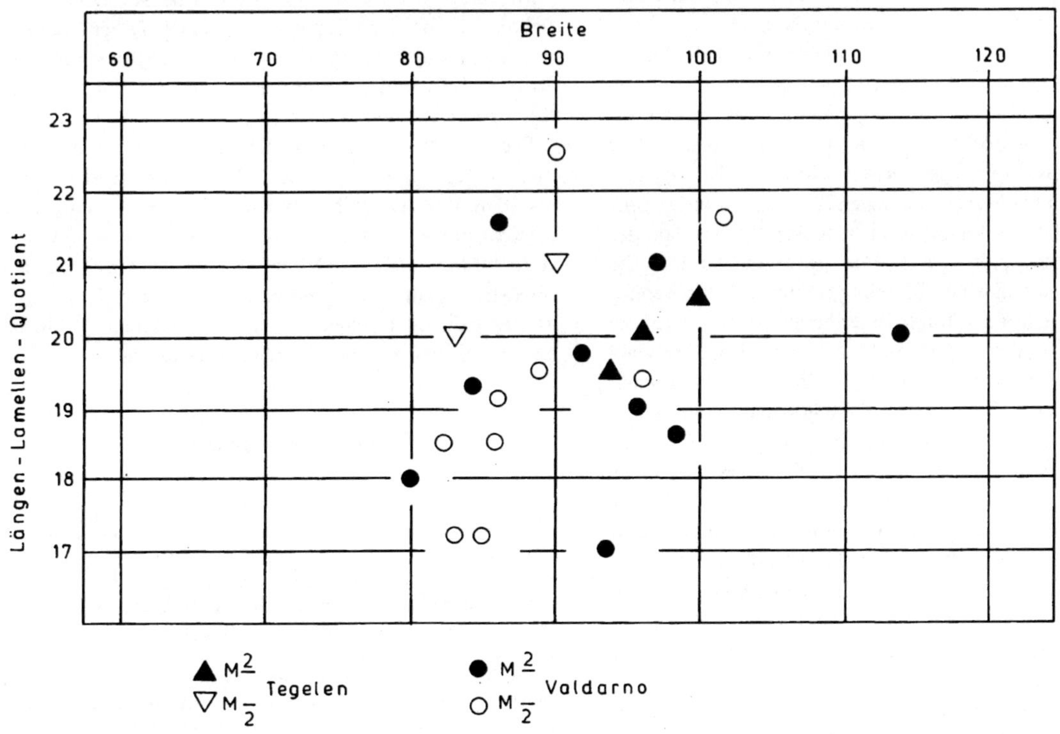

Abb. 3. Zahnbreite und Längen-Lamellen-Quotient der zweiten Molaren aus den Tegelen-Schichten und dem Valdarno

etwas ältere Elefanten der *meridionalis-trogontherii-primigenius*-Reihe.

Es mag sein, daß dieser „phylogenetische Unterschied" eine Folge der räumlich weit auseinanderliegenden Standorte ist. Es könnte vielleicht auch sein, daß die beiden Fundgebiete nicht ganz gleichaltrig sind. Während die italienischen Zähne aus dem oberen Valdarno stammen, könnten die niederländischen Elefanten in einem etwas älteren Abschnitt des Villafranchiano gelebt haben.

6. Die Biberart Trogontherium aus den Tegelen-Schichten

Von Resten des Altbibers *Trogontherium cuvieri* FISCHER aus den Tegelen-Schichten liegt in niederländischen Museen und in weit geringerem Umfange in nichtniederländischen Museen wie Basel, Wiesbaden und Heidelberg, ein reiches Material. Vor allem die in den Niederlanden aufbewahrten Funde hat SCHREUDER (1929) in einer eingehenden Monographie bearbeitet. Eine Überarbeitung erfolgte durch MAI (1977), in die die neueren Funde einbezogen werden konnten. Die Autorin bemühte sich herauszuarbeiten, wie weit die Gebisse von *Trogontherium* aus Schichten von verschiedenalten Fundplätzen Europas sich unterscheiden, ob sich also innerhalb des Alt- und Mittelpleistozäns eine phylogenetische Weiterentwicklung erkennen läßt. Von dieser Arbeit (MAI 1977) sind verschiedene Abschnitte bereits veröffentlicht worden, so das Kapitel über die Biber von Jockgrim in der Rheinpfalz (GUENTHER & MAI 1977), von Mosbach bei Wiesbaden (MAI 1979a), von Mauer an der Elsenz (MAI 1979b) und von Bilzingsleben in Thüringen (GUENTHER 1983). Eine Publikation des Kapitels über die *Trogontherium*-Funde aus den Tegelen-Schichten ist nicht vorgesehen; es sollen daher hier einige der wichtigsten Ergebnisse referiert werden.

Tegelen und Mosbach haben das umfangreichste Material geliefert, so daß diese Fundstellen für Vergleiche besonders geeignet sind. Die Tegelen-Fossilien — es sind neben zahlreichen Einzelzähnen Fragmente des Ober- und Unterkiefers — zeigen keine Spuren einer Abrollung, doch sind sie in mehr oder weniger kleine Stücke zerfallen. So gibt es nur ganz wenige Schädel von *Trogontherium* und die Unterkiefer sind zum großen Teil vor den Prämolaren und unmittelbar hinter den dritten Molaren abgebrochen. Nur in wenigen Fällen befindet sich der Inzisor noch in der Alveole. Der Erhaltungszustand der *Trogontherium*-Reste aus den Tegelen-Schichten ist, obwohl sie mit wenigen Ausnahmen aus schluffig-tonigem Material kommen, nicht besser als derjenige der Mosbacher Funde, die aus Sanden geborgen wurden. Ein Teil der Brüche mag bei der Bergung erfolgt sein, die wesentlichen Beschädigungen erfolgten aber bereits vorher, eventuell schon vor der Einbettung im Sediment, vor allem aber wohl in diesem. Vielleicht mag hierbei die Frosteinwirkung einer Kaltzeit nach dem Tiglium eine Rolle gespielt haben.

H. MAI hat sich weiter bemüht, an Hand von *Trogontherium*-Unterkiefern zu überprüfen, welches Sterbealter die Tiere hatten. Die Tabelle 1 gibt hierzu einen Einblick.

Fast 50% der fossilen Unterkiefer stammen also von alten und sehr alten Individuen. Jüngere Tiere sind mit 19% vertreten und auf die mittleren Altersklassen fällt ein Anteil von 31%. Gute Lebensbedingungen sind Voraussetzung für ein hohes Lebensalter und bei guten Lebensbedingungen sollte auch die Vermehrungsrate hoch liegen.

Man kann bei dem Fundgut von einer natürlichen Sterbegemeinschaft sprechen, wobei allerdings der Prozentsatz der jüngeren, voll ausgewachsenen Tiere etwas überhöht erscheint.

Tabelle 1. Zur Altersverteilung von *Trogontherium*-Unterkiefern der Fundstellen im Raum Tegelen

Anzahl	Kaufläche	Alter	Prozent
3	4 Striide	sehr jung	9,6
3	1 Hypostriid	jung	9,6
9	4 Inseln	voll ausgewachsen	29,0
5	3 Inseln	älter	16,1
10	2 Inseln	alt	32,1
1	1 Insel	sehr alt	3,2

Auf vier Diagrammen hat MAI die auf der Kaufläche gemessenen Längen und Breiten der Zähne dargestellt. Hierbei zeigt sich, daß die Prämolaren des Oberkiefers — weniger deutlich die des Unterkiefers — der Tegelen-Fauna bei geringerer Länge eine größere Breite aufweisen als die entsprechenden Zähne von Mosbach. Die Oberkieferprämolaren von Bilzingsleben liegen in dem Feld der Tegelen-Funde, die Unterkieferprämolaren von Bilzingsleben jedoch zeichnen sich durch eine besondere Länge aus, sind allerdings fast alle etwas schmaler als die entsprechenden Zähne von Tegelen. Die ersten und zweiten Molaren von Tegelen sind bei gleicher Breite zumeist kürzer als die entsprechenden Zähne von Mosbach. Die dritten Molaren von Tegelen sind im mittleren Wert bei gleicher Breite länger als die entsprechenden Zähne von Mosbach. Eine deutliche Trennung mit Hilfe dieser Merkmale ist jedoch nur bei Vorlage eines umfangreichen Materials sinnvoll, da die Felder in den Grenzbereichen sich überdecken.

Bei den Backenzähnen von Tegelen tritt nach MAI (1977) eine Besonderheit auf. Ein Teil der Prämolaren besitzt ein fünftes Striid. Dieses befindet sich am vorderen Teil des Zahns, reicht jedoch bei einem wenig angekauten Zahn noch nicht bis zur Kaufläche und wird erst bei stärkerer Abrasion sichtbar. Es hat eine Länge von ± 7 mm. Es gibt Zähne von sehr jungen Tieren, die kein Anzeichen dieses zusätzlichen Striids zeigen und es finden sich tief abgekaute Zähne, die eine deutliche Kerbe aufweisen.

Die Breiten der Unterkiefer der Tegelen-Trogontherien sind mit 16—18 mm geringer als die von Mosbach mit 18 bis 20 mm. Aber auch hier überschneiden sich die Grenzbereiche etwas, so daß eine Abtrennung nur beim Vorliegen eines umfangreicheren Materials, als es von Tegelen und Mosbach bekannt ist, und die Errechnung eines mittleren Wertes sinnvoll erscheint. In der Ausbildung der Inzisiven zeigen sich keine auswertbaren Unterschiede.

Zusammenfassung

Im Grenzgebiet BRD—Niederlande, in der Nähe von Tegelen und Venlo, werden seit langem altpleistozäne Tone abgebaut. In diesen fand man zahlreiche Reste des Altbibers *Trogontherium*, von *Cervus rhenanus* und *Eucladoceras tegulensis*. Andere Tierarten konnten zum Teil nur durch Einzelfunde nachgewiesen werden. Von *Archidiskodon meridionalis* liegen, außer einigen Knochen in verschiedenen niederländischen Museen, elf Backenzähne, ausschließlich als Fragmente vor. Sie bestehen zumeist nur aus wenigen Lamellen, vielfach vom proximalen Zahnende. Die Bestimmung der Dentitionsstufe ist daher nicht immer leicht. Es handelt sich um Reste von fünf adulten, aber keineswegs alten und um solche von fünf jüngeren und jungen Tieren. Mit zwei fraglichen Ausnahmen ist kein letzter Molar überliefert.

Die Beschreibung des Fundguts und der Vergleich mit Zähnen aus dem Valdarno bringt neue Daten zur Kennzeichnung der Backenzähne von *Archidiskodon meridionalis*. Der Südelefant hat niedrige, ovale bis tetragonale, besonders breite Backenzähne. die im Vergleich zu den Gebissen aus späteren Zeiten der *Archidiskodon-Parelephas-Mammuthus*-Reihe aus weniger, jedoch besonders kräftig ausgebildeten Lamellen zusammengesetzt sind. Der Schmelz ist bei den Molaren im Regelfalle stärker als 3 mm und oft mit groben Zacken, geperlt oder gekörnt, ausgebildet. Das Zement zwischen den Lamellen ist zumeist schmal. Die Lamellen enden auf der Kauflächenseite oft in acht Mamillen, die auffallend kräftig und lang sind, was zur Folge hat, daß bei fortschreitender Abrasion mitunter auf der Kaufläche bis zu acht Lamellen in Mamillen enden. Die Molaren wurden beim Nachschub mit ihrer Vorderseite weniger stark gegen den Gegenzahn vorgeschoben als dies beim Mammut des jüngsten Glazials der Fall ist. Der Längen-Lamellen-Quotient hat beim Südelefanten höhere Werte als bei den späteren Elefanten (*Parelephas* und *Mammuthus*) und erreicht beim dritten Molaren Quotienten um 20.

Bei *Parelephas* und vor allem bei *Mammuthus* sind die zweiten und dritten Molaren fast immer mit Hilfe des Längen-Lamellen-Quotienten und der Breite zu trennen. Beim Südelefanten überdecken sich die beiden Felder in den Grenzbereichen. Alle diese Unterscheidungsmerkmale wurden an Backenzähnen aus dem Valdarno südlich von Florenz überprüft. Dabei zeigte es sich, daß die Tegelen-Zähne im mittleren Wert einen phylogenetisch etwas älteren Charakter aufweisen als die italienischen Zähne.

Von *Trogontherium* liegt aus diesem Raum ein sehr umfangreiches Material vor, das keine Spuren einer Abrollung erkennen läßt. Außer zahlreichen Einzelzähnen wurden Reste von Unterkiefern und seltener von Oberkiefern geborgen. Vollständige Erhaltung findet sich dann, wenn der fossile Rest in einer Konkretion geschützt war, bei ungeschützter Einlagerung im Sediment zerbrachen die Fossilien, vielleicht durch die Bildung von Frostböden einer späteren Kaltzeit.

Fast 50% der Unterkiefer stammen von alten und sehr alten Tieren; die mittleren Alterklassen sind mit 31% und die jüngeren mit 19% vertreten. Aus dieser Altersverteilung kann man auf eine natürliche Sterbegemeinschaft schließen.

Summary

In the area of the German-Dutch frontier near Tegelen and Venlo, Lower Pleistocene clays have been worked for a long time. In these deposits rich fossil remains of the beaver *Trogontherium*, of *Cervus rhenanus*, and *Eucladoceras tegulensis* are found. Other taxa are recorded by few fossils only. Among those *Archidiskodon* is represented — aside from a few isolated lamellae — by eleven fragmentary molars which are kept in several Dutch museums. The elephant molars are represented by fragmentary specimens only, in general showing a few lamellae, mainly of the proximal part of the tooth. The determination of the dentition-stage, therefore, is not easy in every case. All in all remains of five adult but not senile, and five specimens of younger and young individuals are recorded. Two questionable remains of the last molar are included in this material.

The description of the fossil material and the comparison of the teeth with those of the Valdarno has given new date for distinguishing molar-teeth of *Archidiskodon meridionalis*. The Southern Elephant is characterized by low oval to tetragonal, especially broad molar-teeth which are composed in comparison to the later elephants of the *Archidiskodon-Parelephas-Mammuthus* line by fewer but particularly strong lamellae. The enamel in the molar-teeth is regularly thicker than 3 mm and often characterized by rugosities, pearls or grains. The cement between the lamellae is mainly narrow. Towards the wear surface the lamellae often end with eight mamillae which are especially strong leading to the result that within a progressive abrasion sometimes eight lamellae end in mamillae. The molar-teeth are less strongly compacted in a proximal to distal direction than in late glacial mammoth. The length-lamellae-quotient reaches within the Southern Elephant higher values than in the later elephants (*Parelephas* and *Mammuthus*) and reaches in the third molar a quotient around 20.

In *Parelephas* and mainly in *Mammuthus* the second and third molars can always be distinguished by means of the length—lamellae-quotient and the breadth of the tooth. In the Southern Elephant, however, both squares intersect in the boundary area. All these data distinguishing those teeth have been checked on molar-teeth from the Valdarno near Florence. The result was, that the Tegelen-teeth in the middle value represent phylogenetically a somewhat older stage than the Italian teeth.

Trogontherium of this area is represented by rather extensive material showing no traces of rolling. Aside from numerous isolated teeth, remains of mandibles and less often of maxillae are found. Completely preserved fossils have been found enclosed in concretions; when unsheltered deposition has occurred in the sediment the fossils have been found broken, perhaps as a result of the freezing processes of a later glacial period.

Nearly 50% of the lower jaws are attributed to old and very old individuals; the middle-age classes are represented by 31%, and the younger by 19%. By these percentages it is possible to conclude that the animals died a natural death.

Резюме

В районе границы ФРГ и Нидерландов, у Тегелена и Венло, долгое время разрабатывались нижнеплейстоценовые глины. В этих отложениях обнаружены богатые ископаемые остатки трогонтериевого бобра, *Cervus rhenanus* и *Eucladoceros tegulensis*. Другие таксоны представлены лишь немногочисленными остатками. Среди них имеется (кроме нескольких изолированных пластин) одинадцать фрагментарных моляров *Archidiskodon*, хранящихся в нескольких голландзких музеях. Моляры слона представлены только неполными экземплярами, имеющими как правило несколько пластин, преимущественно в проксимальной части зуба. Поэтому определение номера смены зуба в каждом случае было сложным. В итоге определены остатки 5 взрослых (но не дряхлых) особей и 5 особей молодого и более молодого возраста. В этот материал включены два остатка, предположительно принадлежащих последнему моляру.

Описание ископаемого материала и сравнение этих зубов с зубами из Вальдарно дало новые данные для диагностики коренных зубов *Archidiskodon meridionalis*. Южный слон характеризуется низкими овально-четырехугольными и особенно широкими молярами, которые в сравнении с более поздними слонами линии *Archidiskodon-Parelephas-Mammuthus* состоят из меньшего числа особенно мощных пластин. Эмаль коренных зубов постоянно толще 3 мм и часто характеризуется складкамиж бугорками и зернами. Цемент между пластинами как правило узкий. По направлению к жевательной поверхности пластины часто заканчиваются 8 сосочками, особенно мощными и длинными. В результате на прогрессивных стадиях стирания иногда до 8 пластин имеют рисунок, состоящий из сосочков. Моляры не так сильно сжаты в передне-заднем направлении, как у более позднего „ледникового" мамонта. Коэффициент LLQ (длина зуба к числу пластин) достигает у южного слона более высоких значений, чем у более поздних слонов (*Parelephas* и *Mammuthus*), и на третьем моляре может приближаться к 20.

У *Parelephas* и, главным образом, у *Mammuthus* второй и третий моляры всегда можно различить по коэффициентам LLQ и ширине зуба. Однако у южного слона поля обеих этих величин перекрываются в пограничной области. Все признаки, отличающие эти зубы, были проверены на молярах из Вальдарно около Флоренции. В результате показано, что тегеленские зубы по средним значениям представляют филогенетически несколько более раннюю стадию, чем итальянские.

Trogontherium из данного района представлен довольно обширным материалом, не обнаруживающим следов окатанности. Помимо многочисленных изолированных зубов, найдены остатки нижних и реже верхних челюстей. Полностью сохранившиеся экземпляры найдены в конкрециях; если в осадке оказывались незащищенные кости, они обычно сломаны, возможно, в результате мерзлотных процессов, происходивших в течение более позднего оледенения.

Около 50% нижних челюстей принадлежит старым и очень старым особям. К среднему возрастному классу относится

31%, а к более молодому — 19% остатков. По этому соотношению можно заключить, что животные погибали естественной смертью.

Literatur

AZZAROLI, A., 1983: Quaternary mammals and the "End-Villafranchian dispersal event" — a turning point in the history of Eurasia. — In: Palaeogeogr., Palaeoclimatol., Palaeoecol. — Amsterdam **44**. — S. 117—139.

BERNSEN, J. J. A., 1930: Eine Revision der fossilen Säugetierfauna aus den Tonen von Tegelen 1. — In: Natuurhist. Maanbl. — Limburg **19**. — S. 147—152.

DOPPERT, J. W. CHR.; ZONNEFELD, J. I. S., 1955: Over de stratigrafie van het fluviatile Pleistoceen in West-Nederland en Noordbrabant. — In: Meded. Geol. Sticht., Ser. N. — Maastricht **5** (8). — S. 13—30.

FRENZEL, B., 1967: Die Klimaschwankungen des Eiszeitalters. — Braunschweig: Vieweg & Sohn. — 12, 291 S. (Die Wissenschaft; 129).

GUENTHER, E. W., 1981: Die Backenzähne der Mammute von Salzgitter-Lebenstedt. — In: Quartärpaläontologie. — Berlin **4**. — S. 27—41.

—, 1983: Die interglaziale Fundstelle Bilzingsleben in Thüringen und ihre Elefanten *(Palaeoloxodon antiquus)* und ihre Biber (*Trogontherium cuvieri* und *Castor fiber*). — In: Schr. naturwiss. Ver. Schleswig-Holstein. — Kiel **53**. — S. 133—153.

GUENTHER, E. W.; MAI, H., 1977: Die pleistozänen Schichten von Jockgrim in der Rheinpfalz. — In: Schr. naturwiss. Ver. Schleswig-Holstein. — Kiel **47**. — S. 5—24.

KORTENBOUT VAN DER SLUIJS, G., 1960: The fossil tapir of Maalbeek, Netherlands. — In: Publ. Nat. Hist. Gen. Limburg. — Limburg **12**. — S. 12—18.

KORTENBOUT VAN DER SLUIJS, G.; ZAGWIJN, W. H., 1962: An introduction to the stratigraphy and geology of the Tegelen claypits. — In: Meded. Geol. Sticht., N.S. — Maastricht **15**. — S. 31—37.

KURTÉN, B., 1968: Pleistocene mammals of Europe. — London.

LÜTTSCHWAGER, J.; VAN BEMMEL, A. C. V., 1962: Über von Hominiden bearbeitete Knochenfunde aus dem oberen Villafranchien in den Niederlanden. — In: Beaufortia zool. Mus. — Amsterdam **102** (9). — S. 139—144.

MAI, H., 1977: Untersuchung von Gebissen der pleistozänen Biberarten *Trogontherium* und *Castor* und ihre stratigraphische Einordnung. — Kiel, Univ., ungedr. Diss.

—, 1979: Die Biberart *Trogontherium* aus den Mosbacher Sanden bei Wiesbaden. — In: Mainzer naturwiss. Archiv. — Mainz **7**. — S. 41—64, (1979a).

—, 1979: Die Biberarten *Castor* und *Trogontherium* aus den altpleistozänen Schichten von Mauer an der Elsenz. — In: Schr. naturwiss. Ver. Schleswig-Holstein. — Kiel **48**. — S. 35—46, (1979b).

REID, C.; REID, E. M., 1907: The fossil flora of Tegelen-sur-Meuse, near Venlo, in the Province of Limburg. — In: Verh. koninkl. Neederl. Akad. Wet., Sect. 2. — Amsterdam **13** (6). — S. 1—26.

RICHARZ, S., 1921: Neue Wirbeltierfunde in den Tonen von Tegelen bei Venlo. — In: Cbl. Mineral. etc. — Stuttgart.

SCHREUDER, A., 1929: Conodontes *(Trogontherium)* and *Castor* from the Tiglian clay, compared with Castoridae from other localities. — In: Archiv. Mus. Teyler. — Haarlem **3** (6). — S. 99—318.

—, 1931: Conodontes, *Trogontherium* and other Castoridae. — In: Paläont. Z. — Stuttgart **13**. — S. 148—176.

URBAN, B., 1978: The Interglacial of Frechen I, Rheinland. A section of the Tiglian A-type. — In: Geol. en Mijnbouw. — The Hague **57** (3). — S. 401—406, (1978a).

—, 1978: Die Stellung altquartärer Altwasserabsätze in der Niederrheinischen Bucht anhand neuer paläobotanischer Untersuchungen. — In: Kölner geogr. Arb. — Köln **36**. — S. 69—77, (1978b).

—, 1978: Vegetationsgeschichtliche Untersuchungen zur Gliederung des Altquartärs der Niederrheinischen Bucht. — In: Sonderveröff. geol. Inst. Univ. Köln. — Köln 34. — 165 S., (1978c).

—, 1979: Bio- und Magneto-Stratigraphie Plio/Pleistozäner Ablagerungen in der Niederrheinischen Bucht. — In: Acta geol. Acad. Sci. Hung. — Budapest **22** (1—4). — S. 153—160.

—, 1980: Paläökologische Untersuchungen zum Krefeld-Interglazial am Niederrhein. — In: Eiszeitalter u. Gegenwart. — Hannover **30**. — S. 73—88.

ZAGWIJN, W. H., 1957: Vegetation, climate and time-correlations in the Early Pleistocene of Europe. — In: Meded. geol. Sticht., N.S. — s'Gravenhage **9**. — S. 233—244.

—, 1960: Aspects of the Pliocene and Early Pleistocene vegetation in the Netherlands. — In: Meded. geol. Sticht. — Maastricht **53** (5). — S. 785.

—, 1963: Pollenanalytic investigations in the Tiglian of the Netherlands. — In: Meded. geol. Sticht., N.S. — Maastricht **16**. — S. 49—71.

—, 1973: Tegelen. Landshap, flora, fauna en klimat circa 2 milijoen jaar geleden. — In: Naturhist. Maanblad. — **62** (12). — S. 1—12.

—, 1974: Bemerkungen zur stratigraphischen Gliederung der plio-pleistozänen Schichten des niederrheinisch-deutschen Grenzgebietes zwischen Venlo und Brüggen. — In: Z. dt. geol. Gesellsch. — Hannover **125**. — S. 2—16.

TAFEL I

Archidiskodon meridionalis (NESTI), Tegelen-Schichten

Fig. 1: T2. Kaufläche eines zweiten Oberkiefermolaren. Es sind nur der proximale Talon und die anschließende letzte Lamelle erhalten. Das Zwischenzement fehlt. Die letzte Lamelle war nur wenig angekaut, der Talon wahrscheinlich noch nicht. Alle Lamellen enden noch in Mamillen.

Fig. 2: T2. Proximalseite des zweiten Oberkiefermolaren. Die letzte Lamelle, die zum größten Teil von Resten des Talons verdeckt wird, endet in fünf breiten und zwei schmalen Mamillen.

Fig. 3: T3. Kaufläche eines zweiten Oberkiefermolaren. Der erhaltene proximale Zahnteil ist schon derart stark niedergekaut, daß sich in seinem distalen Abschnitt bereits ein Dentinplanum ausgebildet hat. Die breiteren Abschnitte auf der Kaufläche sind letzte Reste des Dentins der Lamellen, die schmalen vom Schmelz umgebenen Inseln sind die Reste des Zwischenzements. In der Mitte der Lamellen hat sich eine Brücke ausgebildet, die diese miteinander verbindet, was dem Normalverhalten von sehr stark niedergekauten Backenzähnen entspricht.

Fig. 4: T3. Innenseite des zweiten Oberkiefermolaren. Die Wurzeln sind bereits zum großen Teil resorbiert. Durch diese starke Resorption wurden zwei Hypoplasien an der proximalen Zahnseite undeutlich gemacht. Da sich an der proximalen Zahnseite die Impression des nachfolgenden M^3 erkennen läßt, handelt es sich um den hinteren Zahnabschnitt. Man kann annehmen, da noch genügend vom Zahn erhalten ist, daß der Molarenrest nicht ausgeworfen wurde.

Fig. 5: T8. Kaufläche, wahrscheinlich eines dritten Milchmolaren. Erhalten sind nur eine Lamelle und Bruchstücke der distal und proximal anschließenden Lamellen. Der eine frühe Dentitionsstufe repräsentierende Zahn (etwa 12 Jahre altes Individuum) zeigt bereits eine erstaunlich grobe Körnung des Schmelzes.

Fig. 6: T9. Kaufläche eines wahrscheinlich zweiten Milchmolaren. Von dem Zahn ist nur eine Lamelle erhalten, so daß die Bestimmung, welchem Kieferteil er entstammt, nicht möglich ist. Der Schmelz ist mit etwa 3 mm für einen so früh in der Dentitionsreihe stehenden Zahn erstaunlich dick. Er zeigt eine kräftige Fältelung.

Alle Abbildungen in nat. Größe

TAFEL I

TAFEL II

Archidiskodon meridionalis (NESTI), Tegelen-Schichten

Fig. 1: T6. Kaufläche eines wahrscheinlich ersten Molaren des Oberkiefers. Der Zahn ist weitgehend abgekaut, möglicherweise handelt es sich um einen Auswurfrest. An der Distalseite hat sich bereits ein Dentinplanum gebildet. An der Proximalseite befindet sich eine starke Impression, die einen großen Teil des proximalen Talons abgepreßt hat.

Fig. 2: T7. Kaufläche des ersten Molaren oder des dritten Milchmolaren des Oberkiefers. Es ist nur der proximale Abschnitt erhalten. Bei den beiden letzten Lamellen sind nur die Mamillen abradiert. Der hintere Talon ist noch nicht in die Abrasion einbezogen. Bei der drittletzten Lamelle ist die Verschmelzung der Innenseite bereits abgeschlossen, während der Externpfeiler von der übrigen Lamelle noch abgesetzt ist.

Fig. 3: T7. Externseite eines ersten Molaren oder eines dritten Milchmolaren. Der proximale Talon liegt noch weit von der Abrasionsfläche entfernt. Deckzement fehlt weitgehend, das Zwischenzement scheint stellenweise herausgebrochen zu sein.

Fig. 4: T5. Kaufläche eines zweiten Molaren des rechten Unterkiefers. Erhalten sind Reste der drei letzten Lamellen und des proximalen Talons. Dieser ist jedoch durch die Impression des nachfolgenden Zahns weitgehend abgepreßt. Eine Schmelzbrücke verbindet, wie es für den wurzelnahen Abschnitt der Lamellen kennzeichnend ist, diese miteinander.

Alle Abbildungen in nat. Größe

TAFEL II

Quartärpaläontologie 6 67—74 Berlin 1986

Zur Geschichte der Erforschung von fossilen Kleinsäugerfaunen aus Interglazialtravertinen in Thüringen

Von Wolf-Dieter Heinrich, Berlin

Inhalt

Abstractum

Der Beitrag vermittelt einen Überblick über die Geschichte der Erforschung von fossilen Kleinsäugerfaunen aus den thüringischen Travertinfundstätten, besonders von Burgtonna, Taubach, Weimar-Ehringsdorf und Bilzingsleben. Es lassen sich drei Forschungsetappen aushalten, die charakterisiert werden.

This paper summarizes the history of research on the fossil small mammal faunas recovered from the travertines of Thuringia, mainly from the localities of Burgtonna, Taubach, Weimar-Ehringsdorf, and Bilzingsleben. Three periods of research are distinguished and discussed.

В статье приводится обзор истории изучения местонахождений мелких млекопитающих в травертинах Тюрингии, главным образом, Бургтонны, Таубаха, Веймара-Эрингсдорфа и Бильцингзлебена. Всего здесь выделено и рассмотрено три периода исследований.

1. Einleitung

Unser Wissen über Aufsammlungen von fossilen Kleinsäugern in den thüringischen Travertinfundstätten reicht mindestens bis in das vorige Jahrhundert zurück (Portis 1878; Pohlig 1881, 1884, 1885, 1891; und andere). Seitdem haben zahlreiche Sammler, Forscher und wissenschaftliche Institutionen zur Vervollständigung des Kenntnisstandes beigetragen. Insgesamt lassen sich drei Forschungsetappen aushalten, die im folgenden in einer knappen forschungsgeschichtlichen Darstellung umrissen werden sollen. Dieser Abriß erhebt keinen Anspruch auf Vollständigkeit. Es werden nur die wichtigsten Etappen berücksichtigt, die zum Verständnis des heutigen Kenntnisstandes und der gegenwärtigen Forschungstendenzen beitragen sollen[1]).

Die inhaltliche Bestimmung des Begriffs „Kleinsäuger" folgt der in der Quartärpaläontologie allgemein üblichen Unterscheidung zwischen Makro- und Mikromammaliern (Jánossy 1961; Thenius 1962). Danach sind unter Kleinsäugern Vertreter aus den Ordnungen Chiroptera, Insectivora, Rodentia und Lagomorpha zu verstehen.

2. Forschungsgeschichtlicher Abriß

2.1. Die erste Forschungsetappe

Die wahrscheinlich älteste wissenschaftliche Bearbeitung fossiler Kleinsäugerreste aus den Ilmtaltravertinen geht auf Portis (1878) zurück, der in der zweiten Hälfte des vorigen Jahrhunderts aus dem Travertin von Taubach die Arten *Cricetus frumentarius* (= *Cricetus major*, zur Revision vgl. Heinrich und Jánossy 1977)[2]) und *Castor fiber* beschreibt. In diesem Zusammenhang weist Portis (1878, S. 144) erstmals auf die ungewöhnliche Größe der Taubacher *Cricetus*-Funde hin, und er betont, daß die Zähne dem *Cricetus frumentarius* (= *Cricetus cricetus*) entsprechen, „doch sind sie, wie auch die übrigen Knochen, mehr wie ein Drittheil grösser als die entsprechenden Skeletttheile eines erwachsenen Individuums dieser Art, das mir zur Vergleichung diente". In Verbindung damit äußerte Portis (1878) ferner, daß die in Taubach nachgewiesene Großsäugerfauna vermutlich mit weiteren, bereits von Nehring (unter anderem 1877, 1878) für verschiedene „diluviale" Fundstellen Deutschlands (Thiede bei Braunschweig, Westeregeln usw.) belegten Kleinsäugerarten vergesellschaftet sein dürfte. Für das bisherige Fehlen entsprechender Funde in Taubach macht Portis (1878, S. 154) Sammler verantwortlich, „welche sich aus dem Sammeln von Knochen ein Geschäft machen", dabei aber das Aufsammeln von Kleinsäugerresten vernachlässigen.

Wenig später teilt Pohlig (1881, 1884, 1885) Faunenlisten mit, in denen Kleinsäugerfunde aus thüringischen Travertinfundstellen berücksichtigt werden. Seine Faunenzusammenstellung, die auf Aufsammlungen aus „den alten Tuffkalken und -Sanden Weimars (bzw. Ehringsdorfs und Taubachs)" beruht, enthält folgende Kleinsäugerarten und -gattungen (Pohlig 1881, S. 32): *Hypudaeus* (= ?), *Myodes* (= ?)[3]) und *Castor* cf. *fiber*.

[1]) Zur Geschichte der geologischen und paläontologischen Erforschung der thüringischen Travertinfundstätten vgl. auch Claus (1928, 1978), Kahlke (1958, 1961), Steiner und Wiefel (1966, 1974, 1977) sowie Wiefel (1974).

[2]) Die heute üblichen taxonomischen Bezeichnungen werden — soweit erforderlich und möglich — in runden Klammern aufgeführt. Eckige Klammern verweisen auf Korrekturen von Gattungsbezeichnungen in der Originalliteratur.

[3]) Es läßt sich heute nicht mehr ermitteln, auf welche Rodentierarten bzw. -gattungen das bei Pohlig (1881, S. 32) unter *Hypudaeus* und *Myodes* (?) aufgeführte Fundgut zu beziehen ist. Zur Synonymie von *Hypudaeus* vgl. Kretzoi (1964), von *Myodes* Kretzoi (1964, 1965).

Von dem gleichen Autor (POHLIG 1884, S. 261 und 262) werden ferner für die Ilmtaltravertine unter der nicht eindeutigen Fundortbezeichnung „Weimar-Taubach" folgende Arten erwähnt: *Castor fiber antiqui*, *Arvicola amphibius* (= *Arvicola* sp.)[4]) und *Arvicola*? *agrestis* (= *Microtus agrestis*?). In der gleichen Veröffentlichung sind für die Fundstätte Taubach *Cricetus* cf. *frumentarius* (= *Cricetus major*) und eine weitere spezifisch nicht näher bestimmte Nagetierart, für die Fundstelle Weimar *Spermophilus* cf. *citillus* (= *Citellus* sp.) und für Tonna (= Burgtonna) *Cricetus* cf. *frumentarius* (= *Cricetus cricetus*) aufgeführt. Außerdem werden Angaben über die Häufigkeit der einzelnen Säugetierarten mitgeteilt und der für Taubach und Weimar bzw. Weimar-Taubach und Tonna (= Burgtonna) belegte paläontologische Fundkomplex stratigraphisch der sogenannten *Antiquus*-Stufe zugewiesen (POHLIG 1884, S. 50, S. 26; 1885, S. 261). Das Vorkommen des Bibers *(Castor fiber)* in den Travertinen von Taubach wird ferner unter anderem von POHLIG (1890, S. 52; 1891, S. 38), von GÖTZE (1892, S. 374), von NEHRING (1895, S. 431) und bei VON LINSTOW (1908, S. 274) erwähnt. Letzterer Autor (VON LINSTOW 1908, S. 274) betont außerdem, daß der Biber nicht nur in Taubach, sondern auch in den „gleichaltrigen Kalktuffen von Ehringsdorf und Weimar" sowie in Bilzingsleben (WÜST 1902) auftritt.

In einer vorläufigen Übersicht über die paläolithischen Fundschichten in der Umgebung von Weimar erwähnen HAHNE und WÜST (1908, S. 200) den Siebenschläfer *Myoxus glis* (= *Glis glis*) aus den Oberen Travertinen. Nähere Fundortbezeichnungen werden nicht mitgeteilt. Von besonderer Bedeutung für die Kleinsäugerforschung sind in diesem Zeitraum horizontierte Fossilaufsammlungen von WÜST (1910), der die Oberen Travertine von Ehringsdorf nach den dort angetroffenen Molluskenvergesellschaftungen in drei Fundschichten (h, i, k) untergliedert, die eine stratigraphische Abfolge darstellen. Aus der Fundschicht h des Fischerschen Steinbruchs von Ehringsdorf werden in dieser Zeit *Myoxus glis* (= *Glis glis*) und *Microtus* [= *Arvicola*] *arvalis* sp. oder *agrestis* sp. (= *Microtus arvalis* vel *agrestis*) gesammelt. Im Zusammenhang damit verweist WÜST (1910, S. 209) auf eine Sammlung des Bauinspektors REBLING in Weimar, in der neben nicht bestimmten Chiropterenresten Funde folgender Kleinsäugerarten vorhanden sind: *Myoxus glis* (= *Glis glis*), *Cricetus cricetus* (= *Cricetus cricetus*), *Mus* sp., *Microtus* [= *Arvicola*] *arvalis* sp. oder *agrestis* sp. (= *Microtus arvalis* vel *agrestis*). WÜST (1910, S. 209) hält es nicht für ausgeschlossen, daß diese Fundstücke aus der Fundschicht h stammen, jedoch darf dies nicht als sicher angesehen werden. Während für die Fundschicht i keine Kleinsäugerreste nachweisbar waren, ließ sich für die Fundschicht k abermals *Myoxus glis* (= *Glis glis*) belegen (WÜST 1910, S. 215). Dagegen ist diese Art für die Fundstätte Bilzingsleben (WÜST 1902) nicht sicher nachgewiesen.

Weitere Nachweise gehen auf WEISS (1910) zurück. Dieser führt in einer tabellarisch zusammengestellten Fossilliste für die Unteren Travertine von Weimar, Taubach und Ehringsdorf neben fraglichen unbestimmten Chiropterenfunden *Cricetus* cf. *frumentarius* (= *Cricetus* sp.)[5]), *Castor fiber*, *Arvicola amphibius* (= *Arvicola* sp.) und *Lepus* sp. an (WEISS 1910, S. 29—30). Für die Oberen Travertine der gleichen Lokalitäten werden dagegen mit Einschränkungen (cf.?) *Spermophilus* cf. *citillus* (= *Citellus* sp.) sowie als gesichert *Arvicola amphibius* (= *Arvicola* sp.) angegeben (WEISS 1910, S. 30). Diese Faunenliste von WEISS (1910) ist für die Kleinsäugerforschung insofern von Bedeutung, als sie zwei Nachweise enthält, die bis heute weder überprüft noch durch Neufunde bestätigt werden konnten. Der eine bezieht sich auf *Lepus* sp., über den WEISS (1910, S. 34) auch in seinen Erläuterungen zu den Fossillisten keine näheren Angaben über die Herkunft mitteilt. Der andere Nachweis betrifft Belege für *Castor fiber* aus dem Werkbanktuff des Steinbruchs E. XII, den WEISS (1910, S. 15) als den — stratigraphisch — höchsten aller Ehringsdorfer Steinbrüche ansieht. In seinem Idealprofil der „Weimaraner, Ehringsdorfer und Taubacher Schichten" wird dieser Werkbanktuff mit dem Schichtenkomplex 10 vereinigt und liegt damit deutlich über dem Schichtenkomplex 12, der den Pariser Horizont von Ehringsdorf repräsentieren soll (WEISS 1910, S. 20—21). Setzt man die Richtigkeit der WEISSschen Parallelisierung voraus, so müßte *Castor fiber* auch im Oberen Travertin von Ehringsdorf vorkommen. Dies ist zwar nicht ausgeschlossen, doch liegen nach KAHLKE (1958 und persönliche Mitteilung) bis heute aus dem Oberen Travertin von Ehringsdorf keine Biber-Funde vor. Erschwerend für die Beurteilung dieser Funde wirkt sich weiterhin aus, daß WEISS (1910, S. 30) selbst in seiner tabellarischen Zusammenstellung *Castor fiber* für den Oberen Travertin von Weimar, Taubach und Ehringsdorf nicht aufführt. Hinzu kommt, daß gegen die Zusammenfassung und Parallelisierung von Schichten in der Form, wie sie WEISS (1910) vornahm, schwerwiegende Bedenken geltend gemacht worden sind (vgl. dazu unter anderem STEINER und WIEFEL 1974, S. 29 sowie ZEISSLER 1975, S. 17). Die angeblichen *Castor*-Funde aus dem Oberen Travertin von Ehringsdorf, die auch von späteren Autoren nicht mehr erwähnt werden, müssen daher vorläufig als nicht gesichert angesehen werden.

Von SCHMIDT (1912) werden ohne genaue Fundortbezeichnungen für die Unteren Travertine von Ehringsdorf, Weimar und Taubach *Cricetus* cf. *frumentarius* (= *Cricetus* sp.), für die Oberen Travertine *Spermophilus* sp. (= *Citellus* sp.), *Arvicola amphibius* (= *Arvicola* sp.), *Glis glis*, *Cricetus cricetus*, *Arvicola arvalis* oder *agrestis* (= *Microtus arvalis* vel *agrestis*) und *Mus* sp. erwähnt (zitiert nach JÁNOSSY 1975, S. 502).

Ungefähr im gleichen Zeitraum beschreibt SCHÄFER (1909, S. 451, 454—457) für den Interglazialtravertin von Burgtonna *Castor fiber antiqui* (= *Castor fiber*), *Cricetus frumentarius* (= *Cricetus cricetus*) und *Sorex vulgaris* (= *Sorex araneus*). Zuvor hatte bereits POHLIG (1884, S. 50; 1885, S. 261) für diese Fundstätte *Cricetus* cf. *frumentarius* (= *Cricetus cricetus*)

---

[4]) Eine eindeutige taxonomische Bestimmung von *Arvicola*-Resten aus dem Zeitraum Saale-Komplex — Eem-Warmzeit ist nur an Hand biometrischer Analysen der Schmelzband-Differenzierung möglich. Nicht mehr verfügbare Altfunde entziehen sich demzufolge einer genauen Determination. Solche Belege für die Gattung *Arvicola* LACÉPÈDE werden daher hier unter der taxonomischen Bezeichnung *Arvicola* sp. geführt.

[5]) Bei WEISS (1910) und auch bei späteren Autoren sind in den Faunenlisten für die Ilmtaltravertine die einzelnen Fundorte nicht getrennt aufgeführt. Inzwischen ist bekannt, daß in der Fundstätte Taubach nur *Cricetus major*, in Weimar-Ehringsdorf nur *Cricetus cricetus* auftritt. Daher ist bei alten Angaben ohne nähere Herkunftsbezeichnungen nicht mehr zu entscheiden, um welche *Cricetus*-Art es sich handelt. Deshalb werden diese Altfunde hier unter der Bezeichnung *Cricetus* sp. geführt.

angeführt. von Linstow (1908, S. 275) erwähnt „zwei zusammenhängende Unterkieferäste vom Biber", die aus den „interglazialen Kalktuffen von Burgtonna stammen und sich im Privatbesitz von H. F. Schäfer (Gotha) befinden". Die von Pohlig (1884, 1885) und von Schäfer (1909) für den Travertin von Burgtonna belegten Kleinsäugerarten werden von Soergel (1912, S. 58, außerdem Tab. auf S. 78) übernommen. Im Zusammenhang damit ist ein Hinweis von Soergel (1912, S. 58, Fußnote 1) von Bedeutung, wonach *Sorex* sp. im „älteren Travertin von Ehringsdorf" auftritt. Allerdings wird diese Angabe in späteren Faunenlisten dieses Autors und anderer Verfasser nicht mehr berücksichtigt. Unter Berufung auf Portis (1878) und Pohlig (1885) zitiert Soergel (1912, S. 57) in der gleichen Veröffentlichung für den Interglazialtravertin von Taubach die beiden Kleinsäugerarten *Castor fiber* (= *Castor fiber spelaeus*, zur Revision vgl. Kretzoi 1977) und *Cricetus frumentarius* (= *Cricetus major*). Beide Arten werden in einer Tabelle berücksichtigt, die alle bis dahin von Taubach, Burgtonna, Gräfentonna, Mauer, Mosbach und Süßenborn bekannten Säugetierarten nach ökologischen Gesichtspunkten klassifiziert wiedergibt (Soergel 1912, S. 79). Allerdings sind in dieser Aufstellung *Cricetus frumentarius* (= *Cricetus cricetus*) für Burgtonna sowie *Castor fiber* für Taubach und Burgtonna nicht angegeben.

Im Jahre 1916 zitiert Werth (1916, S. 121) unter Berufung auf Wüst (1910) für die „Travertine (Kalktuffe) des Ilmtales bei Weimar" unter anderem „Siebenschläfer, Biber, Hamster, Erd- oder Feldmaus und eine echte Maus" (vgl. ferner Werth 1928). Bei der Beschreibung von *Glis glis* mut. *suessenbornensis* (= *Glis glis*; zur Revision vgl. Fejfar 1969) zieht Soergel (1920, S. 68 und 70 sowie Taf. 3, Fig. 3) Funde von *Glis glis* aus dem Oberen Travertin von Ehringsdorf zum Vergleich heran.

In einer Sammelfaunenliste für die „unteren Weimarer Travertine und Rabutz" nennt Wiegers (1920, S. 120) neben verschiedenen Großsäugern offenbar unter Bezug auf Weiss (1910) *Castor fiber*, *Cricetus* cf. *frumentarius* (= *Cricetus* sp.), *Arvicola amphibius* (= *Arvicola* sp.) und *Lepus* sp., später (Wiegers 1928, S. 175) unter der gleichen Herkunftsbezeichnung *Castor fiber*, *Cricetus frumentarius* (= *Cricetus* sp.) und *Myoxus glis* (= *Glis glis*). Die hier aufgeführten Kleinsäugerarten sind auf die Ilmtaltravertine zu beziehen.

In einer Veröffentlichung für die Tagung der Paläontologischen Gesellschaft in Weimar 1925 gibt Soergel (1926, S. 15) für den Bereich der Unteren Travertine von Ehringsdorf, Taubach und Weimar folgende Rodentier an: *Castor fiber*, *Cricetus cricetus* (= *Cricetus* sp.) und *Myoxus glis* (= *Glis glis*). Für den Oberen Travertin von Ehringsdorf und Weimar werden *Myoxus glis* (= *Glis glis*), *Cricetus cricetus* und *Microtus arvalis* oder *Microtus agrestis* erwähnt (Soergel 1926, S. 18). An dieser Zusammenstellung ist bemerkenswert, daß hier für den Bereich der Unteren Travertine von Ehringsdorf, Taubach und Weimar der „Siebenschläfer *Myoxus glis* L. nach Mitteilung von Wüst" angegeben wird.[6]) Diese Angabe konnte allerdings bisher weder für den Interglazialtravertin von Weimar (vgl. Kahlke 1958) und Taubach (vgl. Kahlke 1958, 1961; Heinrich und Jánossy 1977) noch für den Unteren Travertin von Ehringsdorf (Heinrich 1980b, 1981b; vgl. auch Jánossy 1974, 1975) bestätigt und durch Neufunde untermauert werden. Diese Soergelsche Aufstellung findet bei Wiegers (1931, S. 461—462) Berücksichtigung. Schließlich ist Schaub (1935, Fußnote auf S. 581) zu nennen, der bei der Bearbeitung eines Hamsterskeletts von Taubach zu dem Schluß kommt, daß hier die pleistozäne „Riesenform" *Cricetus major* vorliegt. Damit werden nach über fünfzig Jahren die bereits im vergangenen Jahrhundert gemachten Beobachtungen von Portis (1878, S. 144) bestätigt, der seinerzeit mit Nachdruck auf die ungewöhnlichen Größenverhältnisse eines *Cricetus*-Fundes von Taubach hingewiesen hatte.

Während in der ersten Forschungsperiode Kleinsäugerarten aus den Travertinfundstätten Ehringsdorf, Taubach und Burgtonna wiederholt in der Fachliteratur zitiert werden, scheinen in diesem Zeitraum Hinweise auf das Vorkommen von fossilen Kleinsäugern in den Deckschichtenfolgen der genannten Travertinfundstätten fast völlig zu fehlen. Von gewisser Bedeutung ist allerdings eine fossilführende Spaltenfüllung in den Travertinen von Ehringsdorf, in der aus Lößablagerungen neben Vogel- und Großsäugerresten auch Rodentier geborgen werden (Soergel 1926, S. 22), deren Bearbeitung jedoch offenbar noch aussteht (vgl. Kahlke 1958, S. 103, Fußnote 28).

Die erste Forschungsetappe klingt mit dem Ende der dreißiger Jahre dieses Jahrhunderts aus. Der zweite Weltkrieg führt zu einer mehrjährigen Unterbrechung der geologischen und paläontologischen Forschungen in den thüringischen Travertinfundstätten (vgl. dazu Behm-Blancke 1960; Steiner und Wiefel 1974).

2.2. Die zweite Forschungsetappe

Die zweite Forschungsetappe beginnt Anfang der fünfziger Jahre dieses Jahrhunderts mit den Untersuchungen einer Forschungsgruppe des Museums für Ur- und Frühgeschichte Weimar. Sie wird vorbereitet mit ersten Bestandsaufnahmen und Neuaufsammlungen sowie einer Revision der Wirbeltierfaunen aus den Travertinfundstätten Ehringsdorf und Taubach durch Kahlke (1958, 1961). Dabei werden die bis dahin für die Ilmtaltravertine nachgewiesenen und noch auffindbaren Kleinsäugerarten weitestgehend berücksichtigt. Darüber hinaus ermöglichte die Gründung des Instituts für Quartärpaläontologie Weimar im Jahre 1962 nicht nur eine generelle Inventur der über zahlreiche Sammlungen und Institutionen verstreuten Fundstücke, sondern zugleich weitere umfangreiche Neuaufsammlungen. Ausgehend von den Ergebnissen, die die Untersuchungen der fossilen Kleinsäugerfaunen von Voigtstedt (Guenther 1965; Jánossy 1965; Kretzoi 1965) und von Süßenborn bei Weimar (Fejfar 1969; Kretzoi 1969) zeitigten, werden auch für das in Ehringsdorf geborgene Kleinsäugerfundgut auf Veranlassung von H. D. Kahlke gesonderte Bearbeitungen durch Spezialisten vorgesehen. Die erzielten Untersuchungsergebnisse werden auf dem III. Internationalen Paläontologischen Kolloquium „Das Pleistozän von Weimar-Ehringsdorf" in Weimar 1968 vorgelegt und diskutiert sowie in Veröffentlichungen von Jánossy

[6]) Vgl. dazu die Angaben von Wiegers (1928, S. 175), der unter der nicht eindeutigen Herkunftsbezeichnung „untere Weimarer Travertine und Rabutz" *Myoxus glis* (= *Glis glis*) anführt, der auf die Ilmtaltravertine zu beziehen ist.

(1974, 1975) und KRETZOI (1975) dokumentiert. Durch KRETZOI (1975) erfahren die *Castor*-Funde eine eingehende Untersuchung; neben morphologischen Detailanalysen und biometrischen Kennzeichnungen werden paläoökologische und phylogenetische Einzelheiten mitgeteilt. Erstmals wird der Versuch unternommen, ausschließlich an Hand fossiler Kleinsäugerreste die stratigraphische Position des Oberen Travertins von Ehringsdorf zu fixieren. So zieht JÁNOSSY (1975, S. 508) auf Grund seiner Befunde für die Kleinsäugerfauna aus dem Oberen Travertin von Ehringsdorf „die Schwelle des Würms (Altwürm sensu lato)" als stratigraphisches Alter in Betracht. Allerdings wird noch für den gesamten Oberen Travertin von Ehringsdorf ein „zusammenfassendes" Bild der ehemaligen Kleinsäugerassoziationen entworfen. Mit der gesonderten Kennzeichnung von Kleinsäugerfaunen aus den einzelnen lithostratigraphischen Einheiten wird erst begonnen. Ein Beispiel ist der Pariser Horizont (vgl. JÁNOSSY 1974, 1975).

Im Anschluß an diese Tagung beginnt eine internationale Forschungsgemeinschaft mit der Neubearbeitung der fossilen Wirbeltierreste aus den Interglazialtravertinen von Taubach und Burgtonna für das IV. Paläontologische Kolloquium, das im September 1972 unter dem Thema „Das Pleistozän von Taubach und Burgtonna — Zur biostratigraphischen Grenze zwischen Glazial und Interglazial im europäisch-asiatischen Raume" ebenfalls in Weimar stattfindet. Hierfür übernehmen M. KRETZOI, Budapest, D. JÁNOSSY, Budapest, und W.-D. HEINRICH, Berlin, die Bearbeitung der Kleinsäugerfaunen des Interglazialtravertins von Taubach. Von letzteren beiden werden außerdem die Kleinsäugerfunde aus dem Interglazialtravertin von Burgtonna untersucht und revidiert. Die *Castor*-Belege dieser Fundstätte sind Gegenstand einer gesonderten Untersuchung von L. SCHMIDT, Greifswald.

Dank der zwischenzeitlich vom Institut für Quartärpaläontologie Weimar und von W. STEINER (1972) in Taubach vorgenommenen Aufsammlungen steht den Bearbeitern nicht nur ein relativ umfangreiches, sondern auch ein exakt stratifiziertes Fundgut zur Verfügung. Darauf aufbauend entwirft KRETZOI (1977) ein umfassendes Bild über phylogenetische Entwicklungstendenzen an Molaren pleistozäner Castoriden, wobei auf die in Süßenborn bei Weimar (KRETZOI 1969) und Ehringsdorf (KRETZOI 1975) gesammelten Erfahrungen zurückgegriffen werden kann. Darüber hinaus gelingt es in Taubach, einzelne lithostratigraphische Einheiten des Travertinprofils (Bruch VOLLMAR) und die durch sie dokumentierten erdgeschichtlichen Zeiträume erstmals mit Kleinsäugerspektren zu belegen (HEINRICH und JÁNOSSY 1977). Diese Untersuchungen ermöglichen ökologische und faunengeschichtliche Angaben, die durch stratigraphische Aussagen ergänzt werden. Im Zusammenhang damit wird — wenn auch vorerst mit Vorbehalten — eine überregionale Korrelation mit ungarischen Fundstätten versucht (HEINRICH und JÁNOSSY 1977).

Im Unterschied zu Taubach steht aus dem Travertin von Burgtonna — von einem Neufund (*Apodemus sylvaticus*) abgesehen — den Bearbeitern kein neu aufgesammeltes Fundgut für die Untersuchungen zur Verfügung. Da überdies einige der Originale SCHÄFERS (1909) als verschollen angesehen werden müssen, kann vorerst nur ein Teil der bisher von dieser Lokalität bekannten Kleinsäugerfunde untersucht und revidiert werden (HEINRICH und JÁNOSSY 1978a). Allerdings wird während dieser Untersuchungen über dem Interglazialtravertin von Burgtonna eine bis dahin unbekannte Wirbeltierfundstätte entdeckt und komplex bearbeitet.[7]) Von den dort erzielten Untersuchungsergebnissen sei hier der Nachweis eines stratigraphisch bedeutsamen *Lagurus*-Horizonts hervorgehoben, der erste weitreichende überregionale Korrelationen mit mehreren europäischen Fundstätten ermöglicht (HEINRICH und JÁNOSSY 1973, 1978b; HEINRICH und JÄGER 1978), die für die Beurteilung der stratigraphischen Stellung der thüringischen Interglazialtravertine ebenfalls von Bedeutung sind. Demgegenüber können in den Deckschichten des Ehringsdorfer Travertinprofils im Verlauf der zweiten Forschungsperiode keine fossilen Kleinsäuger geborgen werden, obwohl diese Ablagerungen mehrfach Gegenstand feinstratigraphischer Untersuchungen sind (GUENTHER 1958; BEHM-BLANCKE 1960).

2.3. Die dritte — gegenwärtige — Forschungsetappe

Die jüngste Etappe der Erforschung von fossilen Kleinsäugerfaunen aus thüringischen Travertinen steht verstärkt im Zeichen biostratigraphischer Diskussionen. Einen besonderen Schwerpunkt stellt die Wirbeltierfundstätte Weimar-Ehringsdorf dar. Dort sind durch die Bearbeitung fossiler Kleinsäugerfaunen weitere Ergebnisse erbracht worden. Wesentliche Feststellungen ermöglicht ein neuer Fundkomplex aus dem Unteren Travertin dieser Lokalität (HEINRICH 1980b, 1981b). Dort konnten erstmals Kleinsäugerarten nachgewiesen werden, die von anderen thüringischen Travertinfundstätten bisher nicht bekannt sind (z. B. *Microtus gregalis*, *Ochotona* sp.; HEINRICH 1981b). Ein wesentlicher Teil dieses Fundguts wird methodischen Verbesserungen bei der Gewinnung von Kleinsäugerresten aus dem Travertin verdankt. Während in der Vergangenheit Skelettreste von Mikromammaliern vorzugsweise aus lockeren und daher schlämmbaren Travertinlagen gewonnen worden sind, ist es in den letzten Jahren gelungen, solche auch aus festen Travertinbänken freizulegen, die sich der Untersuchung bisher weitgehend entzogen. Grundlage hierfür ist ein Verfahren, Knochen und Zähne mit Hilfe von verdünnter Essigsäure aus den einbettenden Travertinen herauszulösen. Damit wurde ein Aufbereitungsverfahren in die paläontologische Untersuchung pleistozäner Travertine eingeführt, das sich bereits früher bei der Gewinnung von Skelettresten, z. B. von säugetierähnlichen Reptilien und Säugern aus mesozoischen Ablagerungen (siehe HENKEL 1966; HENKEL und KREBS 1969) bewährt hat.

Erweitert wurde nicht nur das Methodenspektrum für die Materialgewinnung, sondern auch die Reihe der untersuchten Travertinfundstätten. Von besonderer Bedeutung ist in diesem Zusammenhang die Bestimmung von Mikromammaliern aus dem Travertinkomplex von Bilzingsleben (HEINRICH 1980a), die zur paläoökologischen und biostratigraphischen Beurteilung der dort aufgeschlossenen Travertinablagerungen beitragen. Bisher sind folgende Arten nachgewiesen (HEINRICH 1980a): *Talpa* sp., *Clethrionomys glareolus* und *Arvicola*

[7]) Über Anlaß und Ablauf der Untersuchungen in dieser Wirbeltierfundstätte informiert ein gesonderter Aufsatz des Verfassers (HEINRICH 1978a).

cantiana. Hinzu kommen (MAI 1978): *Trogontherium cuvieri* und *Castor fiber.* Wesentlich ist dort der bisher erdgeschichtlich älteste Nachweis für *Arvicola* in Thüringen. Darüber hinaus ist es inzwischen gelungen, diese Gattung in Travertinablagerungen der Fundstätte Weimar — Parkhöhlen nachzuweisen (HEINRICH 1982a, 1984). Damit steht aus thüringischen Travertinfundstätten eine Serie von pleistozänen *Arvicola*-Populationen unterschiedlichen stratigraphischen Alters zur Verfügung, die den Zeitraum vom Holstein-Komplex bis zur Weichsel-Kaltzeit umfaßt. Hierzu zählen:

Weichsel-Kaltzeit:	Deckschichtenfolge über dem Travertin von Burgtonna
Eem-Warmzeit:	Travertine von Burgtonna, Taubach und Weimar-Parkhöhlen
Eem oder Warmzeit im Saale-Komplex:	Travertine von Weimar-Ehringsdorf
Warmzeit im Holstein-Komplex:	Travertinkomplex von Bilzingsleben

Diese Folge von Fundstätten bot relativ günstige Voraussetzungen für eine vergleichende biometrische Kennzeichnung von *Arvicola*-Populationen unterschiedlichen erdgeschichtlichen Alters in einem geographisch eng begrenzten Gebiet (Thüringer Becken). Für eine solche Untersuchung erwies sich die Schmelzband-Differenzierung an den ersten unteren Molaren (M_1) als besonders geeignet (HEINRICH 1978b, 1980b, 1982a, b, c). Ihre präzise Erfassung erlaubt sowohl neue Aussagen zur erdgeschichtlichen Altersstellung der Fundstätten als auch Einblick in einen langfristig wirksamen Evolutionstrend bei der Gattung *Arvicola.* Die in Thüringen gewonnenen Ergebnisse konnten inzwischen durch weitere Studien in anderen Teilen Mitteleuropas, z. B. im Karpaten-Becken, grundsätzlich bestätigt werden (HEINRICH 1982a). Die hierbei mit Hilfe von *Arvicola* erzielten Aussagen betreffen im wesentlichen drei Aspekte:

1. Begründung der *Arvicola-Microtus*-Stufe innerhalb der biostratigraphischen Gliederung des Pleistozäns Europas auf der Grundlage von Mikromammaliern (FEJFAR und HEINRICH 1980, 1981a, b, 1982, 1983).
2. An Hand der *Arvicola*-Funde aus den thüringischen Travertinen sowie aus der Deckschichtenfolge, die über dem Travertin von Burgtonna angetroffen wurde, ist es möglich, die Ablösung von *Arvicola cantiana* durch *Arvicola terrestris* zeitlich zu erfassen und biostratigraphisch für eine Zonengliederung der *Arvicola-Microtus*-Stufe zu nutzen: *Arvicola-cantiana-* und *Arvicola-terrestris*-Zone (FEJFAR und HEINRICH 1981a, b; HEINRICH 1982a, b, c).
3. Die evolutiven Veränderungen am Schmelzband der M_1 von *Arvicola* erlauben durch die Berechnung von Schmelzband-Differenzierungs-Quotienten innerhalb der *Arvicola-cantiana*- und *Arvicola-terrestris*-Zone weitergehende Feindatierungen, die bei radiometrischer Eichung zu einem Hilfsmittel der Geochronologie werden können (HEINRICH 1982a, c). Allerdings kommt diesen Indices auch ohne Eichung erhebliche Bedeutung zu, so z. B. bei der Klärung von Altersbeziehungen zwischen verschiedenaltrigen Travertinfundstätten. So konnte mit Hilfe von Schmelzband-Differenzierungs-Quotienten der *Arvicola*-Populationen gezeigt werden, daß die Travertine von Taubach nicht älter, sondern jünger sind als diejenigen von Weimar-Ehringsdorf (HEINRICH 1982a). Diese Aussagen gewinnen zusätzlich dadurch an Bedeutung, daß die betroffenen Interglazialtravertine stammesgeschichtlich wichtige Hominidenfunde geliefert haben (Bilzingsleben, Weimar-Ehringsdorf) und darüber hinaus überregional bedeutsame Fundpunkte für die Archäologie des Paläolithikums in Mitteleuropa (besonders Bilzingsleben, Weimar-Ehringsdorf und Taubach) darstellen.

Zu den untersuchten Fundpunkten gehört mindestens einer, für den Kleinsäuger aus dem warmzeitlichen wie aus dem kaltzeitlichen Abschnitt des letzten glazialklimatischen Zyklus vorliegen. Dies ist Burgtonna, wo zum einen eine letztinterglaziale Kleinsäugerfauna aus dem Travertin und zum anderen eine weichselkaltzeitliche Abfolge von Mikromammalierfaunen im unmittelbaren Hangenden der Travertinablagerungen (Deckschichtenfolge) nachzuweisen ist (HEINRICH und JÁNOSSY 1978b). Damit wird dort der Wandel von Kleinsäugerfaunen im Übergangsbereich vom Interglazial zum Glazial faßbar. Dies war Anlaß, die Travertinfundstätte Burgtonna (Steinbruch nordöstlich vom Ort) als Stratotypus für die *Arvicola-Microtus*-Stufe vorzuschlagen (FEJFAR und HEINRICH 1983). Diese Stufe wurde wegen der Lage des Stratotypus im Thüringer Becken in Analogie zu bereits eingebürgerten biostratigraphischen Begriffen, wie z. B. Villányium oder Biharium als Toringium bezeichnet (FEJFAR und HEINRICH 1980, 1981a, b, 1983).

Zu einer weiterführenden Auswertung von Kleinsäugerfaunen aus thüringischen Interglazialtravertinen hat in den letzten Jahren die Gewinnung von radiometrischen Daten beigetragen. Seit 1963 sind zunächst einzelne Datierungen für Weimar-Ehringsdorf (ROSHOLT und ANTAL 1963; ČERDYNCEV u. a. 1975; ISAAC 1974; KUKLA 1975; SCHWARCZ nach STEINER 1977; SCHWARCZ 1980), Taubach (KUKLA 1975) und Bilzingsleben (GLAZEK u. a. 1980; HARMON u. a. 1980) veröffentlicht worden. Neuerdings steht eine weitere, bedeutend umfangreichere Serie von radiometrischen Datierungen zur Verfügung, an der die hier besprochenen Travertinablagerungen von Bilzingsleben, Weimar-Ehringsdorf, Taubach, Weimar-Parkhöhlen und Burgtonna beteiligt sind (BRUNNACKER u. a. 1983). Damit zeichnen sich weitere Möglichkeiten ab, und zwar sowohl für die geochronologische Eichung der irreversiblen Wertefolge von Schmelzband-Differenzierungs-Quotienten der *Arvicola*-Populationen als auch für Feststellungen zum zeitlichen Ablauf von Evolutionsvorgängen.

Die meisten der bisher erkennbaren Tendenzen in der gegenwärtigen Forschungsetappe erfordern eine weitere Absicherung der Aussagen durch zusätzliche Materialgewinnung. So ist eine umfassendere Kennzeichnung der Kleinsäugerfaunen für Bilzingsleben wie für Weimar-Ehringsdorf besonders wünschenswert. Im ersteren Fall lassen die fortdauernden archäologischen Ausgrabungen, im zweiten Falle der wiederaufgenommene Steinbruchbetrieb neue Funde erwarten. In Burgtonna steht eine Wiederaufnahme des Travertinabbaus ebenfalls in Aussicht. Inzwischen sind weitere Travertinvorkommen in gezielte Untersuchungen einbezogen worden, so z. B. im Stadtgebiet von Mühlhausen. Dabei wird es im Rahmen von künftigen Untersuchungen nicht nur um das Auffinden von bisher nicht nachgewiesenen Arten gehen, sondern auch um die Verbreiterung der statistischen Basis für biometrische Aussagen, so z. B. bei *Arvicola.* Darüber hinaus erscheint es angebracht, auch andere Gattungen auf biostratigraphische und evolutionsbiologische Aussagemöglichkeiten zu überprüfen. Ansätze hierfür sind bereits durch KRETZOI (1975, 1977), z. B. für *Castor,* deutlich gemacht worden.

Zusammenfassung

Die Geschichte der Erforschung von fossilen Kleinsäugerfaunen aus Interglazialtravertinen in Thüringen vollzog sich in drei Etappen:

Die erste Etappe beginnt im vorigen Jahrhundert und reicht etwa bis zur Mitte unseres Jahrhunderts. In diesem Zeitraum werden Kleinsäugerreste nur gelegentlich gesammelt und untersucht. Eine systematische Suche und gezielte Bergung erfolgt nicht. Folglich ist die Zahl der in diesem Zeitraum nachgewiesenen Kleinsäugerarten, die in der Fachliteratur immer wieder unter Berufung auf frühere Autoren zitiert werden, relativ gering. Vielfach werden Artenlisten mit nicht eindeutigen strati- und topographischen Herkunftsbezeichnungen veröffentlicht. Biostratigraphische Rückschlüsse auf die Altersstellung der Travertinfundstätten, die von einer Analyse der Kleinsäugerfunde ausgehen, fehlen in dieser Forschungsetappe.

Die zweite Forschungsetappe reicht von Anfang der fünfziger bis in die zweite Hälfte der siebziger Jahre unseres Jahrhunderts. Sie ist gekennzeichnet durch Neuaufsammlungen in den Travertinvorkommen von Weimar-Ehringsdorf, Taubach und Burgtonna sowie in einer fossilführenden Deckschichtenfolge über dem Travertin von Burgtonna. Dieses Fundgut wurde unter Berücksichtigung der Altfunde von internationalen Forschungsgemeinschaften unter paläoökologischen, biostratigraphischen und faunengeschichtlichen Gesichtspunkten nach dem neuesten Kenntnisstand umfassend bearbeitet.

Die dritte Forschungsetappe geht ohne scharfe Zäsur aus der zweiten hervor. Sie hat zunächst die Anzahl der Travertinfundplätze mit Kleinsäugern erweitert (Bilzingsleben, Weimar-Parkhöhlen, Mühlhausen) und durch neue Untersuchungsmethoden die Aussagemöglichkeiten vergrößert. Hierzu zählen insbesondere der Einsatz des Essigsäure-Aufschlußverfahrens, die verstärkte Anwendung der Biometrie zur Erfassung von Evolutionsvorgängen und zur Präzisierung biostratigraphischer Aussagen sowie deren Verknüpfung mit radiometrischen Datierungen. In diesem Zeitraum wurde Burgtonna als Stratotypus für die *Arvicola-Microtus*-Stufe (Toringium) vorgeschlagen. Schließlich kam die Untersuchung von Kleinsäugerfaunen verstärkt für die biostratigraphische Beurteilung von thüringischen Travertinfundstätten bei interdisziplinären Komplexuntersuchungen zur Geltung. In diesem Rahmen haben sich Nachweise von *Arvicola* als besonders aussagekräftig erwiesen.

Summary

Research on the small mammal faunas from the interglacial travertines of Thuringia is divisible into three phases. The first phase began in the last century and lasted until the middle of the present century. In this period, fossil remains of small mammals were collected and studied only occasionally. Systematical field research and corresponding field collection was not carried out at this time and the number of small mammal species known was accordingly poor. Those few fossil species that were recorded during this period have been cited in subsequent literature but often species lists have been published without reference to exact stratigraphical and locality records. Biostratigraphical correlations of the different travertine deposits arising from an analysis of the small mammal remains are absent from this period of research.

The second phase began in the 1950's and lasted until the late 1970's. It was characterized by new field collections from the travertine deposits of Weimar-Ehringsdorf, Taubach and Burgtonna, and a series of cap-deposits overlying the travertines of Burgtonna. These new collections have been studied by an international team of scientists with respect to their palaeoecological, biostratigraphical, and fauna-history data by the most up-to-date methods available.

The third phase followed the second without any sharp break. The result of this third period of research was that the number of travertine deposits from which fossil small mammals had been recorded was enlarged (Bilzingsleben, Park Caves of Weimar, Mühlhausen) and new methods of research were introduced, increasing the level of information recoverable. In this respect we have to mention the acetic acid method of recovering small mammals from calcified matrices, the extensive use of biometry to recognize evolutionary trends and the combination of precise biostratigraphical information with radiometric dating. In this period of research the fossil locality of Burgtonna has been proposed as the stratotype of the *Arvicola-Microtus* stage (Toringicum). Finally, data on small mammal faunas have been used more and more for biostratigraphical correlations of the Thuringian travertines within the framework of complex interdisciplinary research. In this respect the genus *Arvicola* is of particular significance.

Резюме

История изучения ископаемых фаун мелких млекопитающих из межледниковых травертиновых залежей Тюрингии может быть разделена на три периодач. Первый период начался еще в прошлом веке и продолжался, возможно, вплоть до середины нашего столетия. В этот период ископаемые остатки мелких млекопитающих собирались и изучались лишь случайно. Систематических сборов и исследований в этот период не было.

Как следствие этого, число видов мелких млекопитающих, обнаруженных в этот период и приводимых авторами того времени со ссылками на более ранние публикации, относительно невелико. Часто видовые списки публиковались без точной стратиграфической и топографической документации. Биостратиграфические выводы о стратиграфическом положении травертиновых залежей на основании изучения мелких млекопитающих в этот период исследований также отсутствовали.

Второй период исследований охватывает время от начала 50-х до второй половины 70-х голов нашего века. Этот период характеризуется новыми сборами в травертинах Веймара-Эрингсдорфа, Таубаха и Бургтонны, а также в костеносных покровных осадках на кровле травертинов Бургтонны. Эти новые сборы, также как и более ранние материалы, интенсивно изучались международной группой исследователей в соответствии с наиболее современным уровнем знаний — в палеоэкологическом, биостратиграфическом и эволюционном аспектах.

Третий период исследований непосредственно продолжает второй. В это время число местонахождений костеносных травертинов (с остатками мелких млекопитающих) увеличилось (Бильцингслебен, Парковые пещеры Веймара, Мюльхаузен), и благодаря новым методам исследований достигнута возможность получения новых результатов. К таким методам можно отнести применение растворения в уксусной кислоте, возрастающее использование биометрии для выявления эволюционного процесса и получения биостратиграфических результатов, а также увязку этих результатов с радиометрическими данными. В этот период было предложено считать местонахождение Бургтонна стратотипом яруса *Arvicola-Microtus* (Торингий). Наконец, изучение мелких млекопитающих использовалось для решения биостратиграфических проблем травертинов Тюрингии в рамках междисциплинарных комплексных исследований. Оказалось, что большое значение в этом отношении имеет палеонтологическая летопись *Arvicola*.

Literatur

BEHM-BLANCKE, G., 1960: Altsteinzeitliche Rastplätze im Travertingebiet von Taubach, Weimar, Ehringsdorf. — In: Alt-Thüringen. — Weimar **4**. — 246 S.

CLAUS, H., 1928: Beiträge zur Geschichte der geologischen Forschung in Thüringen. — In: Beitr. Geol. Thüringen. — Jena **1** (2). — S. 1—35.

—, 1978: Die geologisch-paläontologische Erforschung der Burgtonnaer Travertinlagerstätten. — In: Quartärpaläontologie. — Berlin **3**. — S. 9—41.

BRUNNACKER, K.; JÄGER, K.-D.; HENNIG, G. J.; PREUSS, J., 1983: Radiometrische Untersuchungen zur Datierung mitteleuropäischer Travertinvorkommen. — In: Ethnogr.-archäol. Z.— Berlin **24** (2). — S. 217—266.

ČERDYNČEV, V.; SENINA, H.; KUZMINA, E. A., 1975: Die Altersbestimmung der Travertine von Weimar-Ehringsdorf (über das Alter des Riß-Würm-Interglazials). — In: Abh. zentr. geol. Inst. — Berlin 23. — S. 7—14.

FEJFAR, H., 1969: Die Nager aus den Kiesen von Süßenborn bei Weimar. — In: Paläont. Abh., A. — Berlin **3** (3/4). — S. 761 bis 770.

FEJFAR, O.; HEINRICH, W.-D., 1980: Zur biostratigraphischen Abgrenzung und Untergliederung des kontinentalen Quartärs in

Europa an Hand von Arvicoliden (Mammalia, Rodentia). — In: Čas. Mineral. a Geol. — Praha **25** (2). — S. 185—189.

FEJFAR, O.; HEINRICH, W.-D., 1981: Biostratigraphical Delimitation and Division of the Continental Quaternary of Europe on the basis of Arvicolids (Rodentia, Mammalia). — In: IUGS-UNESCO, Internat. Geol. Correlation Progr., Project 73/1/24 "Quaternary Glaciations in the Northern Hemisphere", Rep. No. 6. — Prague: S. 83—86, (1981 a).

FEJFAR, O.; HEINRICH, W.-D., 1981: Zur biostratigraphischen Untergliederung des kontinentalen Quartärs in Europa anhand von Arvicoliden (Rodentia, Mammalia). — In: Ecl. geol. Helv. — Basel **74** (3). — S. 997—1006, (1981 b).

FEJFAR, O.; HEINRICH, W.-D., 1982: A Division of the European Quaternary based on Arvicolid Rodents: A Proposal. — In: IUGS-UNESCO, Internat. Geol. Correlation Progr., Project 73/1/24 "Quaternary Glaciations in the Northern Hemisphere", Rep. No. 7. — Prague: S. 82—83; S. 114.

FEJFAR, O.; HEINRICH, W.-D., 1983: Arvicoliden-Sukzession und Biostratigraphie des Oberpliozäns und Quartärs in Europa. In: W.-D. HEINRICH (Hrsg.): Wirbeltierevolution und Faunenwandel im Känozoikum. — Berlin: Akad.-Verl. — S. 61—109, (Schriftenr. geol. Wiss.; 19/20).

GŁAZEK, I.; HARMON, R. S.; NOWAK, K., 1980: Uranium-series dating of the hominid-bearing travertine deposit at Bilzingsleben, GDR, and its stratigraphical significance. — In: Acta geol. Pol. — Warszawa **30** (1). — S. 1—14.

GÖTZE, A., 1892: Die paläolithische Fundstelle von Taubach bei Weimar. — In: Verh. Berliner Gesellsch. Anthropol. — Berlin 1892. — S. 366—377.

GUENTHER, E. W., 1958: Feinstratigraphische Untersuchungen einer Zwischenschicht und der Deckschichten der Travertine von Ehringsdorf. — In: Alt-Thüringen. — Weimar **3**. — S. 1—15.

—, 1965: Die Biber (*Trogontherium cuvieri* FISCH. und *Castor fiber* L.) der altpleistozänen Fundstelle von Voigtstedt in Thüringen. — In: Paläont. Abh., A. — Berlin **2** (2/3). — S. 565—583.

HAHNE, E.; WÜST, E., 1908: Die paläolithischen Fundschichten der Gegend von Weimar. — In: Cbl. Mineral. etc. — Stuttgart 1908. — S. 197—210.

HARMON, R./S.; GLAZEK, I.; NOWAK, K., 1980: $^{230}Th/^{234}U$ dating of travertine from the Bilzingsleben archeological site. — In: Nature. — London **284**. — S. 132—135.

HEINRICH, W.-D., 1978: Zur Entdeckungs- und Forschungsgeschichte der Wirbeltierfundstätte in der jungpleistozänen Deckschichtenfolge über dem Travertin von Burgtonna (Bezirk Erfurt). — In: Quartärpaläontologie. — Berlin **3**. — S. 185—186, (1978 a).

—, 1978: Zur biometrischen Erfassung eines Evolutionstrends bei *Arvicola* (Rodentia, Mammalia) aus dem Pleistozän Thüringens. — In: Säugetierkdl. Inf. — Berlin **2**. — S. 3—21, (1978 b).

—, 1980: Kleinsäugerfunde aus dem Travertinkomplex von Bilzingsleben. Vorläufige Mitteilung. — In: Ethnogr.-archäol. Z. — Berlin **21**. — S. 36—41, (1980 a).

—, 1980: Biostratigraphische Aspekte einer neuen Kleinsäugerfauna aus dem Unteren Travertin von Weimar-Ehringsdorf. — In: Z. geol. Wiss. — Berlin **8** (7). — S. 923—927, (1980 b).

—, 1981: Systematische Zusammenstellung der in den thüringischen Interglazialtravertinen von Burgtonna, Taubach und Weimar-Ehringsdorf nachgewiesenen Kleinsäugerarten. — In: Quartärpaläontologie. — Berlin **4**. — S. 127—130, (1981 a).

—, 1981: Fossile Kleinsäugerreste aus dem Unteren Travertin von Weimar-Ehringsdorf (Thüringen, DDR) (Vorläufige Mitteilung). — In: Quartärpaläontologie. — Berlin **4**. — S. 131—143, (1981 b).

—, 1981: Zur stratigraphischen Stellung der Wirbeltierfaunen aus den Travertinfundstätten von Weimar-Ehringsdorf und Taubach in Thüringen. — In: Z. geol. Wiss. — Berlin **9** (9). — S. 1031 bis 1055, (1981 c).

—, 1982: Zur Evolution und Biostratigraphie von *Arvicola* (Rodentia, Mammalia) im Pleistozän Europas. — Z. geol. Wiss. — Berlin **10** (6). — S. 683—735, (1982 a).

—, 1982: The Importance of *Arvicola* (Mammalia, Rodentia) for the Biostratigraphy of the European Quaternary. — In: IUGS-UNESCO, Internat. Geol. Correlation Progr., Project 73/1/24 "Quaternary Glaciations in the Northern Hemisphere", Rep. No. 7. — Prague: S. 93—96, (1982 b).

—, 1982: Ein Evolutionstrend bei *Arvicola* (Rodentia, Mammalia) und seine Bedeutung für die Biostratigraphie im Pleistozän Europas. — In: Wiss. Z. Humboldt-Univ. Berlin. Math.-naturwiss. R. — Berlin **31** (3). — S. 155—160, (1982 c).

—, 1984: *Arvicola* sp. (Rodentia, Mammalia) aus dem Pleistozän der Parkhöhlen in Weimar. — In: Quartärpaläontologie. — Berlin **5**. — S. 431—432.

HEINRICH, W.-D.; JÄGER, K.-D., 1978: Zusammenfassende stratigraphische und palökologische Interpretation des Fossilvorkommens in der untersuchten jungpleistozänen Deckschichtenfolge über dem interglazialen Travertin von Burgtonna in Thüringen. — In: Quartärpaläontologie. — Berlin **3**. — S. 269—285.

HEINRICH, W.-D.; JÁNOSSY, D., 1973: *Lagurus lagurus* (PALLAS, 1773) (Rodentia, Mammalia) aus dem Jungpleistozän von Burgtonna (Bezirk Erfurt). — In: Z. geol. Wiss. — Berlin **1** (7). — S. 597—602.

HEINRICH, W.-D.; JÁNOSSY, D., 1977: Insectivoren und Rodentier aus dem Travertin von Taubach bei Weimar. — In: Quartärpaläontologie. — Berlin **2**. — S. 401—411.

HEINRICH, W.-D.; JÁNOSSY, D., 1978: Insectivoren und Rodentier aus dem Travertin von Burgtonna in Thüringen. — In: Quartärpaläontologie. — Berlin **3**. — S. 167—170, (1978 a).

HEINRICH, W.-D.; JÁNOSSY, D., 1978: Fossile Säugetierreste aus einer jungpleistozänen Deckschichtenfolge über dem interglazialen Travertin von Burgtonna in Thüringen. — In: Quartärpaläontologie. — Berlin **3**, — S. 231—254, (1978 b).

HENKEL, S., 1966: Methoden zur Prospektion und Gewinnung kleiner Wirbeltierfossilien. — In: N. Jahrb. Geol. u. Paläont., Monatsh. — Stuttgart 1966 (3). — S. 178—184.

HENKEL, S.; KREBS, B., 1969: Zwei Säugetier-Unterkiefer aus der Unteren Kreide von Una (Prov. Cuenca, Spanien). — In: N. Jahrb. Geol. u. Paläont., Monatsh. — Stuttgart 1969 (8). — S. 449—463.

ISAAC, G., 1974: Stratigraphy and Patterns of Cultural Change in the Middle Pleistocene. — In: Current Anthropology. — Chicago **5** (4). — S. 508—514.

JÄGER, K.-D.; HEINRICH, W.-D., 1981: Correlations between Loess Stratigraphy and Vertebrate Palaeontology in the Weichselian of Thuringia and their Palaeoecological Implications. — In: IUGS-UNESCO, Internat. Geol. Correlation Progr., Project 73/1/24 "Quaternary Glaciations in the Northern Hemisphere", Rep. No. 6. — Prague: S. 128—145.

JÄGER, K.-D.; HEINRICH, W.-D., 1982: The Travertine at Weimar-Ehringsdorf — An Interglacial Site of Saalian Age? — In: IUGS-UNESCO, Internat. Geol. Correlation Progr., Project 73/1/24 "Quaternary Glaciations in the Northern Hemisphere", Rep. No. 7. — Prague: S. 98—113.

JÁNOSSY, D., 1961: Die Entwicklung der Kleinsäugerfauna Europas im Pleistozän (Insectivora, Rodentia, Lagomorpha). — In: Z. Säugetierkd. — Berlin [West]—Hamburg **26**. — S. 40—50.

—, 1965: Die Insectivoren-Reste aus dem Altpleistozän von Voigtstedt in Thüringen. — Paläont. Abh., A. — Berlin **2** (2/3). — S. 663—679.

—, 1974: Die Kleinsäuger-Fauna aus dem Pariser Horizont und den untersten Schichten des Oberen Travertins (Proben M3, M5—8 und d). In: W. STEINER: Zur geologischen Dokumentation des Pariser Horizontes im Travertinprofil von Ehringsdorf bei Weimar. — In: Abh. zentr. geol. Inst. — Berlin 21. — S. 199 bis 247.

—, 1975: Kleinsäugerfunde aus den Travertinen von Weimar-Ehringsdorf. — In: Abh. zentr. geol. Inst. — Berlin 23. — S. 501 bis 511.

KAHLKE, H. D., 1958: Die jungpleistozänen Säugetierfaunen aus dem Travertingebiet von Taubach—Weimar-Ehringsdorf (Vorbericht). — In: Alt-Thüringen. — Weimar **3**. — S. 97—130.

—, 1961: Revision der Säugetierfaunen der klassischen deutschen Pleistozän-Fundstellen von Süßenborn, Mosbach und Taubach. — In: Geologie. — Berlin **10** (4/5). — S. 493—532.

—, 1975: Zur chronologischen Stellung der Travertine von Weimar-Ehringsdorf. — Zur biostratigraphischen Grenze zwischen Interglazial und Glazial im mitteleuropäischen Kontinentalbereich auf der Grundlage der Großsäugerassoziationen. — In: Abh. zentr. geol. Inst. — Berlin 23. — S. 591—596.

KRETZOI, M., 1964: Über einige homonyme und synonyme Säugetiernamen. — In: Vertebrata Hungarica. — Budapest **6** (1/2). — S. 131—138.

—, 1965: Die Nager und Lagomorphen von Voigtstedt in Thüringen und ihre chronologische Aussage. — In: Paläont. Abh., A. — Berlin **2** (2/3). — S. 586—661.

—, 1969: Die *Castor*- und *Trogontherium*-Reste aus den Kiesen von Süßenborn bei Weimar. — In: Paläont. Abh., A. — Berlin **3** (3/4). — S. 771—783.

—, 1975: Die *Castor*-Funde aus dem Travertinkomplex von Weimar-Ehringsdorf. — In: Abh. zentr. geol. Inst. — Berlin 23. — S. 513—532.

—, 1977: Die *Castor*-Reste aus den Travertinen von Taubach bei Weimar. — In: Quartärpaläontologie. — Berlin **2**. — S. 389 bis 400.

KUKLA, G. J., 1975: Loess Stratigraphy of Central Europe. — In: K. W. BUTZER; G. L. ISAAC (Eds.): After the Australopithecines: Stratigraphy Ecology and Culture Change in the Middle Pleistocene. — The Hague—Paris: Mouton Publ. — S. 99—188.

LINSTOW, O. v., 1908: Die Verbreitung des Bibers im Quartär. — In: Abh. u. Ber. Mus. Natur- und Heimatkd. Magdeburg. — Magdeburg **1** (4). — 387 S.

MAI, H., 1978: Untersuchungen von Gebissen der pleistozänen Biberarten *Trogontherium* und *Castor* und ihre stratigraphische Einordung. — In: Schr. naturwiss. Ver. Schleswig-Holstein. — Kiel **48**. — S. 35—39.

MANIA, D., 1975: Zur Stellung der Travertinablagerungen von Weimar-Ehringsdorf im Jungpleistozän des nördlichen Mittelgebirgsraumes. — In: Abh. zentr. geol. Inst. — Berlin 23. — S. 571—589.

NEHRING, A., 1877: Beiträge zur Kenntnis der Diluvialfauna. — In: N. Jahrb. Mineral. etc. — Stuttgart 1877. — S. 556—557.

—, 1878: Die quarternären Faunen von Thiede und Westeregeln. — In: N. Jahrb. Mineral etc. — Stuttgart 1878. — S. 843—846.

—, 1895: Über einen diluvialen Kinderzahn von Předmost in Mähren unter Bezugnahme auf den schon früher beschriebenen Kinderzahn aus dem Diluvium von Taubach bei Weimar. — In: Verh. Berliner anthropol. Gesellsch. — Berlin 1895. — S. 425 bis 433.

POHLIG, H., 1881: Katalog einer Collection von Fossilien, Gesteinen, Mineralien, Zoologischem etc., erstere besonders aus Trias und Pleistocaen Central-Deutschlands. — Weimar: Voigt'sches Officin. 35 S.

—, 1884: Untersuchungen über das Pleistocaen oder Quartär. — In: Sitz.-Ber. niederrhein. Gesellsch. Natur- u. Heilkd. — Bonn **41**. — S. 47—61.

—, 1885: Vorläufige Mittheilungen über das Plistocaen, insbesondere Thüringens. — In: Z. Naturwiss. — Halle **58**. — S. 258 bis 276.

—, 1890: Die großen Säugetiere der Diluvialzeit. — Leipzig: Verl. Freese. — 64 S.

—, 1891: Ueber neue Ausgrabungen von Taubach bei Weimar. — In: Sitz.-Ber. niederrhein. Gesellsch. Natur- u. Heilkd. — Bonn **48**. — S. 38—39.

PORTIS, A., 1878: Ueber die Osteologie von *Rhinoceros Merckii* JÄG. und ueber die diluviale Säugethierfauna von Taubach bei Weimar. — In: Palaeontographica. — Cassel **25**. — S. 141—162.

ROSHOLT, I. N.; ANTAL, P. S., 1963: Evaluation of the Pa^{231}/U to Th^{233}/U method for Dating Pleistocene carbonate rocks. — In: U.S. Geol. Surv. — Prof. Paper. — Washington **450-E**. — S. E108—E111.

SCHÄFER, H. F., 1909: Über die pleistocäne Säugetierfauna und die Spuren des paläolithischen Menschen von Burgtonna in Thüringen. — In: Z. dt. geol. Gesellsch. — Berlin **61**. — S. 445—469.

SCHAUB, S., 1935: Die Hamster vom Roten Berg bei Saalfeld. — In: Ecl. geol. Helv. — Basel **28** (2). — S. 580—590.

SCHMIDT, L., 1978: Der Biber *(Castor fiber)* aus dem Pleistozän von Burgtonna. — In: Quartärpaläontologie. — Berlin **3**. — S. 163—166.

SCHMIDT, R. R., 1912: Die diluviale Vorzeit Deutschlands. — Stuttgart: E. Schweizerbart'sche Verlagsbuchh. — XIII. 281 S.

SCHWARCZ, H. P., 1980: Absolute age determination of archaeological sites by Uranium series dating of travertines. — In: Archaeometry. — Oxford **22** (1). — S. 3—24.

SOERGEL, W., 1912: *Elephas trogontherii* POHL. und *Elephas antiquus* FALC., ihre Stammesgeschichte und ihre Bedeutung für die Gliederung des deutschen Diluviums. — In: Palaeontographica. — Stuttgart **60**. — S. 1—114.

—, 1920: Der Siebenschläfer aus den Kiesen von Süßenborn bei Weimar. — In: Z. dt. geol. Gesellsch. — Berlin **71**. — S. 59—79.

—, 1926: Exkursion ins Travertingebiet von Ehringsdorf. — In: Palaeont. Z. — Berlin **8**. — S. 7—33.

STEINER, W., 1972: Ein neues Profil mit paläolithischem Fundhorizont im Travertin von Taubach bei Weimar. — In: Wiss. Z. Hochsch. Archit. u. Bauw. — Weimar **19** (5/6). — S. 485—492.

—, 1973: Zur Stratigraphie des Oberen Travertins im Quartärprofil von Ehringsdorf bei Weimar. — In: Z. geol. Wiss. — Berlin **1** (5). — S. 515—529.

—, 1974: Zur geologischen Dokumentation des Pariser Horizontes im Travertinprofil von Ehringsdorf bei Weimar. — In: Abh. zentr. geol. Inst. — Berlin 21. — S. 199—247.

—, 1975: Zur stratigraphischen Stellung des Travertinprofils von Ehringsdorf bei Weimar. — In: Alt-Thüringen. — Weimar **13**.

—, 1976: Das neue Standardprofil des Travertinvorkommens von Ehringsdorf bei Weimar. — In: Z. geol. Wiss. — Berlin **4** (5).

—, 1977: Das geologische Profil des Travertin-Komplexes von Taubach bei Weimar. — In: Quartärpaläontologie. — Berlin **2**. — S. 83—118.

STEINER, W.; WIEFEL, H., 1966: Zur Geschichte der geologischen Forschung in Weimar. Teil I: Forscherpersönlichkeiten und Institutionen. — In: Wiss. Z. Hochsch. Archit. u. Bauw. — Weimar **13** (3). — S. 247—273.

STEINER, W.; WIEFEL, H., 1974: Die Travertine von Ehringsdorf bei Weimar und ihre Erforschung (Zur Geschichte der geologischen Forschung in Weimar, Teil II). — In: Abh. zentr. geol. Inst. — Berlin 21. — S. 11—60.

STEINER, W.; WIEFEL, H., 1977: Zur Geschichte der geologischen Erforschung des Travertins von Taubach bei Weimar. — In: Quartärpaläontologie. — Berlin **2**. — S. 9—81.

THENIUS, E., 1962: Die Großsäugetiere des Pleistozäns von Mitteleuropa. Eine Übersicht. — In: Z. Säugetierkd. — Berlin [West]—Hamburg **27**. — S. 65—83.

WEISS, A., 1910: Das Pleistocän der Umgegend von Weimar. — Hildburghausen: Kommissionsverl. Gadow. — 66 S.

WERTH, E., 1916: Über die Paläolith-Fundstätten der Gegend von Weimar. — In: Z. Ethnol. — Berlin **48**. — S. 119—131.

—, 1928: Der fossile Mensch. Grundzüge einer Paläanthropologie. — Berlin: Verl. Gebr. Borntraeger. — X, 398 S.

WIEFEL, H., 1974: Geschichte der geologischen Erforschung. In: W. HOPPE; G. SEIDEL (Hrsg.): Geologie von Thüringen. — Gotha/Leipzig: H. Haack. — S. 13—25.

WIEGERS, F., 1920: Diluvialprähistorie als geologische Wissenschaft. — In: Abh. preuß. geol. Landesanst., N. F. — Berlin 84.

—, 1928: Diluviale Vorgeschichte des Menschen. Bd. 1: Allgemeine Diluvialprähistorie mit einem Beitrag von HANS WEINERT: Die fossilen Menschenreste. — Stuttgart: F. Enke. — VIII, 299 S.

—, 1931: Das geologische Alter des oberen Kalktuffs von Ehringsdorf. — In: Jahrb. preuß. geol. Landesanst. — Berlin **52**. — S. 461—465.

WÜST, E., 1902: Säugetierreste aus dem Kalktuffe von Bilzingsleben bei Kindelbrück. — In: Z. Naturwiss. — Stuttgart **75**. — S. 237—239.

—, 1910: Die plistozänen Ablagerungen des Travertingebietes der Gegend von Weimar und ihre Fossilbestände in ihrer Bedeutung für die Beurteilung der Klimaschwankungen des Eiszeitalters. — In: Z. Naturwiss. — Leipzig **82**. — S. 161 bis 252.

ZEISSLER, H., 1975: Konchylien im Ehringsdorfer Pleistozän. — In: Abh. zentr. geol. Inst. — Berlin 23. — S. 15—90.

Quartärpaläontologie 6 75—81 Berlin 1986

Anmerkungen zur intellektuellen Höherentwicklung des Menschen im Plio-Pleistozän (Primates: Hominidae: Australopithecus und Homo)[1])

Von HELMUT HEMMER, Mainz

Mit einer Abbildung

Inhalt

Abstractum

Stellvertretend für die aus dem Fossilbestand pliozän/pleistozäner Menschen nicht unmittelbar bewertbare intellektuelle Leistungsfähigkeit werden Cephalisationswerte zusammengestellt, wie sie sich aus korrelativen Schätzungen von Hirngewichten und Körpergewichten ergeben. Hieraus resultiert der Befund eines Übergangs von nahezu evolutiver Stasis der Hirnevolution während der intragenerischen *Australopithecus*-Entwicklung zu höchster evolutiver Dynamik in der *Homo*-Linie und einer Trennung dieser Linien nicht später als auf dem *Australopithecus afarensis*-Niveau. Die Abspaltung dieses Hominidenzweigs von afrikanischen Menschenaffen ist deren rezenten Hirnwerten zufolge spätestens im Endmiozän anzusetzen. Die Übereinstimmung mit molekularen Daten ist augenfällig.

The intellectual abilities of Pliocene and Pleistocene man cannot evaluated directly on the basis of available fossils. Cephalization values may be substituted for these abilities, and can be calculated from brain and body weight estimates. The use of this procedure results in a picture of near stasis in brain evolution within the genus *Australopithecus*, but highly dynamic evolution in the genus *Homo*, and points to the divergence of these lineages not later than at the *Australopithecus afarensis* level. The branching off of these hominids from the African apes would have taken place at the end of the Miocene at the latest, according to the cephalization values of the modern species. The correspondence of this dating with molecular data is obvious.

На основании имеющихся ископаемых остатков нельзя непосредственно оценить мыслительные способности плиоценового и плейстоценового человека. Взамен этого можно использовать показатели цефализации, вычисленные по расчетным значениям веса мозга и тела. Использование этой методики дает картину почти полного застоя в эволюции мозга в роде *Australopithecus*, но весьма динамичной эволюции в роде *Homo*, а также указывает на расхождение этих линий не позднее, чем на уровне *Australopithecus afarensis*. Отделение этих гоминид от африканских человекообразных обезьян должно было произойти самое позднее в конце миоцена, судя по показателям цефализации современных видов последних. Эти хронологические оценки находятся в очевидном соответствии с данными молекулярной биологии.

1. Einleitung

Der Verfasser begegnete CONSTANTIN CONSTANTINOVIČ FLEROV erstmals im September 1969 anläßlich des UNESCO-Symposiums: „The Origin of *Homo sapiens*" in Paris (BORDES 1972). FLEROV stellte dort im großen internationalen Rahmen eine repräsentative Sammlung seiner überzeugenden Rekonstruktionsgemälde eindrucksvoller Vertreter der pleistozänen paläarktischen Lebewelt aus, gemeinsam mit einer Präsentation lebensechter plastischer Portraitrekonstruktionen pleistozäner Menschen durch M. M. GERASIMOV (GERASIMOV 1972). Das künstlerische Schaffen dieser beiden eminenten sowjetischen Naturwissenschaftler auf der Basis langer und sorgsamer wirbeltierpaläontologischer und paläoanthropologischer Studien und gründlicher Naturbeobachtung vermittelte damals der Fachwelt und dem interessierten Laienpublikum einen bisher wohl unerreicht konzentrierten Einblick in Erscheinungsbild und Umwelt des Eiszeitmenschen.

Wenn diese vor allem im Kollegenkreis weltweit bekannten künstlerischen Darstellungen aus der Hand FLEROVS auch nur ein Mußestunden-Beiprodukt seiner säugetierpaläontologischen Arbeiten gewesen sein mögen, so bezeugen sie doch in unübertroffener Weise den ganzheitlichen Aspekt, unter dem FLEROV die von ihm bearbeitete Tierwelt in ihren eiszeitlichen Ökosystemen sah. Dieser Umstand kam ihm neben eingehender Detailkenntnis aus gründlicher morphologischer Arbeit beispielsweise sicherlich bei der phylogenetischen Deutung des rezenten kanadischen Waldbisons als eine eigene, vom Präriebison ähnlich weit wie der Wisent getrennte Art und gleichzeitig als Reliktform des pleistozänen holarktischen *Bison priscus* zu Hilfe — eine bislang in der *Bison*-Rezentsystematik noch weitgehend unberücksichtigte Deutung, die in zukünftigen Studien der Isolationsmechanismen zwischen Wald- und Präriebison Bestätigung finden sollte.

Wenn sich auch die Gespräche des Verfassers bei späteren Begegnungen mit FLEROV mehr um verschiedene Aspekte pleistozäner Großsäugetiere denn um den Menschen drehten, so erscheint es in Würdigung des Wirkens von C. C. FLEROV also doch durchaus angebracht, seinem Angedenken erste gesammelte Daten aus einer neuen Blickrichtung auf die mit ihrer Hirngrößenentwicklung aufgezeigte intellektuelle Evolution plio-pleistozäner Menschen zu widmen, die er als Beitrag zum Gesamtverständnis der pleistozänen Lebewelt und der ökologischen Rolle des Menschen in ihr sicherlich besonders begrüßt hätte.

[1]) Dem Andenken an Professor Dr. CONSTANTIN CONSTANTINOVIČ FLEROV.

2. Die relative Hirngröße als Ausdruck der Entwicklungshöhe

Bereits gegen Ende des letzten Jahrhunderts wurde zu vergleichenden Abschätzungen von Hirnevolutionsstufen aus den Grundparametern Hirngewicht und Körpergewicht bei Säugetieren die Allometrieformel $y = b \cdot x^a$ entwickelt, nachdem es sich gezeigt hatte, daß weder absolute Hirnmaße noch einfacher Indexbezug auf ein Maß der Körpergröße relevante Aussagen erbringen können (Snell 1892; Dubois 1897; Lapicque 1898). In den folgenden acht Jahrzehnten entwickelte sich dann allerdings die betreffende Forschung in einer Sackgasse, indem interspezifische Allometrieexponenten (meist zwischen den Werten 0,5 und 0,7 variierend) den Berechnungen zugrunde gelegt wurden (z. B. Stephan und Bauchot 1965; Bauchot und Stephan 1966, 1969; Jerison 1973; Eisenberg und Wilson 1978; Stephan u. a. 1981; Harvey und Bennett 1983). Daß demgegenüber allein die Abschätzung von Cephalisationskonstanten auf der Basis des durchschnittlichen intraspezifischen Exponenten (gewöhnlich mit Werten zwischen 0,2 und 0,3) zu der jeweiligen Hirnleistungsfähigkeit angemessenen Daten führen kann, wurde vom Verfasser für Primaten seit 1971 mehrfach betont; zwischenzeitlich konnten signifikante Zusammenhänge solchermaßen gewonnener Hirngrößenwerte mit unterschiedlichen psychobiologischen Parametern bei Primaten aufgezeigt werden (Hemmer 1971, 1974, 1978, 1979, 1981). Unabhängig davon nutzte Passingham (1975) den intraspezifischen Exponenten zur Hirngrößenkalkulation in der Hominidenevolution, McHenry (1975, 1976) stellte für diese Evolutionsreihe die Resultate verschiedener Kalkulationsmethoden zusammen, während kurz zuvor Pilbeam und Gould (1974) nur teilweise mit intraspezifischem/intragenerischem Ansatz Stufen progressiver Cephalisation diagrammatisch darstellten.

In Entsprechung zur Hirngewichts-Körpergewichts-Allometrie erscheint eine Beurteilung von Hirnevolutionsstufen an Hand von Schädelmaßen allein über Allometrie-Beziehungen vertretbar. Der Verfasser hat auch hierauf seit einer monographischen Bearbeitung dieser Thematik (Hemmer 1967) mehrfach eindringlich hingewiesen (z. B. Hemmer 1972). Der Vergleich lediglich absoluter Hirnschädelkapazitäten ohne Berücksichtigung die Körpergröße zum Ausdruck bringender Bezugsmaße wird nicht dadurch sinnvoller, daß er erstaunlicherweise in der paläoanthropologischen Literatur bis heute am Leben erhalten ist.

Daß die vergleichende Beurteilung der Hirnevolutionshöhe plio-pleistozäner Menschen trotz Nutzung allometrischer Bezugsverfahren für das Hirnraumvolumen dennoch problematisch blieb, beruht auf der entscheidenden Grundschwierigkeit, ein relevantes Maß zu finden, das die als solche nicht unmittelbarer Messung zugängliche Körpergröße über die gesamte menschliche Evolution hin möglichst gleichbleibend gut repräsentiert. Der Verfasser wählte hierfür zunächst das bei zahlreichen Säugetieren lange Zeit als recht stabil empfundene Maß Basallänge (pr-ba-Länge, Gesichtslänge der Anthropologie = Martin Maß Nr. 40) (Hemmer 1967). Dieses Maß schließt jedoch den gesamten Kieferapparat mit ein, für den mittlerweile eng zeitkorrelierte Reduktion seit dem frühen *erectus*-Niveau belegt ist (Roth 1983). Ferner ist es nur an solchen Funden nehmbar, bei denen wesentliche Partien des Gesichtsschädels neben dem Hirnschädel erhalten sind. Dies engt die mögliche Beurteilungsbasis fossiler menschlicher Schädelpopulationen entscheidend ein und führte zunächst dazu, an seiner Stelle die größte Hirnschädellänge (g-op) als Körpergrößenbezugsmaß für die Hirnraumkapazität heranzuziehen (Hemmer 1972). Prinzipielle Erkenntnisänderungen innerhalb des Evolutionsbereichs von frühen *erectus*-Populationen zum heutzeitlichen *sapiens* waren — im Vergleich zum Gesichtslängenbezug — hierbei nicht zu erwarten, da für die betreffende Evolutionsspanne die Allometrieverhältnisse Gesichtslänge/Größte Schädellänge (im Vergleich zur *Australopithecus*-Stufe) recht konstant blieben (Hemmer 1967). Dafür ergab sich deutlicher Erkenntniszuwachs durch die nunmehr gegebene Möglichkeit starker Ausweitung der einbeziehbaren Fundstücke. Für eine weitere Benutzung dieser Bezugsgrundlage muß nunmehr allerdings ein zeitbezogener Korrekturfaktor herangezogen werden, wie er durch den Nachweis zeitkorrelierter Grazilisierung des Kauapparats durch Roth (1983) sowohl zwingend als auch unter Verrechnung der Einzeldaten dieser Autorin insofern möglich wird, als sich eben die Gesichtslängen-Schädellängen-Allometrie kaum änderte. Fehlt eine solche Korrektur, so muß die Hirnevolutionshöhe im Schädellängen-Bezug fälschlicherweise um so geringer erscheinen, je älter ein Fund ist, bzw. einen je größeren Kauapparat er besitzt (diese Zeitkorrelation ist bei der Neubeurteilung der von Hemmer, 1972, publizierten Diagramme zu beachten).

Äußerst problematisch erscheint die Suche nach einem durchgängig brauchbaren Körpergrößenbezugsmaß am Schädel für den zur *erectus*-Stufe führenden Evolutionsabschnitt des Menschen. Wie die von Roth (l.c.) für den Kauapparat zusammengestellten Daten erkennen lassen, setzte die das gesamte Pleistozän hindurch verfolgbare, regelhafte, zeitkorrelierte Wandlung dieses Schädelteils erst etwa mit dem frühen *erectus*-Niveau ein. Die *habilis*-Stufe leitet vom *Australopithecus*-Typ zum Beginn dieser Weiterentwicklung zunächst mit einer Verbreiterung des Kauapparats über. Der *Australopithecus*-Schädelbau, der sich in seinen Gesichtsschädel/Hirnschädel-Allometrieverhältnissen grundlegend vom Bau des späteren *Homo*-Schädels (*erectus-sapiens*-Bereich) unterscheidet (Hemmer 1967), liefert schwerlich einen unmittelbar mit *Homo* gemeinsamen bzw. einen durch einen einfachen Zeitkorrekturfaktor anpaßbaren Körpergrößenbezug.

Damit ist es zumindest für die pliozäne und unterpleistozäne Evolution zur Bewertung der Hirnentwicklungshöhe und damit der intellektuellen Kapazität notwendig, über die Abschätzung des Körpergewichts und die Umsetzung des Hirnraumvolumens des Schädels in Hirngewicht Hirngrößenwerte aus der Hirngewichts/Körpergewichts-Allometrie zu bestimmen. Solche Werte (Methode zur Bestimmung der Cephalisationskonstanten nach Hemmer 1971) bieten den Vorteil, sowohl unmittelbaren Vergleich mit heutigen *Homo sapiens*-Populationen als auch mit anderen Primatenarten zu ermöglichen. Sie sollten daher für die gesamte Evolution zum Menschen geschätzt werden.

Um aus Hirnschädelkapazitäten Schätzwerte für Hirngewichte zu erhalten, ist es notwendig, eine Vorstellung vom Zusammenhang dieser beiden Parameter bei unterschiedlicher absoluter Größe zu entwickeln. Dies ist für Säugetiere allgemein in einem vertretbaren Fehlerrahmen möglich, wenn die Kapazitätswerte und die Werte des Index $\frac{\text{Hirnraumkapazität}}{\text{Hirngewicht}}$ bei verschiedenen Größenstufen verrechnet werden. Hierzu wurden folgende Hirnraumvolumen/Hirngewichts-Daten benutzt: *Pomatogale velox* (Insectivora; Daten aus STEPHAN und KUHN 1982, Kapazität bei 0,0037 cm^3), Mittelwert aus 11 Chiroptera-Arten (Daten für die Kapazitäten aus EISENBERG und WILSON 1978, für das Gewicht aus STEPHAN u. a. 1981, Kapazität bei 0,87 cm^3), Mittelwert aus sechs *Rodentia*-Arten (Daten aus STEPHAN und DIETERLEN 1982, Kapazität bei 1,19 cm^3), Mittelwert der *Lemur*-Arten (Primates; Daten für die Kapazität: eigene Messungen, für das Gewicht aus BAUCHOT und STEPHAN 1966, Kapazität bei 24 cm^3), *Canis lupus* (aus RÖHRS und EBINGER 1983 nach Hirngewicht-Hirnschädelkapazitäts-Formel und mittlerem Hirngewicht bestimmt, Kapazität bei 148 cm^3), sowie *Gorilla gorilla* (nach Angaben bei BAUCHOT und STEPHAN 1969, Kapazität bei 496 cm^3). Logarithmische Verrechnung der Kapazitätswerte erbringt einen Wert des Korrelationskoeffizienten von 0,94 und läßt für den Kapazitätsbereich > 1200 cm^3 (*Homo sapiens*-Bereich) den Indexwert $\frac{\text{Kapazität}}{\text{Gewicht}}$ von 1,08 extrapolieren, der dem von WELCKER (zit. n. MARTIN und SALLER 1959) für den Kapazitätsbereich 1200—1300 cm^3 rezenter Menschen bestimmten Wert von 1,10 nahe kommt und mit anderen für *Homo sapiens* berechneten Werten übereinstimmt (zusammengestellt in MARTIN und SALLER 1959). Damit zeigt sich die zur Vermeidung eines Kreisschlusses *Homo sapiens* nicht einschließende Berechnung für Säugetiere allgemein als für die hier zu bearbeitende Fragestellung durchaus relevant. Zur Hirngewichts-Abschätzung aus Hirnraumkapazitäten wird daher unter Zugrundelegen der betreffenden Regression im folgenden mit dem Hirnkapazität/Hirngewichts-Indexwert 1,08 für Volumina > 1200 cm^3, mit dem Wert 1,07 für Volumina 600—1200 cm^3, mit dem Wert 1,06 für Volumina 300—600 cm^3 gearbeitet.

3. Die relative Hirngröße pliozäner und pleistozäner Hominiden

Körpergewichtsschätzungen für den südafrikanischen *Australopithecus africanus* wurden auf der Grundlage unterschiedlicher Gewichtsregressionen postkranialer Skelettelemente von mehreren Autoren vorgenommen, zusammengestellt und diskutiert (z. B. PILBEAM und GOULD 1974; HEMMER 1974; PASSINGHAM 1975; MCHENRY 1975, 1976; ZIHLMANN und LOWENSTEIN 1983). Einzelwerte solcher Schätzungen schwanken für verschiedene Individuen zwischen 20 und 53 kg, Mittelwerte liegen zwischen 32 und 37 kg. In Anbetracht recht guter Übereinstimmung auf sehr unterschiedliche Weise gewonnener Werte darf ein intersexuell mittleres Gewicht zwischen 30 und 40 kg als real angenommen werden. Ebenso steht ein Hirnraumvolumen grob zwischen 400 und 500 cm^3 außer Frage, wobei auf ein Mittel zwischen 440 und 450 cm^3 zu schließen ist (Bestimmungen n. HOLLOWAY, 1973, von den meisten Autoren — s. oben — übernommen). Die Hirngewichtstransformation führt zu einer Schätzung von im Mittel grob um 420 g. Diesen Daten zufolge sollte der relative Hirngrößenwert (Cephalisationskonstante) von *Australopithecus africanus* im Bereich um 38 gelegen haben.

Nach den bisher publizierten Daten (Schätzwerte 385, 450 und 500 cm^3 für das Hirnraumvolumen, entsprechend im Mittel um 420 g Hirngewicht, vergleichbare Körpergröße/ BOAZ 1983; WHITE u. a. 1983) besteht kein Grund, für den spätpliozänen ostafrikanischen *Australopithecus afarensis* (bzw. *A. africanus afarensis*, je nach taxonomischer Handhabung — in einer den rezenten Steppenpavianen wohl vergleichbaren Situation, wo selbst für lebende Populationen keine allseitige Übereinstimmung über Subspezies-, Semispezies- oder Speziescharakter zu erzielen ist) einen wesentlich anderen Hirngrößenwert anzunehmen. Für den zeitlich auf *Australopithecus africanus* folgenden pleistozänen südafrikanischen *Australopithecus robustus* (oder wiederum *A. africanus robustus*) erscheinen mittlere Körpergewichte zwischen 40 und 50 kg als diskutabel (PILBEAM und GOULD 1974; PASSINGHAM 1975; MCHENRY 1975, 1976). Schätzungen bzw. Messungen des Hirnraumvolumens liegen für diese Form zwischen 500 und 530 cm^3, was Hirngewichten im Bereich zwischen 470 und 500 g entspricht. Die ostafrikanische Spätform *Australopithecus boisei* (bzw. *A. africanus boisei*) schließlich mag bei ähnlichen Hirngewichten wie *A. robustus* im mittleren Körpergewicht wenig schwerer als jener gewesen sein (PILBEAM und GOULD, 1974, geben 47,5 kg als mittleres Körpergewicht, 510 cm^3 als mittleres Hirnschädelvolumen, MCHENRY, 1976, gibt entsprechend 46,5 kg und 534 cm^3). Ein deutlicher Unterschied des zu etwa 41—42 zu bestimmenden Hirngrößenwerts ergibt sich im Vergleich der ostafrikanischen und der südafrikanischen *Australopithecus*-Endformen mit diesen Daten nicht.

Die hohe Variabilität der für *Homo habilis* geschätzten Hirnschädelkapazitäten (509—752, Mittel 638 cm^3/HOLLOWAY 1978) legt Hirngewichte zwischen 480 und 700 g nahe; das Mittel ist im Bereich um 600 g zu suchen. PASSINGHAM (1975) errechnet zwei Körpergewichte zu 43 und 49 kg, MCHENRY (1976) bestimmt zwei Gewichte auf 51 und 54 kg, wobei die jeweils letztgenannten Werte mit unterschiedlichen Ansätzen für ein und denselben Fund (Femur KNM-ER 1481) erhalten wurden. PILBEAM und GOULD (1974) schlossen zuvor als mehr allgemeine Einschätzung auf Körpergewichte von *Homo habilis* nicht unter denen von *Australopithecus robustus* und nicht über denen von *Australopithecus boisei*, für die der Bereich zwischen 40 und 50 kg plausibel erscheint (s. oben; PILBEAM und GOULD nehmen 40—45 kg an). Für 600 g mittleres Hirngewicht resultiert mit einem Körpergewicht von nur 40 kg ein Cephalisationswert von 52, mit einem Gewicht von 45 kg ein solcher von 51, schließlich mit einem Gewicht von 50 kg ein solcher von 50. Somit dürfte ein Wert um 50 der Realität für *Homo habilis* sehr nahe kommen.

Für den zeitlich auf *Homo habilis* folgenden ostafrikanischen *Homo erectus* (bzw. *Homo sapiens*, *erectus*-Gruppe, je nach taxonomischer Auffassung) gibt RIGHTMIRE (1980) drei Schädelkapazitäten zu 727, 850 und 1067 cm^3 (mit den Datierungen um 0,7/1,5/1,2 Mill. J.), was Hirngewichten zwischen 680 und 1000 g, Mittel bei 825 g, entspricht. Ein identischer Mittelwert resultiert aus den Schätzungen der Hirnraumvolumina der frühen javanischen Formen der *erectus*-Gruppe (750—1029 cm^3, Mittel 883 cm^3/PASSINGHAM 1975, entsprechend Hirngewichten zwischen 700 und 960 g, Mittel um 825 g). PASSINGHAM (1975) bestimmt für diesen frühen *erectus* ein Körpergewicht von 56 kg; CRONIN et al. (1981) setzen für *Homo erectus* allgemein ein mittleres Körpergewicht von 48,5 kg. 45 kg Gewicht erbringen mit einem Hirngewicht von 825 g einen Cephalisationswert von 70, 50 kg ergeben mit dem gleichen mittleren Hirngewicht einen Wert von 68,5, 55 kg schließlich einen solchen von 67. Es erscheint damit nicht als verfehlt, für die mittlere Datierung um 1 Mill. J. einen mittleren relativen Hirngrößenwert von 70 zu setzen.

Die früh-mittelpleistozäne chinesische *erectus*-Population von Chou-kou-tien dürfte Hirngewichte zwischen 855 und 1135 g, Mittel um 975 g, besessen haben (Kapazitäten 915 bis 1225 cm^3/PASSINGHAM 1975). PILBEAM & GOULD (1974) schätzen ihr mittleres Körpergewicht auf 53 kg, PASSINGHAM (1975) auf 49 kg. Ein mittlerer Ansatz um 50 kg führt mit dem genannten mittleren Hirngewicht zu einem Cephalisationswert von 81. Für die mittelpleistozänen europäischen „Anteneandertaler" (Heidelberger Rasse von *Homo sapiens*, Funde von Vertesszöllös, Petralona, Steinheim, Swanscombe und Tautavel) begründet HEMMER (1982) einen Cephalisationswert um 90. Damit erreicht diese Gruppe bereits den Hirngrößenbereich heutiger australischer Ureinwohner (HEMMER l. c.). Für das Zeitalter grob um 500000 Jahre läßt sich im gemeinsamen Blick auf diese europäische Population und auf den ostasiatischen *erectus* also eine mittlere Cephalisation um 85 ansetzen.

Für den rezenten *Homo sapiens* zusammengestellte Hirngewichts/Körpergewichts-Daten (BAUCHOT und STEPHAN 1969) und Hirnschädelkapazitäts + Körpergewichtsdaten für unterschiedliche geographische Populationen (MARTIN und SALLER 1959) zeigen eine Gesamtstreuung des heutzeitlichen Menschen um einen relativen Hirngrößenwert bei 100, mit Tiefstwerten bei den australischen Ureinwohnern (Bereich 90—95; HEMMER 1982) und Höchstwerten (Bereich 105—110) bei ostasiatischen Bevölkerungen (Chinesen, Japaner).

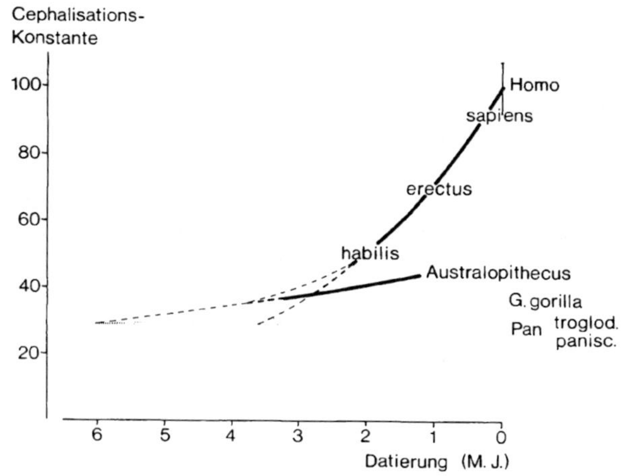

Abb. 1. Hirngrößenevolution (Cephalisationskonstante n. HEMMER 1971) im Zeitkorrelat im Genus *Homo* (dicke Linie, berechnet als Regressionsgerade der Beziehung log Cephalisation/Datierung auf der Basis der im Text angegebenen Daten) und im Genus *Australopithecus* (unter der Annahme entsprechender logarithmischer Beziehung konstruiert aus den Daten für *A. afarensis/africanus* und *A. boisei/robustus*)

Unterbrochene Linien: Extrapolationen der *Homo*- bzw. der *Australopithecus*-Evolutionslinien bzw. mögliche Verbindung der *Australopithecus*- mit der *Homo*-Kurve bei Annahme nicht abrupter Änderung der Evolutionsgeschwindigkeit. Dünne Punktlinien: Denkbarer Verlauf der Hirngrößenentwicklung bei den afrikanischen Menschenaffen *Gorilla gorilla*, *Pan troglodytes* und *Pan paniscus* unter der (infolge Fehlens relevanter Fossilfunde aus der frühen *Pan*-Linie vorläufig unbeweisbaren) Annahme seit der Trennung von der *Australopithecus-Homo*-Linie unveränderter relativer Hirngröße auf dem Weg zum Bonobo (*Pan paniscus*); damit Konstruktion des spätestens zu erwartenden Verzweigungspunktes der beiden großen Hominidenäste unter der (ebenfalls unbewiesenen) Voraussetzung weiterer Extrapolationsmöglichkeit der *Australopithecus*-Kurve

4. Diskussion

Zur formalen Bestimmung einer Entwicklungslinie der relativen Hirngröße im Verlauf der pleistozänen Evolution innerhalb des Genus *Homo* ist von den oben extrahierten Wertepaaren auszugehen: 50 für die Zeit um 2 Mill. J. (*habilis*-Wert), 70 für die Datierung um 1 Mill. J. (*erectus*-Wert), 85 für 0,5 Mill. J. (europäische frühe *sapiens*- und ostasiatische späte *erectus*-Populationen im Mittel) und 100 für den heutzeitlichen Menschen. Logarithmische Verrechnung der Cephalisationswerte resultiert in einem Wert r = —0,9996, d. h., nahezu absoluter Korrelation. Dieser trotz der äußerst geringen Datenzahl klare Zusammenhang erlaubt den Versuch einer Extrapolation der Cephalisationskurve über das *Homo habilis*-Niveau hinaus (Abb. 1). Die aus der Verbindung des *Australopithecus africanus* (+ *afarensis*)-Werts (38 auf 2,8 Mill. J. gesetzt) und des *Australopithecus robustus/boisei*-Werts (42 auf 1,75 Mill. J. gesetzt) erhaltene *Australopithecus*-Hirnevolutionslinie wird von der genannten *Homo*-Linie bei etwa 2,8 Mill. J. geschnitten. Eine noch weitere Rückwärtsverlängerung der *Homo*-Linie ließe für ein mittleres Körpergewicht zwischen 30 und 40 kg (in diesem Zeitabschnitt für *Australopithecus afarensis* angesetzt) bei 3,2 Mill. J. (mittlere Datierung für die Hadar-Serie von *A. afarensis*: BOAZ 1983) eine Hirnschädelkapazität um 390 cm^3 erwarten, die gegenüber den realen Schätzwerten (s. oben) klar zu niedrig liegt.

Geht man nicht von der recht unwahrscheinlich anmutenden Annahme einer zu Beginn der *Homo*-Eigenevolution geringeren relativen Hirngröße als gleichzeitig in der *Australopithecus*-Linie aus, so deutet dieser Befund darauf hin, daß sich um die Pliozän-Pleistozän-Wende die Evolutionsgeschwindigkeit der Hirngröße bei Vor- und Frühmenschen deutlich änderte: aus der etwa 2,5 Mill. J. lang sehr langsam zunehmenden Evolutionshöhe innerhalb des Genus *Australopithecus*, die nahezu im Sinne evolutiver Stasis interpretiert werden mag, erwuchs nun eine hohe evolutive Dynamik, mit der zur Zeit des Erlöschens der *Australopithecus*-Populationen bereits eine 50% höhere relative Hirngröße im Vergleich zu jenen Endformen von *Australopithecus* erreicht wurde. Gleichzeitig zeigt dieser Befund an, daß die *Homo*-Linie spätestens vor etwa 2,8 Mill. J. von der *Australopithecus*-Linie abzweigte — im Falle eines nicht sehr wahrscheinlichen abrupten Änderns der Hirnevolutionsgeschwindigkeit; bei allmählicher Änderung von der *Australopithecus*- zur *Homo*-Linie (Alternativkurve in Abb. 1) jedoch früher. Als gemeinsame Stammform ließe sich diesen Daten zufolge also *Australopithecus afarensis* im derzeit bekannten zeitlichen Umfang ansehen. Von den von WHITE u. a. (1983) diskutierten Alternativ-Hypothesen zur Phylogenie von *Australopithecus* und *Homo* ist damit höchstens die auch von jenen Autoren favorisierte letzte Hypothese (Fig. 18, C) akzeptabel (*A. afarensis* als Basis, Trennung während der Evolution zu *A. africanus*). Es ist aber ebenfalls eine *Homo*-Abspaltung vor etwa 4 Mill. J. diskutabel, wie sie dann als zwingend anzusehen wäre, wenn die Datierung des Kanapoi-Humerus auf 4 Mill. J. stimmig ist (s. hierzu Kritik durch BOAZ 1983), für den SENUT (1979) überzeugend *Homo*-Zugehörigkeit darlegte.

Ist man nicht geneigt, für die Hirnevolution der afrikanischen Hominidengenera *Pan* und *Gorilla* regressive Entwicklung anzunehmen — ein evolutiver Vorgang, der nach derzeitiger Kenntnis bei keinem wildlebenden Säugetier eine Parallele hätte —, so sollte die relative Hirngröße zur Zeit der Trennung dieser Linie(n) von der *Australopithecus-Homo*-Linie im höchsten Fall das Niveau erreicht haben, das bei den rezenten Arten dieser Gattungen als geringstes zu finden ist. Dies ist mit dem mittleren Cephalisationswert um 29 des Bonobos (*Pan paniscus*) gegeben [übereinstimmend aus den Rohdaten von PILBEAM und GOULD (1974) und ZIHLMAN und LOWENSTEIN (1983) resultierend], während Schimpanse (*Pan troglodytes*) und Gorilla (*Gorilla gorilla*) etwas höher liegen (Schimpanse bei 32, Gorilla bei 33, jeweils nach Hirngewichts/Körpergewichts-Einzelwerten offensichtlich adulter Individuen in der Zusammenstellung bei BAUCHOT und STEPHAN, 1969, sowie aus der Wandlung von Hirnschädelkapazitäts- und Körpergewichtsmittelwerten bei PILBEAM und GOULD 1974). Versuchsweise Rückwärtsextrapolation der *Australopithecus*-Hirnevolutionslinie (Abb. 1; aus log Cephalisationskonstante/Alter) führt zum Schnittpunkt mit der nach der Bonobo-Cephalisation gesetzten Abszissenparallelen bei 6 Mill. J. Dieser Zeitpunkt dürfte damit den grundsätzlich spätest möglichen für die phylogenetische Aufzweigung zu den Genera *Pan* und *Gorilla* bzw. *Australopithecus* und *Homo* markieren. Jede geringe Steigerung der relativen Hirngröße für die letztlich zum Bonobo führende Linie seit dieser Verzweigung rückt den Trennungspunkt weiter im oberen Miozän zurück.

Dieses rein von den Hirngrößendaten her erarbeitete Verzweigungsbild deutet ein überraschend hohes Maß an Übereinstimmung mit der derzeitigen Gesamtbeurteilung jenes Aufspaltungspunktes nach molekularen Daten an. CRONIN (1983) erhält bei einer Zusammenstellung entsprechender Werte für die *Homo-Pan-Gorilla*-Aufzweigung eine mittlere Datierung von 5,0 ± 1,5 Mill. J. und gibt der klaren Überzeugung Ausdruck, daß die betreffende Divergenz im groben Zeitraum zwischen 4 und 8 Mill. J. stattfand. Die Deckung dieser Ergebnisse aus total unterschiedlichen methodischen Ansätzen dürfte andeuten, daß tatsächlich kaum eine absicherbare Höherentwicklung des Gehirns in der zu *Pan panicus* führenden Linie seit dem oberen Miozän stattgefunden haben sollte.

Ähnlich enge Übereinstimmung der Ergebnisse jener sehr differenten Methoden zur Bestimmung wesentlicher Abschnitte der menschlichen Evolution ergeben sich für den Endabschnitt. Die geringste in heutigen menschlichen Bevölkerungen gefundene Cephalisation (australische Ureinwohner) liegt auf dem Niveau, das sich auf der generellen Evolutionslinie des Gehirns in der Gattung *Homo* (Abb. 1) bereits für das obere Mittelpleistozän findet. Wiederum unter der Voraussetzung, keine regressive Hirnevolution — hier für die Australier — annehmen zu müssen, bedeutet dies die Abspaltung der Wurzeln dieser Bevölkerung von anderen *Homo sapiens*-Populationen spätestens zu dieser Zeit. Molekulare Daten verweisen auf eine Radiation moderner Bevölkerungen vor etwa 100000—500000 Jahren (CRONIN 1983), also eben auf die Zeit, zu der in Europa die „Anteneandertaler" die entsprechende Hirngröße erreicht hatten (HEMMER 1982).

Die hier für das gesamte Pleistozän wahrscheinlich gemachte streng regelhafte progressive Hirnevolution in der *Homo*-Linie steht prinzipiell für eine entsprechend zeitkorrelierte intellektuelle Höherentwicklung. Wie der Verfasser zeigte (HEMMER 1974), stehen die Cephalisationswerte bei Simiern mit dem mittleren Alter der ♀♀ bei Erreichen der vollen Geschlechtsreife im Zusammenhang, d. h., mit der Dauer der Jugendentwicklung, der in der sozialen Umwelt behüteten Lernzeit der Kinder und Jugendlichen, in der die genetische Potenz der Gehirnorganisation intensiv für die Bereitstellung erlernter Verhaltensprogramme genutzt wird. Ausgehend von der Kurve für die untere Grenze des einfachen mittleren Fehlers der Geschlechtsreifealter (1. Konzeption)/Cephalisationswert-Beziehung, die sowohl für die Extrapolation in den basalen Lemurenbereich (*Microcebus murinus*) als auch für die Extrapolation auf *Homo sapiens* stimmige Werte erbringt, lassen sich für die hier dargestellten Punkte der Hominiden-Hirnevolution folgende Alterswerte für die 1. Konzeption abschätzen: Trennpunkt *Pan-Gorilla*-Linie und *Australopithecus-Homo*-Linie: höchstens $7^1/_2$ Jahre; *Australopithecus afarensis/africanus*: 9 Jahre; Endglieder des Genus *Australopithecus*: $9^1/_2$ Jahre; mittlerer *Homo habilis*-Bereich (vor 2 Mill. J.): $10^1/_2$ Jahre; mittlerer *erectus*-Bereich (vor 1 Mill. J.): 13 Jahre; mittlerer Wert europäisch-ostasiatischer Bevölkerungen vor 0,5 Mill. J.: 15 Jahre, „Anteneandertaler": $15^1/_2$ Jahre; heutiger *Homo sapiens*: zwischen $15^1/_2$ und 18 Jahren. Seit der Abspaltung der *Homo*-Linie von der *Australopithecus*-Linie hat sich die im weitgehend entspannten Feld der Jugendzeit ruhende Lernzeit also nahezu verdoppelt, seit der Abspaltung von der Linie zu den afrikanischen Großaffen mehr als verdoppelt. Die als damit einhergehend zu erwartende ganz beträchtliche Ausweitung und Anreicherung erlernter komplexer Verhaltensprogramme muß zwangsläufig erhebliche Auswirkungen auf die Stellung einer jeweiligen Menschenform in der interspezifischen Hierarchie und Konkurrenzfähigkeit mit sich gebracht haben. Bereits bei *Australopithecus afarensis* ist an eine den afrikanischen Menschenaffen gegenüber wohl fortgeschrittene soziale relative Koordination zu denken, wie sie aus der Beobachtung des vollen Gleichschritts zweier eng nebeneinander hergehenden, körpergrößenmäßig sehr unterschiedlichen Individuen auf der Spurenplatte von Laetolil (s. z. B. Photo in WHITE u. a. 1983) eindringlich nahegelegt wird. Mitten in die aufstrebende Hirnevolution der *Homo*-Linie fällt der Übergang des Menschen zur Großsäugetierjagd. Die ersten konkreten Hinweise auf Großsäuger-Ausschlachtung datieren aus der Zeit vor etwa 2 Mill. J. (BLUMENBERG 1979), ohne daß hier eine Entscheidung darüber möglich wäre, wie die betreffenden Tiere (Elefant, Flußpferd) zu Tode kamen. Dies ist die Zeit eines Hirngrößenwertes um 50. Klare Hinweise auf Großwildjagd durch den Menschen finden sich erst weit später, im Mittelpleistozän (BLUMENBERG l. c.), assoziiert mit Menschen höherer Cephalisation. Dieses Übergangsfeld menschlichen Verhaltens zur Groß-Carnivorie läßt wesentliche, bislang aber kaum diskutierte, noch überhaupt beachtete Verschiebungen in den assoziierten Faunen großer Carnivoren erwarten. Welche generellen Konsequenzen beim Übergang von zunächst Unterordnung in der interspezifischen Carnivoren-Hierarchie unter eine oder mehrere dominante Carnivoren-Arten zur Co-Dominanz und schließlich zur Dominanz zu erwarten sind, führte EATON (1979) aus. Es soll nicht Aufgabe dieser Arbeit sein, aus den bekannten pliozänen, unter- und mittelpleistozänen Carnivorenfaunen mit Anwesenheit von *Australopithecus*- bzw. *Homo*-Formen unterschiedlicher Hirnevolutionshöhe korrelative Beziehungen zu extrahieren. Immerhin erscheint in diesem Zusammenhang bereits jetzt bemerkenswert, daß große Säbelzahnkatzenarten in verschiedenen Bereichen der Holarktis solange regelmäßig oder sogar vergleichsweise häufig im Fundgut auftreten, auch bei gleichzeitiger Existenz anderer Großcarnivoren, wie Arten des

Genus *Panthera*, solange *Homo sapiens* fehlt (in Mitteleuropa also bis zum unteren Mittelpleistozän, in Nordamerika bis zum Jungpleistozän). Ähnliches scheint für die Großgeparden (*Acinonyx pardinensis*) zu gelten. Zukünftige Diskussion dieser Problematik sollte weiterführende Anregungen zum Verständnis pleistozäner Großsäugerfaunen bringen.

Zusammenfassung

Die sich im Verlauf seiner Evolution progressiv ändernde Stellung des Menschen in der interspezifischen Hierarchie pliozän-pleistozäner Großsäugerfaunen und der sich damit stetig wandelnde menschliche Einfluß auf seine Umwelten ist in unmittelbarem Zusammenhang mit seiner intellektuellen Höherentwicklung zu sehen, die ihrerseits prinzipiell aus seiner Hirngrößenevolution abzuschätzen ist. Zu deren vergleichender Beurteilung im Evolutionsabschnitt *Australopithecus—Homo* und im Bezug auf rezente Menschenaffen dürfte allein der Versuch einer Erfassung der Cephalisationswerte aus Hirngewichts-Körpergewichts-Schätzungen zum Erfolg führen können. Um dies zu erreichen, müssen Hirnraumvolumina in Hirngewichte umgerechnet und Körpergewichte aus Regressionen postkranialer Skelettelemente erhalten werden. Aus der Auswertung entsprechender Literaturdaten wird demgemäß die Cephalisationskonstante für verschiedene *Australopithecus*-Formen eingekreist und in der Gattung *Homo* für das *habilis*-Niveau, frühe und späte *erectus*-Stufen und für die europäischen „Anteneandertaler" erörtert. Logarithmische Verrechnung solchermaßen erhaltener Cephalisationsstufen im Zeitbezug erbringt für die Gattung *Homo* von *habilis* bis zum Jetztmenschen nahezu absolute Korrelation, die den Versuch einer Extrapolation über den *habilis*-Evolutionsabschnitt hinaus rechtfertigt. Mit diesem Verfahren zeigt es sich, daß die Abzweigung des Genus *Homo* aus der *Australopithecus*-Stufe mit einem Übergang von auf letzterer nahezu evolutiver Stasis der Hirnevolution zu hoher evolutiver Dynamik verbunden war und nicht später als auf dem *Australopithecus-afarensis*-Niveau erfolgt sein sollte. Als Zeit dieser Aufspaltung ist der Abschnitt zwischen 2,8 und 4 Millionen Jahren diskutabel. Als nach den Kriterien der Hirngrößenevolution spätestmögliche Abzweigungszeit der afrikanischen Menschenaffen *(Pan, Gorilla)* von *Australopithecus-Homo* ist die Endzeit des Miozäns vor etwa 6 Millionen Jahren anzusetzen. Dieser Befund ist ebenso gut mit den molekularen Daten zur Aufspaltung der Hauptlinien der Hominiden in Einklang zu bringen, wie die aus den Hirnwerten zu erschließende erste Trennung zu heutzeitlichen menschlichen Populationen bereits mit dem Beginn des oberen Mittelpleistozäns mit entsprechenden biochemischen Werten übereinstimmt. Abschließend werden die aus der Hirngrößenkorrelation abschätzbaren Geschlechtsreifealter für die Angehörigen aufeinander folgender Abschnitte der Vor- und Frühmenschen-Evolution zusammengestellt und eine Zuordnung zentraler Änderungen des Ernährungsverhaltens zur Hirnevolution versucht.

Summary

Man's progressively changing position in the interspecific hierarchy of Pliocene and Pleistocene mammalian faunas and hence his changing influence on his environments may be understood as directly correlated with the progressive increase of his intellectual abilities, which in their turn depend on his progressive brain size evolution. It is only the attempt to calculate cephalization values by brain and body weight estimates that provides successful comparative evaluation of this process within the evolutionary sequence *Australopithecus — Homo*, compared with the modern African apes. Therefore values for the cephalization constant are calculated for different *Australopithecus* geographic and time populations and discussed for the *habilis* level, early and late *erectus* levels and the European "Anteneandertal" populations within the genus *Homo*, in all cases relying on published body weight calculations and being based on the conversion of braincase volumes into brain weight estimates. Logarithmic transformations of the resulting cephalization values show a nearly absolute correlation with absolute dating within the genus *Homo* for the span from *habilis* to modern *sapiens* populations. This justifies an attempt to extrapolate the brain evolutionary curve beyond the *habilis* stage. This procedure reveals that the branching of the genus *Homo* from its Australopithecine stem has been accompanied by a transition from near stasis of brain evolution to highly dynamic change. It also indicates that this phylogenetic event took place not later than at the *Australopithecus afarensis* level, and dates it plausibly within the time span between 2.8 and 4.0 mill. yrs. The criteria of brain evolution point to the dichotomy of the African ape branch and the *Australopithecus/Homo* line not later than at the end of the Miocene. This agrees well with the latest evaluation of molecular data on hominid ramification. The result that the first radiation of extant *sapiens* populations took place at the early Upper Middle Pleistocene also agrees with biochemical values. Finally the age of female sexual maturity as calculable by the cephalization constant correlation is listed for different successive stages in the course of human brain evolution, and crucial changes in elimentary behaviour are tentatively linked with the brain progression.

Резюме

Можно предполагать, что прогрессировавшее изменение положения человека в межвидовой иерархии плиоценовых и плейстоценовых млекопитающих и, следовательно, изменение его воздействия на окружающую среду было непосредственно связано с прогрессировавшим ростом его мыслительных способностей, которые, в свою очередь, основывались на прогрессировавшей эволюции размера мозга. Успешная сравнительная оценка этого процесса в рамках эволюционной последовательности *Australopithecus-Homo* по сравнению с современными африканскими человекообразными обезьянами возможна только путем попытки вычислить показатели цефализации по расчетным значениям веса мозга и тела. Поэтому для различных географических и хронологических серий *Australopithecus* рассчитаны постоянные показатели цефализации, основанные на опубликованных расчетах веса тела и переводе данных по объему мозговой полости в оценки веса мозга. Эти же показатели рассмотрены для уровня *habilis*, раннего и позднего уровней *erectus* и для европейских „донеандерталоидных" популяций рода *Homo*. Представление полученных показателей в логарифмической форме обнаруживает почти полную корреляцию с абсолютными датировками для рода *Homo* в пределах промежутка времени от *habilis* до современных сапиентных популяций. Это дает основание попытаться экстраполировать кривую эволюции мозга за пределы стадии *habilis*. Применение данного метода показало, что ответление рода *Homo* от его австралопитецинного корня сопровождалось переходом от почти полного эволюционного застоя в развитии мозга к высокой эволюционной динамике. Оно указывает также, что это филогенетическое событие произошло не позднее, чем на уровне *Australopithecus afarensis*, и датирует его, вероятно, хронологическим промежутком между 2,8 и 4 млн. лет. Критерии эволюции мозга свидетельствуют о том, что разделение ветви африканских человекообразных обезьян и линии *Australopithecus/Homo* произошло не позднее, чем в конце миоцена. Это вполне сопоставимо с новейшими оценками дивергенции гоминид по молекулярным данным. В такой же степени результаты первой радиации, приведшей в начале поздней части среднего плейстоцена к сапиентным популяциям современного типа, сопоставимы с оценками по биохимическим данным. Наконец, для различных последовательных стадий в ходе эволюции человеческого мозга приводится возраст достижения женщинами половой зрелости, вычисленный по корреляции константы цефализации, и делается попытка связать решающие изменения в пищевом поведении с прогрессивным развитием мозга.

Literatur

BAUCHOT, R.; STEPHAN, H., 1966: Données nouvelles sur l'encéphalisation des insectivores et des prosimiens. — In: Mammalia. — **30**. — S. 160—196.

BAUCHOT, R.; STEPHAN, H., 1969: Encéphalisation et niveau evolutif chez les simiens. — In: Mammalia. — **33**. — S. 235—275.

BLUMENBERG, B., 1979: The origin of hominid megafaunal carnivory. — In: Carnivore. — **2** (3). — S. 71—73.

BOAZ, N. T., 1983: Morphological trends and phylogenetic relationships from Middle Miocene hominoids to late Pliocene hominids. — In: New interpretations of ape and human ancestry/ed. by R. L. CIOCHON; R. S. CORRUCCINI. — New York: Plenum Press. — S. 705—720.

BORDES, F. (Ed.), 1972: The origin of *Homo sapiens*. — Paris: UNESCO.

CRONIN, J. E., 1983: Apes, humans, and molecular clocks. — In: New interpretations of ape and human ancestry/ed. by R. L. CIOCHON; R. S. CORRUCCINI. — New York: Plenum Press. — S. 115—150.

CRONIN, J. E.; BOAZ, N. T.; SPRINGER, C. B.; RAK, Y., 1981: Tempo and mode in hominid evolution. — In: Nature. — London **292**. — S. 113—122.

DUBOIS, E., 1897: Sur le rapport du poids de l'encéphale avec la grandeur du corps chez les a mammifères. — In: Bull. Soc. Anthrop. — Paris **8**. — S. 337—376.

EATON, R. L.; HUTCHINS, M., 1979: Socio-ecology and diet of carnivores: a model for hominid evolution. — In: Carnivore. — **2** (3). — S. 68—70.

EISENBERG, J. F.; WILSON, D. E., 1979: Relative brain size and feeding strategies in the Chiroptera. — In: Evolution. — Lancaster **32**. — S. 740—751.

GUERASSIMOV, M. M., 1972: Le reconstitution documentée de l'homme fossile et la formation de l'homme moderne. — In: The origin of *Homo sapiens*/ed. by F. BORDES. — Paris: UNESCO. — S. 287—290.

HARVEY, P. H.; BENNETT, P. M., 1983: Brain size, energetics, ecology, and life history patterns. — In: Nature. — London **306**. — S. 314—315.

HEMMER, H., 1967: Allometrie-Untersuchungen zur Evolution des menschlichen Schädels und seiner Rassentypen. — Stuttgart. — G. Fischer Verl.

—, 1971: Beitrag zur Erfassung der progressiven Cephalisation bei Primaten. — In: Proc. 3rd. internat. Congr. Primat./Zürich 1970, Bd. 1: — Basel; Verl. Karger. — S. 99—107.

—, 1972: The evolutionary significance, taxonomy and environmental aspects of the Upper Pleistocene Neanderthal and neanderthaloid men of Europe, Asia, and Africa. — In: The origin of *Homo sapiens*/ed. by F. BORDES. — Paris: UNESCO. — S. 65—71.

—, 1974: Progressive Cephalisation und Dauer der Jugendentwicklung bei Primaten nebst Bemerkungen zur Situation bei Vor- und Frühmenschen. — In: Bevölkerungsbiologie/Hrsg. W. BERNHARD; E. KANDLER. — Stuttgart: G. Fischer Verl. — S. 527—533.

—, 1978: Socialization by intelligence: Social behaviour in carnivores as a function of relative brain size and environment. — In: Carnivore. — **1**. — S. 102—105.

—, 1979: Beobachtungen zur Soziobiologie madagassischer Lemuren. — In: Z. Kölner Zoo. — Köln **22**. — S. 43—51.

—, 1981: Brain, sociobiology, and evolution in primates. — In: Primate behaviour and sociobiology/ed. by A. B. CHIARELLI; R. S. CORRUCCINI. — Berlin [West] u. a.: Springer-Verl. — S. 148—155.

—, 1982: Major factor in the evolution of hominid skull morphology, biological correlates, and the position of the neandertals. — In: Congr. internat. pal. hum., 1st Congr.: L'*Homo erectus* et la place de l'homme de Tauteval parmi les hominidés fossiles, Prétirage. T 1. — Nice. — S. 339—364.

HOLLOWAY, R. I., 1973: Endocranial volumes of early african hominids, and the role of the brain in human mosaic evolution. — In: J. human Evolution. — **2**. — S. 449—459.

—, 1978:

JERISON, H. J., 1973: Evolution of the brain and intelligence. — London, New York: Acad. Press.

LAPICQUE, M. L., 1898: Sur la relation du poids de l'encephale au poids du corps. — In: C. R. Séance Soc. Biol. — **5**. — S. 62—63.

MARTIN, R.; SALLER, K., 1959: Lehrbuch der Anthropologie. Bd. 2. — Stuttgart: G. Fischer Verl.

MCHENRY, H. M., 1975: Fossil hominid body weight and brain size. — In: Nature. — London **254**. — S. 686—688.

—, 1976: Early hominid body weight and encephalization. — In: Amer. J. phys. Anthrop. — **45**. — S. 77—84.

PASSINGHAM, R. E., 1975: Changes in the size and organization of the brain in man and ancestors. — In: Brain Behav. Evolution. — **11**. — S. 73—90.

PILBEAM, D.; GOULD, S. J., 1974: Size and scaling in human evolution. — In: Science. — Washington **186**. — S. 892—901.

RIGHTMIRE, G. P., 1980: In: Current arguments on early man/ed. by L.-K. KÖNIGSSON. — Oxford: Pergamon Press. — S. 70—85.

RÖHRS, M.; EBINGER, P., 1983: Noch einmal: Wölfe mit unterschiedlichen Cephalisationsstufen? — In: Z. zool. Syst. u. Evolutionsforsch. — **21**. — S. 314—318.

ROTH, H., 1983: Comparison statistique de la forme des arcades alveolaire et dentaire des mandibules des hominidés fossiles. — Marseille, Univ. Provence, Thèse.

SENUT, B., 1979: Comparison des hominidés de Gomboré IB et de Kanapoi: Deux piéces du genre *Homo*. — In: Bull. Mém. Soc. Anthrop. — Paris **6**. — S. 111—117.

SNELL, O., 1892: Die Abhängigkeit des Hirngewichtes von dem Körpergewicht und den geistigen Fähigkeiten. — In: Arch. Psychiat., Nervenbr. — **23**. — S. 436—446.

STEPHAN, H.; BAUCHOT, R., 1965: Hirn-Körpergewichtsbeziehungen bei den Halbaffen (Prosimii). — In: Acta Zool. — **46**. — S. 209—231.

STEPHAN, H.; DIETERLEN, 1982: Relative brain size in Muridae with special reference to *Colomys goslingi*. — In: Z. Säugetierkd. — Hamburg **47**. — S. 38—47.

STEPHAN, H.; KUHN, H.-J., 1982: The brain of *Micropotamogale lamottei* HEIM DE BALSAC, 1954. — In: Z. Säugetierkd. — Hamburg **47**. — S. 129—142.

STEPHAN, H.; NELSON, J. E.; FRAHM, H. D., 1981: Brain size comparison in Chiroptera. — Z. zool. Syst. u. Evolutionforsch. — **19**. — S. 195—222.

WHITE, T. D.; JOHANSON, D. C.; KIMBEL, W. H., 1983: *Australopithecus africanus*. Its phylogenetic position reconsidered. — In: New interpretations of ape and human ancestry/ed. by R. L. CIOCHON; R. S. CORRUCCINI. — New York: Plenum Press. — S. 721—780.

ZIHLMAN, A. L.; LOWENSTEIN, J. M., 1983: *Ramapithecus* and *Pan paniscus*. Significance for human origins. — In: New interpretations of ape and human origins./ed. by R. L. CIOCHON; R. S. CORRUCCINI. — New York: Plenum Press. — S. 677—694.

Quartärpaläontologie 6 83—86 Berlin 1986

Biostratigraphical Correlations (Mammals) of Quaternary Continental Deposits of Europe and the Far East[1])

By HANS DIETRICH KAHLKE, Weimar

Contents

Abstractum

On the basis of biostratigraphical data (large mammals) attempts are made to correlate the Quaternary faunistic sequence of Europe and the Far East.

Auf der Grundlage biostratigraphischer Daten (Großsäuger) werden Versuche diskutiert, die quartären Faunenfolgen Europas und die des Fernen Osten zu korrelieren.

На основе биостратиграфических данных по крупным млекопитающим делается попытка корреляции последовательности четвертичных фаун Европы и Дальнего Востока.

1. Palaeogeographical and Palaeozoological Background

In the Late Neogene and Quaternary the palaeofaunistic picture of the Eurasian continent was highly influenced by two landbridges: a composite one in the far west, connecting Africa with Europe and Asia perhaps along a western (Afro-Iberian), a central (Afro-Sicilian), and an East Mediterranean (Afro-Arabian) landbridge, and another in the far north-east, Beringia, connecting Siberia with the Nearctic.

Already in Late Neogene times there was a world-wide interchange of faunistic elements between Africa, Europe, Asia, and North America along these highways distributing biostratigraphic marker-taxa in both directions. With the progress of world-wide climatic deterioration in the Pleistocene, leading to the glaciation of wide areas in the Holarctic, this mainly west-east or east-west directed faunistic exchange was finally overlain by a fluctuating north-south migration of faunistic elements in glacial times, or a south-north migration in interglacial times.

Only against this palaeogeographic-palaeoclimatic background may we understand the different ecological character of Pleistocene association within the same area — in the far west as well as in the far east — and finally the present zoogeographical state of Eurasia.

2. Biostratigraphical Sequences of Terrestrial Deposits of Europe and Continental East Asia

In Europe, in general, we have to distinguish in the Quaternary two different faunistic complexes or facies: an East European faunistic complex of Asian affinities, and a West European faunistic complex of Mediterranean/African affinities (KAHLKE 1961; RUTTE 1967). In the Pleistocene the boundary between both complexes was marked by the easternmost distribution of the genus *Hippopotamus*, a typical Mediterranean/African element.

The Early Pleistocene (Neogene/Quaternary boundary; XIth INQUA Congress, Moscow 1982, and 27th International Geological Congress, Moscow 1984, INQUA- and IGCP-Working Groups: Locus typicus: Vrica, Calabria, Italy; Stratum typicum: basal plane of the claystone overlying bed "e", $\approx$1.6 Mill. yrs.) fauna of Europe is characterized by Late Villafranchian associations of Olivola faunal unit type (AZZAROLI 1983) following the Saint Vallier faunal unit type of uppermost Pliocene age according to the recently adopted N/Q boundary.

In the Far East, continental China, Indochina, and South-East Asia as far as Java, we meet with a similar picture. In general we have to distinguish here two faunistic complexes as well: a North-Tsinling complex of Siberian affinities north of the Tsinling range, and a South-Tsinling complex of Malayan affinities south of the Tsinling range (PEI 1957; KAHLKE 1968) extending southwards as far as Java. In palaeofaunistic respects the South-Tsinling faunistic complex was characterized by the large Pongidae and other taxa of Late Neogene Indo-Malayan affinities (*Gigantopithecus*, *Pongo*).

In continental China, in the Pliocene/Pleistocene intramontane basins — Yangyuan (Nihewan) basin, Yuanmou basin — the Lower Pleistocene continental deposits are characterized by fossil associations of late "Yellow Nihewan" type. According to palaeomagnetic research (CHENG et al.

[1]) Paper presented at the 27th International Geological Congress, Moscow, USSR, 1984, Section C 03.1.1., Quaternary Chronostratigraphy of Eurasia.

1978; LIU 1982) the boundary between the Lower and the Upper ("Yellow") Nihewan is more or less coincident with the Matuyama/Gauss boundary (≈2.5 Mill. yrs.). The fauna of the Lower Yellow Nihewan (TEILHARD DE CHARDIN and PIVETEAU 1930) is correlated with the Saint Vallier faunal unit, now uppermost Pliocene according to the recently adopted N/Q boundary. In southern China the Lower Pleistocene is represented by deposits (cave deposits) with Leng-chai-shan faunal unit associations.

Up to the present day there has been no general agreement about where to place the boundary between the Lower and the Middle Pleistocene. Moreover in accordance with geological principles this boundary has to be determined in a marine environment. Against a palaeoclimatic background we can see this boundary as coincident with the progressive climatic deteriorations within the Pleistocene ("Glacial Pleistocene"). In the area of the unstable Palaearctic ice shields we have an important turning-point in the glacial sequence with the beginning of the Anglian/Elsterian/Mindelian/Oka/Dagu period, causing fundamental changes in environment and fauna in vast areas of the Holarctic.

In the far west, in Europe, we have to record, that with the beginning of the so called "Glacial Pleistocene" — Anglian/Elsterian/Mindelian/Oka — large mammals of North Asiatic/Beringian affinities: *Praeovibos*, *Ovibos*, and *Rangifer*, elements of steppe and cold steppe environment, were enabled for the first time to reach Central Europe and finally Western Europe as shown by the fossil faunae of the well known localities of Süssenborn, Bad Frankenhausen, Mosbach, and Tautavel, to name only a few localities. At the same time faunistic elements of "praeglacial" European character and of Mediterranean/African affinities (*Hippopotamus*) were retreating to the south.

A similar picture is met with in the Far East, continental China, Dagu period. At this time the distribution of southern elements of Late Neogene Indo-Malayan affinities retreated more and more to the south, indicating the very end of the Late Neogene Indo-Malayan faunistic predominance in continental South-East Asia, and foreshadowing the extinction of the large Pongidae *Gigantopithecus* and *Pongo* on the continent in the Pleistocene and Holocene.

At the same time faunistic elements of northern affinities: Sino-Siberian faunistic complex (North-Tsinling complex) were extending far to the south, intermingling with elements of the Sino-Malayan complex (South-Tsinling complex) in a relatively southern location as is shown by the recently discovered Long-tan cave fauna (HUANG et al. 1982) with *Sinomegaceros* and *Bison* of northern, *Ailuropoda*, *Stegodon*, *Tapirus*, and *Megatapirus* of southern affinities. A similar picture is found at Tautavel, South-West France, where elements of northern character (*Praeovibos*, *Rangifer*) are mixed in a southern geographical location with a local fauna of West European/Mediterranean character (CRÉGUT 1979; KAHLKE 1981).

The following interglacial period of Europe, the Holsteinian, is characterized in biostratigraphical respects by the retreat of continental/northern faunistic elements and a corresponding immigration of faunistic elements of southern/Mediterranean affinities: the early *Palaeoloxodon antiquus/Dicerorhinus kirchbergensis* associations, enriched by *Bubalus* of South Asiatic affinities, extending as far as Central Europe: Schönebeck (SCHERTZ 1937), Steinheim an der Murr (BERCKHEMER 1927), and perhaps the northern part of the Rhine valley-graben (FRANZEN and VON KOENIGSWALD 1979).

Again we find the same picture in the Far East, continental China, where faunistic elements of southern origin, the Sino-Malayan faunistic complex, including *Palaeoloxodon*, the merckoid *Dicerorhinus choukoutiensis* as well as *Bubalus* reached in this interglacial the area around Beijing as recorded in the famous Zhoukoudian deposits, *Homo erectus pekinensis* too is a southern element in a northerly geographical position at this time.

Up to this geological horizon the biostratigraphical correlation of the Pleistocene sequences seems to be fairly well recorded in both areas under discussion. After this time, however, a problematical gap follows in the continental Far Eastern fossil record.

In Europe, the following glacial period, the Saalian/Rissian/Dnepr-Moscow glaciation, is characterized by a new invasion of continental- northern faunistic associations, the early *Mammuthus primigenius/Coelodonta antiquitatis* fauna, extending finally over nearly all Europe.

Knowing of the Late Pleistocene (Tali glacial) invasion by Siberian faunistic elements into Central China, we have to expect a similar invasion in the Chinese equivalent of the far western Saalian (Lushan) as in the Tali glaciation. But this faunistic horizon is not known, perhaps not identified in the Far East. In the most recent papers on this question (ZHOU and GUO 1980; ZHOU 1982) the fossil association for example of Tingtsun (PEI et al. 1958) has been quoted to be correlated with the Lushan or western Saalian/Rissian glacial period. But this correlation seems to be quite impossible because we have here — Tingtsun — a mixed fossil association including southern ("interglacial") elements like Indian elephant, *Dicerorhinus* sp., and *Bubalus* (WOO 1980) in a relatively northern geographical position. This association is — in western terms — an association of the Eemian, perhaps Late Eemian, last Pleistocene interglacial period.

The only possible explanation for this problem, the apparent lack of a faunistic complex in the continental Far East (North and Central China) correlated in ecological respects to the far western Saalian/Rissian glacial period is, that it was not possible to distinguish within the Pleistocene *Mammuthus primigenius/Coelodonta antiquitatis* faunistic complex between an earlier one — in western terms a Saalian/Rissian — and a later one — in western terms a Weichselian/Würmian — association. In Europe, in the extra-glacial area it is sometimes a problem too, especially if the fossil record is poor and no other stratigraphical data are available.

In Europe, the next interglacial period, the Eemian, is characterized by the late *Palaeoloxodon antiquus/Dicerorhinus kirchbergensis*-associations as known from different classical localities of Europe.

In China there are also Late Pleistocene faunistic associations of large mammals — for example the already mentioned association of Tingtsun, Shanxi — with southern types in a relatively northern geographical position. According to the present state of knowledge no other interpretation is possible than to see in these associations fossil faunae of the last interglacial period of China, when southern elements again reached a more northerly geographical distribution.

As in Europe, the continental deposits of the last glacial period in China (Tali glaciation) were characterized by a more or less uniform faunistic block — differentiated by endemic elements in both areas only — extending from Western Europe as far as Southern Central China: the Late Pleistocene *Mammuthus primigenius/Coelodonta antiquitatis*-associations known from numerous fossil localities of this vast region.

To conclude we may say, that an intensive study of the ecological and evolutionary character of Quaternary fossil associations will provide more and more details leading towards a synthesis of biostratigraphical correlations of the continental deposits of bothEurope and the Far East, proving or disproving the hypothesis of glacial events and corresponding faunal movements, occurring simulteneously in the Palaearctic.

Summary

Biostratigraphical correlations of Quaternary continental deposits of Europe and the Far East have a long history. Tracing the question of European/Far Eastern Quaternary correlations by means of biostratigraphy we may point out, that during the Pleistocene in Europe as well as in the Far East a north-south, and a south-north fluctuation of faunistic elements is known, documented by alternating mixed fossil associations in the contact zones s.l. of both areas. The changing direction of these migrations is unquestionably the result of changing climatic conditions. Other factors are to be excluded. On the basis of these faunistic migrations recorded by mixed fossil associations of southern affinities in northerly geographical positions, or of northern affinities in southerly geographical positions respectively, and in connection with the evolutionary status of the different fossil associations, it seems possible to add further details to the study of European-Far Eastern biostratigraphical correlations.

Zusammenfassung

Biostratigraphische Korrelationen quartärer Kontinentalablagerungen zwischen Europa und dem Fernen Osten haben eine lange Geschichte. Verfolgt man die Frage solcher Korrelationsversuche so kann man aufzeigen, daß während des Pleistozäns in Europa wie auch im Fernen Osten eine Nord-Süd- und eine Süd-Nord-Fluktuation von Faunenelementen bekannt ist, die sich durch einander abwechselnde Mischfaunen in den Kontaktzonen s. l. beider Räume belegen lassen. Die jeweils wechselnde Richtung dieser Faunenbewegungen ist zweifellos das Ergebnis wechselnder klimatischer Bedingungen; andere Faktoren scheiden offensichtlich aus.

Auf der Basis dieser Faunenbewegungen, die durch Mischfaunen südlicher Affinität in einer relativ weit nördlichen geographischen Position oder aber nördlicher Affinität in einer relativ weit südlichen geographischen Position angetroffen werden, und in Verbindung mit dem evolutionären Status der verschiedenen fossilen Assoziationen erscheint es möglich, weitere Details zur Frage biostratigraphischer Korrelationen zwischen Europa und dem Fernen Osten aufzuzeigen.

Резюме

Попытки биостратиграфических сопоставлений четвертичных континентальных отложений Европы и Дальнего Востока предпринимались давно. Анализируя проблему корреляции четвертичных отложений Европы и Дальнего Востока методами биостратиграфии, можно подчеркнуть, что как в Европе, так и на Дальнем Востоке на протяжении плейстоцена известны смены северных фаунистических элементов на южные и южных на северные. Эти смены документированы чередующимися смешанными фаунистическими ассоциациями, существовавшими в контактных (в широком смысле) зонах обеих областей. Изменение направлений этих миграций несомненно является результатом изменений климатических условий. Прочие факторы следует исключить. На основе этих фаунистических миграций, устанавливаемых по смешанным ассоциациям ископаемых форм южной принадлежности, найденных в географически более северных местонахождениях, или северной принадлежности, найденных в географически более южных местонахождениях, а также на основе эволюционного положения различных ассоциаций ископаемых форм, представляется возможным детализировать биостратиграфические корреляции между Европой и Дальним Востоком.

References

AZZAROLI, A., 1983: Quaternary mammals and the "End-Villafranchian" dispersal event — a turning point in the history of Eurasia. — In: Palaeogeogr., Palaeoclimat., Palaeooecol. — Amsterdam **44** — P. 117—139.

BERCKHEMER, F., 1927: *Buffelus murrensis* n. sp. Ein diluvialer Büffelschädel von Steinheim a. d. Murr. — In: Jahresh. Ver. vaterländ. Naturkd. Württ. — Stuttgart **83**. — P. 146—158.

CRÉGUT, E., 1979: La faune de mammifères du Pleistocène moyen de la Caune de l'Arago á Tautavel, Pyrénées orientales. — Marseille, Univ. Provence, These, 381 pp.

FRANZEN, J. L.; KOENIGSWALD, W. v., 1979: Erste Funde von Wasserbüffel *(Bubalus murrensis)* aus pleistozänen Schottern des nördlichen Oberrhein-Grabens. — In: Senckenberg. leth. — Frankfurt (Main) **60** (1/3). — P. 253—263.

HUANG, W. B., 1979: On the age of the cave-fauna of South China — In: Vertebrata Palasiatica. — Beijing **17** (4). — P. 327—343. — In Chinese, english summary.

HUANG, W. B.; et al., 1982: Preliminary study of the fossil hominid skull and fauna of Hexian, Anhui. — In: Vertebrata Palasiatica. — Beijing **20** (3). — P. 248—256. — In Chinese, english summary.

KAHLKE, H. D., 1961: On the Complex of the *Stegodon-Ailuropoda* Fauna of South China and on the chronological position of *Gigantopithecus blacki* v. KOENIGSWALD. — In: Vertebrata Palasiatica. — Peking **2**. — P. 83—108. — In Chinese, english summary.

—, 1968: Zur relativen Chronologie ostasiatischer Mittelpleistozän-Faunen und Hominoidea-Funde. — In: Evolution und Hominisation/Hrsg. G. KURTH. — Stuttgart. G. Fischer Verl. — P. 91—118.

—, 1981: Die biostratigraphische Stellung der fossilen Faunen der Caune de l'Arago in Tautavel. — In: Colloque internat. C.N.R.S., Prétirage. — Paris. — P. 245—251.

LI, H. M.; WANG, J. D., 1982: Magnetostratigraphic study of several typical geological sections in North China. — In: Quaternary geology and environment of China/ed. by T. S. LIU. — Beijing. — P. 33—38.

LIU, T. S., 1982: Pleistocene stratigraphy and Plio/Pleistocene boundary in China. — In: Quaternary geology and environment in China/ed. by T. S. LIU. — Beijing. — P. 1—6.

PEI, W. C., 1957: The zoogeographical divisions of Quaternary mammalian faunas in China. — In: Vertebrata Palasiatica. — Peking **1** (1). — P. 9—24.

PEI, W. C.; et al., 1958: Report of the excavation of the palaeolithic site at Tingtsun, Hsaihfen Hsien, Shansi. — Peking. — 111 pp.

RUTTE, E., 1967: Die Cromer-Wirbeltierfundstelle Würzburg-Schalksberg. — In: Abh. naturwiss. Ver. Würzburg. — Würzburg **8**. — P. 1—26.

SCHERTZ, E., 1937: Ein neuer Wasserbüffel aus dem Diluvium Mitteldeutschlands (*Buffelus wanckeli* nov spec.). — In: Palaeont. Z. — Berlin **19**. — P. 57—72.

SONG, Z. C.; et al., 1982: The lower boundary of the continental Quaternary in some areas of China based on palynological data. — In: Quaternary geology and environment of China/ed. by T. S. LIU. — Beijing. — P. 7—12.

TEILHARD DE CHARDIN, P.; PIVETEAU, J., 1930: Les mammifères de Nihowan (Chine). — In: Ann. Paléont. — Paris **19**. — 134 pp.

Wang, N. W.; He, X. X., 1982: Biostratigraphical succession of the marine and terrestrial Quaternary in North China. — In: Papers Chinese geologists submitted to the XI INQUA Congress. — Beijing. — Vol. 9, p. 1—8.

Woo, J. K., 1980: Palaeanthropology in the New China. — In: Current argument on early man/ed. by L. K. Königsson. — Oxford etc. — P. 182—206.

Zhou, M. L., 1982: A review of studies on the Quaternary glacial and interglacial sequences in China. — In: Quaternary geology and environment in China/ed. by T. S. Liu. — Beijing. — P. 20.

Zhou, M. L.; Guo, Y. Z., 1980: The Quaternary of China and that of the Netherlands. — In: Bull. Chin. Acad. Geol. Sci. — Beijing **6** (1, 2). — P. 96—105. — In Chinese, english summary.

Quartärpaläontologie 6 87—92 Berlin 1986

Zwei neue Stangenfragmente der verticornis-Gruppe (Mammalia, Cervidae) aus dem Pleistozän des Thüringer Raums (DDR)

Von RALF-DIETRICH KAHLKE, Weimar

Mit 3 Abbildungen und einer Tafel

Inhalt

Abstractum

In der vorliegenden Arbeit werden zwei neue Stangenfragmente aus der Gruppe der Steppenhirsche (*verticornis*-Gruppe) von Kalbsrieth (Kreis Artern) und von Kutzleben (Kreis Bad Langensalza) beschrieben. Die stratigraphische Stellung der Funde wird diskutiert.

In this paper two new antler-fragments of steppe-deer (*verticornis*-group) are described from Kalbsrieth (Kreis Artern) and from Kutzleben (Kreis Langensalza). The stratigraphical position of the fossils is discussed.

В данной статье описываются два новых фрагмента рогов оленей, отнесенных к группе *verticornis* (степные олени). Они происходят из Кальбсрита (район Артерн) и из Кутцлебена (район Лангензальца). Рассматривается стратиграфическое положение этих ископаемых остатков.

1. Geologisch-stratigraphischer Rahmen und Fundumstände

1.1. Kalbsrieth (Kreis Artern)

Unmittelbar südöstlich des Dorfes Kalbsrieth (Kreis Artern), am Rote-Berg, finden sich als Nordwestteil des Bottendorfer Höhenzugs in maximal 8 m aufgeschlossener Mächtigkeit (Herbst 1982) pleistozäne Unstrut-Kiese, die gegenwärtig durch den VEB Meliorationsgenossenschaft Kalbsrieth abgebaut werden. Der von elsterzeitlichem Geschiebemergel überdeckte Kieskörper wurde in einer Subrosionssenke akkumuliert, so daß Gesamtmächtigkeiten bis zu 40 m erbohrt werden konnten (persönliche Mitteilung von Dipl.-Geol. H. GESANG, Erfurt). Der Kies lagert in einer Höhe von +100—140 m NN.

Bereits DAMES (1882, S. 15) kannte die Kiese von Kalbsrieth. Er kennzeichnete sie als dem unteren Diluvium zugehörig und beschrieb ihre petrographische Zusammensetzung. Neben einheimischen Komponenten aus Thüringen und dem Harz nannte er vereinzelt auftretende nordische Geschiebe. Bei letzteren dürfte es sich allerdings um aus dem Hangenden sekundär eingetragenes Material handeln. WÜST (1900, S. 149) betonte ausdrücklich das Fehlen einer nordischen Komponente und ordnete die Kiese einer Kaltzeit vor der ersten Vereisung der zweiten Eiszeit, also dem Frühelster zu. Zur Zusammensetzung der präglazialen Kiese im Helme- und Unstrutgebiet äußerte sich weiterhin JANKOWSKI (1961, S. 54—55). Quantitativ-petrographische Analysen von Sedimenten des Kalbsrieth-Roßlebener Schotterzugs (WÜST 1900, S. 146ff.; 1901, Karte) führte schließlich STEINMÜLLER (1972a, S. 13, Abb. 5) durch.

Eine erste paläontologische Untersuchung legte ZEISSLER (1965, S. 278—279) mit einer vorläufigen Liste der Malako-Fauna vor, die später (ZEISSLER 1968) erweitert wurde sowie eine stratigraphische, paläoklimatische und palökologische Ausdeutung erfuhr. Der von ZEISSLER (1968, S. 402, Abb. 1) in der Grube von Kalbsrieth beprobte, 1 m mächtige Profilausschnitt wurde von ihr als warmzeitlich bezeichnet. Auf Grund des Auftretens von *Fagotia* (= *Melanopsis*) *acicularis* (FÉRUSSAC) konnte cromerisches Alter angegeben werden (ZEISSLER 1968, S. 407). Weitere Malako-Faunenlisten finden sich bei MANIA (1972, S. 20ff., Abb. 7; 1973, S. 84, Tab. 2), der die Faunengemeinschaft der unteren Schotter der Borntal-Warmzeit, die der oberen Schotter dagegen der Älteren Voigtstedter Warmzeit zuordnete (MANIA 1973, S. 18). Das bedeutet, letztere wären dem sogenannten Muschelton von Voigstedt (vgl. KRUTZSCH 1965, S. 241; STEINMÜLLER 1977, S. 5) synchron. Eine Vertebratenfauna war bisher von Kalbsrieth nicht bekannt. Umso erfreulicher ist es, daß im Herbst 1982 neben einigen anderen Fundstücken (vgl. Abschnitt 3, Altersstellung) ein Stangenrest, der dem *verticornis*-Kreis zuzuordnen ist, geborgen werden konnte.

Das Fundstück [IQW 1983/18688 (Kalbsr. 18206)] wurde aus dem südlichen Abschnitt der mit 0° streichenden Ostwand der Grube entnommen. Es fand sich 1 m über der Grubensohle unmittelbar oberhalb einer 0,2 m mächtigen, graugrünen Siltlage in schräggeschichteten, stark eisenschüssigen Grobkiesen, die bis zur Unterkante des hier durchschnittlich 2 m mächtigen hangenden Geschiebemergels in 7 m Höhe monoton ausgebildet sind. Der Grobkies des gesamten Aufschlußbereichs führt einen hohen Sandanteil, relativ häufig sind in den liegenden Partien Unionidenreste zu beobachten. Gelegentlich sind subrosiv bedingte Lagerungsstörungen erkennbar (vgl. LEHMANN 1922, S. 97).

Der Erhaltungszustand des Geweihrestes (Abwurfstange) erklärt sich aus seiner Lage im Sediment. Basale Bruchstücke von *verticornis*-Stangen finden sich in Kiesen weitaus häufiger als Schaufelpartien, weil in den meisten Fällen die Stange so eingelagert wird, daß die distale Geweihpartie vertikal steht und demzufolge in der Strömung abgestoßen wird (vgl. KAHLKE 1969, S. 571, Abb. 17).

1.2. Kutzleben (Kreis Bad Langensalza)

In der Thüringen-Sammlung des Instituts für Quartärpaläontologie Weimar befindet sich ein weiteres, bisher nicht beschriebenes Stangenfragment (schädelechte Stange) des *verticornis*-Kreises [IQW 1969/11263 (Kutzl. 12315)]. Es wurde aus den Sammlungsbeständen des Museums für Ur- und Frühgeschichte Thüringens zu Weimar übernommen. Außer der Beschriftung „Kutzleben" sind keine weiteren Angaben vorhanden. Aus dem Erhaltungszustand läßt sich vermuten, daß das Fossil sandigen oder kiesigen Ablagerungen entstammt.

Eine erste Erwähnung solcher Vorkommen zwischen Kutzleben (Kreis Bad Langensalza) und Lützensömmern findet sich bei SPEYER (1883, S. 11). WÜST erkannte erstmals das präglaziale Alter der von nordischem Gestein freien Kiese und ordnete sie einem „Tonna-Griefstedter Schotterzug" zu (WÜST 1900, S. 130; 1901, S. 5, Karte), dessen einheitliche petrographische Zusammensetzung KUGLER (1961, S. 187) bestätigen konnte. UNGER und ZIEGENHARDT (1961) zeigten den Deltacharakter der Kiese von Lützensömmern auf. Das Schottermaterial wurde vor der Elster-Maximalvereisung von SW in den aus dem Kindelbrücker Eisstausee hervorgegangenen Nausißer Eisstausee (Kannawurfer Halt) geschüttet (UNGER und ZIEGENHARDT 1961, S. 477). Eine Profilbeschreibung der Grube nordöstlich Lützensömmern gibt GESANG (1962, S. 220). Die Zweiteilung der Kiese (Älterer und Jüngerer Grobschotter) durch die „Lehmzone von Lützensömmern" zeigten später UNGER und GESANG (1969, S. 59) auf. Die durch Kartierungsbohrungen mehrfach durchteufte Zone ist nach Pollenbestand (ERD, in UNGER und GESANG 1969, S. 59) und Ostrakodenfauna (DIEBEL, in UNGER und GESANG 1969, S. 59) warmzeitlicher Entstehung. STEINMÜLLER (1972b, S. 160) korrelierte sie mit dem warmzeitlichen Teil der Lehmzone von Voigtstedt (vgl. RUSKE 1965), so daß Ältere bzw. Jüngere Grobschotter von Lützensömmern den Unteren bzw. Oberen Kiesen von Voigtstedt entsprechen. Nach neueren Untersuchungen vertritt STEINMÜLLER (1977, S. 13) die Ansicht, daß die unteren Kiese von Voigtstedt im Sinne von KRUTZSCH und REICHSTEIN (1956) bzw. KRUTZSCH (1965) in zwei genetisch zu trennende Folgen zu gliedern sind, was eine Korrelation mit dem Profil von Lützensömmern erschwert. STEINMÜLLER (vgl. UNGER 1974a, S. 759) ist weiterhin der Auffassung, daß das Profil von Lützensömmern in Süßenborner Unteren und Oberen Schottern ein Äquivalent findet. In Anlehnung an die Nomenklatur von WOLDSTEDT und DUPHORN (1974, S. 168) ordnete UNGER (1976, Tabelle) die Lehmzone von Lützensömmern der Rhume-Warmzeit zu, wodurch die Älteren Grobschotter in die Helme-Kaltzeit gestellt werden und die Jüngeren ein präglaziales Frühelster repräsentieren.

Für die stratigraphische Herkunft der *verticornis*-Stange kommen neben dem Jüngeren Grobschotter von Lützensömmern Sander- bzw. Sanderdelta-Bildungen bei Kutzleben in Frage (vgl. Profil Lützensömmern/SW bei UNGER 1974b, S. 793). Beide wurden in der Vergangenheit abgebaut. Bei den Sanderbildungen handelt es sich um Sedimente, die aus Richtung der Stirn des Elstereises, also aus N bzw. NE in die gleiche Geländedepression geschüttet wurden wie die Schotter. Ein nennenswerter Altersunterschied zwischen den Jüngeren Grobschottern und der Sanderbildung ist daher nicht anzunehmen (persönliche Mitteilung von Dr. K.-P. UNGER, Jena).

2. Beschreibung

2.1. Abwurfstangenbruchstück von Kalbsrieth
IQW 1983/18688 (Kalbsr. 18206)

Bei dem vorliegenden Fundstück (Abb. 1) handelt es sich um das Basisfragment eines rechten Abwurfs mittlerer Größe. Auch der Eissproß wurde vor der endgültigen Einbettung tief abgebrochen. Er entspringt der Stange etwa im rechten Winkel, ist also zunächst nach vorn gerichtet und biegt dann in charakteristischer Weise bis zum Abbruch nach unten. Während die Kompakta des vorderen und des unteren Be-

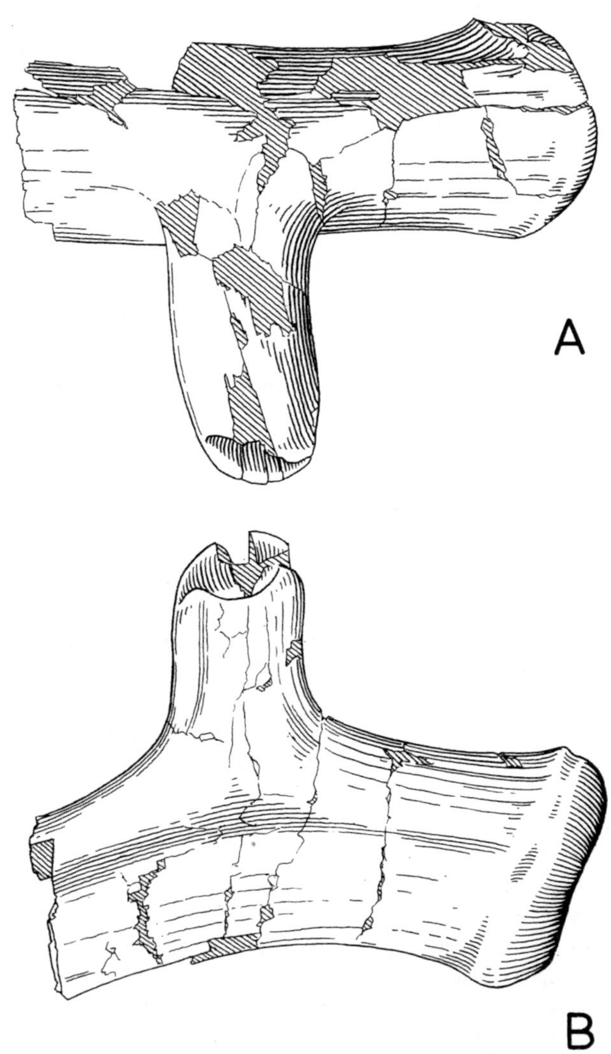

Abb. 1. Rechtes Abwurfstangenbruchstück eines Hirschs der *verticornis*-Gruppe; Kalbsrieth (Kreis Artern), IQW 1983/18688 (Kalbsr. 18206), etwa 1/2 nat. Größe
A — Ansicht von oben; B — Ansicht von vorn

reichs geschlossen erhalten ist, weisen der obere und hintere Bereich Lücken auf. Die Spongiosa im Inneren ist nur noch in Resten vorhanden. Eine insgesamt starke Abrollung der Stange zeigt sich besonders an der Rose, an der keine Perlung mehr erhalten ist, sowie an der Abbruchfläche des Eissprosses. Gefäßrillen sind lediglich im hinteren Bereich in Spuren zu erkennen.

Die Rose steht zur Stangenhauptachse in einem Winkel von etwa 75°. Ihre Basisfläche ist leicht konkav eingezogen. Zwischen Rose und Eissproß weist die Stange einen ovalen Querschnitt auf, der sich oberhalb des Sprosses verflacht. Auffällig ist das relativ hohe Ansetzen des Eissprosses über der Rose bei Fehlen eines Augsproßrudiments. Vergleichbare Stücke der *verticornis*-Populationen von Voigtstedt und Süßenborn setzen zumeist den Eissproß etwas tiefer an oder besitzen einen sogenannten „Augsproßkopf". Nur in wenigen Fällen liegen von den genannten Fundstellen ähnlich dimensionierte Stücke vor.

Maße [in mm]	
Rosenumfang	202
Stangenumfang zwischen Rose und Eissproß (Mitte)	170
Abstand des Eissproßansatzes (unten) von der Rose	62
Durchmesser des Eissprosses (basal, horizontal)	42
Länge des Eissprosses bis zum Abbruch	69
Abwurflänge (Rose bis Endbruchstelle)	123

2.2. Schädelechtes Stangenbruchstück von Kutzleben

IQW 1969/11263 (Kutzl. 12315)

Die rechte, schädelechte Stange (Abb. 2) stammt von einem relativ kräftigen Hirsch. Auch dieses Fossil weist starke Abrollungsspuren auf, die besonders an den Rändern des Os frontale, im Bereich der wohl primär bereits schwach ausgebildeten Rose und an der Abbruchstelle des Eissprosses erkennbar sind. Gefäßrillen sind nur auf der Oberseite der Stange und an der Eissproßabzweigung vorhanden. Die Rose steht in einem Winkel von etwa 65° zu Rosenstock und Stangenhauptachse. Insgesamt weist die Stange einen ovalen Querschnitt auf. Der leicht nach vorn unten gebogene Rest des Eissprosses ist im Querschnitt ebenfalls oval gestaltet. Er setzt tief an, ein Augsproßrudiment fehlt. Die Hauptstange ist oberhalb des Eissproßansatzes tief abgebrochen.

Maße [in mm]	
Rosenstockumfang	205
Rosenumfang	233
Stangenumfang zwischen Rose und Eissproß (Mitte)	212
Abstand des Eissproßansatzes (unten) von der Rose	32
Durchmesser des Eissprosses (basal, horizontal)	54
Länge des Eissprosses bis zum Abbruch	29
Stangenlänge (Rose bis Endbruchstelle)	105

3. Altersstellung

Neben dem oben beschriebenen Stangenrest des in seiner Nomenklatur generisch nicht eindeutig zugeordneten *verticornis*-Hirschs (vgl. KAHLKE 1965, 1969, dort ausführliches Literatur-Verzeichnis) fand sich in den Kiesen von Kalbsrieth bisher folgende Begleitfauna:

Bison sp. IQW 1974/14255 (Kalbsr. 14084), Apex des Processus cornualis sin.;

Dicerorhinus etruscus (FALCONER), IQW 1983/18585 (Kalbsr. 18205), Fragment Mandibula sin. mit Resten von D_3—M_1;

Mammuthus sp., IQW 1983/18684 (Kalbsr. 18208), Fragment I.

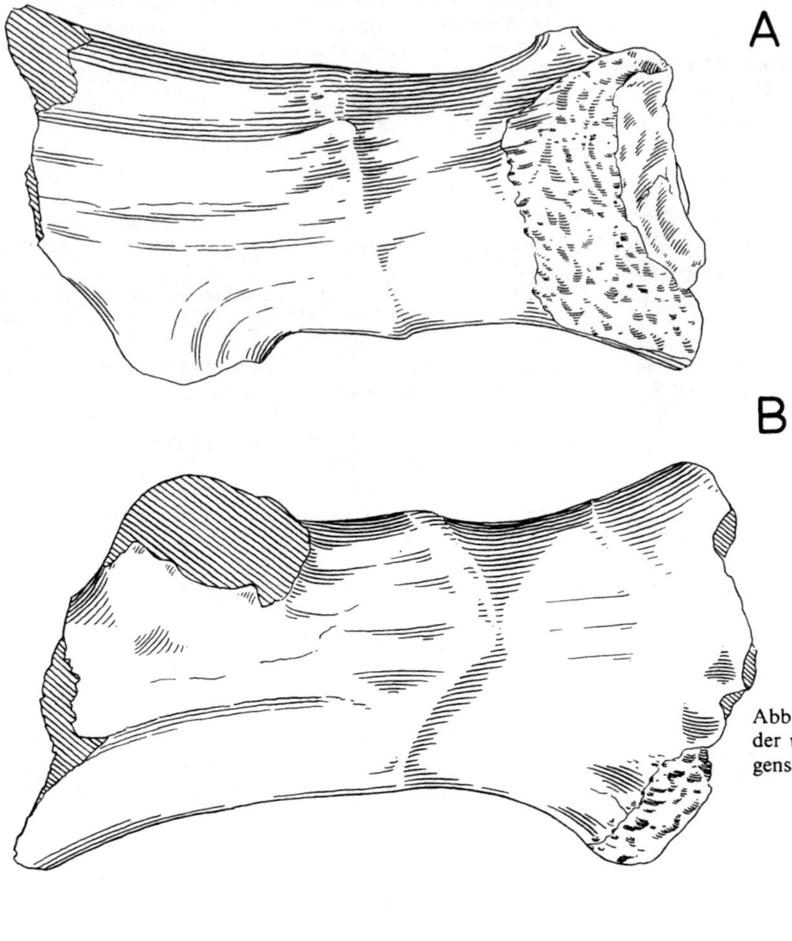

Abb. 2. Rechte, schädelechte Stange eines Hirschs der *verticornis*-Gruppe; Kutzleben (Kreis Bad Langensalza), IQW 1969/11263 (Kutzl. 12315), etwa 1/2 nat. Größe

A — Ansicht von oben; B — Ansicht von vorn

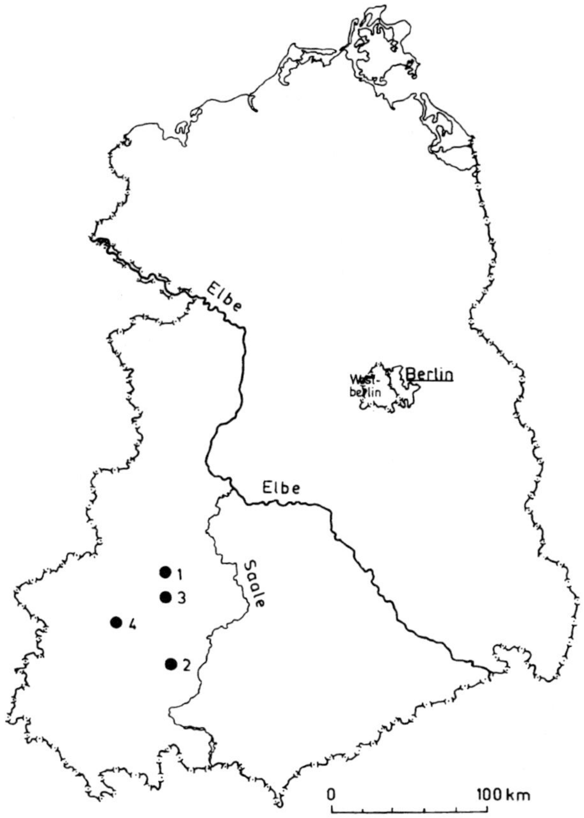

Abb. 3. Fundstellen mit Fossilresten der *verticornis*-Gruppe auf dem Gebiet der DDR

1 — Voigtstedt (Kreis Sangerhausen); 2 — Süßenborn (Kreis Weimar); 3 — Kalbsrieth (Kreis Artern); 4 — Kutzleben (Kreis Bad Langensalza)

Die obere Zeitgrenze des Auftretens der *verticornis*-Gruppe ist im Thüringer Raum durch die Liegendgrenze des Elster-Geschiebemergels gekennzeichnet. Reste des Steppenhirschs fanden sich bisher in Voigtstedt (Kahlke, zuletzt 1969) und Süßenborn (Kahlke, zuletzt 1965).

Für das Fundstück von Kutzleben kann mit relativ hoher Wahrscheinlichkeit die präglaziale Elsterzeit angegeben werden, unabhängig davon, ob das Fossil den Jüngeren Grobschottern oder den Kutzlebener Sanderbildungen entstammt, da die präelsterzeitlichen Älteren Grobschotter bisher nur aus Bohraufschlüssen bekannt sind.

Weil der Zeitpunkt des frühesten Auftretens der *verticornis*-Hirsche im Cromerium s. l. zur Zeit noch nicht fixierbar ist, darf für den Stangenfund von Kalbsrieth mit Rücksicht auf die Ergebnisse von Zeissler (1968) und Mania (1973) lediglich gesagt werden, daß er wohl einer relativen Warmperiode des späten Cromer-Komplexes angehörte.

Insgesamt passen sich die neuen Lokalitäten gut in das räumlich-zeitliche Verbreitungsmodell des *verticornis*-Kreises ein, das von Kahlke (1969, S. 586) aufgezeigt wurde.

Zusammenfassung

Neben die bisher aus dem Thüringer Raum/südlichen Harzvorland bekannten pleistozänen Fundstellen Voigtstedt und Süßenborn mit Fossilresten der *verticornis*-Gruppe treten zwei weitere Lokalitäten. Das Fragment einer Abwurfstange aus den Kiesen von Kalbsrieth (Kreis Artern) darf unter Beachtung der malakozoologischen Untersuchungen von Zeissler (1968) und Mania (1973) in eine relative Warmperiode des späten Cromer-Komplexes gestellt werden. Eine von Kutzleben (Kreis Bad Langensalza) stammende schädelechte Stange ist mit hoher Wahrscheinlichkeit der präglazialen Elsterzeit zuzuordnen.

Summary

Aside from the already known fossil record of steppe-deer (*verticornis*-group) discovered at Voigtstedt and Süßenborn, two further fossil localities are added. The fragment of a shed-antler from the gravels of Kalbsrieth (Kreis Artern) may be correlated according to malacozoological research by Zeissler (1968) and Mania (1973) with a relatively warm period within the late Cromerian-Complex. A second antler-fragment with preserved parts of the os frontale from Kutzleben (Kreis Langensalza) is correlated with high probability with the preglacial Elsterian.

Резюме

К уже известным плейстоценовым местонахождениям Фойгтштедт и Зюссенботн, в которых найдены ископаемые остатки оленей группы *verticornis*, следует добавить два новых местонахождения этих оленей. Фрагмент сброшенного рога из галечников Кальбсрита (район Артерн) может быть отнесен, судя по данным малакологических исследований Цайсслер (1968) и Маньиа (1973), к относительно теплому периоду в пределах позднего кромерского комплекса. Второй фрагмент рога с частями лобных костей из Кутцлебена (район Лангензальца) весьма уверенно относится к доледниковому эльстерию.

Literatur

Dames, W., 1882: Erläuterungen zur geologischen Specialkarte von Preussen und den benachbarten Thüringischen Staaten. No. 37, Bl. Ziegelroda. — Berlin: Preuß. geol. Landesanst. — 23 S.

Gesang, H., 1962: Pleistozäne Lagerstätten der Steine und Erden im Thüringer Becken. — In: Exkursionsf.: Das Pleistozän im sächsisch-thüringischen Raum. — Berlin: Geol. Gesellsch. DDR. — S. 219—224.

Jankowski, G., 1961: Quartäre Ablagerungen im Ried des mittleren Helme- und Unstrutlaufes. — In: Geologie. — Berlin **10** (1). — S. 50—66.

Kahlke, H. D., 1965: Die Cerviden-Reste aus den Tonen von Voigtstedt in Thüringen. — In: Paläont. Abh., A. — Berlin **2** (2/3). — S. 379—426.

—, 1969: Die Cerviden-Reste aus den Kiesen von Süßenborn bei Weimar. — In: Paläont. Abh. A. — Berlin **3** (3/4). — S. 547 bis 610.

Krutzsch, W., 1965: Das geologische Profil von Voigtstedt in Thüringen. — In: Paläont. Abh. A. — Berlin **2** (2/3). — S. 235 bis 248.

Krutzsch, W.; Reichstein, H., 1956: Das Pleistozän-Profil von Voigtstedt-Edersleben (Bezirk Halle). — In: Geologie. — Berlin **5** (4/5). — S. 327—349.

Kugler, H., 1961: Bemerkungen zur Gliederung und Altersstellung der pleistozänen Flußterrassen der Unstrut. — In: Geogr. Ber. — Berlin **6** (3/4). — S. 183—196.

Lehmann, R., 1922: Das Diluvium des unteren Unstruttales. — In: Jahrb. hall. Verb. — Halle **3** (3). — S. 89—124.

Mania, D., 1972: Altpleistozäne Molluskenfaunen im unteren Unstruttal. — In: Exkursionsf.: Aktuelle Probleme der Quartärgeologie in der Deutschen Demokratischen Republik. — Berlin: Dt. Gesellsch. geol. Wiss. — S. 19—25.

—, 1973: Paläoökologie, Faunenentwicklung und Stratigraphie des Eiszeitalters im mittleren Elbe-Saalegebiet auf Grund von Molluskengesellschaften. — Berlin: Akad.-Verl. — 175 S., (Beih. Geologie; 78/79).

RUSKE, R., 1965: Zur petrographischen Ausbildung und Genese der „Lehmzone" von Voigtstedt in Thüringen. — In: Paläont. Abh., A. — Berlin **2** (2/3). — S. 249—258.

SPEYER, O., 1883: Erläuterungen zur geologischen Specialkarte von Preussen und den benachbarten Thüringischen Staaten. No. 52, Bl. Gebesee. — Berlin: Preuß. geol. Landesanst. — 25 S.

STEINMÜLLER, A., 1972: Präglaziale Bildungen des Kalbsrieth-Roßleben-Wendelsteiner Schotterzuges. — In: Exkursionsf.: Aktuelle Probleme der Quartärgeologie in der Deutschen Demokratischen Republik. — Berlin: Dt. Gesellsch. geol. Wiss. — S. 12—17, (1972a).

—, 1972: Die Schichtenfolgen von Süßenborn und Voigtstedt und die Gliederung des Mittelpleistozäns. — In: Geologie. — Berlin **21** (2). — S. 149—165, (1972b).

—, 1977: Die präglazialen Bildungen und elsterkaltzeitliche glaziäre Sedimente im Gebiet Voigtstedt-Edersleben und am Kyffhäuser. — In: Exkursionsf. Arbeitstagg. INQUA-Subkommiss. europ. Quartärstratigraphie (SEQS) zur Stratotypen-Diskussion in der Deutschen Demokratischen Republik. — Berlin: Zentr. geol. Inst. — S. 1—24.

UNGER, K. P., 1974: Quartär. In: Geologie von Thüringen/hrsg. von W. HOPPE; G. SEIDEL — Gotha/Leipzig: H. Haack. — S. 742—781, (1974a).

—, 1974: Die Elstervereisung des Zentralen Thüringer Keuperbeckens. — In: Z. geol. Wiss. — Berlin **2** (7). — S. 791—800, (1974b).

—, 1976: Gliederungstabelle der quartären Flußablagerungen im südlichsten Randbereich des skandinavischen Vereisungsgebiets. — In: Z. geol. Wiss. — Berlin **4** (5). — S. 797—799.

UNGER, K. P.; GESANG, H., 1969: Quartäre Ablagerungen im Zentralen Thüringer Becken und im nördlichen Vorland des Thüringer Waldes und ihre lagerstättenkundliche Bedeutung. — In: Exkursionsf.: Ergebnisse quartärgeologischer Untersuchungen in Thüringen. — Berlin: Dt. Gesellsch. geol. Wiss. — S. 57 bis 82.

UNGER, K. P.; ZIEGENHARDT, W., 1961: Periglaziale Schotterzüge und glazigene Bildungen der Mindel-(Elster-)Eiszeit im zentralen Thüringer Becken. — In: Geologie. — Berlin **10** (4/5). — S. 469—479.

WOLDSTEDT, P.; DUPHORN, K., 1974: Norddeutschland und angrenzende Gebiete im Eiszeitalter. — Stuttgart: Verl. F. K. Koehler. — 500 S.

WÜST, E., 1900: Untersuchungen über das Pliozän und das älteste Pleistozän Thüringens nördlich vom Thüringer Walde und westlich von der Saale. — Stuttgart: — 352 S.

—, 1901: Beiträge zur Kenntnis des Flußnetzes Thüringens vor der ersten Vereisung des Landes. — In: Mitt. Ver. Erdkd. — Halle. — S. 1—17.

ZEISSLER, H., 1965: Konchylien im Pleistozän von Voigtstedt in Thüringen. — In: Paläont. Abh., A. — Berlin **2** (2/3). — S. 273 bis 290.

—, 1968: Mollusken aus dem altpleistozänen Unstrutkies von Kalbsrieth. — In: Ber. dt. Gesellsch. geol. Wiss., A. — Berlin **13** (3). — S. 401—413.

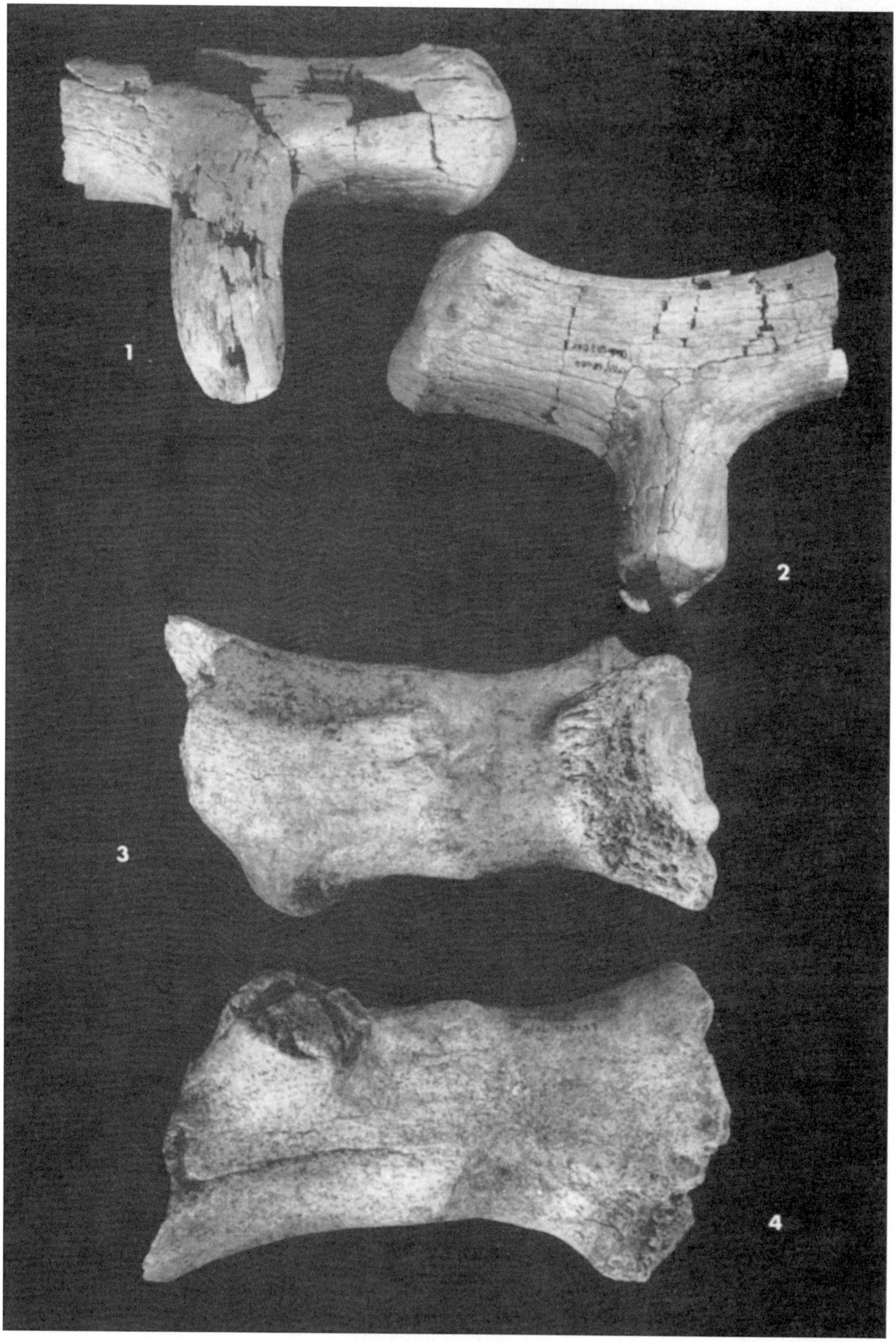

TAFEL I

Rechtes Abwurfstangenbruchstück eines Hirschs der *verticornis*-Gruppe; Kalbsrieth (Kreis Artern): IQW 1983/18688 (Kalbsr. 18206), etwa 1/2 nat. Größe: Fig. 1. Ansicht von oben, Fig. 2. Ansicht von vorn

Rechte schädelechte Stange eines Hirschs der *verticornis*-Gruppe; Kutzleben (Kreis Bad Langensalza); IQW 1969/11263 (Kutzl. 12315), etwa 1/2 nat. Größe: Fig. 3. Ansicht von oben, Fig. 4. Ansicht von vorn

Quartärpaläontologie 6 93–97 Berlin 1986

The Late Cenozoic Two-Humped (Bactrian) Camels of Asia

By Baldirgan Seralievna Kozhamkulova, Alma-Ata

With 2 Figures

Contents

Abstractum

Five species of Camelidae: *Gigantocamelus longipes*, *Paracamelus gigas*, *Paracamelus praebactrianus*, *Camelus knoblochi*, and *Camelus ferus* have existed in Asia since the middle Pliocene.

Vom mittleren Pliozän bis heute existierten in Asien fünf Cameliden-Arten: *Gigantocamelus longipes*, *Paracamelus gigas*, *Paracamelus praebactrianus*, *Camelus knoblochi* und *Camelus ferus*.

В Азии, начиная со среднего плиоцена и до наших дней существовало пять видов: *Gigantocamelus longipes*, *Paracamelus gigas*, *Paracamelus praebactrianus*, *Camelus knoblochi* и *Camelus ferus*.

Fossil History of the Two-Humped Camels of Asia

The earliest fossil remains of Camelidae from Asia are of middle Pliocene age (Belyaeva 1937; Aubekerova 1974). The fossil Camelids of Asia are members of the genera *Gigantocamelus*, *Paracamelus*, and *Camelus*.

In 1974 *Gigantocamelus longipes* Aubekerova of middle Pliocene age was discovered as a member of the late Hipparion-fauna from the Tekes-basin. The fossil fauna of this locality includes the following species: *Anancus kazachstanensis*, *Hipparion houfenense*, *Hipparion* sp., *Dicerorhinus orientalis*, *Gigantocamelus longipes*, *Sinomegaceros*, *Cervavitus flerovi*, *Palaeotragus* sp., *Samotherium* sp., *Tragocerus* sp., *Gazella* ex. gr. *dorcadoides*, *Gazella sinensis*, *Antilospira* sp. From the evidence of a number of morphological differences, Aubekerova (1974) suggested that in contrast to recent camels, *Gigantocamelus* browsed on a relatively soft vegetation, mainly trees and bushes.

The representatives of the genus *Paracamelus* were widely distributed in the late Pliocene. In central Asia *Paracamelus gigas* and *Paracamelus praebactrianus* are characteristic. The first species was described by M. Schlosser from the north-eastern part of China (Henan, Shansi, and Nihowan-basin). Subsequently remains of this camel were described by E. I. Belyaeva from north-western Mongolia, Oschi, and from Tadzhikistan, and finally B. S. Kozhamkulova recorded this camel from Tira, Kuruksaj, and Kazakhstan where it is especially numerous in the north-west, and less so in the southern part of the region.

It is, however, interesting to note, that *Gigantocamelus* lived in Kazakhstan not only in the late Pliocene but also throughout the early Pleistocene.

Paracamelus praebactrianus from the Beteka-deposits was described by Yu. A. Orlov. This camel was widely distributed in Kazakhstan: Turgai-Tobolsk, the southern Altai, the Isim area, and the middle Irtysh.

It was also characteristic of the Ili faunal complex. Isolated remains have been reported from Tadzhikistan (Kozhamkulova 1977) and from Chinese Turkestan (Y. Khaveson 1954). In Chinese Turkestan fossils of *Paracamelus praebactrianus* were found together with remains of bovids, gazelles, and cervids.

In eastern Kazakhstan this camel was found together with *Antilospira gracilis* and *Cervus (Rusa) ubensis*. It is also interesting to note that *Paracamelus praebactrianus* has been recorded from Tadzhikistan (Sharapov 1978), together with numerous representatives of Lagomorpha, Rodentia, Cercopithecidae, Carnivora, Hyaenidae, Felidae, Ursidae, Elephantidae, Rhinocerotidae, Equidae, Cervidae, Giraffidae, Bovidae, and Gazellinae (Sharapov 1978).

Gigantocamelus has been reported from P.R. Mongolia together with fossil remains of the *Hipparion*-fauna, dated by E. I. Belyaeva to the middle Pliocene.

Paracamelus gigas of late Pliocene age has been reported from the Irtysh-area (Podpusk near Pavlodar) as well as from the Kuruksaj faunal-complex.

Khisarova (1963) described early Pleistocene *Paracamelus gigas* from Koshkurgan (15 km north of the town Turkestan). Fossil remains of *Equus caballus mosbachensis*, *Asinus hydruntinus*, *Dicerorhinus kirchbergensis*, and *Bison priscus* have been reported from the same deposits.

In the valley of the River Tira (Tadzhik SSR) bones of this camel have been excavated along with fossil remains of *Canis* sp., *Equus* sp., and *Cervus* sp.

The main distribution area of *Paracamelus*, therefore, was central Asia, whereas the distribution area of *Gigantocamelus* extended from China, across the central Asiatic region, to Kazakhstan as far as the northern Caucasus, with a few localities in the Asov area.

Paracamelus praebactrianus is known mainly from Kazakhstan and lived partly in Tadzhikistan. It is suggested therefore

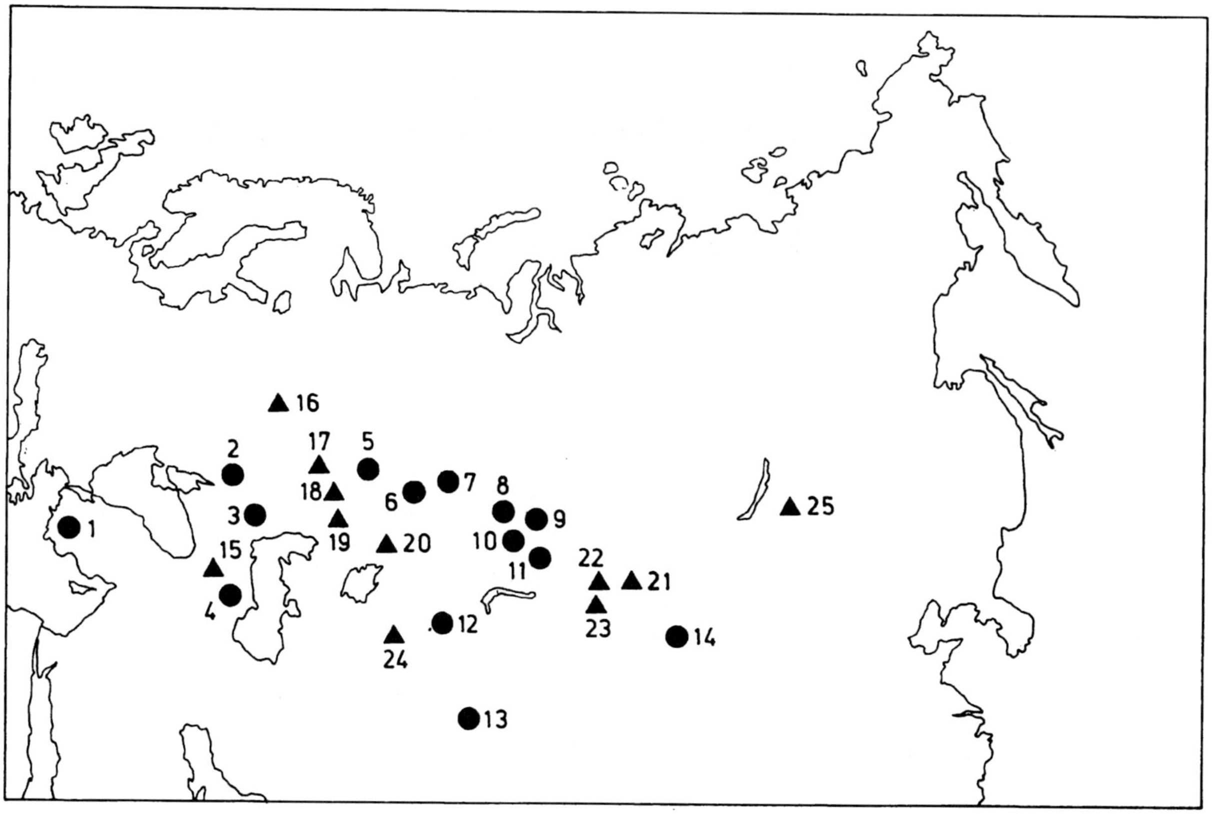

Fig. 1. Fossil localities with *Paracamelus gigas* SCHLOSSER, 1903 (1—14) and with *Camelus bactrianus knoblochi* POLJAKOV, 1880 (15—25)

1 — Turkey; 2 — Liventsovka; 3 — Vinodelnoe; 4 — Bozdag; 5 — Ufa; 6 — Beteke; 7 — Ilinka; 8 — Pavlodar; 9 — Maraldy; 10 — Lebyanzhe; 11 — Podpusk; 12 — Koshkurgan; 13 — Tira; 14 — Mongolia
15 — Leninakan; 16 — Syromyatnikov; 17 — Sarepta; 18 — Rubezhka; 19 — Uralsk; 20 — Martuk; 21 — Ust-Bukhtarma; 22 — Zyryanovsk; 23 — Kamyshinka; 24 — Samarkand; 25 — Transbaikalia

that *Paracamelus gigas* and *Paracamelus praebactrianus* were Asiatic endemics, living under more extreme conditions than *Gigantocamelus* and the European *Paracamelus*.

The direct ancestor of *Paracamelus* is unknown, but it is suggested (KHAVESON 1954) that this species was probably living in America in the middle Pliocene. On the other hand it is obvious that we must derive the genus *Camelus* from *Paracamelus*. In the territory of Tadzhikistan (Lachuti Settlement) fossil remains of *Camelus* cf. *knoblochi* have been discovered, dating from the interval between the Taman and Tiraspol stages (VANGENGEJM et al. 1977).[1])

In the middle Pleistocene in Kazakhstan and North China (Ordos, cannyon of the River Sara-osso-gol, Huanghe) this camel reached the peak of its fortunes. In the west it occurred as far as the southern Urals, the Volga area and the Lesser Caucasus (Armenia). It is interesting to note, that *Camelus knoblochi* was present in Kazakhstan not only in the middle Pleistocene, but in the late Pleistocene as well. Numerous fossil remains of this camel have been found in Soviet Middle Asia (Uzbek SSR).

In Kazakhstan fossil remains of *Camelus knoblochi* are known from the right bank of the River Uba, two kilometres south of Kamyshinka (East Kazakhstan), as well as from the River Rubezhka in the Ural area. Here (Rubezhka) *Camelus knoblochi* was found associated with fossil remains of *Arvicola terrestris*, *Equus caballus*, *Saiga tatarica*, and *Bison priscus*.

In the surroundings of the town of Zyryanovsk, *Camelus knoblochi* was found together with fossil remains of *Clethrionomys rufocanus*, *Allactaga jaculus*, *Myospalax myospalax*, *Cricetus cricetus*, *Citellus* sp., *Ochotona pusilla*, *Talpa altaica*, *Canis lupus*, *Panthera spelaea*, *Hyaena spelaea*, *Ursus* cf. *spelaeus*, *Mammuthus primigenius*, *Equus caballus*, *Coelodonta antiquitatis*, *Bison priscus*, and *Bos primigenius* (KOZHAMKULOVA 1961).

VERESHCHAGIN and MEL'NIKOVA (1958) reported fossil remains of mammals which have been collected by S. S. CHERNIKOV when excavating a late palaeolithic cave near the mouth of the River Buchtarma. The author suggested that at least 14 species of mammals, including *Camelus knoblochi*, were hunted by the ancient Buchtarma people. From the rare stone implements, archaeologists have tentatively dated this ancient living-floor to the Upper Palaeolithic period. Further on, on the right bank of the River Ilek near the Martuk Settlement, in the Aktyubinsk area, a few fossil remains of *Camelus knoblochi*, *Equus caballus*, *Coelodonta antiquitatis*, and *Mammuthus primigenius* have been collected (KOZHAMKULOVA and ZHYLKIGAEV 1976). In the centre of

[1]) Late Cromerian Complex (Taman) to Early Elsterian (Tiraspol).

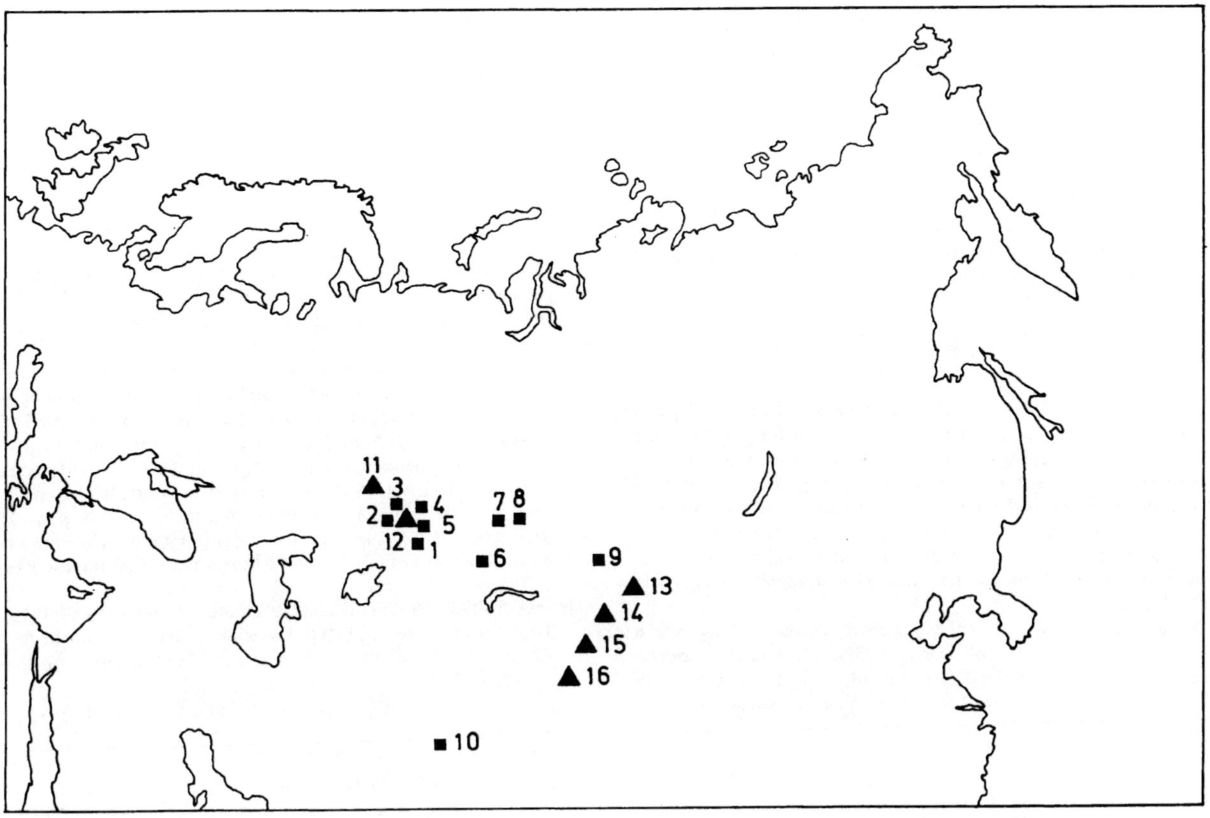

Fig. 2. Fossil localities with *Paracamelus praebactrianus* (ORLOV, 1927) (1—10), and *Camelus ferus* PRZEWALSKI, 1883 (11—16)

1 — Tasty; 2 — Tobol; 3 — Zhukovsky; 4 — Presnovka; 5 — Beteke river, Selim-Dzhevar; 6 — Zavyalovka, Sovkhoz Rostovsky; 7 — Pavlodar; 8 — Sovkhoz Kirova; 9 — Ostraya Sopka; 10 — Kuruksay; 11 — Ushkatta; 12 — Alekseevskoe; 13 to 16 — South-Eastern Kazakhstan

the town of Samarkand, on the right bank of Komsomal Lake, fossil remains of the following mammals have been found together with *Camelus knoblochi*: *Canis lupus*, *Equus caballus*, *Equus hydruntinus*, *Equus hemionus*, *Sus scrofa*, *Cervus elaphus*, *Saiga tatarica*, *Ovis* cf. *arcal*, and *Bos primigenius*. Many of the bones showed signs of having been burnt and smashed — perhaps traces of palaeolithic men (BATYROV 1967, 1969).

The Asiatic *Camelus knoblochi* perhaps evolved at the beginning of the early Pleistocene, as suggested by YA. I. KHAVESON, and in the middle Pleistocene was distributed as far as China, Kazakhstan, South Ural, the River Volga area, as far as Armenia. Most probably this camel vanished in Eastern Europe at the period of maximum glaciation, because at no site with the late-palaeolithic faunal complex have remains of this camel been found (ALEKSEEVA 1977). From the present fossil record of late Pleistocene remains of *Camelus knoblochi*, it seems possible that the main distribution area of this species in the late Pleistocene extended from the Baikal as far as Aktjubinsk.

It is possible to discuss the early distribution of Camelus ferus in Kazakhstan from both skeletal remains and rock-pictures. In the Holocene this camel was numerous in western and south-eastern Kazakhstan and was hunted together with marals, mountain-sheep, gazelles, saigas, and wolves, which are also known from rock-pictures (MARIKOVSKIY 1952, 1953). *Camelus ferus* of the Andronov-period has been discovered in the archaeological horizons of Alekseevsk in the Kustanaj area (KRIVCOVA-GRAKOVA 1948)[2]) as well as in East Kazakhstan (VERESHCHAGIN 1956). The discovery in the Ata-su Settlement, in a grave of Aksu-Ajuly 2 group Central Kazakhstan, as published by AKISHEV et al. in 1956, is well known. In another paper (KUZMINA 1963) the discovery of a statuette of a two-humped camel was reported, which was excavated at the Ushkatta Settlement, in the southern Ural area.

VERESHCHAGIN (1956), who discovered remains of wild-living two-humped camels within the materials of the Andronov-civilization of East Kazakhstan (Ust-Naryn), suggested that wild camels became extinct in Soviet Middle Asia and western Siberia as late as within our own epoch. According to KHAKHLOV (1928) camels disappeared in the area around 1883. It is, however, not certain, if he is referring to wild or domesticated camels in his report.

In prehistoric times the distribution area of the Holocene wild-living two-humped camel increased, as was proved by remains discovered in the east Kazakhstan area as far as central and north Kazakhstan and in the southern Ural area.

From the evidence of rock-pictures, wild camels were living in the Bronze-Age in the valley of the Zaravshan River (near the town of Navoi) and in the area of the Kaunchi and

[2]) The two-humped camels seem to have been not numerous within the Andronov-civilization and do not play an important oeconomic part like cattle- and sheep-breeding of which bones are numerous among the ancient remains.

Dalvarzin Settlements. In Mesolithic times camels were living in the north-east area of Uzbekistan (sea cavern Obi-shir).

"The native area of the wild two-humped camels is in general the deserts between China and India, as well as the area around the northern boundaries of India. They are living in the deserts of Songarh and in the mountain-steppes (that means possibly the hilly area of Kazakhstan, the "Kazakh-threshold" of East-Kazakhstan) as well as around Lake Balash and the River Ili flowing into the Balash" (GEPTNER 1966).

Summary

In Pliocene-Pleistocene times several species of camels lived in Asia. In the Pliocene camels were moving into the eastern parts and in the Pleistocene and Holocene up to modern times they reached their peak, especially in Kazakhstan, with the species *Gigantocamelus longipes*, *Paracamelus gigas*, *Paracamelus praebactrianus*, *Camelus knoblochi*, and *Camelus ferus*. With the exception of *Gigantocamelus longipes* and *Camelus ferus*, all other species lived also in Tadzhikistan. In Mongolia and China *Gigantocamelus* was absent at different times.

In the southern Ural area only a few fossil remains of *Paracamelus* sp., *Camelus knoblochi*, and *Camelus ferus* are known. *Camelus ferus* became extinct in the Asiatic part of the USSR at the end of the 19th century. At present some 400 individuals are still living in the Gobi Desert (SOKOLOV et al. 1978).

Zusammenfassung

Aus dem Pliozän und Pleistozän Asiens sind verschiedene fossile Cameliden bekannt. Während des Pliozäns wanderten Kamele bis in die östlichen Gebiete Asiens, im Pleistozän-Holozän bis in die modernen Zeiten hinein erreichten sie mit den Spezies *Gigantocamelus longipes*, *Paracamelus gigas*, *Paracamelus praebactrianus*, *Camelus knoblochi* und *Camelus ferus* den Höhepunkt ihrer Entwicklung besonders in Kasachstan. Mit Ausnahme von *Gigantocamelus gigas* und *Camelus ferus* kamen die Kamele bis nach Tadshikistan. In der Mongolei und in China fehlte zu verschiedenen Zeiten *Gigantocamelus*.

Aus dem südlichen Ural-Gebiet sind nur wenige Fossilfunde von *Paracamelus* sp., *Camelus knoblochi* und *Camelus ferus* bekannt. *Camelus ferus* starb gegen Ende des 19. Jahrhunderts im asiatischen Teil der UdSSR aus. In der Gegenwart leben noch etwa 400 Individuen in der Wüste Gobi (SOKOLOV u. a. 1978).

Резюме

Таким образом, в Азии существовали различные двугорбые верблюды, которые в Восточное полушарие проникли в плиоцене, а в плейстоцене и до наших дней они получили расцвет, особенно в Казахстане, где были представлены *Gigantocamelus longipes*, *Paracamelus gigas*, *P. praebactrianus*, *Camelus knoblochi* и *Camelus ferus*. Почти все эти виды, кроме первого и последнего встречены и на территории Таджикистана, а в Монголии и Китае пока отсутствует только гигантокамелюс.

На Южном Урале были обнаружены единичные костные остатки *Paracamelus* sp., *Camelus knoblochi* и *Camelus ferus*. Последний вид на Азиатской территории СССР исчез в конце 19 в., а в настоящее время он (при численности 400 особей) обитает в Гобийской пустыне (Соколов и др. 1978).

References

AKISHEV, K. A.; AGEEVA, E. I.; PACEVICH, G. I.; MARGULAN, A. KH., 1956: Osnovnye itogi archeologičeskich rabot na territorii Kazachstana v 1955 godu (Fundamental results of archaeological research in Kazakhstan in the year 1955). — In: Izvest. Akad. Nauk Kaz. SSR, Ser. Istor., Ekon., Filosof. prava. — Alma-Ata (3).

ALEKSEEVA, L. I., 1977: Teriofauna rannego antropogena Vostočnoj Evropy (The Therio-fauna of the Early Anthropogene of East Europe). — Moskva.

AUBEKEROVA, P. A., 1974: Pozdneneogenovaja fauna Tekesskoj vpadiny (The late Neogene fauna of the Tekes-basin). — Alma-Ata, Avtoref. dis. soisk. učen. step. kand. biol. nauk.

BATYROV, B. KS., 1967: Ostatki mlekopitajuščich iz sloev Samarkandskoj verchnepaleolitičeskoj stojanki (Fossil mammalian remains from the upper palaeolithic site of Samarkand). — In: Materialy XXIV naučn. Konf. — Samarkand.

—, 1969: Materialy po istorii teriofauny Južnogo Uzbekistana v verchnem antropogene (Materials and history of the Therio-fauna from South Usbekistan from the upper Anthropogene). — Samarkand, Avtoref. dis. soisk. učen. step. kand. biol. nauk.

BELYAEVA, E. I., 1937: Materialy i charakteristika verchnetretičnoyj fauny mlekopitajuščich severo-zapadnoj Mongolii (Materials and characteristics of the late Tertiary mammalian fauna of north-western Mongolia). — In: Tr. mong. Kom. — Moskva **33**.

BOGOLYUBSKIJ, S. N., 1959: Proischoždenie i preobrazovanie domašnich životnych (Origin and transformation of domesticated animals). — Moskva.

GEPTNER, V. G., 1966: O dikom verbljude (*Camelus ferus* PRZEW., 1883) na territorii SSSR (On the wilde Camel — *Camelus ferus* PRZEW., 1883 — on the territory of the USSR). — In: Zool. Ž. — Moskva **45** (1).

KHAKHLOV, V. A., 1928: Zajzanskaja Kotlovina i Tarabagataj. — Tomsk.

KHAVESON, YA. I., 1954: Tretičnye verbljudy Vostŏcnogo polušatija (rod *Paracamelus*) (The Tertiary Camels of the East — *Paracamelus*). — In: Tr. paleont. Inst. Akad. Nauk SSSR. — Moskva **47** (2).

KHISAROVA, G. D., 1963: Iskopaemye kosti mlekopitajuščich iz Koškurganskogo grifona (Južnyj Kazachstan) (Fossil mammal-bones from Koshgurkan, South Kazakhstan). — In: Materialy po istorii fauny i flory Kazachstana. Vol. 4. — Alma-Ata.

KHUDAJBERDIEV, T. N., 1973: Materialy po teriofaune Severo-Vostočnogo Uzbekistana v pozdnem antropogene (Materials of the Theriofauna of North-East Usbekistan in the late Anthropogene). — Taškent, Avtoref. dis. soisk. učen. step. kand. biol. nauk.

KOZHAMKULOVA, B. S., 1961: Mestonachoždenie iskopaemych mlekopitajuščich antropogena, vskryvaemye gidromonitorami (New locality of fossil mammals of Anthropogene age discovered by hydromonitors). — In: Mater. istorii fauny i flory Kazachstana. Vol. 3. — Alma-Ata.

—, 1969: Antropogenovaja iskopaemaja teriofauna Kazachstana (The Anthropogene fossil Therio-fauna of Kazakhstan). — Alma-Ata: Akad. Nauk Kaz. SSR.

—, 1977: Zoogeografičeskij analiz sostava i raspredelenija antropogenovoj fauny Kazachstana v prostranstve i vremeni (Zoogeographical analysis of the amount and the distribution of the Anthropogene fauna of Kazakhstan in time and space). — In: Mezokajnozojskaja fauna i flora severo-zapadnogo Kazachstana. — Alma-Ata.

KOZHAMKULOVA, B. S.; ZHYLKIGAEV, K. ZH., 1976: Izučenie mamontovoj fauny na territorii Kazachstana (Study of the Mammouth fauna on the territory of Kazakhstan). — In: Izvest. Akad. Nauk Kaz. SSR, Ser. Geol. — Alma-Ata (6).

KRIVCOVA-GRAKOVA, O. A., 1948: Alekseevskoe poselenie i mogil'nik (Settlement and graveyard in Alekseevsk). — In: Tr. gos. istor. Muzeja. — **17**.

KUZ'MINA, E. E., 1963: Drevnejšaja figurka verbljuda iz Orenburgskoj oblasti i problema domestikacii baktriana (The ancient camelstatuette from the Orenburg area and the problem of domestication of *Camelus bactrianus*). — In: Sov. archeol. — Moskva (2).

MARIKOVSKIY, P. I., 1956: O naskal'nych izobraženijach v gorach Čulak (On the rock-pictures in the Chulak-mountains). — In: Vestn. Akad. Nauk Kaz. SSR. — Alma-Ata **63** (6).

SHARAPOV, SH., 1978: Kuruksajskij kompleks posdnepliocenovych mlekopitajuščich Tadžikškoj depressii (The Kuruksaj-Komplex of late Pliocene mammals from the Tadzhik basin). — Kiev, Avtoref. dis. soisk. učen. step. kand. biol. nauk.

SOKOLOV, V. E.; KHOTOLCHU, N.; ORLOV, V. N., 1978: Redkie i ochranjaemye vidy kopytnych pustyn' Mongolii (Rare and protected Ungulata of the Mongolian desert). — In: II. Congr. Theriologicus Internationalis. — Brno, June 20—27 (Abstracts of papers).

VANGENGEJM, E. A., 1977: Paleontologičeskoe obosnovanie stratigrafii antropogena Severnoj Azii (Palaeontological definition of the stratigraphy of the Anthropogene of northern Asia). — Moskva.

VANGENGEJM, E. A.; ALEKSEEVA, L. I.; VISLOBOKOVA, I. A.; GODINA, A. YA.; DMITRIEVA, A. L.; ZHEGALLO, BV. I.; ZAZHIGIN, V. S.; SOTNIKOVA, M. B.; TROFIMOV, B. A.; SHARAPOV, S., 1977: Fauna mlekopitajuščich pozdnego pliocena i eoplejstocena Južnogo Tadžikistana (The mammal fauna of the late Pliocene and the Eopleistocene of South Tadzhikistan). — In: Tezisy dokladov Meždunarodnogo simposiuma po probleme „Granica neogena i četvertičnoj sistemy". 3—13 Oct. 1977 g. Dušanbe, SSSR (Thesis to the International Symposium "Boundary between the Neogene and the Quaternary", 3—13 Okt. 1977 in the city of Dushanbe, SSSR). — Moskva: Nauka.

VERESHCHAGIN, N. K., 1956: O preždnem rasprostranenii nekotorych kopytnych v rajone smykanija Evropejsko-Kazachstanskich i Central'no-Aziatskich stepej (On the earlier distribution of some Ungulata in the contact-area between the European-Kazakh- and Central-Asiatic steppes). — In: Zool. Ž. — Moskva **10** (35).

VERESHCHAGIN, N. K.; MEL'NIKOVA, N. N., 1958: Zoologičeskie otkrytija archeologov v Vostočnom Kazachstane i v Altajskom krae (Zoological discoveries by archaeologists in the East-Kazakhstan and the Altai area). — In: Izvest. VGO. — **90** (4).

Quartärpaläontologie 6 99—100 Berlin 1986

Crocuta (Hyaenidae) from the Pleistocene of Voigtstedt, Thuringia (G.D.R.)

By BJÖRN KURTÉN, Helsinki

With one Figure and one Table

Contents

Abstractum

The spotted hyena, probably the subspecies *Crocuta crocuta praespelaea*, is recorded for the first time from the early Mid-Pleistocene fauna of Voigtstedt near Sangerhausen/Thuringia, G.D.R.

Die Tüpfelhyäne, wahrscheinlich *Crocuta crocuta praespelaea*, wird erstmals aus der frühen Mittelpleistozän-Fauna von Voigtstedt bei Sangerhausen, Thüringen, DDR, nachgewiesen.

В фауне Фойгтштедт (ГДР), относящейся к ранней стадии среднего плейстоцена, впервые установлено присутствие пятностой гиены, возможно принадлежащей к подвиду *Crocuta crocuta praespelaea*.

1. Introduction

The Carnivora from the early mid-Pleistocene clays of Voigtstedt have been described by THENIUS (1965). The faunal list (nine taxa) includes diagnostic forms of the earlier mid-Pleistocene (e.g. *Canis lupus mosbachensis, Ursus deningeri*). Although the main excavations were finished by the time of THENIUS's study, there are a few later additions, mostly representing taxa already put on record by THENIUS. There are, however, two specimens of the spotted hyena (Catalogued in 1966—1967), a taxon not previously recorded in this fauna. The purpose of the present paper is to put this material on record.

I wish to express my thanks to Dr. H. D. KAHLKE and to the staff of the Institut für Quartärpaläontologie, Weimar, for permission to study the material and for many courtesies.

2. Description and Discussion

Crocuta crocuta praespelaea SCHÜTT, 1971

Material: IQW 1966/7367 (Voi. 3693), right lower canine; IQW 1967/9210 (Voi. 3751), right P^3

The material is in the collections of the Institut für Quartärpaläontologie, Weimar.

The canine tooth (Fig. 1A) is completely preserved. The curvature of the root and crown, the position of the internal keel, and the contour of the external enamel base show it to be a lower canine. The crown is heavily worn on the tip and on the posterior face, and there is also some abrasion of the basal portion of the internal keel. The basal internal cingulum is weakly developed and does not form a distinct triangle at its juncture with the internal keel, as in *Hyaena*. Reference to *Crocuta* is indicated.

The third upper premolar (Fig. 1B) also belonged to a fairly old animal, possibly the same as the canine. The tip of the crown is worn flat. The base of the tooth is broken away so that the cingular portion is missing, but the small posterior cusp is formed directly by the cingulum as normally in *Crocuta*, and does not intervene between the cingulum and the posterior face of the main cusp, as in *Hyaena*. As preserved the tooth fragment has a length of 21.2 and a width of 16.0 mm, but the original dimensions, especially the length, were somewhat greater.

The common early mid-Pleistocene (Biharian) spotted hyena of Europe is *Crocuta crocuta praespelaea* SCHÜTT, 1971. Near the end of the Biharian there arose a very large and aberrant subspecies, *C. c. petralonae* KURTÉN, at present known only from the top levels of Petralona Cave, Greece (see KURTÉN and POULIANOS 1977).

In Table 1, the dimensions of the Voigtstedt teeth (tentative estimates for the P^3) are compared to those of a sample of *C. c. praespelaea* from the main deposits of Petralona Cave, and the type of *C. c. petralonae*.

It is evident that the Voigtstedt P^3 is much smaller than its homologue in *C. C. petralonae*, and it also lacks the morphological peculiarities of the latter. The canine tooth is also much smaller than that of *C. c. petralonae* (the crown is broken in the latter but the remaining root testifies to its enormous size).

On the other hand, the Voigtstedt specimens are close to or within the lower part of the observed size range in *C. c. praespelaea*. The fact that both specimens are small need not constitute a significant deviation, since they could have belonged to a single individual.

The Voigtstedt hyena may thus be classified as a small individual of *Crocuta crocuta praespelaea*. The taxon occurs at a number of early mid-Pleistocene sites, including Mosbach (the type locality), Süssenborn, Gombaszög, the Forest Bed, Westbury-sub-Mendip, and the Crenian and Petralonian levels of Petralona Cave (SCHÜTT 1971; KURTÉN 1968, 1969; BISHOP 1982; KURTÉN and POULIANOS 1977, 1981).

Table 1. *Crocuta crocuta*, comparative dimensions of teeth [in mm]

Voigtstedt			*C. c. praespelaea*				*C. c. petralonae* (Type)
			N	OR	M	SD	
C_{inf}	Length	16.5	14	16.7—19.5	17.89 ± 0.26	0.96	—
	Width	13.3	14	12.5—15.4	14.14 ± 0.20	0.75	—
P^3	Length	22—23	8	23.6—26.8	25.34 ± 0.42	1.18	26.8
	Width	16—17	9	17.8—20.5	18.73 ± 0.35	1.06	23.3

N — number of specimens; OR — observed range; M — mean with standard error; SD — standard deviation. The comparative samples are Petralona cave; data from KURTÉN and POULIANOS (1977, 1981), with addition of specimens in the Department of Geology, University of Thessaloniki. Dimensions of Voigtstedt P^3 are tentative estimates.

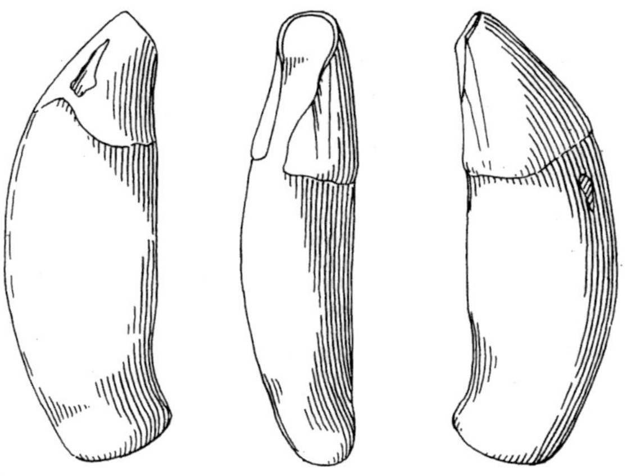

Fig. 1. *Crocuta crocuta praespelaea* SCHÜTT, nat. size

A — right lower canine, IQW 1966/7367 (Voi. 3693); B — P^3 dext., IQW 1967/9210 (Voi. 3751)

Summary

The taxon *Crocuta crocuta praespelaea* SCHÜTT is represented by two isolated teeth from the fossiliferous clays of Voigtstedt near Sangerhausen, GDR. The teeth may represent a single, small, rather aged individual. The taxon is new to this fauna.

Zusammenfassung

Das Taxon *Crocuta crocuta praespelaea* SCHÜTT ist mit zwei isolierten Zähnen aus den fossilführenden Tonen von Voigtstedt bei Sangerhausen, DDR, vertreten. Die Zähne zeigen wahrscheinlich ein einziges, kleines, ziemlich altes Individuum an. Das Taxon ist für diese Fauna neu.

Резюме

Подвид *Crocuta crocuta praespelaea* SCHÜTT представлен двумя изолированными зубами, происходящими из костеносных глин Фойгтштедт около Зангерхаузена (ГДР). Зубы возможно принадлежат одной и той же особи, мелкого размера, довольно старой по индивидуальному возрасту. Этот таксон впервые установлен в данной фауне.

References

BISHOP, M. J., 1982: The mammal fauna of the early middle Pleistocene cavern infill site of Westbury-sub-Mendip, Somerset. — In: Brit. Mus. Nat. Hist., Spec. Pap., Palaeont. — London 28. — 108 pp.

KURTÉN, B., 1968: Pleistocene mammals of Europe. — London: Weidenfeld & Nicolson. — 317 pp.

—, 1969: Die Carnivoren-Reste aus den Kiesen von Süßenborn bei Weimar. — In: Paläont. Abh., A. — Berlin **3** (3/4). — S. 735 bis 756.

KURTÉN, B.; POULIANOS, A. N., 1977: New stratigraphic and faunal material from Petralona Cave, with special reference to the Carnivora. — In: Anthropos. — Athen **4**. — P. 47—130.

KURTÉN, B.; POULIANOS, A. N., 1981: Fossil Carnivora of Petralona Cave, status of 1980. — In: Anthropos. — Athens **8**. — P. 9—56.

SCHÜTT, G., 1971: Die Hyänen der Mosbacher Sande (Altpleistozän, Wiesbaden/Hessen) mit einem Beitrag zur Stammesgeschichte der Gattung *Crocuta*. — In: Mainzer naturwiss. Arch. — Mainz **10**. — S. 29—76.

THENIUS, E., 1965: Die Carnivoren-Reste aus dem Altpleistozän von Voigtstedt bei Sangerhausen in Thüringen. — In: Paläont. Abh., A. — Berlin **2** (2/3). — S. 537—564.

Quartärpaläontologie 6 101—117 Berlin 1986

Die quartären Vertebraten-Faunen in der SFR Jugoslawien

Von MIRKO MALEZ, Zagreb

Mit 4 Abbildungen und einer Tabelle

Inhalt

Abstractum

Auf der Grundlage geologischer und paläontologischer Forschungen zahlreicher Autoren wird eine Synthese zum modernen Forschungsstand der Verbreitung stratigraphischer Fundpunkte quartärer Säugetiere Jugoslawiens vorgelegt. Die Ablagerungen des frühesten Pleistozäns sowie des Unterpleistozäns und des Oberpleistozäns sind sehr reich an fossilen Faunenresten. Sie wurden eingehend studiert, und über sie in verschiedenen Abhandlungen publiziert. Ablagerungen des Mittelpleistozäns sind hingegen in diesem Gebiet selten und dazu arm an fossilen Säugetierresten.

On the basis of geological and palaeontological research by numerous authors, a synthesis of the most recent states of our knowledge on the distribution and stratigraphical position of vertebrates in the Quarternary deposits of Yugoslavia has been made. The deposits of the Earliest and the Lower Pleistocene as well as those of the Upper Pleistocene are very rich in fossil vertebrates. Those deposits have been thoroughly studied and the results have been published in several proceedings. Deposits of the Middle Pleistocene, however, are limited in this area, and in general are poor in fossil vertebrate remains.

На основании геологических и палеонтологических исследований многих авторов проведен обзор современного состояния наших знаний по распространению и стратиграфическому положению позвоночных в четвертичных отложениях Югославии. Отложения древнейшего и нижнего плейстоцена, а также верхнего плейстоцена очень богаты ископаемыми остатками позвоночных. Эти отложения всесторонне изучены, а результаты опубликованы в нескольких трудах. Отложения среднего плейстоцена, однако, имеют в этом районе ограниченное распространение и вообще бедны ископаемыми остатками позвоночных.

1. Einleitung

In Jugoslawien sind die pleistozänen Faunengemeinschaften heute relativ gut bekannt und für einige Lokalitäten und geographische Räume im Detail bearbeitet. Eine Übersicht und Synthese der Quartärfauna des gesamten Territoriums Jugoslawien wurde in ihren Grundzügen erst in neuester Zeit gegeben (MALEZ 1979, 1982).

Die ersten wissenschaftlichen Bearbeitungen unserer Quartärfaunen stammen aus dem 19. Jahrhundert (PARTSCH 1827; WOLDŘICH 1874, 1882, 1897; NEUMAYR 1882, GORJANOVIČ-KRAMBERGER 1883, 1884; HOCHSTETTER 1881; KIŠPATIĆ 1885; und andere). Später, Anfang des 20. Jahrhunderts begann eine Periode intensiverer Erforschung der Quartärfaunen Jugoslawiens, die ihren Höhepunkt nach dem zweiten Weltkrieg erreichte. Es muß jedoch betont werden, daß nicht in allen Teilen Jugoslawiens die Quartärfaunen gleichermaßen gut erforscht und untersucht wurden; das hing in erster Linie von den wissenschaftlichen Interessen einzelner Institute oder Forscher ab.

Für die Kenntnis der pleistozänen Faunen Sloweniens sind Arbeiten von HOCHSTETTER (1881), SEIDL (1912), BATTAGLIA (1929), BRODAR (1938), ANELLI (1941, 1947), KOS (1923, 1939, 1944), ADAM (1958), THENIUS (1958), DROBNE (1964, 1975), RIEDEL (1977) und POHAR (1981) von großer Bedeutung, ganz besonders aber die etwa dreißig Abhandlungen von RAKOVEC (1935—1975).

Die pleistozänen Faunen Kroatiens sind ebenfalls gut erforscht und in Arbeiten von WOLDŘICH (1874, 1882), GASPERINI (1885, 1887), KIŠPATIĆ (1885), GORJANOVIČ-KRAMBERGER (1883, 1884, 1912, 1913), KORMOS (1912, 1931, 1933), TOULA (1907), LAMBRECHT (1915, 1933), LEONARDI (1934), GIROMETTA (1935), HERAK (1947), KOWALSKI (1958), MALEZ-BAČIĆ (1975, 1979), RABEDER (1983) sowie in zahlreichen Beiträgen und Monographien von MALEZ (1956—1982) dargestellt worden.

Angaben über die Quartärfaunen von Bosnien und der Herzegowina finden wir unter anderem in Abhandlungen von FIALA (1893), WOLDŘICH (1897), GRIMMER (1898), MALEZ (1957, 1964, 1968a, 1968b, 1968c, 1969, 1970, 1973, 1979, 1980), MALEZ und PEPEONIK (1969), MALEZ, RUKAVINA und SLIŠKOVIĆ (1972), MALEZ, RUKAVINA, SLISKOVIC und KAPEL (1974), MALEZ, RUKAVINA und SLIŠKOVIĆ (1978), V. MALEZ (1983).

Für Serbien ist in erster Linie die systematische Bearbeitung der pleistozänen Säugetierfauna der Risovača-Höhle bei Arandjelovac durch RAKOVEC (1965) zu erwähnen sowie die Bearbeitung der Vertebraten der prähistorischen Fundstelle Lepenski Vir durch BÖKÖNYI (1969). Ferner liegen noch eine Anzahl kürzerer Artikel über einzelne Funde oder Fundorte von quartären Säugern vor, wie z. B. von BABIĆ (1952), CVIJIĆ (1891, 1912), ĆIRIĆ (1952), JOVANOVIĆ (1891), LASKAREV (1922, 1938), MARKOVIĆ-MARJANOVIĆ (1971), MILIČEVIĆ (1900), STEPANOVIĆ (1953), VESELINOVIĆ-ČIČULIĆ (1952), VUJISIĆ (1958), ŽUJOVIĆ (1900, 1905).

Die Kenntnis der Quartärfauna von Crna Gora (Montenegro) gründet sich hauptsächlich auf die Bearbeitung des osteologischen Materials aus Crvena Stijena bei Petrovići (RAKOVEC 1958; MALEZ 1962, 1967, 1975; MALEZ und MALEZ-BAČIĆ 1974; PAUNOVIĆ 1983), während die Untersuchungen von GAREVSKI (1957, 1964, 1969, 1970) für die Kenntnis der Quartärfaunen Makedoniens eine Grundlage bilden.

2. Die Faunen des ältesten Pleistozäns

Als ältestes Pleistozän betrachten wir den Zeitraum vom Ende des Pliozäns bis zum Ende des Günz-Glazials (Villafranchium), das wir in das untere (Donau-Glazial), das mittlere (Donau/Günz-Interglazial) und das obere (Günz-Glazial) unterteilen.

In der Fauna des unteren Villafranchiums sind noch Pliozän-Relikte, z. B. Mastodonten, Tapire und einige Nashornarten vorhanden (Abb. 1). Fossilführende Ablagerungen dieses Zeitraums sind in Slowenien und Kroatien bekannt. Aus dem Hangenden der Kohlenserie von Velenje (Slowenien) stammen Skelettreste zweier Mastodonten-Arten: *Anancus arvernensis* und *Mastodon borsoni*. Weiter konnten Reste von *Anancus arvernensis* in bituminösen Bildungen im Liegenden der Tongrube Dronjkova Glavica in Strmica nördlich von Knin in Dalmatien gefunden werden, ferner in den basalen Ablagerungen der Tongrube Grmoščica westlich von Zagreb, in den sandigen Sedimenten vom Berge Štrigovčak bei Štrigova, in Medjimurje, in Rastušje bei Slavonski Brod.

Die Ablagerungen des mittleren und oberen Villafranchiums in Jugoslawien sind besser erforscht. Aus dieser Zeit stammt der größte Teil der limnischen Sedimente der Tongruben von Strmica. Die bedeutendsten Vertreter dieses Faunenhorizonts sind *Archidiskodon meridionalis, Dicerorhinus etruscus*, verschiedene Gattungen bzw. Arten früher Cerviden *(Anoglochis ramosus, Euctenoceros senezensis, Eucladoceros dicranios, Cervus philisi), Panthera* cf. *toscana, Ursus etruscus* usw. Die Knochenbrekzien von Šandalja I bei Pula in Istrien wurden wahrscheinlich teilweise während des mittleren und teilweise im oberen Villafranchium gebildet. Die fossile Fauna besteht aus folgenden Arten: *Macaca florentina, Hypolagus brachygnathus, Oryctolagus* cf. *lacosti, Allocricetus bursae, Mimomys* sp., *Canis etruscus, Ursus etruscus, Ursus thibetanus mediterraneus, Euryboas lunensis, Crocuta perrieri, Equus stenonis, Dicerorhinus etruscus, Sus strozii, Euctenoceros senezensis, Anoglochis ramosus, Cervus acoronatus, Cervus philisi, Dama nestii, Leptobos stenometopon, Francolinus* sp., *Columba* sp., *Turdus* sp. und *Testudo* sp. Diese Skelettreste sind als Überbleibsel der Jagdbeute altpaläolithischer Jäger anzusehen, die diese Tiere einstmals in die Höhlen brachten. Neben dieser fossilen Fauna wurden weitere Spuren des ältesten Paläolithikums in Jugoslawien — wie auch an anderen Stellen Südeuropas — entdeckt, die beweisen, daß der fossile Mensch bereits im ältesten Pleistozän den europäischen Kontinent besiedelte. Das Klima war damals

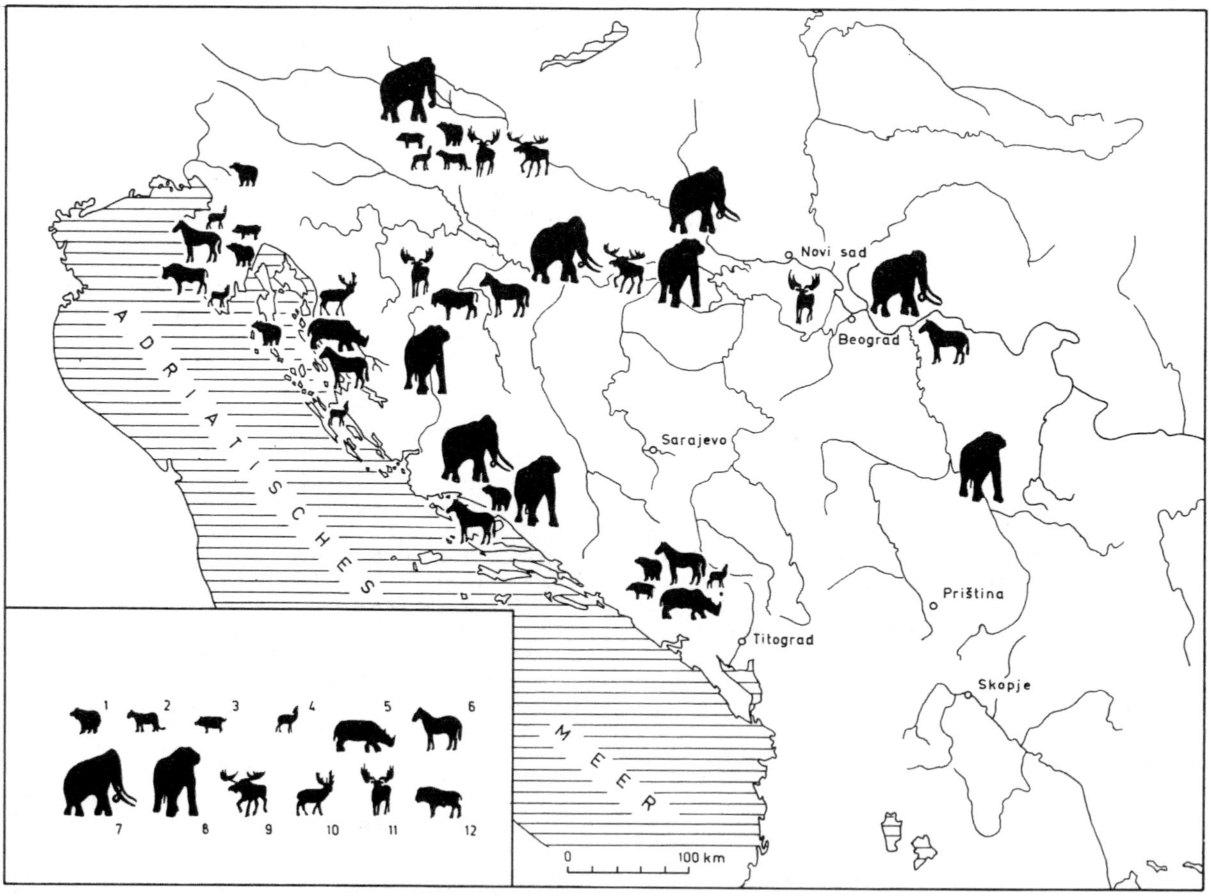

Abb. 1. Die Fauna des ältesten und unteren Pleistozäns in der SFR Jugoslawien (Villafranchium bis Mindel)

1 — *Trogontherium cuvieri*; 2 — *Hystrix refossa*; 3 — *Canis etruscus*; 4 — *Ursus etruscus*; 5 — *Ursus thibetanus mediterraneus*; 6 — *Hyaena perrieri*; 7 — *Acinonyx* sp.; 8 — *Felis toscana*; 9 — *Praealces gallicus*; 10 — *Euctenoceros* sp.; 11 — *Eucladoceros* sp.; 12 — *Orthogonoceros* sp.; 13 — *Cervus philisi*; 14 — *Sus strozzii*; 15 — *Hemitragus stehlini*; 16 — *Macaca florentina*; 17 — *Camelus* sp.; 18 — *Equus stenonis*; 19 — *Leptobos stenometopon*; 20 — *Mastodon borsoni*; 21 — *Anancus arvernensis*; 22 — *Archidiskodon meridionalis*; 23 — *Tapirus arvernensis*; 24 — *Dicerorhinus etruscus*

bedeutend wärmer als heute und das Landschaftsbild unseres Raums wurde durch Steppen und Savannen, durchsetzt von kleineren Waldgebieten, bestimmt.

Im obersten Villafranchium Sloweniens lebte *Praealces* aff. *gallicus*, von dem ein Unterkieferfragment mit vier Zähnen in den limnischen Ablagerungen der Ziegelei Vič bei Ljubljana gefunden wurde. Aus diesem Zeitraum stammen auch die fossilen Faunenreste der Knochenbrekzien des Zadar-Archipels (Iž, Rutnjak, Rava) mit folgenden Formen: *Arvicola* sp., *Acinonyx* cf. *pardinensis*, *Equus stenonis*, *Cervus* cf. *philisi*, *Euctenoceros* sp., *Dama* sp., *Capreolus süssenbornensis*, *Leptobos* cf. *stenometopon*, *Lacerta* cf. *viridis*.

3. Die Faunen des unteren Pleistozäns

In das untere Pleistozän stellen wir die Ablagerungen des Günz/Mindel-Interglazials (= Cromer) und des Mindel-Glazials, die von mehreren Fundstellen Jugoslawiens bekannt wurden (Abb. 1).

In Slowenien ist aus diesem Zeitraum *Ursus thibetanus mediterraneus* nachgewiesen worden (Brekzie von Vrhovlje, in der Nähe der italienischen Grenze, ein Zahn), ferner wahrscheinlich *Dicerorhinus etruscus* (Ulna aus der Tongrube der Ziegelei Zalog unweit Novo Mesto). Aus Ablagerungen des Günz/Mindel-Interglazials stammen Reste von *Hippopotamus antiquus* (Grotte Postojnska Jama, Geröllablagerungen). Wahrscheinlich aus gleichaltrigen Bildungen kommt der Flußpferd-Schädel von Čepićko Polje und der Nachweis von *Hemitragus* cf. *stehlini* aus Knochenbrekzien von Tar in Istrien.

In Kroatien wurden Faunengemeinschaften des unteren Pleistozäns an zahlreichen Fundstellen beobachtet, hauptsächlich in den Knochenbrekzien von Razvodje auf dem Berge Promina, auf der Halbinsel Marjan bei Split, in Podumci bei Unešić, von Tatinja Draga bei Karlobag, von Dubci über Luka Vrulja bei Makarska und zahlreichen anderen Fundstellen. Die Knochenbrekzien enthalten dabei hauptsächlich Skelettreste von Mikromammaliern, seltener Knochen und Zähne der großen kontinentalen Säuger wie Nashörner, Pferde, Hirsche, Rinder und von Raubtieren. Die Knochenbrekzien des Günz/Mindel-Interglazials enthalten eine typisch warmzeitliche Faunengemeinschaft, die darauf hindeutet, daß zu dieser Zeit in unserem Gebiet noch immer warme Steppen, durchsetzt von größeren und kleineren Waldungen, verbreitet waren. Im Verlauf des Mindel-Glazials wurde die Seekreide von Kninsko Polje abgelagert, die zahlreiche Skelettreste der primitiven Form des Steppenelefanten (*Parelephas trogontherii*) erbrachte, ebenfalls Reste der Waldelefanten der Gattungen *Palaeoloxodon* und *Hespero-*

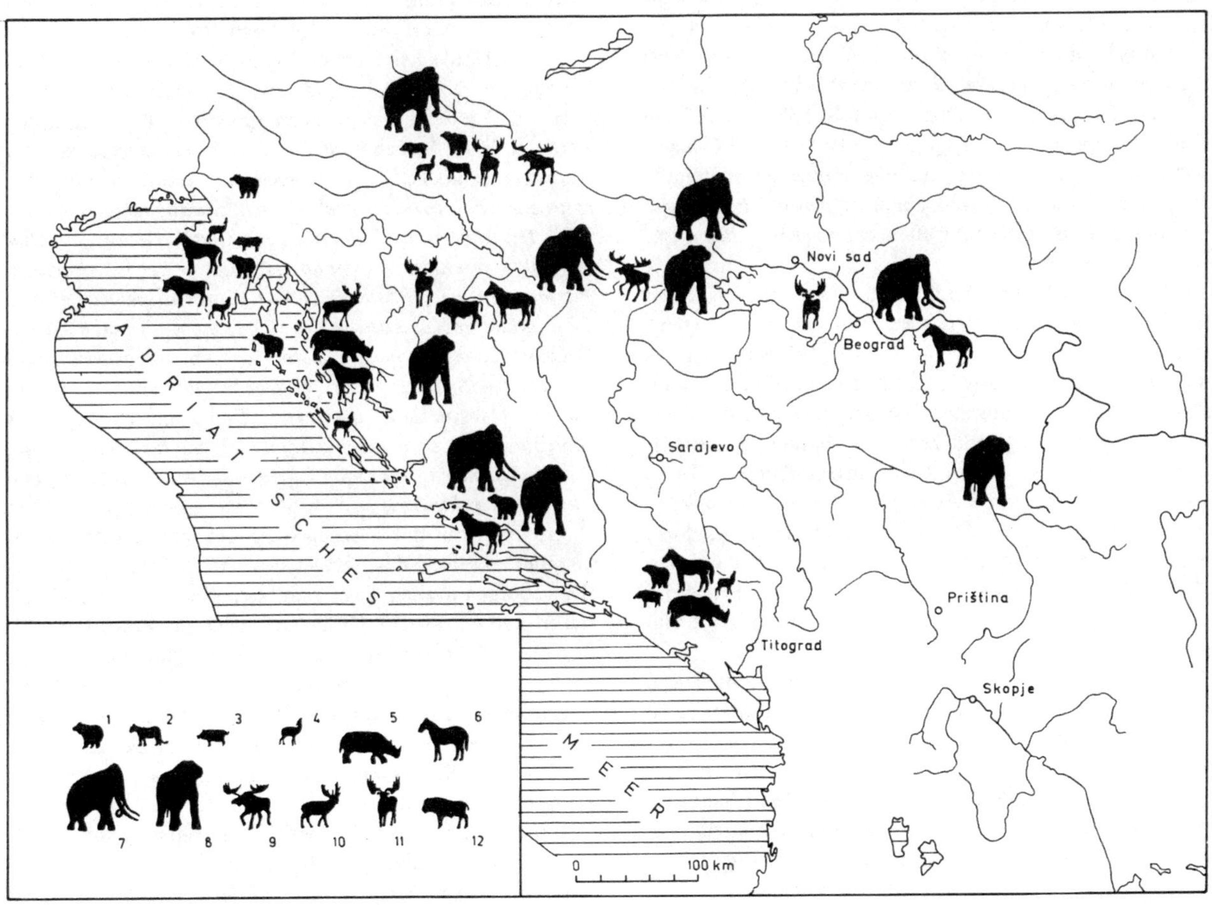

Abb. 2. Die Fauna des mittleren Pleistozäns in der SFR Jugoslawien (Mindel/Riß bis Riß)

1 — *Ursus deningeri*; 2 — *Panthera leo* sp.; 3 — *Sus scrofa*; 4 — *Capreolus süssenbornensis*; 5 — *Dicerorhinus hemitoechus*; 6 — *Equus mosbachensis*; 7 — *Parelephas trogontherii*; 8 — *Palaeoloxodon antiquus*; 9 — *Alces latifrons*; 10 — *Cervus acoronatus*; 11 — *Megaloceros giganteus*; 12 — *Bison schoetensacki*

loxodon. Diese großen kontinentalen Pflanzenfresser deuten darauf hin, daß das Klima in unserem Raum während des Mindel-Glazials nicht so kalt war wie in Mitteleuropa.

4. Die Faunen des mittleren Pleistozäns

Die Ablagerungen des mittleren Pleistozäns umfassen das Mindel/Riß-Interglazial bis zum Ende des Riß-Glazials. Die Bildungen dieses Zeitraums sind in Jugoslawien bislang ungenügend erforscht, und nur an einzelnen Stellen wurden fossile Vertebraten-Reste entdeckt (Abb. 2).

Im Tal der Drau, bei Janezov Vrh unweit Vuhred, wurde eine Elephantiden-Mandibel gefunden, die eine Übergangsstufe zwischen *Parelephas trogontherii* und *Mammuthus primigenius* darstellt. Ebenfalls zu einer frühen Evolutionsstufe des *Mammuthus primigenius* ist ein Skelettfund zu stellen, der aus limnischen Ablagerungen der Ziegelei Bobovk unweit Kranj stammt. Für das Gebiet von Slowenien muß auf zwei bedeutende Funde verwiesen werden: in Ablagerungen des Riß-Glazials von Betalov Spodmol bei Postojna konnte noch *Ursus deningeri* nachgewiesen werden und aus einem interstadialen Stratum des Riß-Komplexes von Jama bei Črni Kal oberhalb Kopar stammt *Capreolus* cf. *süssenbornensis*.

Die bisher vollständigste Fauna aus der Endphase des Mindel/Riß-Interglazials in Kroatien enthielten die Knochenbrekzien von Milovac bei Starigrad südlich des Velebitgebirges. Diese fossile Fauna umfaßt noch primitive Vertreter des *Dicerorhinus kirchbergensis*-Formenkreises, ferner Pferde der Gruppe des *Equus mosbachensis*, dazu Cerviden, Rodentier und Vögel. Die liegenden Horizonte des Komplexes der Höhle Vindija bei Donja Voća wurden im Verlauf des Riß-Glazials sedimentiert und enthalten folgende Faunengemeinschaft: *Canis lupus*, *Ursus deningeri*, *Panthera pardus*, *P. spelaea*, *Trogontherium* sp., *Capreolus* cf. *süssenbornensis*, *Bison* sp., *Capra ibex* und weitere.

In Bosnien und in der Herzegowina konnte bis jetzt noch keine Faunengemeinschaft aus dem mittleren Pleistozän entdeckt werden. Dagegen sind in Serbien Ablagerungen des mittleren Pleistozäns relativ häufig und von mehreren Fundstellen liegen fossile Säugetier-Assoziationen vor, die allerdings noch nicht umfassend bearbeitet wurden. So enthält das mittlere Pleistozän dieses Raums neben *Corbicula fluminalis* auch Skelettreste von *Parelephas trogontherii*, *Megaloceros giganteus*, *Alces alces*, *Bison priscus*. Die Hauptfundorte dieses Faunenkomplexes sind das Flußbett der Donau unterhalb der Brücke von Pančevo, Grabrovac an der Peripherie von Obrenovac, die Mündung der Kolubara in die Save und das Flußbett der Theiß.

Die liegenden Horizonte der Crvena Stijena (Montenegro) wurden im mittleren Pleistozän gebildet, die Straten XXV bis XXXI im Verlauf des Riß-Glazials. Diese Horizonte enthielten Reste folgender Taxa: *Canis lupus*, *Ursus arctos priscus*, *Ursus thibetanus mediterraneus*, *Panthera pardus*, Wildpferde der Gruppen *Equus caballus germanicus* und *Equus mosbachensis abeli*, *Megaloceros giganteus*, *Rupicapra* sp., *Capra ibex*, *Testudo hermanni*, *T. graeca* usw. In Makedonien stammt die Knochenbrekzie beim Dorfe Manastir wahrscheinlich aus dem mittleren Pleistozän; sie wurde im Verlauf des Mindel/Riß-Interglazials abgelagert.

Die mittelpleistozänen Faunenreste von Betalov Spodmol bei Postojna, Vindija bei Donja Voća und Crvena Stijena (Montenegro) gelangten vorwiegend als Jagdbeute des prähistorischen Menschen in die Höhlensedimente, da darin noch andere Spuren seiner Anwesenheit enthalten sind (Herdstellen und paläolithische Werkzeuge).

5. Die Faunen des oberen Pleistozäns

In Jugoslawien sind die Faunengemeinschaften des oberen Pleistozäns die häufigsten und vollständigsten. Sie sind gut erforscht, da ihnen unter anderem als Existenzgrundlage des Neandertalers, dem Träger der mittelpaläolithischen Kulturen, und der späteren sapienten Hominiden-Populationen des jüngeren Paläolithikums großes Interesse zugewendet wird. Das obere Pleistozän umfaßt den Zeitraum vom Riß/Würm-Interglazial bis zum jüngsten Abschnitt der Dryas im Spätglazial. In diesem Zeitraum setzten bedeutende klimatische Schwankungen ein, die eine Folge von Veränderungen in der Zusammensetzung von Flora und Fauna auslösten. Für das Riß/Würm-Interglazial ist eine typische Warmzeit-Fauna charakteristisch (Abb. 3), wogegen die Würm-Interstadiale durch warm-gemäßigte Assoziationen gekennzeichnet sind. Während aller Stadiale des Würm-Glazials herrschten ausgesprochen kaltzeitliche Assoziationen vor (Abb. 4).

Aus den letztinterglazialen Ablagerungen der Betalov Spodmol bei Postojna, Jama bei Črni Kal nördlich von Kopar, Kamnitnik bei Škofja Loka, Dolarjeva Jama bei Logatec, des Abris von Hušnjakovo Brdo in Krapina, der Veternica-Höhle bei Zagreb, Velika Pećina (Ravna Gora), Vindija bei Donja Voća und Crvena Stijena (Montenegro) liegen die Reste warmzeitlicher Faunen vor. Die wichtigsten Vertreter dieser Assoziationen sind: *Talpa europaea*, *Erinaceus europaeus*, *Rhinolophus ferrumequinum*, *Myotis myotis*, *Plecotus auritus*, *Miniopterus schreibersii*, *Sciurus vulgaris*, *Castor fiber*, *Hystrix cristata minor*, *Glis glis*, *Clethrionomys glareolus*, *Arvicola terrestris*, *Apodemus sylvaticus*, *Lepus europaeus*, *Panthera spelaea*, *Panthera pardus*, *Crocuta spelaea*, *Martes foina*, *Meles meles*, *Lutra lutra*, *Vulpes vulpes*, *Ursus spelaeus*, *Palaeoloxodon antiquus*, *Dicerorhinus kirchbergensis*, *Equus germanicus*, *Sus scrofa*, (?) *Hippopotamus antiquus*, *Cervus elaphus*, *Megaloceros giganteus*, *Dama dama*, *Capreolus capreolus*, *Bos primigenius*, *Aquila chrysaetus*, *Gyps fulvus*, *Tetrastes bonasia*, *Anas* sp., *Turdus viscivor*, *Garrulus glandarius*, *Emys orbicularis*.

Assoziiert mit den Faunenresten des letzten Interglazials konnten Artefakte des Levalloisien und des frühen Moustérien geborgen werden. Der Fossilbestand einiger dieser Fundstellen ließ statistische Untersuchungen zur Frage der häufigsten Jagdtiere dieser paläolithischen Hominiden-Gruppe zu. So überwiegen z. B. im Abri von Krapina in den Interglazial-Straten (Riß/Würm) die Reste von *Dicerorhinus kirchbergensis*, gefolgt von *Castor fiber*, *Dama dama*, *Cervus elaphus* und *Bos primigenius*. Demnach war *Dicerorhinus kirchbergensis* das häufigste Beutetier der Neandertaler von Krapina, was Ausdruck spezifischer geomorphologischer und paläoökologischer Verhältnisse ist.

Ähnlich wie in Krapina steht in der Jagdbeute der letztinterglazialen Neandertaler von Vindija *Dicerorhinus kirchbergensis* an erster Stelle, dann folgen die Wildrinder, mehrere Hirscharten, Höhlenbär und weitere. Von anderen Fundstellen liegen unterschiedliche Jagd-Statistiken vor. So erbeuteten z. B. in der Endphase des Riß/Würm-Interglazials

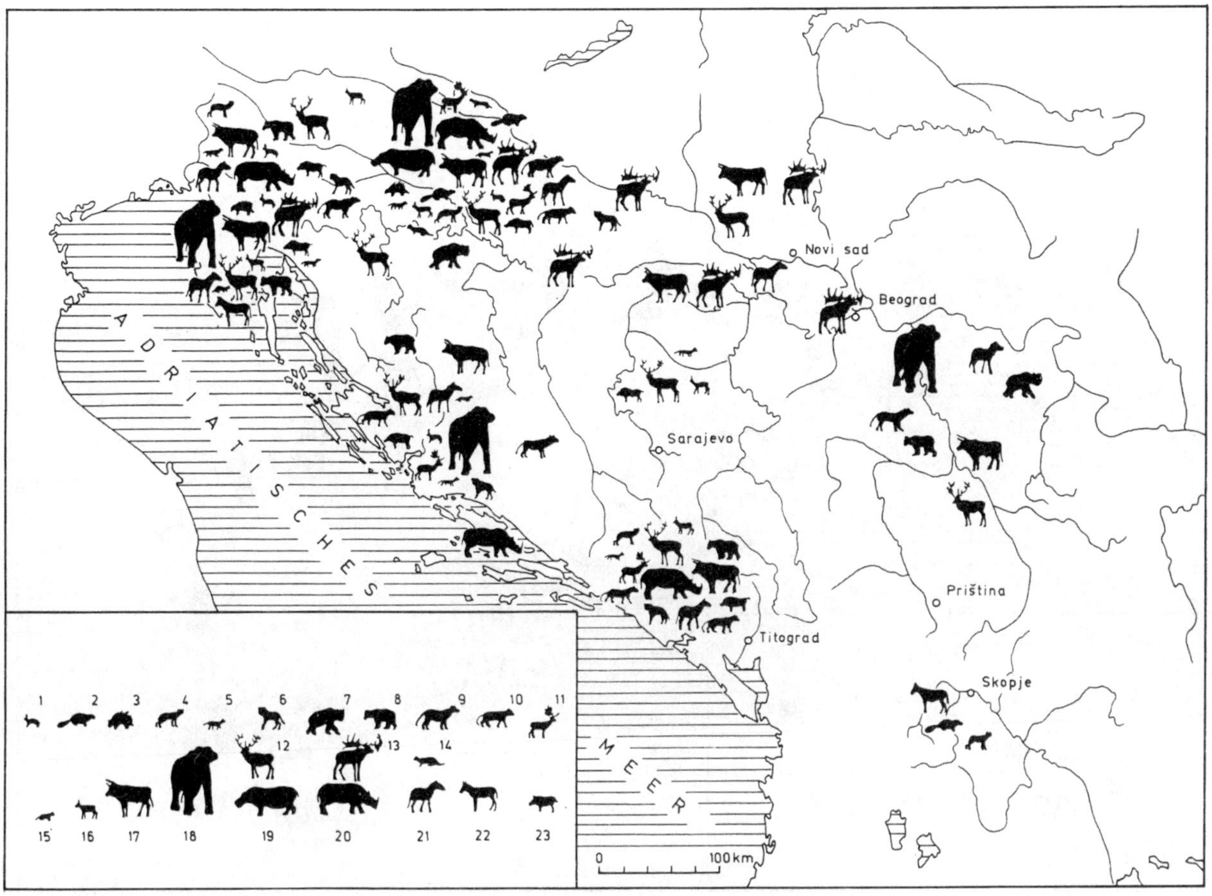

Abb. 3. Warmzeitliche Faunen im oberen Pleistozän der SFR Jugoslawien (Riß/Würm, Würm 1/2, Würm 2/3)

1 — *Lepus europaeus*; 2 — *Castor fiber*; 3 — *Hystrix cristata*; 4 — *Canis lupus*; 5 — *Vulpes vulpes*; 6 — *Crocuta spelaea*; 7 — *Ursus spelaeus*; 8 — *Ursus arctos priscus*; 9 — *Pathera spelaea*; 10 — *Pathera pardus*; 11 — *Dama dama*; 12 — *Cervus elaphus*; 13 — *Megaloceros giganteus*; 14 — *Lutra lutra*; 15 — *Meles meles*; 16 — *Capreolus capreolus*; 17 — *Bos primigenius*; 18 — *Palaeoloxodon antiquus*; 19 — *Hippopotamus antiquus*; 20 — *Dicerorhinus kirchbergensis*; 21 — *Equus germanicus*; 22 — *Asinus hydruntinus*; 23 — *Sus scrofa*

die Träger der Moustérien-Kultur der Höhle Veternica (Medvednica-Gebirge) hauptsächlich den Höhlenbären und nur ein kleiner Teil der Jagdbeute waren Nashörner, Wildrinder, Hirsche, Stachelschweine und andere.

In der Jagdtier-Statistik des letztinterglazialen Menschen der Crvena Stijena stehen die Wildpferde aus den Gruppen *Equus caballus*, *E. germanicus* und *E. mosbachensis* an erster Stelle. Dann folgen — in bedeutend geringerer Anzahl — Wildrinder, Hirsche, Rehe, Wildschweine und andere.

Mit dem Würm-Glazial veränderte sich die Zusammensetzung der Faunengesellschaften Jugoslawiens von Grund auf. Es begannen die großen Faunenverschiebungen als Folge der Klimadepression und der Vergletscherung weiter Gebiete Nord- und Mitteleuropas sowie der Alpen. Der größte Teil des jugoslawischen Territoriums lag während des Würm-Glazials in der subalpinen bzw. in der periglazialen Zone mit relativ günstigen Bedingungen für die Existenz kaltzeitlicher Faunengesellschaften und paläolithischer Besiedlung (Höhlen, Abris usw.).

Die wichtigsten Vertreter der letztglazialen Faunengesellschaften unseres Raums waren *Lepus timidus timidus* und *L. timidus varronis*, *Ochotona pusilla*, *Dicrostonyx torquatus*, *Marmota marmota*, *Cuon alpinus europaeus*, *Alopex lagopus*, *Gulo gulo*, *Coelodonta antiquitatis*, *Mammuthus primigenius*, *Rangifer tarandus*, *Alces alces*, *Bison priscus*, *Ovibos moschatus*, *Rupicapra rupicapra*, *Capra ibex*, *Saiga tatarica* sowie *Lagopus lagopus* und *L. mutus*.

Die Untersuchung dieser letztglazialen Assoziationen, die zugleich die Basis für eine paläolithische Besiedlung dieses Raums bildeten, ermöglicht es, die lokalen paläoklimatischen und paläoökologischen Verhältnisse zu rekonstruieren, die Dynamik und Richtung der Migrationen zu studieren sowie die Paläogeographie ihres Verbreitungsareals aufzuzeigen.

Zur Zeit des Würm-Glazials waren in Südosteuropa zwei Unterarten des Schneehasen verbreitet: die polare Unterart *Lepus timidus timidus* und die alpine Unterart *Lepus timidus varronis*. Die polare Unterart war vorwiegend in den Ebenen anzutreffen. Die bekanntesten Fundstellen Jugoslawiens sind: Betalov Spodmol, Parska Golobina und Ovčja Jama in Slowenien, Romualdo-Höhle, Šandalja II bei Pula, Pećina u Brini bei Drniš, Velika Pećina (Ravna Gora), Bukovac-Höhle bei Lokve, Vindija bei Donja Voća und Veternica-Höhle bei Zagreb in Kroatien, die obere Bijambara-Höhle und Zelena Pećina bei Mostar in Bosnien und in der Herzegowina. Die alpine Unterart des Schneehasen war mehr in der gebirgigen Landschaft des Dinarischen Karstes verbreitet, ebenfalls im Gebiet der Alpen, wo diese Subspezies heute noch in Höhen oberhalb 1300 m anzutreffen ist. Die alpine Unterart wurde bisher in den oberpleistozänen Ablagerungen von Ovčja Jama in Slowenien festgestellt, ferner in Šandalja II bei Pula,

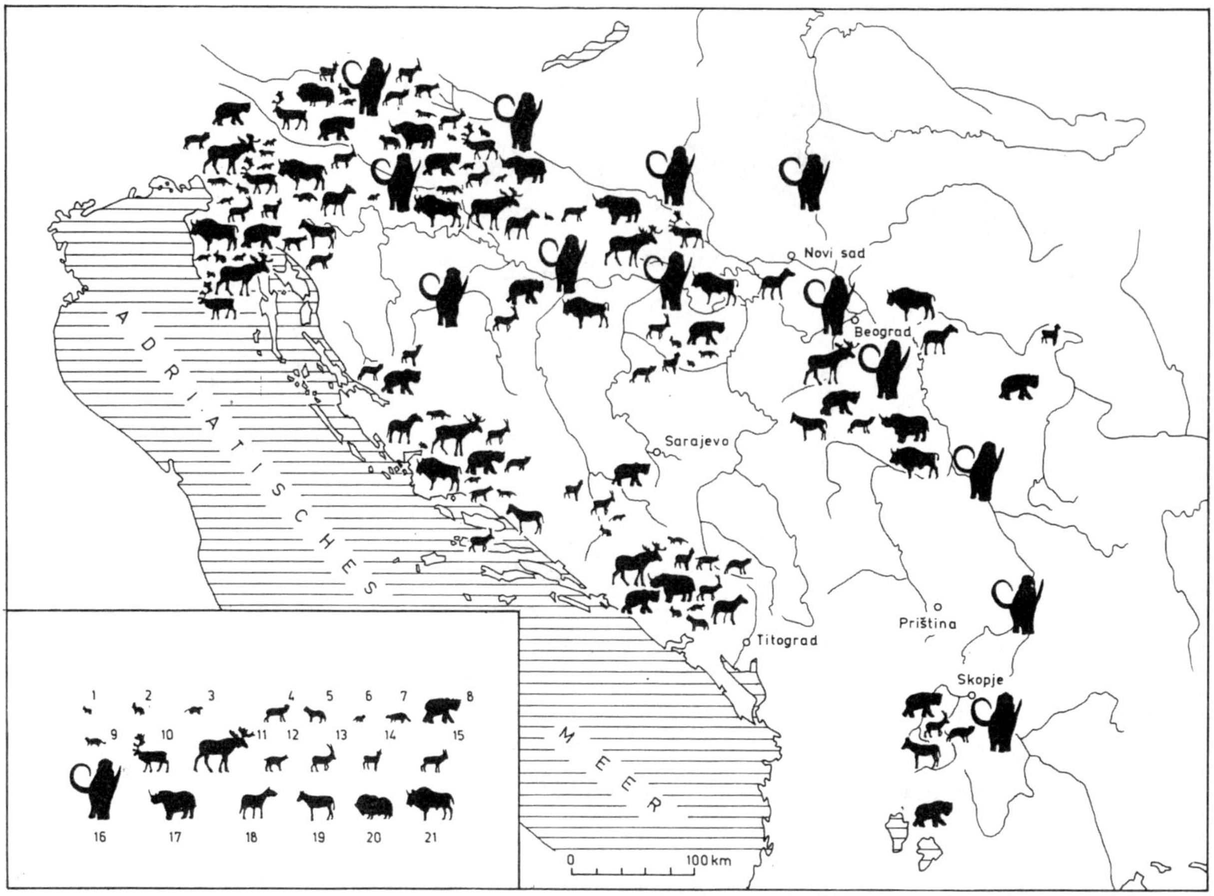

Abb. 4. Kaltzeitliche Faunen im oberen Pleistozän in der SFR Jugoslawien (Würm 1, Würm 2, Würm 3)

1 — *Ochotona pusilla*; 2 — *Lepus timidus*; 3 — *Marmota marmota*; 4 — *Canis lupus*; 5 — *Cuon alpinus europaeus*; 6 — *Alopex lagopus*; 7 — *Gulo gulo*; 8 — *Ursus spelaeus*; 9 — *Felis silvestris*; 10 — *Rangifer tarandus*; 11 — *Alces alces*; 12 — *Lynx lynx*; 13 — *Capra ibex*; 14 — *Rupicapra rupicapra*; 15 — *Saiga tatarica*; 16 — *Mammuthus primigenius*; 17 — *Coelodonta antiquitatis*; 18 — *Equus caballus fossilis*; 19 — *Asinus hydruntinus*; 20 — *Ovibos moschatus*; 21 — *Bison priscus*

Kupići Pećina (Planik) in Kroatien, Zelena Pećina bei Mostar in der Herzegowina und Crvena Stijena bei Petrovići (Montenegro); letzterer ist zugleich der südlichste Fundpunkt seiner spätpleistozänen Verbreitung in Südosteuropa.

Reste von *Ochotona pusilla* waren bis vor kurzem nur aus spätpleistozänen Ablagerungen einiger Fundstellen in Österreich, der Ungarischen VR und der SR Rumänien bekannt, das heißt, aus den nördlichsten Gebieten Südosteuropas. Neuere Forschungen haben nun gezeigt, daß diese Spezies des Steppenbiotops im Spätpleistozän ein viel größeres Areal Südosteuropas einnahm, als bisher angenommen wurde. Fossile Reste von *Ochotona pusilla* wurden in der Veternica-Höhle gefunden, in Kamenika bei Srednji Lipovac in Slawonien, Vindija bei Donja Voća, Šandalja II bei Pula und in der oberen Höhle von Bijambarska in Mittelbosnien. In neuester Zeit wurden gut erhaltene Reste dieser Spezies in Istočna Pećina in Brina bei Drniš nachgewiesen, dies ist der südlichste Verbreitungspunkt dieser Form in Südosteuropa.

Fossile Reste von *Dicrostonyx torquatus* wurden bislang nur an einer Fundstelle Jugoslawiens entdeckt und zwar in Babja Jama in der Nähe von Dob bei Domžale in Slowenien. Die Reste fanden sich im liegenden Horizont des Kulturstratums, das der Endphase des Würm-Glazials entspricht. *Dicrostonyx torquatus* ist ein wichtiges klimatisches und ökologisches faunistisches Element, da diese Form im letzten Glazial die periglazialen Tundren bewohnte und als Indikator für extreme Klimaverhältnisse anzusehen ist. Der Fund dieses Nagers in Babja Jama bezeichnet die südlichste Grenze seines Areals im Würm-Glazial, was sich gut mit den Verbreitungsarealen zweier anderer arktischer Formen deckt — des Polarfuchses und des Moschusochsen (MALEZ 1972, Abb. 5 und 12).

Die spätpleistozänen südosteuropäischen Fundorte von *Marmota marmota* liegen im Gebiet des Dinarischen Karstes und in den Ausläufern der Ostalpen bis hin zu den Karpaten. In Slowenien wurden Reste dieses Nagers an zahlreichen Fundstellen entdeckt, von denen die wichtigsten Potočka Zijalka, Pod-Hom, Zagorice, Trbovlje, Mornova Zijalka, Špehovka, Hude Luknje, Ajdovska Jama, Parska Golobina, Križna Jama bei Lož, Ovčja Jama und Betalov Spodmol sind. In Kroatien wurden ebenfalls zahlreiche Reste des Alpenmurmeltiers gefunden; die bekanntesten sind Šandalja II, Druška Peć am Berg Učka, Pećina na Brehu in der Ćićarija, Pećina na Gradini und Pećina na Šaftici in Kastavština, Velika Pećina (Ravna Gora), Vindija, Krapina, Pećina u Brini bei Drniš und Močiljska Pećina bei Dubrovnik. Ein interessanter *Marmota*-Fundort ist auf der Insel Žirje (Šibernik-Archipel) an der heutigen Meeresküste und die südlichsten Vorkommen liegen nahe Zelena Pećina bei Mostar in der Herzegowina sowie bei Crvena Stijena in Montenegro.

Der Rotwolf (*Cuon alpinus europaeus*) gehört zu den seltensten oberpleistozänen Carnivoren Europas. Er ist als typisch asiatisches Element in diesen Assoziationen anzusehen. In Jugoslawien sind drei Fundpunkte mit *Cuon*-Resten bekannt: Veternica- und Vindija-Höhle in Kroatien sowie Crvena Stijena in Montenegro.

Reste des Polarfuchses (*Alopex lagopus*) wurden in spätpleistozänen Ablagerungen Jugoslawiens mit Sicherheit an sieben Fundstellen registriert. Drei Fundorte liegen in Slowenien (Betalov Spodmol, Križna Jama bei Lož, Ovčja Jama), und vier in Kroatien (Šandalja II, Veternica-Höhle, Velika Pećina, Vindija-Höhle). Alle genannten Fundorte liegen im nördlichen Teil Südosteuropas. Daraus ergibt sich, daß der Polarfuchs zur Zeit der maximalen Vereisung des Würm-Glazials in Südosteuropa nicht südlich von Save und Donau anzutreffen war, das heißt, nicht die Balkanhalbinsel bewohnte.

Etwas häufiger sind Fundstellen mit *Gulo*-Resten in Jugoslawien. In Kroatien wurden deren neun entdeckt: Veternica-Höhle, Vindija-Höhle, Velika Pećina, Krapina, Šandalja II, Steinbruch bei Pula (Spaltenfüllung), Zapadna Pećina in Brina, Trogir und die Knochenbrekzie von Hvar. In Slowenien wurde dieser große nordische Marder bis jetzt nur in Križna Jama bei Lož gefunden, während im benachbarten Triester Karst mehrere Fundstellen bekannt sind (Pytina Jama bei Grabrovica, die Höhle Kostanjevica bei Nabrežina usw.). Die Südgrenze der *Gulo*-Verbreitung während der Maximalvereisung des Würm-Glazials wird durch Fundstellen in Dalmatien markiert.

Spätpleistozäne *Coelodonta*-Reste wurden an zahlreichen Punkten in Jugoslawien entdeckt. In Kroatien konnten Schädel, Mandibeln, isolierte Zähne und einzelne Extremitätenknochen in Ferdinandovac, Varaždinske Toplice, Ludbreg, Županja, Osijek, Vukovar, Grubišno Polje, Virje und in der Vindija-Höhle gefunden werden. In Serbien wurde *Coelodonta antiquitatis* in den paläolithischen Straten der Risovača-Höhle bei Arandjelovac nachgewiesen, ebenfalls in Crvena Stijena (Montenegro). Bisher sind keine Funde aus Slowenien, Bosnien, der Herzegowina und aus Makedonien bekannt. Das südlichste *Coelodonta*-Vorkommen in Südosteuropa liegt in Megalopolis in Griechenland (MELENTIS 1966). Als Ergebnis kann somit festgestellt werden, daß *Coelodonta antiquitatis* im Spätpleistozän in großen Teilen Südosteuropas verbreitet war und vornehmlich die Ebenen bewohnte.

Das Mammut ist ebenfalls von zahlreichen südosteuropäischen Lokalitäten bekannt. In Slowenien sind die bedeutendsten Fundorte Bobovk bei Kranj, Nevlje, Limbuš bei Maribor, Kostanjevica an der Krka und Postojnjska Jama. In Kroatien sind heute mehr als dreißig Fundpunkte bekannt. Weitere Fundstellen liegen in Nordbosnien sowie in den Alluvionen der Flüsse Theiß, Donau, Save, Drau, Morava und ihrer Nebenflüsse, in denen Mammut-Reste geborgen wurden. Die südlichsten *Mammutus*-Fundorte Südosteuropas liegen in Makedonien beim Dorf Orešane unweit von Skopje und in der Höhle Saida in Megalopolis/Griechenland. Zusammenfassend kann man aufzeigen, daß das Mammut im Spätpleistozän die Ebenen von ganz Südosteuropa bewohnte. Skelettreste des Mammuts liegen ferner von zahlreichen paläolithischen Fundstellen vor, so von Nevlje und Kostanjevica in Slowenien, Samobor und Zarilac in Kroatien, Risovača-Höhle bei Arandjelovac in Serbien.

Skelettreste des Ren (*Rangifer tarandus*) wurden nur in den nördlichen Gebieten Jugoslawiens gefunden. Die bekanntesten Fundstellen in Slowenien sind Betalov Spodmol, Ovčja Jama, Križna Jama bei Lož, Vrhnika und Nevlje. In Kroatien konnten Skelettreste von *Rangifer* mit Sicherheit in den Ablagerungen des jüngeren Pleistozäns von Kanegra nachgewiesen werden, ebenfalls von Šandalja II in Kroatien. Dazu liegen *Rangifer*-Reste aus paläolithischen Horizonten von Velika Pećina und aus der Vindija-Höhle vor sowie aus pleistozänen Alluvionen der Drau bei Ferdinandovac. Einen interessanten Fund stellt das Rengeweih dar, das zusammen mit einem solchen vom Elch in Lößablagerungen des Würm-III-Stadials bei Vučedol unweit Vukovar gefunden wurde. Somit stimmt die Verbreitung des spätpleistozänen Ren in Südosteuropa mit dem Verbreitungsareal des gleichalterigen Polarfuchses überein (MALEZ 1972, Abb. 9).

Eine weit größere Verbreitung als das Ren hatte der Elch (*Alces alces*) im Spätpleistozän Südosteuropas. Seine Skelettreste wurden bisher von zahlreichen paläolithischen Fundstellen Kroatiens und Sloweniens bekannt sowie aus den pleistozänen Alluvionen von Save, Drau und Donau. Die bekanntesten Punkte in Slowenien sind Betalov Spodmol, Parska Golobina, Tomažkova Jama bei Sežana, Losova Jama bei Gabrovnica, das Ljubljanaer Moor und Jama bei Glažuta unweit von Ribnica. In Kroatien werden Elch-Reste von mehreren paläolithischen Fundorten genannt, so von Šandalja II, Romualdo-Höhle, Kanegra, Samobor, Velika Pećina, Veternica-Höhle, Vindija-Höhle, Krapina, Pećina u Brini bei Drniš, Pisana Stina am Opor-Gebirge und Zarilac im Becken von Slavonska Požega. Aus Lößablagerungen in Vučedol sind Elch-Reste bekannt, ebenso aus Alluvionen der Save bei Prelošćica, Stara Gradiška und Rajevo Selo. Ein gut erhaltenes Elch-Geweih wurde in Vrbovec bei Križevci entdeckt, sowie in Bosanska Kostajnica, Sremska Mitrovica, Šabac, an der Mündung der Kolubara in die Save, in der Theiß, an mehreren Stellen des Donaubetts. Der südlichste Fund stammt aus den paläolithischen Horizonten von Crvena Stijena in Montenegro. Im südosteuropäischen Spätpleistozän erschienen die Elche zu Beginn des Würm-Glazials. Damals besiedelten sie die nordwestlichsten Gebiete, in die sie vom Norden her über die Ostalpen eingedrungen waren. Zur Zeit der Maximalvereisung im Würm-III-Stadial war die Expansion des Elches nach Südosteuropa am stärksten.

Spätpleistozäne *Bison*-Reste (*Bison priscus*) sind von mehreren paläolithischen Fundstellen Jugoslawiens bekannt. Eine weit größere Anzahl zumeist vollständiger erhaltener Skelettreste stammt aus den oberpleistozänen Ablagerungen von Save, Drau, Donau und ihren Nebenflüssen. Die bekanntesten Lokalitäten in Slowenien sind Črni Kal, Križna Jama bei Lož und Ovčja Jama. In Kroatien kommen spätpleistozäne *Bison*-Reste von folgenden paläolithischen Fundpunkten: Krapina, Vindija-Höhle, Velika Pećina, Samobor, Kamenika bei Srednji Lipovac, Zarilac, Šandalja II, Kanegra und Pećina u Brini. Ferner wurden *Bison*-Reste im paläolithischen Stratum der Risovača-Höhle bei Arandjelovac gefunden. Der südlichste Fundpunkt aber ist Megalopolis (MELENTIS 1966). Bisonten, Mammuts und *Coelodonta*-Nashörner, die alle von Megalopolis nachgewiesen werden konnten, drangen im Spätpleistozän entlang Morava und Vardar weit nach Süden auf die Balkanhalbinsel vor.

Der typischste Vertreter der hochglazialen Fauna, der Moschusochse (*Ovibos moschatus*), konnte bisher in Jugoslawien nur in den oberpleistozänen Ablagerungen der Höhle Potočka Zijalka in den östlichen Karawanken gefunden werden. Wie die Lemminge vermochten die Moschusochsen während der spätpleistozänen Maximalvereisung nicht weiter nach Süden vordringen.

Die Skelettreste von *Rupicapra rupicapra* sind lediglich aus spätpleistozänen Ablagerungen paläolithischer Fundstellen in Kroatien, Slowenien, Bosnien und Montenegro bekannt. Im einzelnen handelt es sich um folgende Lokalitäten: Krapina, Veternica-Höhle, Velika Pećina, Vindija-Höhle, Gornja Cerovačka Pećina, Baba-Höhle auf Biokovo und Romualdo-Höhle in Kroatien, Potočka Zijalka in Slowenien, Gornja Bijambarska Pećina in Bosnien und Crvena Stijena in Montenegro. Letztere ist zugleich der südlichste Expansionspunkt im Verbreitungsareal der Gemse im oberen Pleistozän Südosteuropas. Als Pleistozän-Relikt bewohnen geringe Gemsen-Populationen noch heute sehr begrenzte Gebiete auf einigen Bergen des Dinarischen Karstes (Prenj, Čvrsnica, Ilica, Biokovo und weitere).

Etwas zahlreicher sind die Fundorte mit Skelettresten von *Capra ibex*. In Slowenien konnten sie von folgenden Fundstellen nachgewiesen werden: Mornova Zijalka bei Šoštanj, Jama oberhalb von Rašice bei Črnuče, Črni Kal und Jama treh bratov an der Kupa. In Kroatien sind verschiedene Lokalitäten bekannt: Romualdo-Höhle, Druška Peć auf Učka, Velika Pećina auf Ravna Gora, Vindija-Höhle, Veternica-Höhle, Pećina u Brini bei Drniš, Pisana Stina am Opor-Berge, Jama unterhalb Koštak bei Klis, Baba-Höhle auf Biokovo und Brusje auf der Insel Hvar. In Bosnien und in der Herzegowina wurden *Capra ibex*-Reste in Gornja Bijambarska Pećina bei Olovo, Zelena Pećina bei Mostar und in der Höhle Očauša bei Teslič gefunden. Der südlichste Fundpunkt in Südosteuropa ist die Höhle Makarovec in Makedonien.

Die *Saiga*-Antilope (*Saiga tatarica*) — heute ein typischer Bewohner zentralasiatischer Steppen — dehnte im oberen Pleistozän ihr Verbreitungsareal weit bis nach Westeuropa hin aus. Im Sedimentkomplex G (Stratum G_3) der Höhle Vindija konnten mehrere Skelettreste der Saiga entdeckt werden, wobei dieser Ort zugleich den südwestlichsten Punkt im Verbreitungsareal dieses Tiers im späten Pleistozän Südosteuropas darstellt.

Das Moorschneehuhn (*Lagopus lagopus*) und das Alpenschneehuhn (*Lagopus mutus*) treten an zahlreichen Fundstellen gemeinsam auf. Ihre Reste fanden sich weiterhin zusammen mit denen anderer kaltzeitlicher Formen oder solchen des Hochgebirges. An Orten geringerer Meereshöhe überwiegen die Reste des Moorschneehuhns. Im einzelnen handelt es sich um folgende Lokalitäten: Romualdo-Höhle, Šandalja II, Velika Pećina, Veternica-Höhle und Vindija-Höhle in Kroatien, Gornja Bijambarska Pećina und Zelena Pećina in Bosnien und in der Herzegowina und Crvena Stijena in Montenegro.

Das Alpenschneehuhn konnte an folgenden Fundstellen nachgewiesen werden: Potočka Zijalka, Betalov Spodmol und Hude Luknje in Slowenien, Šandalja II, Druška Peć auf Učka und Bukovac Pećina in Gorski Kotar in Kroatien, Gornja Bijambarska Pećina und Zelena Pećina in Bosnien und in der Herzegowina. Das spätpleistozäne Verbreitungsgebiet des Alpenschneehuhns deckt sich gut mit den gleichaltrigen Arealen von Schneehase, Murmeltier, Gemse und Steinbock (MALEZ 1972, Abb. 1, 3, 13, 14).

Im späten Glazial/frühen Postglazial, das heißt im Zeitabschnitt von der ältesten bis zur jüngsten Dryas, kam es wiederum zu großen Veränderungen in der Zusammensetzung der Faunengesellschaften Jugoslawiens. Durch die langsame Erwärmung und Verbesserung der klimatischen Verhältnisse setzte mit Rückzug der Gletscher eine Transgression des Meeres in die nördliche Adriatische Bucht ein, Inseln lösten sich vom Festlande und es formierte sich ein Relief, das dem heutigen ähnlich war.

In Slowenien bestand im späten Glazial eine sogenannte Mischfauna. In dieser Faunengesellschaft herrschten Alpenmurmeltier, Alpenschneehase, Rothirsch, Ren und Elch vor, die alle auf das Vorkommen von subarktischen Nadelwäldern hinweisen. Biber und Hamster kamen gegen Ende des Spätglazials auf, als das Klima bereits günstiger war.

Im Verlauf des Spätglazials waren Istrien, Nordkroatien, Slawonien, Syrmien, ganz Nordbosnien, Vojvodina und ein Teil von Serbien von Steppen — teilweise typische Lößsteppen — und Parksteppen bedeckt. Die für diesen Biotop charakteristische Faunengesellschaft zeigt besonders folgende Arten: *Ochotona pusilla*, *Spalax leucodon*, *Citellus citellus*, *Cricetus cricetus*, *Cricetulus migratorius*, *Equus caballus fossilis*, *Bison* sp. Den Steppencharakter der Landschaft kennzeichneten besonders die typischen Lößgastropoden wie *Pupilla muscorum*, *Columella columella*, *Vallonia tenuilabris*, *Vertigo parcendata*. Ab älterer Dryas ging das Steppenareal langsam zurück und Laubwälder breiteten sich aus, so daß in den Endphasen des Spätglazials (Allerød und jüngste Dryas) in unserem Gebiet eine typische Waldfauna vorherrschte, deren Hauptvertreter waren: *Cervus elaphus*, *Capreolus capreolus*, *Bos primigenius*, *Sus scrofa*, *Ursus arctos*, *Felis silvestris*, *Lyrurus tetrix*, *Tetrao urogallus*, *Tetrastes bonasia*.

6. Die Faunen des Holozäns

Mit Beginn des Holozäns setzte sich in unserem Raum eine typische Waldfauna durch, deren wichtigste Vertreter waren: Rothirsch, Reh, Wildschwein, Wildrind, Braunbär, Wildkatze, Luchs, Wolf, Hase, Biber. Im Verlauf des Mesolithikums begann die Domestikation einiger Arten, zuerst der Hund, später Ziege, Schaf, Rind und Schwein. Im Neolithikum und in der Bronzezeit wurden Pferd, Esel und Kamel gezähmt. Die Wildfauna des frühen Holozäns umfaßte noch Elch und *Megaloceros*. Durch den allmählichen Übergang zu festen Ansiedlungen, zur Landwirtschaft und Zucht von Haustieren verloren Jagd und die Beutetiere ihre Bedeutung für die Ernährung der Menschen des Neolithikums, der Bronze- und Eisenzeit. In dieser Zeit wurde der Lebensraum des Wildes immer mehr eingeengt, es setzte die Rodung der Wälder und die Umwandlung großer Flächen in Feld- und Weideland ein. Durch die rapide Vermehrung der Menschheit wurde das natürliche Faunenbild in Südosteuropa gestört, so daß die heutige Fauna in gewissem Sinne nur eine Reliktfauna des frühen Holozäns darstellt.

Zusammenfassung

Vom Territorium Jugoslawiens liegen genetisch unterschiedliche Bildungen aller pleistozänen Klimaphasen vor. Höhlensedimente, Knochenbrekzien und lakustrine Ablagerungen sind dabei besonders reich an Säugetierresten. Die bekanntesten Fundstellen sind Velenje, Strmica, Šandalja I, Šandalja II, Podumci, Dubci, Karlo-

Tabelle 1. Die quartäre Vertebratenfauna in der SFR Jugoslawien und ihre Position in der geochronologischen Gliederung

Gattung / **Art** / **Unterart**	Quartär: Pleistozän: Donau — Günz (Villafranchium)	Günz/Mindel — Mindel	Mindel/Riß — Riß	Riß/Würm — Jüngere Dryas	Quartär: Holozän: Präboreal — Atlantikum	Subboreal — Subrezent
1	2	3	4	5	6	7
Mammalia						
Insectivora						
Erinaceus europaeus			+	+	+	+
Sorex minutus		+	+	+	+	+
Sorex araeneus			+	+	+	+
Sorex alpinus				+	+	+
Sorex runtonensis		+	+			
Neomys fodiens			+	+	+	+
Suncus etruscus				+	+	+
Beremendia fissidens		+	+			
Crocidura leucodon				+	+	+
Crocidura russula				+	+	+
Talpa fossilis		+	+			
Talpa minor		+	+			
Talpa gracilis		+	+			
Talpa europaea				+	+	+
Talpa caeca				+	+	+
Talpa episcopalis		+	+			
Chiroptera						
Rhinolophus ferrumequinum		+	?	+	+	+
Rhinolophus euryale				+	+	+
Rhinolophus hipposideros				+	+	+
Myotis aemulus		+				
Myotis daubentoni			+	+	+	+
Myotis mystacinus				+	+	+
Myotis emarginatus				+	+	+
Myotis myotis				+	+	+
Plecotus crassidens		+	+			
Plecotus auritus				+	+	+
Miniopterus schreibersi		+	?	+	+	+
Pipistrellus pipistrellus				+	?	+
Verspertilio murinus				+	?	+
Verspertilio nilsoni			+	+	+	+
Nyctalus noctula				+	?	+
Eptesicus serotinus				+	+	+
Barbastella barbastella				+	?	+
Primates						
Macaca florentina	+	+				
Homo sp.	+					
Homo sapiens neanderthalensis				+		
Homo sapiens fossilis				+		
Homo sapiens sapiens					+	+
Carnivora						
Canis etruscus	+	+				
Canis lupus			+	+	+	+
Canis aureus				+	+	+
Canis familiaris intermedius					+	
Canis familiaris palustris					+	
Cuon alpinus europaeus				+		
Vulpes vulpes				+	+	+
Vulpes crucigera				+	+	+
Vulpes corsae				+	+	+
Alopex lagopus				+		

Tabelle 1. (Fortsetzung)

1	2	3	4	5	6	7
Ursus etruscus	+	+				
Ursus deningeri			+			
Ursus spelaeus				+		
Ursus arctos priscus			+	+		
Ursus arctos arctos					+	+
Ursus thibetanus mediterraneus	+	+	+			
Euryboas lunensis	+					
Crocuta perrieri	+	+				
Crocuta spelaea			+	+		
Felis silvestris				+	+	
Felis lunensis		+				
Felis lynx pardina				+		
Lynx lynx				+	+	+
Panthera toscana	+	+				
Panthera spelaea			+	+		
Panthera pardus			+	+		
Acinonyx pardinensis	+	+				
Gulo gulo				+	+	
Martes martes				+	+	+
Martes foina				+	+	+
Mustela putorius				+	+	+
Mustela eversmanni				+	?	+
Mustela erminea				+	+	+
Mustela nivalis				+	+	+
Meles meles				+	+	+
Lutra lutra				+	+	+
Proboscidea						
Anancus arvernensis	+					
Mastodon borsoni	+					
Archidiskodon meridionalis	+	+				
Palaeoloxodon antiquus		+	+	+		
Hesperoloxodon sp.			+			
Parelephas trogontherii			+			
Mammuthus primigenius				+		
Perissodactyla						
Tapirus arvernensis	+					
Dicerorhinus etruscus	+	+				
Dicerorhinus kirchbergensis			+	+		
Dicerorhinus hemitoechus			+	+		
Coelodonta antiquitatis				+		
Equus stenonis	+	+				
Equus cf. *altidens*		+				
Equus mosbachensis-abeli			+			
Equus germanicus				+		
Equus caballus fossilis				+		
Asinus hydruntinus				+	+	
Artiodactyla						
Sus strozzii	+	+				
Sus scrofa			+	+	+	+
Hippopotamus antiquus		+	+	?		
Camelus sp.		+				
Anoglochis ramosus	+	+				
Euctenoceros ctenoides	+	+				
Euctenoceros senezensis	+	+				
Eucladoceros dicranios		+				
Cervus acoronatus	+	+				
Cervus philisi	+	+				
Cervus elaphus			+	+	+	+
Megaceros giganteus			+	+	+	
Dama nestii		+	+			
Dama dama				+	?	+
Capreolus süssenbornensis		+	+			
Capreolus capreolus			+	+	+	+
Praealces gallicus	+	+				
Alces latifrons			+			
Alces alces				+	+	
Rangifer tarandus				+		

Tabelle 1. (Fortsetzung)

1	2	3	4	5	6	7
Saiga tatarica				+		
Rupicapra rupicapra				+	+	
Ovibos moschatus				+		
Ovis musimon				+		
Ovis aries					+	+
Capra ibex				+	+	+
Capra hircus					+	+
Hemitragus stehlini		+				
Leptobos stenometopon	+	+				
Leptobos etruscus		+				
Bison priscus			+	+		
Bison schoetensacki			+			
Bison bonasus					+	
Bos primigenius				+	+	
Bos taurus brachyceros					+	+
Rodentia						
Sciurus vulgaris				+	+	+
Marmota marmota				+	+	+
Citellus citellus				+	?	+
Castor fiber				+	+	
Trogontherium cuvieri	+	+				
Trogontherium sp.			+			
Muscardinus avellanarius				+	?	+
Hystrix refossa	+	+				
Hystrix cristata minor				+		
Spalax leucodon				+	?	+
Glis glis				+	+	+
Glis minor	+					
Cricetus cricetus				+	?	+
Cricetus cricetus major				+		
Allocricetus bursae	+	+				
Cricetulus migratorius				+		
Dinaromys dalmatinus		+				
Dinaromys episcopalis		+				
Dinaromys milleri			+			
Dinaromys bogdanovi					+	+
Pliomys bolkayi		+				
Pliomys cf. *hollitzeri*		+				
Mimomys cf. *pusillus*		+				
Mimomys cf. *savini*		+				
Mimomys malezi		+				
Microtus cf. *praehintoni*		+				
Mimomys intermedius		+				
Arvicola terrestris				+	+	+
Arvicola schermani				+		
Lagurus pannonicus		+				
Allophaiomys pliocaenicus		+				
Microtus arvalis				+	+	+
Microtus arvalinus		+				
Microtus agrestis				+	+	+
Microtus gregalis				+		
Microtus nivalis				+	?	+
Microtus nivalinus		+				
Pitymys hintoni		+				
Pitymys subterraneus				+	+	+
Dicrostonyx torquatus				+		
Apodemus sylvaticus		+	?	+	+	+
Apodemus flavicolis		+			+	
Apodemus mystacinus		+		+	+	
Rattus rattus				+	?	+
Mus musculus				+	?	+
Lagomorpha						
Ochotona pusilla				+		
Hypolagus brachygnathus	+	+				
Oryctolagus lacosti	+					
Oryctolagus cuniculus					+	+
Lepus timidus timidus				+		
Lepus timidus varronis				+	+	
Lepus europaeus				+	+	+

Tabelle 1. (Fortsetzung)

1	2	3	4	5	6	7
Aves						
Podiceps auritus				+		
Podiceps ruficollis				+	+	
Ardea sp.				+		
Ciconia ciconia				+		
Cygnus cygnus				+	+	
Anser anser				+	+	
Anas strepera				+		
Anas crecca				+	+	
Anas platyrhynchos				+		
Aythya ferina				+		
Neophron percnopterus				+		
Gyps fulvus				+		
Aegypius monachus				+		
Aquila chrysaëtos				+	+	
Buteo buteo				+	+	
Accipiter nisus				+	+	
Accipiter cf. *gentilis*				+		
Haliaëtus albicilla				+		
Circus aëruginosus				+		
Falco cf. *subbuteo*				+		
Falco cherrus				+		
Falco columbarius				+		
Falco vespertinus				+		
Falco tinnunculus				+	+	
Lagopus lagopus				+		
Lagopus mutus				+	+	
Lyrurus tetrix				+	+	
Tetrao urogallus				+	+	+
Tetrastes bonasia				+		
Perdix perdix				+	+	+
Coturnix coturnix				+		
Phasianus colchicus					+	+
Francolinus sp.	+					
Grus grus				+	+	
Rallus aquaticus				+		
Porzana sp.				+		
Gallinula chloropus				+		
Fulica atra				+	+	
Otis tarda				+	+	+
Otis tetrax				+		
Vanellus vanellus				+		
Capella gallinago				+		
Scolopax rusticola				+	+	
Larus argentatus				+	+	
Larus ridibundus				+		
Larus minutus				+		
Sterna hirundo				+		
Columba livia	+			+	+	
Tyto alba				+		
Bubo bubo				+		
Nyctea scandiaca				+		
Strix aluco				+		
Asio otus				+		
Asio flammeus				+		
Coracias garrulus				+		
Picus canus				+		
Picus viridis				+		
Dendrocopos medius				+		
Dendrocopos minor				+		
Galerida cristata				+		
Hirundo rustica				+		
Hirundo cf. *daurica*				+		
Delichon urbica				+		
Motacilla cinerea				+		
Motacilla alba				+		
Motacilla stora				+		
Anthus trivialis				+		
Lanius excubitor				+		
Bombycilla garrulus				+		
Troglodytes troglodytes				+		

Tabelle 1. (Fortsetzung)

1	2	3	4	5	6	7
Erithracus rubecula				+		
Turdus viscivorus				+		
Turdus pilaris				+		
Turdus merula	+			+	+	
Parus major				+	+	
Sitta europaea				+		
Emberiza citrinella				+		
Fringilla coelebs				+		
Chloris chloris				+		
Coccothraustes coccothraustes				+	+	
Pyrrhula pyrrhula				+		
Sturnus vulgaris				+	+	
Garrulus glandarius				+	+	+
Pica pica				+	+	
Pyrrhocorax pyrrhocorax				+		
Pyrrhocorax graculus				+	+	
Coloeus monedula				+		
Corvus corone				+	+	
Corvus corax				+	+	
Amphibia						
Bufo bufo				+		
Bufo vulgaris				+	+	
Rana escudenta				+	+	
Rana agilis				+		
Rana mehelyi				+		
Salamandra macullosa				+	+	
Reptilia						
Lacerta viridis		+		+	+	
Emys orbicularis				+	+	
Testudo graeca		+	+	+		
Testudo hermanni		+	+	+		
Testudo sp.	+					
Pisces						
Cyprinus carpio				+	+	+
Esox lucius				+	+	+
Salmo trutta				+		
Rutilus rutilus				+	+	
Tinca cf. *tinca*				+	+	
Abramis sp.					+	
Silurus sp.					+	+
Perca fluviatilis				+	+	+
Lucioperca sp.					+	
Chrysophris aurata					+	+

bag, Razvodje, die Marjan-Halbinsel, Milovac, Crvena Stijena, Krapina, Vindija, Veternica, Pećine u Brini, Cerovacer Höhle, Bijambare-Höhle, Zelena Pećina, Risovača, Makarovec.

Die folgenden Tertiär-Säuger sind noch in Ablagerungen des unteren Villafranchiums vorhanden: *Mastodon borsoni, Anancus arvernensis, Tapirus arvernensis* und *Dicerorhinus megarhinus*. Das mittlere und obere Villafranchium enthält die charakteristische Faunengemeinschaft warmer savannenähnlicher Steppen mit folgenden Formen: *Archidiskodon meridionalis, Dicerorhinus etruscus, Anoglochis ramosus, Euctenoceros senezensis, Eucladoceros dicranius, Cervus philisi, Cervus acoronatus, Dama nestii, Praealces gallicus, Capreolus süssenbornensis, Leptobos stenometopon, Canis etruscus, Ursus etruscus, Euryboas lunensis, Acinonyx pardinensis, Crocuta perrieri, Equus stenonis, Macaca florentina, Hypolagus brachygnathus*.

Die Faunengemeinschaft des Cromer-Interglazials (Günz/Mindel) zeigt das Vorherrschen eines warmen Steppen-Biotops mit zahlreichen Wäldern, Seen und Wasserwegen an. Aus lakustrinen Kreide-Ablagerungen, aus Schottern und Sanden wie aus Knochenbrekzien liegen fossile Reste von Großsäugern vor: *Hippopotamus antiquus, Parelephas trogontherii, Palaeoloxodon antiquus, Dicerorhinus hemitoechus, Equus mosbachensis, Alces latifrons, Hemitragus stehlini, Bison schoetensacki*. Die wichtigsten Reste fossiler Mikromammalier aus diesen Ablagerungen sind: *Sorex runtonensis, Beremendia fissidens, Talpa episcopalis, Plecotus crassidens, Trogontherium cuvieri, Allocricetus bursae, Dinaromys dalmatinus, D. episcopalis, Pliomys bolkayi, Mimomys malezi, M. savini, M. pusillus, Microtus praehintoni, Lagurus pannonicus, Allophaiomys pliocaenicus, Pitymys hintoni*.

Die Fauna der mittelpleistozänen Ablagerungen ist bisher noch unzureichend erforscht. Die wichtigsten Vertreter sind: *Sorex minutus, Talpa minor, Myotis daubentoni, Dinaromys milleri, Canis lupus, Ursus deningeri, Dicerorhinus kirchbergensis, Equus abeli* und *Megaceros giganteus*.

Die Bildungen des oberen Pleistozäns, die bisher am umfassendsten erforscht wurden, erbrachten eine sehr reiche Fauna. Von einigen Fundstellen (Krapina, Šandalja, Vindija, Veternica) liegen alternierende Faunenfolgen vor, das heißt Faunengesellschaften interglazialen bzw. glazialen Typus, wobei die Reste der großen Kontinental-Säuger — die Jagdbeute paläolithischer Jäger — dominieren. Die fossile Fauna des oberen Pleistozäns (Riß/Würm bis Dryas) ist durch mehr als 130 Säugetierarten vertreten und erlaubt die Rekonstruktion klimatischer Verhältnisse und Biotope wie auch ihren Wechsel und die Migration einzelner Arten oder ganzer Faunengemeinschaften in Südosteuropa.

Summary

In Yugoslavia genetically different deposits of all Pleistocene climatic phases are known. Cave sediments, bone breccia, and lacustrine layers are particularly rich in mammalian skeletal remains. The best known fossil sites are Velenje, Strmica, Šandalja I, Šandalja II, Podumci, Dubci, Karlobag, Razvodje, the Marjan Peninsula, Milovac, Crvena Stijena, Krapina, Vindija, Veternica, Pećine u Brini, Cerovacer Cave, Bijambare Cave, Zelena Pećina, Risovača, Makarovec, etc.

The following Tertiary mammals are present in the layers of the Lower Villafranchian: *Mastodon borsoni*, *Anancus arvernensis*, *Tapirus arvernensis*, and *Dicerorhinus megarhinus*. The layers of the Middle and Upper Villafranchian contain the characteristic faunal community of a warm savannah-like steppe, where the most important representatives are *Archidiskodon meridionalis*, *Dicerorhinus etruscus*, *Anoglochis ramosus*, *Euctenoceros senezensis*, *Eucladoceros dicranios*, *Cervus philisi*, *Cervus acoronatus*, *Dama nestii*, *Praealces gallicus*, *Capreolus süssenbornensis*, *Leptobos stenometopon*, *Canis etruscus*, *Ursus etruscus*, *Euryboas lunensis*, *Acinonyx pardinensis*, *Crocuta perrieri*, *Equus stenonis*, *Macaca florentina*, *Hypolagus brachygnathus*, etc.

The faunal community of the Cromerian interglacial (Günz/Mindel) indicates the existence of a warm steppe biotope with extensive forests, lakes, and water courses. In the lacustrine chalk layers, in the sand- and pebble-deposits, as well as in bone breccia layers the fossil remains of following species have been discovered: *Hippopotamus antiquus*, *Parelephas trogontherii*, *Palaeoloxodon antiquus*, *Dicerorhinus hemitoechus*, *Equus mosbachensis*, *Alces latifrons*, *Hemitragus stehlini*, *Bison schoetensacki*, etc. The most significant micromammal remains are *Sorex runtonensis*, *Beremendia fissidens*, *Talpa episcopalis*, *Plecotus crassidens*, *Trogontherium cuvieri*, *Allocricetus bursae*, *Dinaromys dalmatinus*, *D. episcopalis*, *Pliomys bolkayi*, *Mimomys malezi*, *M. savini*, *M. pusillus*, *Microtus praehintoni*, *Lagurus pannonicus*, *Allophaiomys pliocaenicus*, and *Pitymys hintoni*.

The fauna of the Middle Pleistocene layers has not been sufficiently studied, but the most significant representatives are *Sorex minutus*, *Sorex araneus*, *Talpa minor*, *Myotis daubentoni*, *Dinaromys milleri*, *Canis lupus*, *Ursus deningeri*, *Dicerorhinus kirchbergensis*, *Equus abeli*, and *Megaceros giganteus*.

The Upper Pleistocene layers, containing a very rich mammalian fauna, have been most thoroughly studied. At some sites (Krapina, Šandalja, Vindija, Veternica) faunal communities of warm and cold periods alternate, while the remains of big continental mammals — the prey of palaeolithic hunters — predominate. The Upper Pleistocene fauna (Riss/Würm up to Dryas) is represented by more than 130 species of mammals and enables the reconstruction of climatic conditions and biotopes, as well as their changes and the migrations of individual species or whole communities in South-Eastern Europe.

Резюме

В Югославии известны генетически различные отложения всех климатических фаз плейстоцена. Пещерные отложения, костеносные брекчии, и озерные слои особенно болашы скелетными остатками млекопитающих. К наиболее известным местонахождениям относятся Веленье, Стрмица, Шандалья I, Шандалья II, Подумцы, Дубцы, Карлобаг, Разводье, полуостров Марьян, Миловац, Црвена Стийена, Крапина, Виндийа, Ветерница, Пецине у Брини, пешеры Церовацер и Бийамбаре, Зелена Пецина, Рисовача, Макаровец и др.

В слоях нижнего виллафранка присутствуют остатки следующих третичных млекопитающих: *Mastodon borsoni*, *Anancus arvernensis*, *Tapirus arvernensis*, *Dicerorhinus megarhinus*. Слои среднего и верхнего виллафранка содержат характерное фаунистическое сообщество теплой саваннoподобной степи. Наиболее важными представителями его являются *Archidiskodon meridionalis*, *Dicerorhinus etruscus*, *Anoglochis ramosus*, *Euctenoceros senezensis*, *Eucladoceros dicranios*, *Cervus philisi*, *Cervus acoronatus*, *Dama nestii*, *Praealces gallicus*, *Capreolus süssenbornensis*, *Leptobos stenometopon*, *Canis etruscus*, *Ursus etruscus*, *Euryboas lunensis*, *Acinonyx pardinensis*, *Crocuta perrieri*, *Equus stenonis*, *Macaca florentina*, *Hypolagus brachygnathus*, и т. д.

Фаунистическое сообщество кромерского межледниковья (гюнц/миндель) указывает на присутствие биотопа теплой степи с многочисленными лесами, озерами и водотоками. В озерных меловых слоях, в песчаных и галечных отложениях, а также в прослоях костяной брекчии обнаружены ископаемые остатки следующих видов: *Hippopotamus antiquus*, *Parelephas trogontherii*, *Palaeoloxodon antiquus*, *Dicerorhinus hemitoechus*, *Equus mosbachensis*, *Alces latifrons*, *Hemitragus stehlini*, *Bison schoetensacki*, и т.д.

Среди остатков млекопитающих наиболее важны *Sorex runtonensis*, *Beremendia fissidens*, *Talpa episcopalis*, *Plecotus crassidens*, *Trogontherium cuvieri*, *Allocricetus bursae*, *Dinaromys dalmatinus*, *D. episcopalis*, *Pliomys bolkayi*, *Mimomys malezi*, *M. savini*, *M. pusillus*, *Microtus praehintoni*, *Lagurus pannonicus*, *Allophaiomys pliocaenicus*, *Pitymys hintoni*, и т. д.

Фауна среднеплейстоценовых слоев недостаточно изучена, но ее наиболее важными представителями являются *Sorex minutus*, *Talpa minor*, *Myotis daubentoni*, *Dinaromys milleri*, *Canis lupus*, *Ursus deningeri*, *Dicerorhinus kirchbergensis*, *Equus abeli*, *Megaceros giganteus*.

Верхнеплейстоценовые слои, содержащие очень богатую фауну млекопитающих, изучены наиболее тщательно. В некоторых местонахождениях (Крапина, Шандалья, Виндийа, Ветерница) чередуются фаунистические сообщества теплых и холодных периодов; преобладают остатки крупных континентальных млекопитающих — добычи палеолитических охотников. Фауна позднего плейстоцена (рисс/вюрм — дриас) представлена более чем 130 видами млекопитающих и дает возможность реконструировать климатические условия и биотопы Юго-Восточной Европы, а также их изменения и миграции отдельных видов или целых сообществ.

Literatur

ADAM, D. K., 1958: *Dicerorhinus kirchbergensis* (JÄGER) aus einer Karsthöhle bei Črni Kal (Istrien, Jugoslawien). — In: Raspr. SAZU. — Ljubljana **4** Cl. 4. — S. 435—440.

ANELLI, F., 1933: Il pozzo ossifero delle Cave di Cà Negra presso Punta Salvore nel Vallone di Sicciole. — In: Atti I. Congr. speleol. nazionale. — Trieste.

—, 1941: Un importante reperto di *Gulo gulo* L. nella Grotta di Castagnavizza. — In: Grotte d'Italia. — Trieste **14**.

—, 1947: Su un reperto di *Castor fiber* L. delle Grotte di Postumia. — In: Boll. Soc. Adriat. Sci. Nat. Trieste. — Milano **43**.

BABIĆ, N., 1952: Kostur mamuta u Vojvodanskom muzeju. — In: Zb. Matice srpske, Ser. Prir. Nauke. — Novi Sad **2**. — S. 172.

BATTAGLIA, R., 1929: La "*Hyaena crocuta spelaea*" delle Grotte di Postumia. — In: Grotte d'Italia. — Milano **3**.

BÖKÖNYI, S., 1969: Kičmenjaci. In: D. SREJOVIĆ: Lepenski Vir — nova praistorijska kultura u Podunavlju. — Beograd. — S. 224—228.

—, 1978: The Vertebrata Fauna of Vlasac. In: Vlasac (A Mesolithic Settlement in the Iron Gates), Bd. 2. — Beograd. — S. 35—65.

BRODAR, S., 1938: Das Paläolithikum in Jugoslawien. — In: Quartär. — Berlin **1**. — S. 140—172.

ĆIRIĆ, A., 1952: Fosilni ostaci pećinskog medvjeda (*Ursus spelaeus* ROSENMÜLLER) u Prirodnjačkom muzeju srpske zemlje. — In: Glasn. Prir. Muzeja. — A. — Beograd **5**. — S. 277—292.

CVIJIĆ, J., 1891: Prekonoška pećina. — In: Geol. anali. — Beograd **3** (1). — S. 272—299.

—, 1912: Petnička pećina. — In: Glasn. geogr. društva. — Beograd **1**. — S. 105—109.

DROBNE, K., 1964: Živalske kosti iz Resnikovega prekopa. — In: Poročilo o razsk. neolita in eneolita v Sloveniji. — Ljubljana **1**.

—, 1975: Fauna koliščarskih naselbin na Ljubljanskem barju. — In: Arheol. Vestn. — Ljubljana **24**. — S. 217—224.

FIALA, F., 1893: Höhlenforschungen in Bosnien. — In: Wiss. Mitth. Bosnien u. Herzegowina. — Sarajevo **1**. — S. 29—34.

GAREVSKI, R., 1956: Geologische und paläontologische Forschungen in den Höhlen von Mazedonien. — In: Fragm. Balcanica. — Skopje **1**. — S. 34.

—, 1957: Fund der Molaren eines fossilen Elefanten (*Mammonteus primigenius* BLUM.) im Dorfe Orešane im Becken von Skopje. — In: Acta Mus. Maced. Sci. Nat. — Skoje **5** (4) (45). — S. 61—70.

—, 1964: Ostaci od glodar vo pešterata Makarovec II vo okolinata na T. Veles. — In: Fragm. Balcanica. — Skopje **5** (10) (120).

—, 1969: Die stratigraphische und paläontologische Bedeutung der pleistozänen Fauna aus der Höhle Makarovec in der Babuna Schlucht (Umgebung von Titov Veles). — In: Pos. izd. Prir. Muzeja. — Skopje **6**. — S. 1—73.

—, 1970: Ostatoci na mečka od pešterata "Jaorec" vo okolinata na selo Velmej (Ohridsko). — In: Peti jugosl. speleol. Kongr. — Skopje. — S. 223—235.

GASPERINI, R., 1885: Contributo alla conoscenza geologica del diluviale Dalmato. — In: Ann. Dalmato. — Zadar **2**. — S. 3—15.

—, 1887: Secondo contributo alla conoscenza geologica del diluviale Dalmato. — Split: Progr. scuola reale sup. anno 1886/87.

GIROMETTA, U., 1935: Špiljski nalazi u srednjoj Dalmaciji. — In: Hrvatski planinar. — Zagreb **31** (10/11). — S. 3—16.

GORJANOVIC-KRAMBERGER, D., 1883: Ostanci kvaternarnih sisara gore zagrebačke. — In: Rad jugosl. Akad. znan. umjet. — Zagreb **64**. — S. 108—111.

—, 1884: Fosilni sisari Hrvatske, Slavonije i Dalmacije. — Rad jugosl. Akad. znan. umjet. — Zagreb **69**. — S. 1—36.

—, 1912: Die fossilen Proboscider aus Kroatien-Slavonien. — In: Djela jugosl. Akad. znan. umjet. — Zagreb **21**. — S. 5—23.

—, 1913: Über fossile Rhinocerotiden Kroatiens-Slavoniens. — In: Djela jugosl. Akad. znan. umjet. — Zagreb **22**. — S. 1—70.

GRIMMER, J., 1898: Fosilni ostaci sisavaca iz Save. — In: Glasn. Zem. Muzeja. — Sarajevo **10**. — S. 377—386.

HERAK, M., 1947: Geological age and taxonomical characters of the Cave-Bear of Croatia. — In: Geol. Vjestn. — Zagreb **1**. — S. 12—47.

HOCHSTETTER, F., 1881: Die Kreuzberghöhle bei Laas in Krain und der Höhlenbär. — In: Denkschr. Akad. Wiss. Math.-naturwiss. Cl. — Wien **43**. — S. 293.

JOVANOVIĆ, D., 1891: Ljudska lubanja pod Beograda nadjena sa ostacima mamuta. — In: Geol. Anali. — Beograd **4**.

KIŠPATIĆ, M., 1885: Kosti iz Baračeve spilje kod Kršlja. — In: Viest. hrvat. arkeol. Družtva. — Zagreb **7** (2). — S. 33—37.

KORMOS, T. 1912: Die ersten Spuren des Urmenschen im kroatischen Karstgebirge. — In: Földt. Közlöny. — Budapest **40**. — S. 97—104.

—, 1931: Über eine neuentdeckte Forestbed-Fauna in Dalmatien. — In: Palaeobiologica. — Wien **4**. — S. 113—136.

—, 1933: Nuove tracce di una „Forest Bed" fauna nella regione Adriatica. — In: Grotte d'Italia. — Trieste **7** (2). — S. 69—71.

KOS, F., 1923: Evropski los, *Alces alces* L. iz jame „pri Glažuti". — In: Glasn. Muzeja Društva Slovenije, B. — Ljubljana 2—3.

—, 1939: Der Quartärfund von Nevlje. — Quartär. — Berlin **2**. — S. 150—151.

—, 1944: Pleistozäne Proboscidier des Naturwissenschaftlichen Museums in Laibach. — In: Prirod. Izvest. — Ljubljana **1**. — S. 23—92.

KOWALSKI, K., 1958: Altpleistozäne Kleinsäugerfauna von Podumci in Nord-Dalmatien. — In: Palaeont. jugosl. — Zagreb 2. — S. 5—30.

LAMBRECHT, K., 1915: Die pleistozäne Vogelfauna von Krapina. — In: Barlangkutatás. — Budapest **3**. — S. 116—120.

—, 1933: Handbuch der Palaeornithologie. — Berlin. 1024 S.

LASKAREV, V., 1922: Sur le loess des environs de Beograd. — In: Geol. Anali. — Beograd **7** (1). — S. 13.

—, 1938: Treća beleška o kvartarnim naslagama u okolini Beograda. — In: Geol. Anali. — Beograd **15**. — S. 1—40.

LEONARDI, P., 1934: *Elephas* (*Palaeoloxodon*) *antiquus* FALC. di Cittanova nell Istria. — In: Atti Mus. Civ. Stor. Nat. — Trieste **12**. — S. 153—166.

MALEZ, M., 1956: Geologische und paläontologische Forschungen in der Höhle Veternica. — In: Acta geol. — Zagreb **1**. — S. 83—88.

—, 1957: Höhlenhyänenfund in Slawonien. — In: Bull. sci. Cons. Acad. Yougosl. — Zagreb **3** (3). — S. 68—69.

—, 1958: Einige neue Resultate der paläontologischen Erforschung der Höhle Veternica. — In: Palaeont. Jugosl. — Zagreb **1**. — S. 1—24.

—, 1959: Das Paläolithikum der Veternicahöhle und der Bärenkult. — In: Quartär. — Bonn **10/11** (1958/1959). — S. 171—188 (1959a).

—, 1959: Ein Beitrag zur Kenntnis der Höhlenbären der Ćićarija in Istrien. — In: Geol. Vjesn. — Zagreb **12** (1958). — S. 95—113 (1959b).

—, 1959: A new Old-Pleistocene Fauna in Dalmatien. — In: Bull. sci. Cons. Acad. Yougosl. — Zagreb **5** (1). — S. 13—14, (1959c).

—, 1960: Etruskisches Nashorn — *Dicerorhinus etruscus* (FALCONER) — aus altpleistozänen Brekzien der Halbinsel Marjan bei Split (Dalmatien). — In: Mammalia pleistocaenica. — Brno **1**. — S. 115—125, (1960a).

—, 1960: Erster Fund des Südelefanten (*Archidiskodon meridionalis* NESTI) in Jugoslawien. — In: Bull. sci. Cons. Acad. Yougosl. — Zagreb **5** (3). — S. 65—67, (1960b).

—, 1961: Die altpleistozäne Brekzienfauna der Halbinsel Marjan bei Split. — In: Palaeont. Jugosl. — Zagreb **4**. — S. 5—43, (1961a).

—, 1961: Der Fund zweier pleistozäner Säugetiere bei Zagreb und eine Übersicht der benachbarten Fundstellen. — In: Geol. Vjesn. — Zagreb **14** (1960). — S. 63—88, (1961b).

—, 1961: Die Höhlenhyäne aus der Höhle Vindija bei Voća in NW-Kroatien. — In: Geol. Vjesn. — Zagreb **14** (1960). — S. 221—244, (1961c).

—, 1961: *Megaceros giganteus* (BLUMENBACH) aus der Saveanschwemmung bei Sisak (Kroatien). — In: Geol. Vjesn. — Zagreb **14** (1960). — S. 345—354, (1961d).

—, 1961: Oberpleistozäne Schachtfauna in Pisana Stina auf dem Opor-Gebirge (Dalmatien). — In: Bull. sci. Cons. Acad. Yougosl. — Zagreb **6** (1). — S. 2—4, (1961e).

—, 1962: *Gulo gulo* L. aus dem oberen Pleistozän Dalmatiens. — In: Bull. sci. Cons. Acad. Yougosl. — Zagreb **7** (3). — S. 57—58, (1962a).

—, 1962: Erster Fund des Rotwolfs (*Cuon alpinus europaeus* BOURGUIGNAT) im oberen Pleistozän der Balkanhalbinsel. — In: Bull. sci. Cons. Acad. Yougosl. — Zagreb **7** (4/5). — S. 97—98, (1962b).

—, 1962: Romualdo cave — a new significant Pleistocene site in Istria. — In: Bull. sci. Cons. Acad. Yougosl. — Zagreb **7** (6). — S. 159—160, (1962c).

—, 1963: Die quartäre Fauna der Höhle Veternica (Medvednica — Kroatien). — In: Palaeont. Jugosl. — Zagreb **5**. — 200 S.

—, 1964: Der erste Fund der großen Höhlenkatze — *Panthera spelaea* (GOLDF.) — im Pleistozän von Bosnien und Herzegowina. — In: Geol. Glasn. — Sarajevo **10**. — S. 265—274.

—, 1964: Šandalja bei Pula — ein neuer und wichtiger paläolithischer Fundort in Istrien. — In: Bull. sci. Cons. Acad. Yougosl. — Zagreb **9** (6). — S. 154—155, (1964b).

—, 1965: Neue allgemeine Schwankungsbreite der Art *Ursus spelaeus* (ROSENM. & HEINROTH). — In: Geol. Vjesn. — Zagreb **18** (1964). — S. 133—139, (1965a).

—, 1965: Die Höhlen von Cerovac. — In: Speleol. dr. Hrvat. — Zagreb **1**. — S. 1—44, (1965b).

—, 1965: Druška Peć auf der Učka — ein neuer Fundort der Fauna des oberen Pleistozäns in Istrien. — In: Bull. sci. Cons. Acad. Yougosl., (A). — Zagreb **10** (3). — S. 65—66, (1965c).

—, 1965: Crvena Stijena in Montenegro — eine bedeutende paläontologische Lokalität auf der Balkanhalbinsel. — In: Bull. sci. Cons. Acad. Yougosl., (A). — Zagreb **10** (5). — S. 145—147, (1965d).

—, 1965: Der altpleistozäne Fundort Dubci in Mitteldalmatien. — In: Bull. sci. Cons. Acad. Yougosl., (A). — Zagreb **10** (12). — S. 418—420, (1965e).

—, 1966: Die Gattung *Ochotona* LINK, 1795 (Lagomorpha BRANDT, 1855) in Jugoslawien. — In: Bull. sci. Cons. Acad. Yougosl., (A). — Zagreb **11** (1/2). — S. 5—6.

—, 1967: Das Paläolithikum der Höhle Velika Pećina auf der Ravna Gora (NW Kroatien). — In: Arheol. Rad i Raspr. JAZU. — Zagreb **4/5**. — S. 7—68, (1967a).

—, 1967: Die altpleistozäne Knochenbrekzienfauna beim Dorfe Dubci in Dalmatien. — In: Rad jugosl. Akad. znan. umjet. — Zagreb **345**. — S. 55—100, (1967b).

—, 1968: Pfeifhasenreste aus dem oberen Pleistozän in Jugoslawien. — In: Geol. Vjesn. — Zagreb **21** (1967). — S. 147—156, (1968a).

—, 1968: Die altpleistozänische Vertebratenfauna in dem Gebiete des Dinarischen Karstes. — In: Bull. sci. Cons. Acad. Yougosl., (A). — Zagreb **13** (5/6). — S. 151—152, (1968b).

—, 1969: Diffusione del genere *Ochotona* nel Pleistocene superiore dell' Europa sud-orientale. Scritti sul Quaternario in onore di Angelo Pasa. — In: Museo Civico Storia Nat. — Verona. — S. 67—74, (1969a).

—, 1969: Die altpleistozänen Wirbeltierfaunen aus den Höhlen von Kroatien. — In: V. Internat. Kongr. Speläol., Stuttgart, 1969, (Abh.), 4. Sekt. Biospeläol. — München B 21/1—B 21/6, (1969b).

—, 1970: Die Ergebnisse der Revision der pleistozänen Fauna aus Krapina. Krapina 1899—1969. — Zagreb. — S. 45—56, (1970a).

—, 1970: The extinct mammals of Slavonia. — In: Zb. rad. I. znan. sabora Slavonije i Baranje. — Osijek. — S. 101—105, (1970b).

—, 1971: Die Jungwürmfauna und die Spuren des Paläolithikums in der Oberen Bijambarehöhle bei Olovo in Bosnien. — In: Rad jugosl. Akad. znan. umjet. — Zagreb **358**. — S. 209—248, (1971a).

—, 1971: The colonization of Dinaric Karst in the Pleistocene. — In: Simp. o Zaštiti Prir. u našem Kršu, JAZU. — Zagreb (1970). — S. 63—80, (1971b).

—, 1972: Über die Verbreitung kaltzeitlicher Tiere im jüngeren Pleistozän Südost-Europas. — In: Rad jugosl. Akad. znan. umjet. — Zagreb **364**. — S. 133—180.

—, 1973: Spätwürmfauna in der Zelena Pećina bei Mostar in der Herzegowina. — In: Bull. sci. Cons. Acad. Yougosl., (A). — Zagreb **18** (1—3). — S. 7—8.

—, 1975: On the significance of the genus *Homo* discovery in the Villafranchian sediments of Šandalja I near Pula. — In: Rad jugosl. Akad. znan. umjet. — Zagreb **371**. — S. 181—201, (1975a).

—, 1975: Hyänenhöhlen aus dem Pleistozän in Kroatien. — In: Rad jugosl. Akad. znan. umjet. — Zagreb **371**. — S. 307—316, (1975b).

—, 1975: Die Ornithofauna aus Quartärablagerungen der Höhle Velika Pećina im Ravna Gora-Gebirge in Nordwestkroatien. — In: Larus. — Zagreb **26—28** (1972—1974). — S. 45—54, (1975c).

—, 1975: Die Entdeckung von fünf Skeletten des fossilen Braunbären in der Banić-Höhle auf der Insel Cres. — In: Bull. sci. Cons. Acad. Yougosl., (A). — Zagreb **20** (1/2). — S. 5—7, (1975d).

—, 1979: Quartärgeologische und paläolithische Verhältnisse in der Höhle Gospodska Pećina oberhalb der Quelle der Cetina (Dalmatien). — In: Krš Jugosl. — Zagreb **10** (2). — S. 45—76, (1979a).

—, 1979: Kvartarna fauna Jugoslavije. In: Praeistorija jugoslavenskih zemalja, I, Paleolitsko i mezolitsko doba. — Sarajevo: S. 55—79, (1979b).

—, 1979: Beitrag zur Kenntnis der Verbreitung des Pfeifhasen im oberen Pleistozän in Kroatien. — In: Rad jugosl. Akad. znan. umjet. — Zagreb **383**. — S. 345—361, (1979c).

—, 1980: Die Höhle Bezdanjača bei Vrhovine in der Lijka und ihre Quartärfauna. — In: Vesn. arheol. Muzeja, 3 ser. — Zagreb **12/13**. — S. 1—26, (1980a).

—, 1980: Speleological Chronology in Some Caves in Bosnia and Herzegovina. — In: Naš Krš. — Sarajevo **6** (9). — S. 3—32, (1980b).

—, 1981: Fossile Vertebrata im Gebiet von Biokovo und ihre paläoökologische Bedeutung. — In: Acta biokovica. — Makarska **1**. — S. 39—70.

—, 1982: Sequences of the Pleistocene Mammals in Yugoslavia. — In: XI. INQUA Congr., Moscow 1982, Abstr., 2. — Moscow. — S. 171.

MALEZ, M.; PEPEONIK, Z., 1969: Entdeckung des ganzen Skelettes eines fossilen Leoparden in der Vjetrenica-Höhle auf dem Popovo Polje (Herzegowina). — In: Bull. sci. Cons. Acad. Yougosl., (A). — Zagreb **14** (5/6). — S. 144—145.

MALEZ, M.; RUKAVINA, D.; SLIŠKOVIĆ, T., 1972: Eine neue paläontologische und paläolithische Fundstelle in Westbosnien. — In: Bull. sci. Cons. Acad. Yougosl., (A). — Zagreb **17** (7/8). — S. 229—231.

MALEZ, M.; MALEZ-BAČIĆ, V., 1974: Upper Pleistocene Ornithofauna of Crvena Stijena near Petrovići in Montenegro. — In: Geol. Vjesn. — Zagreb **27**. — S. 113—125.

MALEZ, M.; RABEDER, G., 1984: Neues Fundmaterial von Kleinsäugern aus der altpleistozänen Spaltenfüllung Podumci 1 in Norddalmatien (Kroatien, Jugoslawien). — In: Beitr. Paläont. Österr. — Wien 11. — S. 439—510.

MALEZ, M.; RUKAVINA, D.; CRNOLATAC, I., 1974: Discovery of the Pleistocene "Hyena Cave" in northern Istria. — In: Bull. sci. Cons. Acad. Yougosl., (A). — Zagreb **19** (1/2). — S. 8—9.

MALEZ, M.; RUKAVINA, D.; ŠERCELJ, A., 1975: Paläontologische Merkmale des Villafranchium der Lokalität Dronjkova Glavica in Norddalmatien. — In: Bull. sci. Cons. Acad. Yougosl., (A). — Zagreb **20** (3/4). — S. 69—71.

MALEZ, M.; RUKAVINA, D.; SLIŠKOVIĆ, T., 1978: Quartärgeologische und paläontologische Verhältnisse in der Höhle Rastuša bei Teslić. — In: Glasn. Zem. Muz. BiH, N.S. — Sarajevo **16**. — S. 5—22.

MALEZ, M.; RUKAVINA, D.; SLIŠKOVIĆ, T.; KAPEL, A., 1974: Two new sites of Pleistocene Fauna and of the Palaeolithic in Bosnia. — In: Bull. sci. Cons. Acad. Yougosl., (A). — Zagreb **19** (3/4). — S. 80—83.

MALEZ-BAČIĆ, V., 1975: Die oberpleistozäne Ornithofauna aus den Höhlen Nordwestkroatiens. — In: Rad jugosl. Akad. znan. umjet. — Zagreb **371**. — S. 317—324.

—, 1979: Die pleistozäne Ornithofauna aus Šandalja in Istrien und ihre stratigraphische und paläökologische Bedeutung. — In: Palaeont. jugosl. — Zagreb **21**. — S. 1—46.

MALEZ, V., 1983: Die Ornithofauna aus dem oberen Pleistozän in Bosnien und der Herzegovina. — In: Rad jugosl. Akad. znan. umjet. — Zagreb **404**. — S. 75—107.

MARKOVIĆ-MARJANOVIĆ, J., 1966: The Morava-Vardar Valley as a climatogenous factor during the glacial epoch and its part in the determination of the lower limit of the Quaternary. — In: VI. Savjet. geol. u Ohridu. — Beograd **1**. — S. 300—322.

—, 1970: Data concerning the stratigraphy and the fauna of the Lower and Middle Pleistocene of Yugoslavia. — In: Palaeogeogr. Palaeoclimatol., Palaeoecol. — Amsterdam **8**. — S. 153 to 163.

—, 1971: Coast Lines, Pleistocene Sediments and Fauna of the Eastern Part of Adriatic in Yugoslavia. — In: Quaternaria. — Roma **15**. — S. 187—195.

—, 1972: Pleistocene inferieur de la region Danubienne avec *Equus* cf. *süssenbornensis* — Serbie. — In: VII. Kongr. geol. Jugoslavije. — Zagreb. — S. 183—192.

MATEJIĆ, B., 1957: Fosilni ostaci mamuta kod sela Draža u Baranji. — In: Zap. Srp. geol. Društva za 1956 god. — Beograd. — S. 51 bis 52.

MELENTIS, J., 1966: Studien über fossile Vertebraten Griechenlands. 4. Die pleistozänen Nashörner des Beckens von Megalopolis im Peleponnes (Griechenland). — In: Ann. Géol. Helléniques, 1 ser. — Athènes **16**.

MILIČEVIĆ, M., 1900: Zubi od mamuta u Varvarinu. — In: Zap. Srp. geol. Društva. — Beograd **57** (1).

NEUMAYR, M., 1882: Die diluvialen Säugetiere der Insel Lesina. — In: Verh. k. u. k. geol. Reichsanst. — Wien **161**.

PARTSCH, P., 1827: Nachrichten über die Knochen-Breccie von Dalmatien. — In: Leonard's Z. — Frankfurt (Main). — S. 524 bis 528.

PAUNOVIĆ, M., 1983: Les tourtues terrestres du genre *Testudo* Linné, 1758, trouvées dans les couches de pleistocéne a Crvena Stijena (Petrovići, Monténégro). — In: Rad jugosl. Akad. znan. umjet. — Zagreb **404**. — S. 109—123.

POHAR, V., 1976: La grotte de Marovška zijalka. — In: Geologija. — Ljubljana **19**. — S. 107—119.

—, 1981: La faune pléistocène de la cavité de Jama pod Herkovimi pečmi. — In: Geologija. — Ljubljana **24**. — S. 241—284.

RABEDER, G., 1983: *Mimomys malezi* n. sp., ein neuer Arvicolide (Rodentia) aus dem Altpleistozän von Dalmatien. — In: Beitr. Paläont. Österr. — Wien 10. — S. 1—13.

RAKOVEC, I., 1933: *Coelodonta mercki* JÄG. iz Dolarjeve jame pri Logatcu. — In: Prir. Razpr. — Ljubljana **2**. — S. 5—41.

—, 1935: Über diluviale Murmeltiere aus den Südostalpen. — In: Prir. Raspr. — Ljubljana **2** (5). — S. 245—292.

—, 1938: Ein Moschusochs aus der Höhle Potočka Zijalka (Ostkarawanken). — In: Prir. Razpr. — Ljubljana 3. — S. 253 bis 262.

—, 1939: O diluvialnem kozorogu iz neke kočevske jame. — In: Zb. Prir. Društva. — Ljubljana 1. — S. 59—62.

—, 1940: Über die diluvialen Steinböcke aus Slowenien und Dalmatien. — In: Razpr. SAZU. Mat.-prir. Razpr. — Ljubljana 1. — S. 55—112.

—, 1942: Über die Nashornreste vom Hügel Kamnitnik bei Bischoflack. — In: Razpr. SAZU. Mat.-prir. Razpr. — Ljubljana 2. — S. 241—262.

—, 1949: New finds of pleistocene marmots in Slovenia (NW. Yugoslavia.). — In: Razpr. SAZU. Razr. Prir.-med. vede. — Ljubljana 4. — S. 205—228.

—, 1951: On *Felis spelaea* Goldf. from the Postojna Cavern. — In: Razpr. SAZU. — Ljubljana 1., Cl. 4. — S. 127—172.

—, 1952: On Bison of the Lacustrine Age from the Ljubljansko barje in Slovenia (NW Yugoslavia). — In: Arheol. Vestn. — Ljubljana 3 (2). — S. 257—287, (1952a).

—, 1952: O fosilnih sesalskih ostankih iz Betalovega spodmola. — In: Letopis SAZU. — Ljubljana 4 (1950—1951). — S. 205—225, (1952b).

—, 1954: On Elephants Remains in Slovenia. — In: Razpr. SAZU. — Ljubljana 2, Cl. 4. — S. 215—275, (1954a).

—, 1954: *Libralces* aff. *gallicus* Azzaroli from the Vić Terrace near Ljubljana. — In: Razpr. SAZU. — Ljubljana 2, Cl. 4. — S. 277—295, (1954b).

—, 1954: *Hippopotamus* from the Postojna Basin. — In: Razpr. SAZU. — Ljubljana 2, Cl. 4. — S. 297—317, (1954c).

—, 1955: The Pleistocene Bovidae in Slovenia. — In: Razpr. SAZU. — Ljubljana 3, Cl. 4. — S. 301—328, (1955a).

—, 1955: The new findings of Proboscidea in the North-Eastern Slovenia. — In: Razpr. SAZU. — Ljubljana 3, Cl. 4. — S. 329—358, (1955b).

—, 1956: The Remnants of the Elk (*Alces alces* L.) in Yugoslavia. — In: Geol. Anali. — Beograd 24. — S. 1—13, (1956a).

—, 1956: The Development of Pleistocene in Slovenia (NW Yougoslavia). — In: I. jugosl. geol. kongr., Bled 1954. — Ljubljana. — S. 59—72, (1956b).

—, 1957: O jamskem medvedu. — In: Proteus. — Ljubljana 19 (1956/57) 10. — S. 265—272.

—, 1958: The Beavers of the Lacustrine Age from the Ljubljana Moor and from other Holocene Find Spots in Slovenia. — In: Razpr. SAZU. — Ljubljana 4, Cl. 4. — S. 209—267, (1958a).

—, 1958: The Pleistocene Mammalia from the Cave Črni Kal in Northern Istria. — In: Razpr. SAZU. — Ljubljana 4, Cl. 4. — S. 365—433, (1958b).

—, 1958: Les Mammiferès Pleistocenes de l'Abri „Crvena Stijena" prés de Petrovići (Monténégro). — In: Glasn. Zem. Muzeja, N. S. — Sarajevo 13. — S. 65—75.

—, 1959: The Quaternary Mammalian Fauna from the Cave Betalov Spodmol near Postojna. — In: Razpr. SAZU. — Ljubljana 5, Cl. 4. — S. 287—348.

—, 1960: Über die Wanderungen der pleistozänen Säugetiere im Gebiete südlich der Alpen. — In: Geogr. Vestn. — Ljubljana 32. — S. 191—200.

—, 1961: The Upper Pleistocene Fauna from the Cave Parska Golobina (in the Pivka Basin). — In: Razpr. SAZU. — Ljubljana 6, Cl. 4. — S. 271—349, (1961a).

—, 1961: The new Marmot Remains from Slovenia. — In: Razpr. SAZU. — Ljubljana 6, Cl. 4. — S. 351—365, (1961b).

—, 1963: The Late Würmian Fauna from the Caves Jama v Lozi and Ovčja Jama in Slovenia (NW Yugoslavia). — In: Arheol. Vestn. — Ljubljana 13/14 (1962—1963). — S. 241—272.

—, 1965: Pleistocene Mammalian Fauna from Risovača near Arandjelovac (Serbia). — In: Razpr. SAZU. — Ljubljana 8, Cl. 4. — S. 223—317.

—, 1967: The Cave Bear from the Mokrica Cave in the Savinja Alps (Slovenia, Yugoslavia). — In: Razpr. SAZU. — Ljubljana 10, Cl. 4. — S. 121—203, (1967a).

—, 1967: The Mammal Fauna of Slovenia in the Pleistocene Epoch. — In: Arheol. Vestn. — Ljubljana 18. — S. 291—299, (1967b).

—, 1968: The Mastodons from the Šalek Valley. — In: Razpr. SAZU. — Ljubljana 11, Cl. 4. — S. 299—350, (1968a).

—, 1968: Über das älteste Pleistozän Jugoslawiens. — In: Anz. math.-naturwiss. Kl. österr. Akad. Wiss. — Wien 8. — S. 169—176, (1968b).

—, 1969: Su nuovi resti di *Panthera* (*Leo*) *spelaea* (Goldf.) rinvenuti in Slovenia (Jugoslavia). — In: Scriti sul Quaternario in onore di Angelo Pasa, Mus. Civ. Stor. Nat. — Verona. — S. 53—65.

—, 1975: Über quartäre Säugetierfaunen Sloweniens (NW Jugoslawien). — In: Arheol. Vestn. — Ljubljana 24 (1973). — S. 225—270.

Riedel, A., 1977: I resti animali della Grotta delle ossa (Škocjan). — In: Atti Mus. Civ. Stor. Nat. — Trieste 30, 2, 7. — S. 125—208.

Seidl, F., 1912: Širokočelni los (*Alces latifrons*) v starejški diluvijalni naplavini Ljubljanskega barja. — In: Carniola, N. S. — Ljubljana 3. — S. 261—274.

Stepanović, J., 1953: O nalasku ostataka mamuta (*Elephas primigenius* Blumenb.) u Nišu. — In: Nauka i Prirod. — Beograd 6. — S. 171—172.

Thenius, E., 1958: Über einen Kleinbären aus dem Pleistozän von Slowenien nebst Bemerkungen zur Phylogenese der plio-pleistozänen Kleinbären. — In: Razpr. SAZU. — Ljubljana 4, Cl. 4. — S. 631—646.

Toula, F., 1907: *Rhinoceros Mercki* Jäger in Österreich. — In: Jahrb. geol. Reichsanst. — Wien 57 (3). — S. 445—454.

Veselinović-Čičulić, M., 1952: Nov prilog za poznavanje kvartarnih sisavaca u okolini Beograda. — In: Zb. rad SAN Geol. Inst. — Beograd 22 (3). — S. 121—122.

Vujisić, T., 1958: Još jedan nalazak mamuta u lesu okoline Beograda. — In: Zb. rad Geol. Inst. „J. Žujović". — Beograd 10.

Woldřich, J. N., 1874: Mitteilungen aus Dalmatien. — In: Verh. geol. Reichsanst. — Wien 8. — S. 185—189.

—, 1882: Beiträge zur Fauna der Breccien und anderer Diluvialgebilde Österreichs mit besonderer Berücksichtigung des Pferdes. — In: Jahrb. geol. Reichsanst. — Wien 32 (4). — S. 435 bis 470.

—, 1896: Wirbeltierfauna des Pfahlbaues von Ripač. — In: Wiss. Mitth. Bosnien u. Herzegowina. — Wien 5 (1897). — S. 79—113.

—, 1903: Wirbeltierfauna des Pfahlbaues Donja Dolina in Bosnien. — In: Wiss. Mitth. Bosnien u. Herzegowina. — Wien 9. — S. 156—164.

—, 1904: Wirbeltierfauna des Pfahlbaues von Dolnja Tuzla in Bosnien. — In: Wiss. Mitth. Bosnien u. Herzegowina. — Wien 11.

Žujović, J., 1900: Zähne und zerfallene Knochen von Mammut im Löss von Višnjica. — In: Geol. Anali. — Beograd 5 (2). — Annexe S. 4.

—, 1905: Novi nalasci mamutovih zuba u Beogradu. — In: Zap. Srp. geol. Društva. — Beograd 3 (60).

Quartärpaläontologie 6 119—122 Berlin 1986

Methodische Erwägungen zum Problem des Aussterbens

Von Rudolf Musil, Brno

Inhalt

Abstractum

In vorliegendem Beitrag wird das Problem des Aussterbens ganz allgemein betrachtet, nicht in Beziehung auf irgendeine Art oder Gattung.

In this paper the problem of extinction in general is discussed, not that of certain species or genera.

В настоящей статье проблема вымирания рассматривается в целом, а не в отношении каких-то определенных видов или родов.

1. Allgemeine Probleme zur Frage des Aussterbens

Das Aussterben einzelner Taxa oder ganzer Gemeinschaften ist ein allgemein gültiges, in den Eigenschaften der lebenden Materie verankertes Gesetz, eine immanente Voraussetzung der Evolution, denn wo Neues entsteht, muß Bestehendes vergehen. Das Aussterben äußert sich selbstverständlich bei verschiedenen Formen in unterschiedlicher Weise; es handelt sich aber immer um ein und dieselbe Gesetzmäßigkeit.

In der vorliegenden Arbeit wird nicht der Versuch unternommen, die Frage des Aussterbens bestimmter Taxa oder Gruppen konkret zu beantworten, sie will vielmehr theoretische Grundlagen zur Lösung der Problematik vorlegen. Leider haben sich mit der Frage des Aussterbens bedeutend weniger Arbeiten befaßt als mit der Frage der Entwicklung, die deshalb auch wesentlich eingehender erforscht ist.

Die Aussicht, die Frage des Aussterbens objektiv zu beantworten, hängt in hohem Maße davon ab, ob und inwiefern die Vorstellung, von der man bei seinen Überlegungen ausgeht, richtig ist. Richtig heißt vor allem, daß sie nicht einseitig fixiert wird, weil das die Arbeit von Anfang an als fragwürdig erscheinen ließe. Deshalb dürfen auch die zur Lösung unseres Problems gewählten Arbeitsalternativen nicht statisch sein, sondern müssen sich im Verlauf der Tatsachenfindung dynamischer Änderungen und Umgruppierungen fähig erweisen.

Die das Aussterben bewirkenden Faktoren sind komplex; es handelt sich um ein kompliziertes System biotischer und abiotischer Komponenten und ihrer Wechselbeziehungen. Die erfolgreiche Bewältigung unserer Aufgabe erfordert deshalb einen umfassenden Blick auf das gesamte Problem; es ist unumgänglich notwendig, die Grenzen der einzelnen Wissensdisziplinen zu überschreiten und alle von den verschiedenen Aspekten gewonnenen Erkenntnisse als Ganzes zu betrachten. Nur im Lichte einer solchen Synthese kann man mit einer Lösung der gegebenen Problematik rechnen.

Soweit sich das Studium des Aussterbens auf die Untersuchung von Fossilien des Känozoikums stützt, kann es auf die Anwendung des Prinzips des Aktualismus nicht verzichten. Dieses dient gewissermaßen als Regulator paläoökologischer Schlußfolgerungen.

Die erfolgreiche Bearbeitung unserer Problematik beruht demnach auf folgenden Prämissen:

— Ausbildung alternativer Ausgangsvorstellungen, das heißt Bestimmung der Probleme, Ziele und Art ihrer Bearbeitung, die sich im Verlauf der Tatsachenfindung jedoch variieren lassen;
- Komplexes und systematisches Herangehen an die Lösung des Problems;
— Anwendung des Aktualismusprinzips.

2. Arbeitsmethode zur Beantwortung der Frage des Aussterbens

Jedes Systemstudium setzt eine bestimmte Arbeitsmethode voraus. Bei fossilen Gemeinschaften handelt es sich um offene Systeme, die vom Gesichtspunkt der Menge ihrer Komponenten, das heißt der Subsysteme, und ihrer komplexen wechselseitigen Bindungen als höchst kompliziert zu klassifizieren sind. Ein solches System ist niemals stabil, sondern befindet sich in einem autoregulativen und adaptiven dynamischen Gleichgewicht, das ständig von äußeren und inneren Faktoren angegriffen wird. Den Weg zur Lösung des Problems möchte ich in großen Zügen folgendermaßen festlegen:

— Beschreibung des Systemverhaltens,
— Klärung des Verhaltens,
— eventuelle Voraussage des künftigen Verhaltens.

Das Studium des Ganzen geht notwendigerweise vom Studium der einzelnen Komponenten aus. Während aber diese Komponenten früher mehr oder weniger als isolierte

Phänomene untersucht wurden und man eher ihre additiv-summarische Auffassung unterstrich, geht man heute davon aus, daß das Spezifische der komplizierten und vielschichtig verflochtenen Organisation des Ganzen vor allem in den wechselseitigen Beziehungen der einzelnen Komponenten seiner Struktur insgesamt verankert ist (JABLONSKIJ 1978).

Selbstverständlich ist die Anzahl der Subsysteme und damit auch die aller ihrer Beziehungen überaus groß. In der Praxis bedeutet dies, daß es ausgeschlossen ist, alle gleichmäßig in die Betrachtungen einzubeziehen und richtig zu erfassen. Das führt notwendigerweise dazu, die in der Zielsetzung unseres Studiums dominanten Komponenten herauszustellen und damit auch der Bestimmung der angemessenen Methoden höchste Aufmerksamkeit zu widmen. Eines der wichtigsten Momente des Studiums beruht in der Wahl jener Komponenten und Methoden, die bei der Analyse unserer Problematik entscheidend sind.

Die wechselseitigen Beziehungen können, allgemein gesprochen, im Rahmen des zu lösenden Problems äußerlicher oder innerlicher Art sein, sie können auf einer oder mehreren Ebenen verlaufen. Wesentlich ist aber, daß sie niemals einen linearen Verlauf nehmen, mit anderen Worten, daß man aus dem Komplex nicht etwa zwei Erscheinungen herauslösen und sie in ihrer Beziehung von Ursache und Wirkung studieren darf, sondern daran zu denken hat, daß bei jeder einzelnen Komponente in einer bestimmten Zeiteinheit ein Netz äußerer und innerer Funktionsbeziehungen abläuft. Lineare Kausalbeziehungen wären deshalb kein Abbild der echten Beziehungsstruktur oder zumindest allzu stark simplifiziert.

Eine zweite Eigenschaft des studierten Systems ist es, daß nicht alle diese gegenseitigen Beziehungen kategorischen Charakter tragen; manche sind nur wahrscheinlich und bieten demzufolge eine unendliche Menge potentieller Varianten der Weiterentwicklung, von denen sich allerdings nur wenige verwirklichen.

Diese Tatsachen erschweren die Untersuchung des Untergangs und Aussterbens beliebiger Taxa oder ganzer Artengemeinschaften; sie stellen an die Kenntnisse des wissenschaftlichen Bearbeiters hohe Anforderungen (RAUP u. a. 1973).

Natürlich muß man sich dessen bewußt sein, daß dabei wechselseitige Rückbeziehungen vorkommen. Sie können vom Standpunkt der studierten Problematik positiv sein, das heißt den untersuchten Vorgang positiv beeinflussen und beschleunigen, oder sie wirken negativ, wenn sie im Widerspruch zur Entwicklung stehen und diese verzögern. Wenn die negative Rückbindung zu überwiegen beginnt, wird das System stabil.

Beim Studium von Systemen ist vor allem das Ziel begrifflich klarzustellen, das man erreichen will. Seine konkrete Bestimmung vereinfacht die Untersuchungen. Wenn alle Subsysteme demselben Ziel zustreben und ihr Verhalten in einer bestimmten Zeiteinheit unterschiedlich ist, bedeutet das, daß sie dieses Ziel auf verschiedene Weise erreichen können.

Das Aussterben ist vom Standpunkt der zahlreichen Subsysteme, ihrer Schichtung und gegenseitigen Bindungen eine komplizierte Systemstruktur. Deshalb erfordert seine methodische Untersuchung unbedingt:

— die Bestimmung der Zwischenziele und des Endziels;
— die Klärung der dominanten Subsysteme und Bezüge vom Standpunkt der zu lösenden Problematik;
— das Studium in gegenseitigen nicht-linearen Bezügen;
— die Bestimmung der potentiellen Möglichkeiten einer gegebenen Entwicklung sowie — auf der Grundlage von Tatsachen — die Ausschaltung jener, die nicht in Betracht kommen.

3. Evolution und Aussterben

In Abschnitt 2 wurde der theoretische Zutritt dargelegt, den eine objektive Beantwortung der Frage des Aussterbens voraussetzt. Er entspricht unserer heutigen Erkenntnis, obwohl man seine konkreten Einstellungen nicht verabsolutieren darf. Grundsätzlich gestattet er aber, alle äußeren und inneren Bindungen in ihren gegenseitigen Dimensionen zu erfassen und zu beschreiben. Zugleich ist zu betonen, daß die Ergebnisse einer derartigen Bearbeitung niemals die mechanische Summierung analytischer Teile eines Ganzen sind, sondern aus der Synthese der gewonnenen Erkenntnisse hervorgehen.

Das Aussterben ist eine der am weitesten verbreiteten und wichtigsten Erscheinungen der Evolution. Man hat sie immer im Zusammenhang mit der Kehrseite der Medaille, der Entstehung von Taxa und Gemeinschaften zu sehen. Der Frage nach der Entstehung und folgenden Evolution pflegt man höhere Aufmerksamkeit zu schenken. Vielleicht nicht zuletzt deshalb, weil sie derzeit vor allem Neontologen beantworten und das Aussterben eher im Arbeitsbereich der Paläontologen liegt.

Es ist heute eindeutig klar, daß das Aussterben von einem ganzen Komplex wechselseitig zur Wirkung kommender Faktoren verursacht wird, die man in zwei Kategorien einteilen kann. Es ist die Kategorie der inneren, jedem Individuum, jeder Population und jeder Art eigenen Faktoren, ob es sich nun um genotypische oder phänotypische Merkmale handelt. Zur zweiten Kategorie der äußeren Faktoren gehören jene, die außerhalb der Individuen zur Wirkung kommen. Rein theoretisch gesehen, müssen beide Kategorien im Rahmen der Entwicklung nicht dieselbe Rolle spielen, ihr Rang kann sich sogar im Verlauf der Zeit ändern, so daß ihre Dominanz nur eine bestimmte Zeit stabil bleibt. Sind die äußeren Beziehungen dominant, sprechen wir von Ektosystemen, wenn es die inneren sind, von Endosystemen. Bei den Ektosystem kommt es zur gegenseitigen Substitution oder gar Vertauschbarkeit; typisch ist bei ihnen die Konkurrenz und infolge dieser das Phänomen der Auswahl. Bei den Endosystemen sind die Beziehungen zwischen den äußeren und inneren Faktoren größtenteils schwach, es dominieren die Beziehungen zwischen den inneren Faktoren.

Damit ist zugleich gesagt, daß sich die Frage des Aussterbens nicht isoliert betrachten läßt; man hat vielmehr im Rahmen höherer Zusammenhänge zu untersuchen, mit anderen Worten: nicht nur ihre Subsysteme zu zergliedern, sondern das System des Aussterbens selbst als Subsystem übergeordneter Systeme zu betrachten.

Wie ich bereits früher feststellen konnte (MUSIL 1983), spielen Analogien beim Studium unseres Themas eine Rolle, die auf den ersten Blick zu denselben Ergebnissen führen — erst ihre Analyse weist auf andere Ursachen hin. Das bedeutet, daß man nicht nur die Frage des „Wie“, sondern auch die des „Weshalb“ beantworten muß. Bei der Analyse des Aussterbens hat man bedingungslos die für diese Erscheinung charakteristischen Grundstrukturen und Erscheinungen aufzugliedern. Von der Genauigkeit ihrer Bestimmung hängt die Erfüllung der Aufgabe ab. Die tiefere Analyse unseres Problems zeigt, daß es verschiedene Schichten besitzt, deren

Wechselbeziehungen schrittweise zu klären sind. Jede höhere Schicht enthält im Vergleich mit der vorhergehenden, tieferen Schicht etwas Neues, das sich in ihren Termini nicht restlos ausdrücken läßt. Jede dieser Schichten verlangt deshalb einen ihr entsprechenden Zutritt.

Beim Studium des Aussterbens kann man mit zwei Alternativen arbeiten. Die erste Alternative befaßt sich mit den einzelnen Populationen, Arten, Gemeinschaften, usw. und den Ursachen, die zur Verringerung ihrer Bestandzahl, eventuell zu ihrem Aussterben geführt haben. Die zweite Alternative betrifft die Ursachen des Aussterbens auf allgemeiner Ebene; ihr Gegenstand sind deshalb nicht einzelne Formen und Assoziationen des Lebens, sondern das Leben als Ganzes. Das Studium im Rahmen dieser zweiten Alternative bewegt sich auf theoretischem Niveau, man analysiert die Grundprinzipien der Einheit im Bereich der lebenden und der leblosen Natur und ist bemüht, die dominanten und wesentlichen äußeren und inneren Beziehungen aufzudecken. Das Hauptziel bleibt allerdings, die innere Einheit der einzelnen Komponenten aufzudecken, und zwar einschließlich solcher, die sich auf den ersten Blick den Zusammenhängen zu entziehen scheinen.

Jede Komponente, sei sie nun organischer oder anorganischer Natur, besitzt natürlich in verschiedenen Gebieten, manchmal auch zu verschiedenen Zeiten, verschiedene Variationsbreiten. Daraus läßt sich folgern, daß die einzelnen Komponenten jeweils in ein breites Spektrum von Bedingungen eingebaut sind, was bis zu einem gewissen Grade ihre Rekonstruktion erschwert.

Bestimmend ist weiterhin die Art des methodischen Zutritts. Bei jedem Prozeß — und bei dem Aussterben handelt es sich zweifellos um einen Prozeß — lassen sich in dieser Hinsicht zwei Hauptaspekte unterscheiden, die man als strategische und taktische bezeichnen könnte, wobei es sich um zwei Betrachtungsweisen ein und desselben Prozesses handelt. Die Strategie bestimmt das Ziel, das wir uns stellen, hier also die Klärung der allgemeinen Ursachen und Umstände des Aussterbens, die Taktik betrifft die Art und Weise, wie wir dieses Ziel erreichen.

Diesen Abschnitt zusammenfassend kann man feststellen:

- Die Ursachen des Aussterbens sind nicht absolut, sondern dynamisch stabil. Offenbar sind sie imstande, sich unter bestimmten Bedingungen zu verändern. Sie sind komplexer Natur; in erster Linie handelt es sich um Veränderungen der Außenwelt, doch darf man innere Faktoren nicht vernachlässigen.
- Der Prozeß des Aussterbens ist ein kompliziertes System, das aus einer Reihe von Subsystemen besteht, deren gründliches Studium mit der Untersuchung des Aussterbens als Subsystem eines übergeordneten Systems zu verbinden ist.
- Dieselben Ergebnisse, das heißt das Aussterben im Endstadium, können auf verschiedene Ursachen zurückgehen.
- Allgemein gesprochen, ist das Aussterben ein Ausdruck der biologischen Entropie, das Entstehen neuer Taxa ihre Überwindung.

4. Probleme der Gesetzmäßigkeit des Aussterbens

Die Entwicklung des Organismus schreitet über eine Steigerung der Wahlmöglichkeiten des Umweltbiotops fort. Das Maß dieser Wahlmöglichkeit hängt von der Organisationsstufe ab. Je höher diese ist, desto größer ist die Wahlmöglichkeit. Wir können deshalb nach der Organisationsstufe die Existenzfähigkeit beurteilen.

Sowohl Populationen einzelner Arten als auch die Gesamtheit der eine Gemeinschaft bildenden Arten sind in Raum und Zeit als komplexe Systeme anzusehen. Ihre Haupteigenschaften lauten:

- Es liegt eine bestimmte Stabilität vor, so daß die gegebene Gemeinschaft imstande ist, kurzfristige Änderungen der Umwelt zu überstehen und relativ rasch in ihren ursprünglichen Zustand zurückzukehren. Der Grad dieser Stabilität ist bei den einzelnen Gemeinschaften verschieden. Jedenfalls ist die jedem System eigene Stabilitätskomponente eine äußerst wichtige Voraussetzung der Systemorganisation. Das System bleibt so lange bestehen, als es fähig ist, Mechanismen hervorzubringen und zu erhalten, die die Einwirkungen neuer Einflüsse zu überwinden vermögen. Störungen können nicht nur von der Umwelt ausgehen, sondern ebenso gut von disproportionalen Entwicklungen der einzelnen Taxa eines Systems.
- Alle Komponenten und Elemente dieses breit verzweigten Systems stehen ununterbrochen in wechselseitiger Korrelation, wirken ständig aufeinander ein und beeinflussen sich gegenseitig. Ihr Stellenwert ist nicht gleichrangig, manche sind dominant und deshalb für diese Untersuchungen hier besonders wichtig.
- Die Stabilität eines bestimmten Systems ist natürlich nicht statisch fixiert, sondern dynamisch. Deshalb ist das System imstande, sich zu ändern, zu entwickeln. Die Ursachen eventueller Änderungen sind ebenfalls komplexer Natur. Jede Gemeinschaft ist demnach ein dynamisches und offenes System gegensätzlich wirkender Kräfte als Hauptquelle seiner Entwicklung, die allerdings verderblich werden, wenn sie die Regulationsmöglichkeiten des betreffenden Systems überschreiten. Änderungen von Gemeinschaften verlaufen entweder in linearer Richtung oder sie sind zyklischen Charakters.

Die linearen führen zu einer besseren Anpassung an die Umwelt, die zyklischen sind dialektische Pole im ununterbrochenen Prozeß von Entstehen und Vergehen. Der Umfang der Anpassungsmöglichkeiten der Taxa oder Gemeinschaften steht in direktem Verhältnis zu der Intensität der Änderungen und dem Tempo ihres Eintretens. Wenn ein beliebiges Taxon oder eine beliebige Gemeinschaft nicht mehr imstande ist, sich Änderungen anzupassen, mit anderen Worten: wenn der Stabilitätsfaktor zu überwiegen beginnt, gehen das Taxon oder die Gemeinschaft zumeist unter. Im entgegengesetzten Fall erfolgt eine Entwicklung, die von der quantitativen Zunahme der Individuen, der Ausweitung des Siedlungsgebiets und — bei genügend langer Dauer — von der möglichen Entstehung weiterer systematischer Gruppen begleitet wird.

Die einzelnen Arten einer Gemeinschaft sind durch zahlreiche Bindungen miteinander verknüpft, die bei manchen Gruppen, beispielsweise Produzenten, Konsumenten und ähnlichem, besonders stark sind. Die Änderung der einen Art ruft deshalb notwendigerweise bestimmte Änderungen anderer Arten hervor, die manchmal sogar stärker sind als die Variabilität infolge abiotischer Einflüsse. Das Erscheinen einer neuen Art in der bestehenden Gemeinschaft kompliziert die Existenzbedingungen der übrigen Arten dieser Gemeinschaft.

Die Vergrößerung einer Gemeinschaft steht im Verhältnis zur Vergrößerung ihrer Artenzahl. Die mathematische Erfassung der Konkurrenz der einzelnen Arten einer Gemeinschaft läßt erkennen, daß ihre gegenseitigen Beziehungen vom Standpunkt des trophischen Niveaus umso verschiedenartiger sind, je komplizierter die Gemeinschaft organisiert ist. Eine solche Gemeinschaft ist stabiler, was bei unveränderten oder nur wenig veränderten Komponenten ein Vorteil, beim Eintreten stärkerer Änderungen jedoch ein Nachteil ist.

Das Studium einer Gemeinschaft, ihrer Entwicklung und ihres Aussterbens muß von der Untersuchung ihrer Teile ausgehen. Auch in diesem Falle gibt es zwei Möglichkeiten, an die Lösung heranzutreten: entweder geht man von den einzelnen Teilen zum Ganzen, oder in umgekehrter Richtung, vom Ganzen zu den einzelnen Teilen, vor. Konkret bedeutet das: von den Populationen der einzelnen Arten zur gesamten Gemeinschaft oder von der Gemeinschaft zu den einzelnen Arten. Der analytische Zutritt stützt sich dabei faktisch auf die Vorstellung der Isoliertheit und Unabhängigkeit der einzelnen Komponenten und ihre gegenseitigen Beziehungen werden übergangen.

Während man früher mehr oder weniger die additiv-summarische Auffassung der Gemeinschaft als Ganzes herausstellte, gehen wir heute davon aus, daß das Spezifische einer so komplizierten und vielschichtigen System-Organisation in der Charakteristik der Bindungen und Beziehungen zwischen ihren einzelnen Komponenten beruht.

Selbstverständlich ist die Zahl dieser Bindungen und Beziehungen groß und es ist notwendig, jene auszuwählen, die sich von der Zielrichtung der Untersuchungen aus gesehen, als am wichtigsten erweisen. Die Bestimmung der für die Lösung unseres Problems entscheidenden Relationen stellt den Kern unserer Arbeiten dar und ist die Voraussetzung richtiger Schlußfolgerungen (MUSIL 1980).

Fassen wir kurz zusammen:

- Das Studium des Aussterbens von Arten geht von ihrer Gemeinschaft aus. Es umfaßt alle zwischen den einzelnen Taxa und zwischen den Organismen und der Umwelt im weitesten Sinne des Wortes herrschenden Beziehungen.
- Die Gemeinschaft ist Änderungen im Verlauf der Zeit und des Waltens ihrer biotischen oder abiotischen Komponenten angepaßt. Sie stellen ein offenes, adaptives System dar, das innerhalb bestimmter Bereiche organisiert ist und mit seiner Umwelt nicht nur Materie, sondern auch Energie und Informationen austauscht und sich selbst reguliert. Das sind seine Hauptunterschiede gegenüber dem geschlossenen System.
- Trotz aller dieser Tatsachen sind Gemeinschaften zeitlich begrenzt und keiner unbegrenzten Selbsterneuerung fähig; ihre Entwicklung stößt also auf zeitliche Grenzen. Von diesem Blickpunkt aus gesehen könnte man sagen, daß eine Gemeinschaft eigentlich kein absolut offenes System ist, weil sie ebenfalls Erscheinungen unterliegt (Intensitätsschwund, allmähliche Verkleinerung, Untergang), die für geschlossene Systeme charakteristisch sind.
- Das System zielt ständig auf die Stabilisierung als einer Art von Idealziel. Sein Fortschreiten wird jedoch immer wieder von antagonistischen Kräften gestört. Einerseits sind diese der Motor seiner Entwicklung, können jedoch andererseits unter bestimmten Voraussetzungen auch den Untergang des Systems herbeiführen.

Der Untergang und das Aussterben von Taxa und Gemeinschaften ist eines der interessantesten Kapitel der Paläontologie, ob man nun die Problematik auf konkreter oder theoretischer Ebene zu lösen versucht. Alle bisher geleisteten Arbeiten lassen aber erkennen, daß es an der Zeit ist, vor allem auf theoretischem Niveau Teamforschung unter Beteiligung von Fachleuten der verschiedensten Disziplinen einzuleiten. Nur auf diesem Wege wird es gelingen, weitere Tatsachen und Erkenntnisse, vielleicht auch Gesetzmäßigkeiten zu erkennen, dis bisher unerkannt geblieben sind.

Zusammenfassung

Das Aussterben ist ein Grundgesetz der natürlichen Evolution. Die Arbeit bringt allgemeine methodologische Richtlinien zum Studium dieses Phänomens, befaßt sich aber nicht mit konkreten Taxa, sondern erörtert mehr die theoretischen Bestimmungsgrundsätze der zu beantwortenden Fragen und zu erreichenden Ziele, die Möglichkeiten ihrer Lösung, die Arbeitsmethodik im Rahmen der einzelnen Systeme und ihre strategischen und taktischen Aspekte. Das Spezifische dieses Studiums liegt in der Kompliziertheit der zu untersuchenden Systeme und der Notwendigkeit, sie im Lichte der wechselseitigen Beziehungen ihrer Komponenten zu analysieren.

Summary

Extinction represents one of the most common and important processes from an evolutionary point of view. The article deals with various methodological aspects necessary for the study of the extinction of taxons and communities. The author is not trying to explain the extinction of concrete taxons but refers to some theoretical basic principles necessary for a solution of these problems. Therefore the basic principles for the determination of the problems and aims, for the way of their solution, the methods of working in various systems, the strategic and tactical aspects of this work and the specificness of this study consisting in the complexity of the given system and in the necessity of its study from the point of reciprocal relations of various components are analysed.

Резюме

Вымирание с эволюционной точки зрения представляет собой один из наиболее обычных и важных процессов. В статье рассматриваются различные методологические аспекты, связанные с изучением вымирания таксонов и сообществ. Автор не пытается объяснить вымирание конкретных таксонов, а рассматривает некоторые основные теоретические принципы, необходимые для решения этих проблем. Поэтому здесь анализируются основные принципы постановки проблем и задач, путей их решения, методы работы в различных системах, стратегические и тактические аспекты этой работы, а также специфика такого исследования, состоящая в сложности данной системы и необходимости изучения ее с точки зрения взаимоотношений различных компонентов.

Literatur

JABLONSKIJ, A. I., 1978: Rozvoj vědy jako otevřeného systemu (Die Entwicklung der Wissenschaft als offenes System). — In: Sistemnye issledovanija. — Moskva: Nauka. — S. 86—109.

MUSIL, R., 1980: Fázovité vymírání savců v pleistocénu (Das Phasenaussterben der Säugetiere im Pleistozän). — In: Problematik der Entstehung und des Erlöschens. — In: Scripta. — Brno **10**. — S. 351—353.

—, 1983: Probleme und Aufgaben der Säugetierpaläontologie des Pleistozäns. In: W. D. HEINRICH (Hrsg.): Wirbeltier-Evolution und Faunenwandel im Känozoikum. — Berlin: Akad.-Verl. — S. 303—332, (Schriftenr. geol. Wiss.; 19/20).

RAUP, D.; GOULD, ST. J.; SCHOPF, TH. J. M. et al., 1973: Stochastic models of phylogeny and the evolution of diversity. — In: J. Geol. — Chicago **81**. — P. 525—542.

Quartärpaläontologie 6 123—129 Berlin 1986

Vegetative Charophytenreste aus dem Quartär Thüringens (DDR)

Von Tilo Nötzold, Berlin

Mit 2 Abbildungen und 2 Tafeln

Inhalt

Abstractum

Es werden neue Taxa vegetativer Reste von Charophyten aus quartären Travertinen Thüringens, DDR, beschrieben.

New taxa of fossil vegetative charophyte-remains are described from Quaternary travertines of Thuringia, German Democratic Republic.

Описываются новые таксоны ископаемых вегетативных остатков харофитов из четвертичных травертинов Тюрингии, Германская Демократическая Республика.

1. Forschungsgeschichte und Einleitung

Seit Jahrzehnten werden mächtige Charophytentravertine zur Gewinnung von Kalk sowie als mächtige Bau- und Dekorationssteine abgebaut. Als Beispiele seien hier nur die Wandverkleidung der Haupthalle des Bahnhofs Berlin-Friedrichstraße sowie die Basilika von Königslutter bei Braunschweig genannt. Bereits J. W. v. Goethe bewahrte in seiner Sammlung Handstücke von Charophytentravertin verschiedener Herkunft. Während aber die fossilen weiblichen Fortpflanzungsorgane der Charophyten, die „Gyrogonite" und „Oogonien" seit 1804, als Lamarck seine „*Gyrogona medicaginula*" publizierte, zunehmend das Interesse der Paläobotaniker fanden, wurden die fossilen vegetativen Organe der Armleuchteralgen vernachlässigt. Lediglich in der Mitte des 19. Jahrhunderts wurden von Unger, Heer, Saporta, v. Ettingshausen, Engelhardt und Fliche insgesamt sieben Organtaxa über fossile Thallusreste von Charophyten aufgestellt. Davon kann man fünf Taxa praktisch streichen, da sie entweder von den Autoren selbst als fraglich Reste bezeichnet wurden und die Zuordnung zu den Charophyten nur unter Vorbehalt geschah, oder die Angaben der morphologischen Verhältnisse zu dürftig sind, um einen Vergleich mit anderen Fossilien zuzulassen. Allen bisher beschriebenen Taxa ist gemeinsam, daß sie dem Tertiär entstammen. Als legitime Taxa können nur „*Chara schubleri* Unger 1850" und „*Chara langeri* v. Ettingshausen 1872" betrachtet werden, wobei letztere Art auf ihre Gyrogonite begründet wurde und schon deshalb umbenannt werden muß. Denn die Fortpflanzungsorgane wurden nicht im organischen Verband mit den Thallusresten gefunden. Auf diese Spezies wird im Zusammenhang mit den vegetativen Charophytenresten aus dem Thüringer Quartär näher eingegangen.

Daily and Durham (1964) publizierten das einzige dem Verfasser bekannte Fossil, an dem sich die vegetativen Thallusreste und die weiblichen Fortpflanzungsorgane im organischen Verband fanden. Leider unterliefen den Autoren dabei einige Verstöße gegen die Vorschriften des Internationalen Code der Botanischen Nomenklatur (ICBN) und auch die Altersdatierung des bemerkenswerten Fossils muß angezweifelt werden (Nötzold 1977). Trotzdem wurde vom Verfasser bewußt nicht die Konsequenz gezogen, das Taxon „*Tectochara ixtapensis* Daily & Durham 1966" für illegitim zu erklären, um nicht dieses einmalige Fossil der wissenschaftlichen Diskussion zu entziehen. Seit über 20 Jahren untersucht der Autor fossile Charophytenreste und ist bestrebt, in sämtlichen Materialien fossile vegetative und generative Organe im organischen Verband zu finden, um die Gyrogonite und Oogonien bestimmten Thallusresten zuordnen zu können. Aber dieses gelang bei der Untersuchung weder der einzelnen Fortpflanzungsorgane noch der verschiedensten Charophytentravertine und -sande. Im Probenmaterial mit Oogonien und Gyrogoniten fanden sich zwar häufig Thallusreste, aber nie im organischen Verband miteinander. Das gleiche gilt für die reinen Travertine. Da die Organtaxa mit ihren Diagnosen für die fossilen Fortpflanzungsorgane aufgestellt wurden, ist es nach dem ICBN unmöglich, die nomina für getrennte Thallusreste zu übernehmen; ja für letztere ist sogar ein neues Taxon zu finden, das den Thalluscharakter zum Ausdruck bringt. Sollte es später hoffentlich gelingen, Organe im Verband vereinigt zu finden, dann hat zwingend eine Revision zu erfolgen.

Nun taucht noch ein weiteres nomenklatorisches Problem auf. Die rezenten Charophyten werden nach ihren vegetativen Organen diagnostiziert. Die früheren Autoren ließen bei der Aufstellung ihrer Taxa die Oogonien bei der Diagnose aus. Sie nannten wohl die Größenvariation der Oogonien und

Oosporenmembranen mit der Anzahl der seitlich sichtbaren Umgänge der Hüllzellen, da sich diese Eigenschaften bei den einzelnen rezenten Taxa indes weitgehend decken bzw. überschneiden, gelten sie nicht als diagnostisches Merkmal. Zur Diagnose wird der „Stengelquerschnitt" (Thallus) herangezogen, ob er „unberindet" oder „berindet" ist, das heißt, ob nur eine einzige Zellröhre vorhanden ist oder ob diese von schwächeren Röhrchen, von Hüllzellen, umgeben ist. Bei berindeten Arten werden dann „einfach berindete" mit einem Kranz mehr oder weniger gleichstarker Hüllzellen und „mehrfach berindete", an denen Hüllzellen größeren Durchmessers mit solchen kleineren Durchmessers abwechselnd, unterschieden. Außerdem gelten die Beblätterungsverhältnisse und eventuelle Ausbildung von Stacheln als diagnostische Merkmale. Beide waren jedoch bisher bei Fossilmaterial nicht zu erkennen. Die Charophytenthalli haben einen „schachtelhalmähnlichen" Aufbau. Man unterscheidet daher zwischen „Knotenzellen" (Nodien) und den dazwischen stehenden „Internodien", wobei sowohl die Dicke der Stengel als auch die Länge der Internodien bei den rezenten Arten keine diagnostische Bedeutung besitzen. Wir kennen heute aus der Boddenkette südlich des Darß und der Halbinsel Zingst Charophytenrasen von über einem Meter Höhe. Die Größe dürfte also von den ökologischen Verhältnissen abhängig sein.

Wenn der Autor trotzdem zur Unterscheidung des vorliegenden Fossilmaterials (s. unten) und zur Findung verschiedener Taxa die Größe der Querschnitte einzelner Thallusreste und ihrer Rindenzellen heranzieht, so geschieht dies, um überhaupt erst einmal gewisse Anhalts- bzw. Unterscheidungskriterien der einzelnen Fossilien zu haben, um auf diesem „Neuland der Charophytenkunde" eine Arbeitsgrundlage zu gewinnen.

An den einzelnen Knoten der Stengel sitzen bei den rezenten Arten je nach Anzahl der Rindenzellen Blättchen und Triebe höherer Ordnung. Die Blättchen sind beim vorliegenden Fossilmaterial nicht erhalten geblieben. Die Frage nach Haupt- oder Nebentrieb ist demzufolge nicht zu beantworten, da ja nur einzelne Stengelglieder vorliegen. Lediglich in zwei Fällen gelang der Querschnitt durch einen Knoten, so daß Haupt- und Nebentriebe gleichzeitig zu erkennen sind.

Nun ergibt sich aber für die Fossilien noch ein weiteres Problem. Bei den Fortpflanzungsorganen beginnt die Kalkinkrustierung erst nach der Befruchtung der Organe, während die Thalli bei den Arten, die Kalk einlagern, bereits vorher damit beginnen. Alle Fossilfunde stammen aus basischen Gewässern und die organische Substanz hat sich bei den Resten nicht erhalten. Weder im Travertin noch in den Travertinsanden beobachtete man bisher inkohlte Reste der Thalli, selbst wenn in der gleichen Probe ganze Oogonien mit erhaltenen Oosporenmembranen aufzufinden waren. Die kalkigen Thalli bilden offensichtlich beim Entstehen der mächtigen Charophytentravertine die Kristallisationskerne. Es ist daher oft schwer zu entscheiden, wie dick ein Thallusrest primär gewesen ist und wieviel Kalk sekundär angelagert wurde.

Bei der Untersuchung von Quartärmaterial kann auf einen gewissen Rezentvergleich nicht verzichtet werden und alle Charophytentravertine entstammen dem Quartär. Nun hat WOOD (1962) die von den früheren Autoren aufgestellten ca. 30 rezenten Charophytenarten Europas, die verkalken, mit außereuropäischen Taxa zu sieben Arten mit zahlreichen infraspezifischen Taxa vereinigt. Da, wie betont, die rezenten Characeen nach ihren Thalli diagnostiziert werden, der Paläobotaniker aber diese Untersuchungen an rezenten Pflanzen nicht durchführen oder kontrollieren kann, ist WOOD notgedrungen zu folgen. Es muß jedoch eingeschränkt werden, daß sich bei der Untersuchung fossiler Fruktifikationsorgane, bei weitgehender Zustimmung, durchaus einige Zweifel am WOODschen System ergaben (NÖTZOLD 1985). Dennoch soll beim Rezentvergleich hier das neue System zugrunde gelegt werden. Allerdings wird, da fossile quartäre Fossilien zur Diskussion stehen, bei den zahlreichen Fragen, die ohnehin bei der Bearbeitung eines völligen „Neulands" offen bleiben müssen, auf die rezenten Charophyten anderer Erdteile verzichtet. Lediglich das außergewöhnliche Fossil „*Tectochara ixtapensis*" muß mit erwähnt werden. Betrachtet man die morphologischen Eigenschaften der von WOOD vereinigten Arten nach ihren alten Diagnosen, die ja nun für die neuen Taxa zu übernehmen sind, so überschneiden sich diese derartig, daß es praktisch nicht möglich ist, auch nur die rezenten Spezies zu unterscheiden. Als Beispiel diene die verbreiteste Charophytenart der Welt, *Chara vulgaris* L., zu ihr gehören nun folgende ältere Arten (mit den jeweiligen Eigenschaften):

Chara vulgaris L. (syn. *Ch. foetida* A. BR.) — Berindung des Stengels vollkommen, Anzahl der Rindenzellen doppelt so hoch wie die der Blätter „Diplostichae" (was bei Fossilien eine große Anzahl bedingen würde) Mittelreihen von den Zwischenreihen mehr oder weniger überwölbt, Stengel und Blätter vollkommen berindet. (Von diesen diagnostischen Merkmalen weichen die der folgenden Taxa teilweise ab.)

Chara vulgaris L. (syn. *Ch. imperfecta* A. BR.) — Berindung unvollkommen.

Chara vulgaris L. (syn. *Ch. dissoluta* A. BR.) — Anzahl der Rindenzellen gleich der der Blätter am darüberstehenden Quirl „Isostichae"

Chara vulgaris L. (syn. *Ch. jubata* A. BR.)
(syn. *Ch. contraria* A. BR.)
(syn. *Ch. gymnophylla* A. BR.)
(syn. *Ch. crassicaulis* SCHLEICHER)
alle wie *Ch. vulgaris* L. (syn. *Ch. foetida* A. BR.).

Daraus ersieht man, daß die diagnostischen Eigenschaften zum Teil innerhalb einer Art bei WOOD sich gegenseitig aufheben. Alle diese grundsätzlichen Probleme mußten den speziellen Erörterungen über die fossilen vegetativen Charophytenreste vorangestellt werden.

2. Das Fossilmaterial

Charophytentravertine liegen vor von den Fundstellen Weimar-Ehringsdorf, Bilzingsleben und Thomas-Müntzer-Stadt Mühlhausen. Einzelne Stengelreste fanden sich im Oogonienmaterial der Fundstellen: Bilzingsleben, Weimar — Parkhöhlen, Herbsleben, Bad Langensalza, Greußen und Königsaue, die veröffentlicht werden (NÖTZOLD 1985).

UNGER und v. ETTINGSHAUSEN nannten ihre Fossilien „*Chara*". Das ist nach dem ICBN überholt. Es muß ein eigenes Taxon für fossile Organe gewählt werden, denn man weiß nicht, ob die anderen Organe der Mutterpflanzen der Diagnose der rezenten Gattung *Chara* entsprechen. Für die fossilen vegetativen Charophytenreste existiert noch kein Organgattungsname, aber sie müssen beschrieben und bekannt werden; die nominae *Charites* und *Characeitas* sind für fossile Fruktifikationsorgane vergeben. Der Name einer rezenten Gattung kommt aus den genannten Gründen nicht in Betracht. Außerdem soll nach den Empfehlungen des ICBN im Namen eines Taxons sein Charakter möglichst zum Ausdruck kommen. Daher wird für die fossilen Charophytenorgane der Gattungsname

2.1. **Charathallus**

gewählt und das Taxon „*Charathallus weimarensis* n. gen. et n. sp." als Genotypus benannt (siehe unten).

DAILY and DURHAM (1966) verbinden unter *Tectochara ixtapensis* ein Oogonium mit Thallusrest. Das berechtigt jedoch nicht, andere fossile Thalli als „*Tectochara*" zu bezeichnen, denn *Tectochara* ist eine Gyrogoniten-Organgattung, deren Thalli außer von *T. ixtapensis* unbekannt sind. Sie werden daher auch von der Gattungsdiagnose nicht erfaßt.

Die beiden älteren Taxa müssen aus obigen Gründen umbenannt werden, was nachfolgend, um den Text über die quartären Reste nicht zu zerreißen, geschehen soll.

2.1.1. **Charathallus schubleri** (UNGER 1850) n. comb.

Synonym: *Chara schubleri* UNGER.

Diagnose: Fossile Thallusreste der Gattung *Charathallus*, Thallusdurchmesser 400—800 µm, diplostich (zweizeilig) berindet von 16 bis 20 Rindenreihen, größere und kleinere alternierend, einschließlich Wirtel von 8 bis 10 Zweigchen.

Locus typicus: Steinheim, bei Darmstadt (Hessen)/BRD.
Stratum typicum: Tertiär, Miozän.

2.1.2. **Charathallus langeri** (v. ETTINGSHAUSEN) n. comb.

Synonym: *Chara langeri* v. ETTINGSHAUSEN.

Diagnose: Fossile Thallusreste der Gattung *Charathallus*, Thallusdurchmesser 900—1000 µm, 10 erhaltene Rindenzellen, sechs Seitenzweige.

Locus typicus: Sagor und Savine, Slowenien (Krain)/SFR Jugoslawien.
Stratum typicum: Tertiär, Oligozän.

v. ETTINGSHAUSEN schließt aus den erhaltenen Rindenzellen, daß deren reale Anzahl 12 und die Berindung diplostich, zweizeilig war. Außerdem verbindet er mit dem Thallusrest als „*Chara langeri*" an der gleichen Lokalität gefundene, abgetrennte Gyrogonite, die inzwischen in *Rhabdochara langeri* (v. ETT.) MÄDLER 1955 umbenannt wurden.

2.1.3. **Charathallus weimarensis** n. gen. et n. sp.

Taf. I, Fig. 1—8

Diagnose der Gattung: Fossile, kalkinkrustierte, vegetative Charophytenorgane.

Diagnose der Art: Fossile, kalkinkrustierte vegetative Charophytenreste, dreizeilig (triplostisch) berindet, 12 bis 21 Rindenzellen. Innendurchmesser der Thalli: 165—800 µm, Außendurchmesser 600—1500 µm.

Holotypus: Präparat W-Ph Nr. 1, Taf. I, Fig. 1, 2.
Locus typicus: Charophytentravertin, Weimar — Parkhöhlen/DDR.
Stratum typicum: Mindel/Riß-(Holstein-)Interglazial.
Derivatio nominis: Nach dem Fundort benannt.
Weitere Fundstellen: Weimar-Ehringsdorf, Mühlhausen.

Eventuelle Beziehungen zu tertiären und rezenten Charophyten: Nach dem Thallusdurchmesser und der hohen Anzahl der Rindenzellen tendiert das Taxon zu *Charathallus schubleri*, wenn UNGER auch nur auf Diplostichie schließt, aber bei 16 bis 20 Rindenzellen dürfte die Art triplostisch gewesen sein. Deshalb wird für die vorliegenden Fossilien auf dreizeilige Berindung geschlossen, wenn dieses auch nicht an allen Exemplaren deutlich zu erkennen ist, denn die Stengelglieder sind oft zusammengedrückt. In der graphischen Darstellung der Innen- und Außendurchmesser der Thalli ergeben sich Abweichungen vom Glockenform-Idealbild, das dürfte daher kommen, daß Hauptstengel und Nebenäste verschieden hoher Ordnung vorliegen und daher zwangsläufig die Nebenäste überwiegen, außerdem kann die sekundäre Verkalkung unterschiedlich sein. Trotzdem werden die Häufigkeitskurven hier gegeben, um bei der Schwierigkeit der Materie eventuellen späteren Bearbeitern möglichst viele Anhaltspunkte zu vermitteln. Die Darstellung der Innendurchmesser der Thalli zeigt weniger Abweichungen von der Glockenform. Das dürfte daran liegen, daß wohl an die Außenseiten der Thalli noch kalkhaltiges Wasser gelangt, durch die Leitungsbahnen abgestorbener Thalli aber keines mehr fließt.

Bei der Triplostichie des *Charathallus weimarensis* kommt als Rezentvergleich nur die Art *Chara globularis* THUILL., in der WOOD die europäischen Arten *Chara aspera* (DET.) WILLDENOW, *Ch. galioides* DC., *Ch. connivens* SALZMANN, *Ch. tenuispina* A. BR., *Ch. fragilis* DESVAUX und *Ch. delicatula* AGARDH als infraspezifische Taxa vereinigt, in Frage. Die älteren Spezies werden genannt, da WOOD damit offensichtlich sämtliche triplostichen Charen Europas zu einer einzigen, außereuropäischen Art stellt. Dieses dürfte wahrscheinlich den Widerspruch der Rezentbotaniker hervorrufen.

2.1.4. **Charathallus thuringensis** n. sp.

Taf. II, Fig. 1—12

Diagnose: Fossile, kalkinkrustierte, vegetative Charophytenreste der Gattung *Charathallus*. Zweizeilig berindet, Anzahl der Rindenzellen 12 (in zwei Ausnahmefällen 14 und 15). Thallus: Innendurchmesser 100—365 µm, Außendurchmesser 235—730 µm.

Da nur einzelne Thallusstücke ohne Verzweigung als Charophytensand vorliegen, können keine Angaben über Nebenäste erfolgen.

Holotypus: Präp. Bil. 1; Taf. II, Fig. 1.
Locus typicus: Bilzingsleben, Bezirk Halle/DDR.
Stratum typicum: Mindel/Riß-(Holstein-)Interglazial.
Derivatio nominis: Der Artname bezieht sich auf die Verbreitung in Thüringen.
Weitere Fundstellen: Weimar — Parkhöhlen, Herbsleben, Greußen, alle Bezirk Erfurt; Königsaue, Bezirk Magdeburg. Alle Fundstellen DDR.
Stratigraphische Verbreitung: Mindel/Riß- (Holstein-) Interglazial bis Holozän.

Am Fossilmaterial läßt sich nicht erkennen, ob die Mittelreihen der Rindenröhrchen überwölbt werden oder nicht. Auf jeden Fall sind die Mittelreihen stärker entwickelt. Die Stengel sind vollständig berindet und es handelt sich um mehr oder

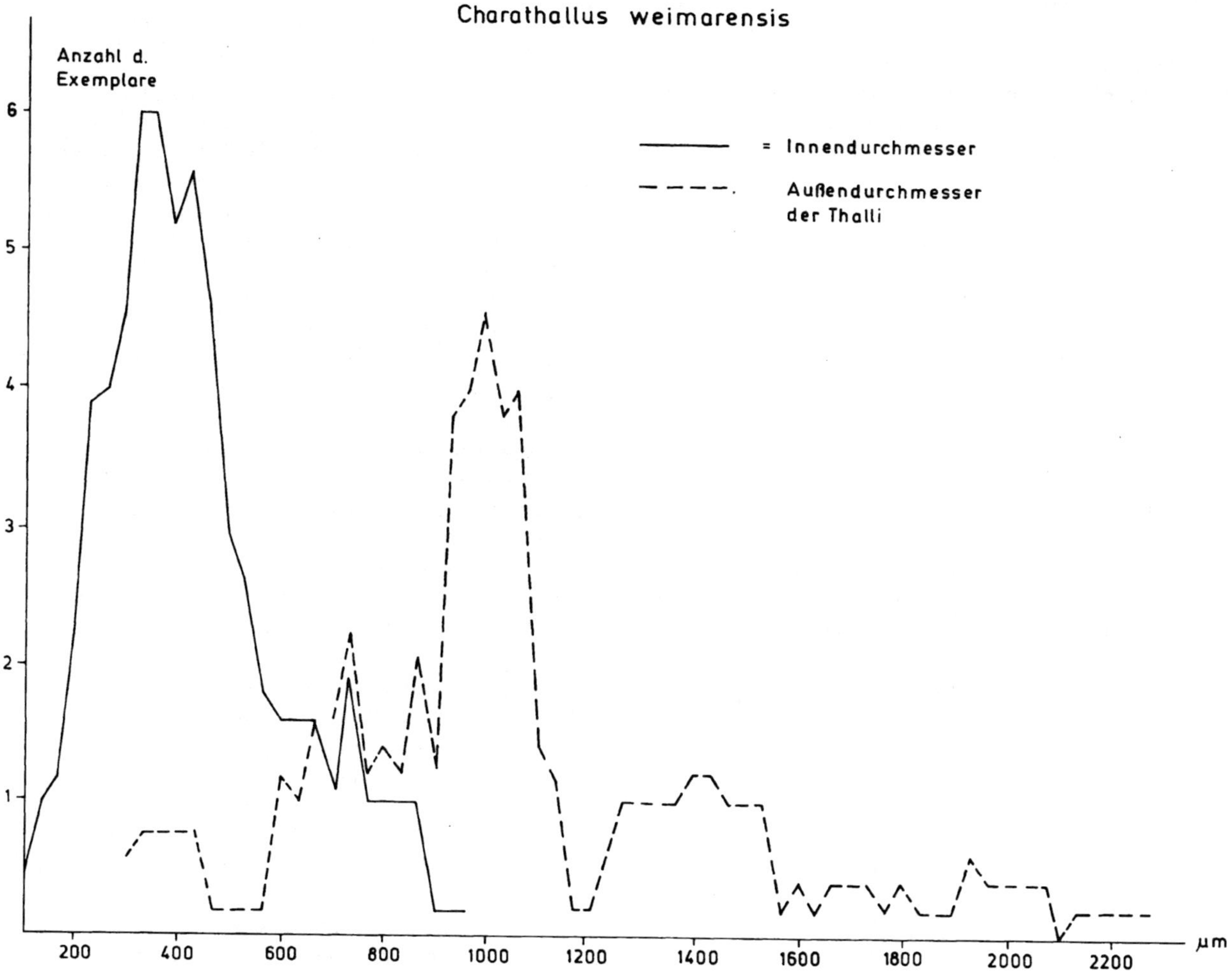

Abb. 1. *Charathallus Weimarensis* n. sp. Häufigkeitskurven der Innen- und Außendurchmesser, bezogen auf die Anzahl der gefundenen Exemplare

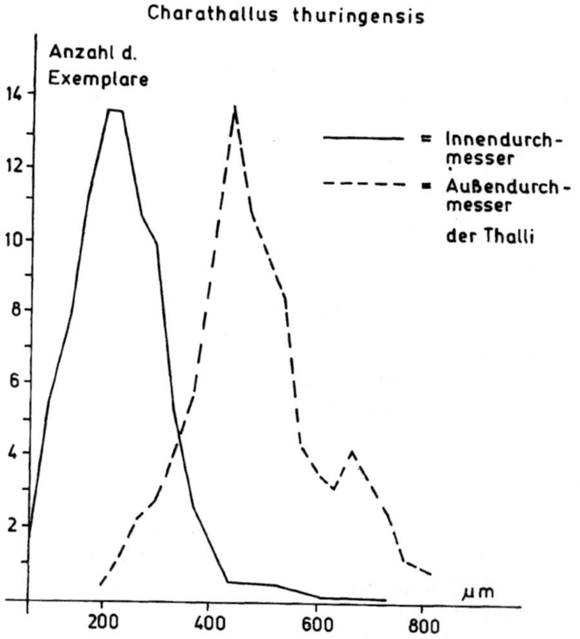

Abb. 2. *Charathallus thuringensis* n. sp.

weniger alternierend stehende, stärkere und schwächere Rinzellen.

Nach diesem Befund können für den Rezentvergleich alle unberindeten und mangelhaft berindeten Arten ausgeschlossen werden. Am ehesten ist *Chara vulgaris* L. zum Vergleich heranzuziehen. Welche älteren Taxa WOOD als infraspezifische Taxa in ihr vereinigt, wurde bereits gesagt. Es kann allerdings ein Vergleich mit der ebenfalls heute in Europa häufigen *Chara hispida* L. nicht ausgeschlossen werden. Zu ihr zählen nach WOOD als infraspezifische Taxa die älteren Arten *Chara polyacantha* A. BR., *Ch. intermedia* A. BR., *Ch. rudis* A. BR. und *Ch. horrida* WAHLSTEDT, eine Neukombination, die völlig zu Recht erfolgte.

Der Bearbeiter hat eine größere Menge Charophytentravertin in Bilzingsleben entnommen. Es wurden über 20 Proben davon geschnitten und angeschliffen, ohne daß auch nur eine einzige Thallusstruktur untersucht werden konnte. Der Travertin von Bilzingsleben ist der älteste des vorliegenden Materials, denn er gehört in das Holstein-Interglazial. Die Diagene des Kalksteins ist soweit fortgeschritten, daß die Anatomie der Charophytenstengel nicht mehr festgestellt werden kann. Es sind nur noch die runden Hohlräume der Thalli als feine Löcher im Travertin zu erkennen. Zum Teil ist dieses auch beim Charophytentravertin aus Mühlhausen der Fall.

Zusammenfassung

Es wurden Charophytentravertine und Travertinsande aus dem Quartär im Südwesten der DDR (Thüringen) untersucht, um die fossilen vegetativen Organe (Thalli) zu beschreiben. Im Gegensatz zu den weiblichen Fortpflanzungsorganen fossiler Characeen, für die in den letzten Jahrzehnten ein gesichertes taxonomisches System erarbeitet wurde, wird hier für die Thallusreste ein Anfang gemacht, wobei zahlreiche Fragen über die Anatomie, ihre Systematik und die Beziehungen zu rezenten Charophyten offen bleiben. Die variationsstatistische Analyse der Größe der Stengelquerschnitte und das Auszählen der erhaltenen Hüllzellenröhrchen im Thallus ließ die Aufstellung des Organgenus *Charathallus* n. gen. mit den zwei Organspezies *Charathallus weimarensis* n. sp. und *Charathallus thuringensis* n. sp. als angebracht erscheinen. Der Charophytentravertin von Bilzingsleben zeigt keine anatomischen Verhältnisse der fossilen Thalli mehr.

Summary

Quaternary charophyte travertines and charophyte travertinesands of Thuringia have been studied to describe the fossil vegetative organs (thalli). Contrary to the female sexual organs of fossil Characeae of which an effectual taxonomic system was founded in the last decades, in this paper a first step is made to reach a similar taxonomic system for the fossil thallus-remains, whereby questions of anatomy, systematics, and relations to recent charophytes are still left open.

A variation-statistic analysis of the size of stem-diameters and the number of preserved cover cellular-tubes in the thallus was making possible to found an "organ-genus" *Charathallus* nov. gen. with two "organ-species" *Charathallus weimarensis* nov. spec., and *Charathallus thuringensis* nov. spec.

In the charophyte travertines of Bilzingsleben due to a progressive secundary travertine recrystallisation it is impossible to recognize any anatomical details of the thalli.

Резюме

Проведено изучение четвертичных харофитовых травертинов и харофитовых травертиновых песков Тюрингии с целью описания ископаемых вегетативных органов (талломов). Для женских половых органов ископаемых Characeae эффективная таксономическая система была создана в последние десятилетия; в этой работе сделан первый шаг к разработке такой же системы для ископаемых остатков талломов, для которых вопросы анатомии, систематики и взаимоотношений с современными харофитами все еще остаются открытыми.

Вариационно-статистический анализ величины диаметра стволиков и числа сохранившихся покровных клеточных трубочек в талломе дал возможность выявить новый „орган-род" *Charathallus* nov. gen. с двумя „орган-видами" *Charathallus weimarensis* nov. sp. и *Charathallus thuringensis* nov. sp.

В харофитовых травертинах Бильцингслебена невозможно различить какие-либо детали анатомического строения талломов из-за прогрессировавшей вторичной рекристаллизации травертина.

Literatur

BRONGNIART, A., 1822: Sur la classification et la distribution des végétaux fossiles en général, et sur ceux des terrains de sédiment supérieur en particulier. — In: Mém. Mus. nat. Hist. Natur. — Paris **8**. — S. 202—240.

DAILY, F. K.; DURHAM, J. W., 1966: Miocene Charophytes from Ixtapa, Chiapas, Mexico. — In: J. Paleont. — Tulsa (Okla.) **40** (5). — P. 1191—1199.

ETTINGSHAUSEN, C. v., 1872: Die fossile Flora von Sagor in Krain. — In: Denkschr. Akad. Wiss. Wien. — Wien **32**. — S. 159—202.

FLICHE, P., 1886: Note sur les flores tertiaires des environs de Mulhouse. — In: Bull. Soc. Indust. — Mulhouse **56**. — S. 353—362.

GROVES, J., 1933: Charophyta. — Fossilium Catalogus II, Plantae, Pars 19 (In: W. JONGMANN (Hrsg.). — Berlin.

HEER, O., 1855: Flora tertiaria helvetiae I. — Winterthur.

LAMARCK, J. B. A. P. M. DE, 1804: Suite des Mémoires sur les Fossiles des environs de Paris. — In: Ann. Mus. nat. Hist. Natur. — Paris **5**. — S. 349—357.

MIGULA, W., 1890—1897: Die Characeen. — In: Kryptogamenflora von Deutschland, Österreich und der Schweiz. V. RABENHORST. — Leipzig.

—, 1925: Charophyta (Charales). — In: Die Süßwasserflora Deutschlands, Österreichs und der Schweiz. 11.: hrsg. A. PASCHER. — Jena.

NÖTZOLD, T., 1962: Bestimmungsmöglichkeiten der *Tectochara*-Oogonien aus dem Pleistozän. — In: Ber. naturforsch. Gesellsch. Freiburg i. Br. — Freiburg i. Br. **52**. — S. 173—183: 3 Abb., 2 Taf., (1962a).

—, 1962: Fossile Charophytenreste vom Tüllinger Berg bei Wei/Rh. — In: Monatsber. dt. Akad. Wiss. Berlin. — Berlin **4** (10). — S. 663—669: 6 Taf., 2 Tab., (1962b).

—, 1966: Die pleistozänen Charophyten-Oogonien Mitteleuropas und ihre Beziehungen zu tertiären Gyrogoniten und rezenten Charophyten-Arten. — In: Quaternaria. — Roma **8**. — S. 217 bis 224: 1 Taf., 5 Tab.

—, 1974: Charophyten-Fruktifikationen aus dem Ilmtravertin von Weimar. — In: Abh. zentr. geol. Inst. — Berlin 21. — S. 327 bis 332: 2 Abb., 1 Taf.

—, 1975: Charophytenreste aus dem Neophytikum Mitteleuropas. — In: Abh. staatl. Mus. Mineral. u. Geol. Dresden. — Dresden **23**. — 265 S.: 3 Abb., 14 Taf., 46 graph. Darst., 7 Tab. (Diese Arbeit enthält eine ausführliche Literaturzusammenstellung von über 300 Titeln.).

—, 1977: *Nitellopsis obtusa* (DESVAUX) J. GOVES und *Lychnothamnus barbatus* (MEYEN) v. LEONHARDI — lebende Fossilien? — In: Ber. naturforsch. Gesellsch. Freiburg i. Br. — Freiburg i. Br. **67**, S. 217—234: 2 Taf.

—, 1980: Die Charophytenreste aus dem Bilzingslebener Travertinkomplex. — In: Ethnograph.-archäol. Z. — Berlin **21**. — S. 20—25.

—, 1986 (s. diesen Band, S. 131—156).

SAPORTA, L. C. J. G. DE, 1862: Études sur la végétation du Sud-Est de la France à l'époque tertiaire, III. — In: Ann. Sci. Natur. Bot. Ser. 4. — Paris **17**. — S. 191—311.

UNGER, F., 1850: Genera et Species Plantarum fossilium. — Wien.

WOOD, R. D., 1962: New combinations and taxa in the revision of Characeae. — In: Taxon. — Utrecht **11**. — S. 7—25.

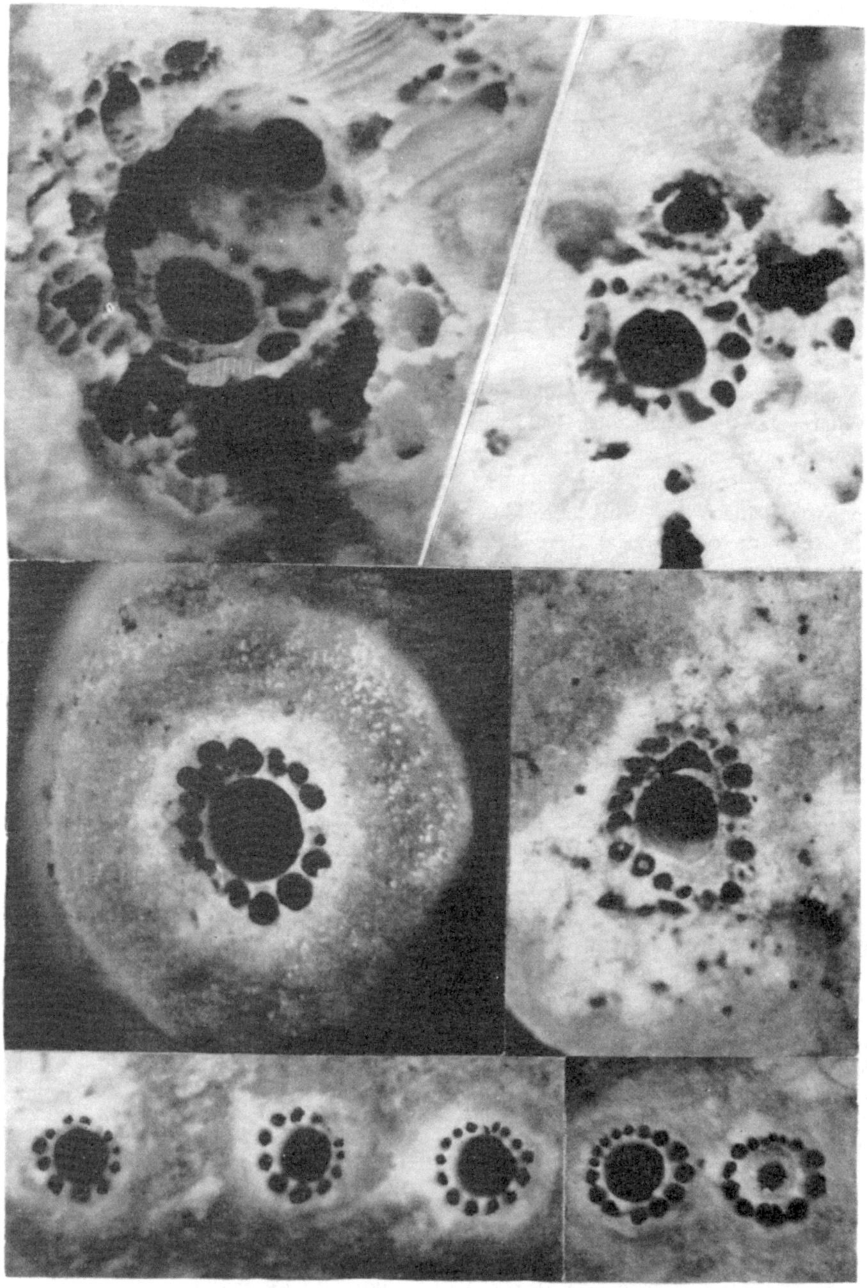

1 2

3 4

5 6 7 8 9

TAFEL I

Fig. 1—9: *Charathallus weimarensis* n. gen. et. n. sp.
Weimar — Parkhöhlen, Bezirk Erfurt/DDR
Alle Stücke Vergr. × 20
Fig. 1. Holotypus der Gattung und Art, Querschnitt durch Nodien mit Haupt- und Nebenästen
Fig. 2. Querschnitt durch Nodien mit Haupt- und Nebenästen
Fig. 3, 4. Starke sekundäre Verkalkung
Fig. 5—9. Querschnitte durch Nebenästchen

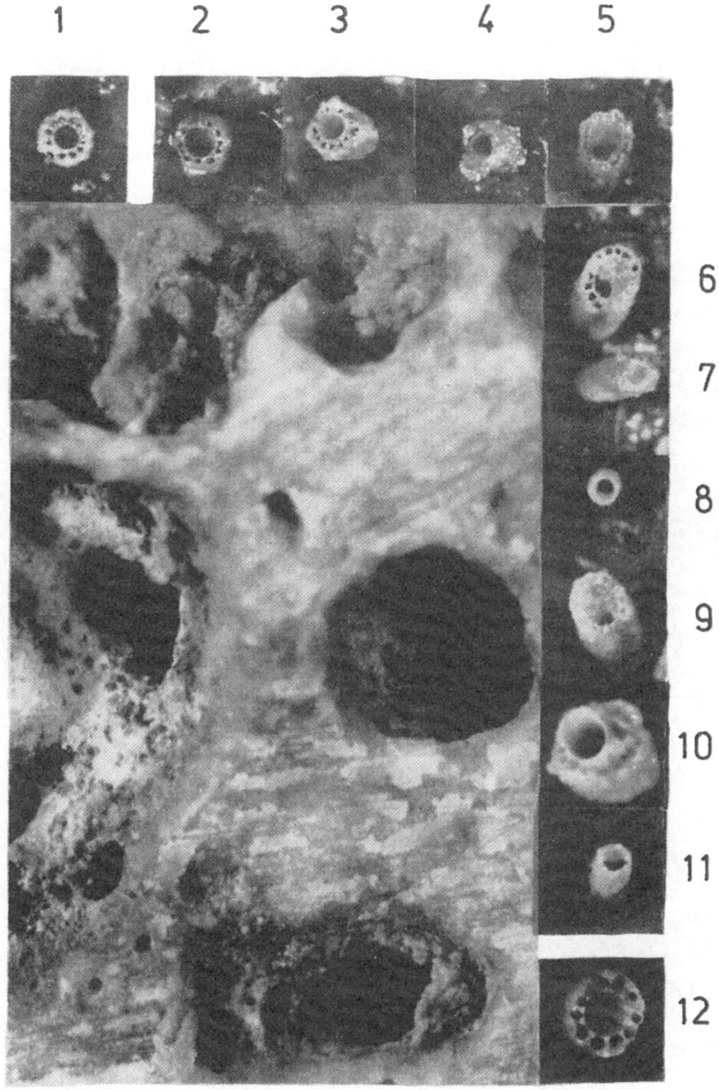

13

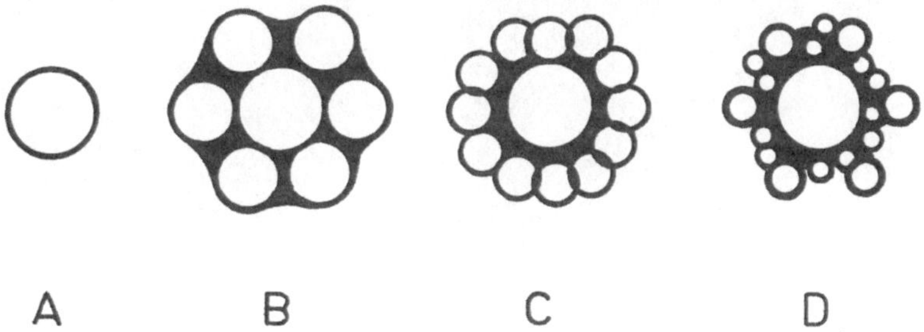

TAFEL II

Fig. 1—12: *Charathallus thuringensis* n. sp.
Fig. 1—11. Bilzingsleben, Charophytensand; Fig. 12. Königsaue
Alle Stücke Vergr. × 20
Fig. 1. Holotypus der Art
Fig. 13. Charophytentravertin von Bilzingsleben

A bis D. Schema der verschiedenen Berindungsverhältnisse der Charophytenthalli im Querschnitt
A — unberindet; B — monostiche, einzeilige Berindung; C — diplostiche, zweizeilige Berindung; D — triplostiche, dreizeilige Berindung

Quartärpaläontologie 6 131—156 Berlin 1986

Neue Charophytenfruktifikationen aus dem Thüringer Quartär (DDR)

Von TILO NÖTZOLD, Berlin

Mit 21 Tabellen und 4 Tafeln

Inhalt

Abstractum

Neufunde von Charophytenfruktifikationen vor allem aus Thüringen/DDR ermöglichten die Aufstellung und Beschreibung der Gattung *Charites*. Elf neue Arten der Gattung *Charites* werden beschrieben. Es werden die phylogenetischen Entwicklungslinien zwischen den tertiären und rezenten Organen verglichen.

Following the recent discovery of fossil charophyte reproductive structures, mostly from Thuringia, GDR, a new genus *Charites* is described, containing eleven new species. The phylogeny and evolution of these charophytes are discussed.

Новые находки репродуктивных органов харофитов, прежде всего из Тюрингии, ГДР, дали возможность установления и описания рода *Charites*. Описано одиннадцать новых видов этого рода. На основании изучения репродуктивных органов проведены сравнения линий филогенетического развития третичных и современных харофитов.

1. Einleitung

Bei der Bearbeitung und Revision der Charophytenreste aus dem Neophytikum Mitteleuropas (NÖTZOLD 1975) werden die Fruktifikationen aus quartären Fundschichten, die dem Bearbeiter bis 1970 vorlagen, in offener Nomenklatur geführt. Es wurden „Charophytenformenkreise“ postuliert, weil dem Autor das Material nicht ausreichend erschien, um taxonomische Einheiten im Sinne des Internationalen Code der Botanischen Nomenklatur (ICBN) festzulegen. Dazu waren zwar Materialien aus plio-pleistozänen Grenzschichten und vor allem aus dem Riß/Würm (Eem)-Interglazial und dem frühen Postglazial ausreichend vorhanden, aber speziell aus dem Mindel/Riß (Holstein)-Interglazial und dem Holozän lagen nur wenige Fossilien vor. Zumindest reichten diese für eine variationsstatistische Analyse nicht aus. Diese Lücke wurde durch Proben aus Thüringen geschlossen. Nun können die Organtaxa der Fruktifikationen der Armleuchteralgen des Quartärs festgelegt werden, wobei die „Charophytenformenkreise“, bis auf diejenigen, die sich nur auf Oosporenmembranen beziehen, eliminiert werden.

Da es dem Autor aus Gesundheitgründen nicht mehr möglich ist, umfangreiche Aufsammlungen im Gelände selbst vorzunehmen, gilt sein Dank seinen befreundeten Kollegen Dr. W. STEINER, Weimar, und Dr. D. MANIA, Jena, für die Überlassung des größten Teils der Fossilien, die diese Arbeit überhaupt erst ermöglichten. Gedankt sei auch Herrn Dr. P. LANGE, Weimar, der die Aufnahmen der Originalia mit dem Rasterelektronenmikroskop anfertigte und Herrn Dr. Dr. H. D. KAHLKE, Weimar, für die Drucklegung dieser Arbeit. Meiner Tochter Ina danke ich für die Anfertigung der graphischen Darstellungen.

Die Charophytenfruktifikationen werden nach der bei NÖTZOLD (1975) beschriebenen Methode variationsstatistisch ausgewertet. Die dort gemachten Ausführungen über die anatomischen und morphologischen Eigenheiten der weiblichen Fortpflanzungsorgane und ihre zunehmende Kalkinkrustation mit fortschreitendem Lebensalter, bedürfen hier nicht der Wiederholung. Wenn auch nach wie vor in jüngeren Publikationen über Charophytenreste zum Teil an der veralteten Methode festgehalten wurde, die vom Lebensalter abhängigen, inkrustationsbedingten Eigenschaften zu diagnostischen Zwecken zu verwenden.

Um den Text zu straffen, sind die Angaben über Größe, Anzahl der seitlich sichtbaren Spiralwindungsumgänge der Hüllzellen (kurz Windungszahl genannt), das Verhältnis Länge/Breite = Q, die Anzahl der ausgewerteten Fossilien und die Mittelwerte (die sich mit den jeweiligen Maxima decken), den Tabellen bzw. graphischen Darstellungen für die einzelnen Fundstellen zu entnehmen.

Die Originale und das Belegmaterial befinden sich in der Charophytensammlung des Staatlichen Museums für Mineralogie und Geologie zu Dresden.

2. Fundstellen quartärer Charophytenreste

Soweit es sich um Fundstellen handelt, die bereits 1975 für „Charophytenformenkreise" genannt wurden, werden die Lokalitäten hier nicht noch einmal behandelt.

2.1. Bilzingsleben/Bezirk Halle

Tab. 1—4, 17—20; Taf. I, Fig. 1—4

Aus vier Proben wurden jeweils 250 Fossilien ausgewertet. Dabei galt es, eine weitere morphologische Definition zu geben. Die fossilen weiblichen Fortpflanzungsorgane von Charophyten aus Bilzingsleben und einiger anderer thüringischer Lokalitäten, unterscheiden sich von den als „Gyrogonit" definierten Resten, also mineralisch ausgefüllten Hüllzellen. Es sind aber keine vollständigen Oogonien im Sinne der botanischen Morphologie, wozu die Oospore mit ihrer Membran gehört. Beim Bilzingsleben-Material (alle Proben) sind nur die leeren, kalkigen Hüllzellen erhalten geblieben. Auch die Oosporenmembranen liegen nicht mehr vor. Letztere bestehen aus Suberin, einer dem Pollenin der Sporomorphen verwandten, korkähnlichen Substanz, oder ganz allgemein chemisch, einem Fett. Da sich die Sporomorphen in kalkigen Sedimenten ebenfalls nur schwer erhalten, sei erst einmal generell gesagt, daß nicht allein die Mutterpflanzen derartiger Reste in basischen Gewässern wuchsen, sondern darüber hinaus die Sedimentation in diesem Bereich erfolgt sein muß. Dem Bearbeiter liegen nach etwa zwanzigjähriger Praxis in der Bearbeitung fossiler Charophytenreste erstmalig derartige hohle Organe, die naturgemäß sehr zerbrechlich sind, vor.

Daher sei im Unterschied zum „Gyrogonit" und „Oogonium" dieser Erhaltungszustand als „Kalkhülle" bezeichnet und der Holotypus der *Charites cava* n. sp. (s. unten) möge dafür als Beispiel dienen.

Da jedoch eine Oosporenmembran wesentlich dicker als eine Pollen- oder Sporenwand ist, bedarf es einer beachtlichen Basizität, um sie in Lösung zu bringen. Eine Frage des stratigraphischen Alters dürfte es nicht sein, denn es liegen aus plio-pleistozänen Grenzschichten Reste von Oosporenmembranen vor. Diese Grenze wird zugleich als stratigraphische Grenze für die Erhaltungsmöglichkeit von Oosporenmembranen angesehen. Als Gradmesser für die Basizität des sedimentierenden Gewässers von Bilzingsleben sei noch der Hinweis gegeben, daß sich in der „Seekreide" von Königsaue nur Oosporenmembranen fanden, Pollen nach MÜLLER (1953) in dieser Schicht jedoch höchst selten nachzuweisen waren.

Die Analyse der Proben Bilzingsleben (Bil.) 1, Bil. 2, (vom 10. Juli 1974), Bil.-Steinrinne = 3 (vom 28. August 1974), Bil. 4 (vom 30. Juni 1974) ergab, daß es sich bei allen vier Proben um das gleiche Fossilmaterial handelt und daß ein einheitliches Taxon vorliegt (s. unten). Die Kalkhüllen gleichen den Fossilien des ehemaligen „Formenkreises Eem" und werden hier *Charites cava* n. sp. benannt.

2.2. Mühlhausen/Bezirk Erfurt

Tab. 5, 15—21

Die Probe Mühlhausen — Klippe wurde den Charasanden zwischen festen Travertinbänken entnommen. MANIA stellt die Fundschicht in das Eem oder Mittelpleistozän? Aus dem Material konnten lediglich 100 Kalkhüllen vermessen werden. Dennoch ergaben die eindeutigen Gaußschen Glockenkurven bei der graphischen Darstellung, daß es sich um ein einheitliches Fossilmaterial handelt. Die Reste gehören zu *Charites thuringica* n. sp.

Tab. 1. Größenbereiche und Windungszahlen des Materials von Bilzingsleben 1

Legende für die Tab. 1—14: Quadrat mit Linien — 7 Windungen; Quadrat, schwarz-weiß — 8 Windungen; Dreieck, auf die Spitze gestellt — 9 Windungen; Quadrat, auf die Spitze gestellt — 10 Windungen; Kreis — 11 Windungen; Quadrat — 12 Windungen; Dreieck, auf Basis stehend — 13 Windungen

Die im Zeichen angegebene Zahl bedeutet die Anzahl der gefundenen Exemplare

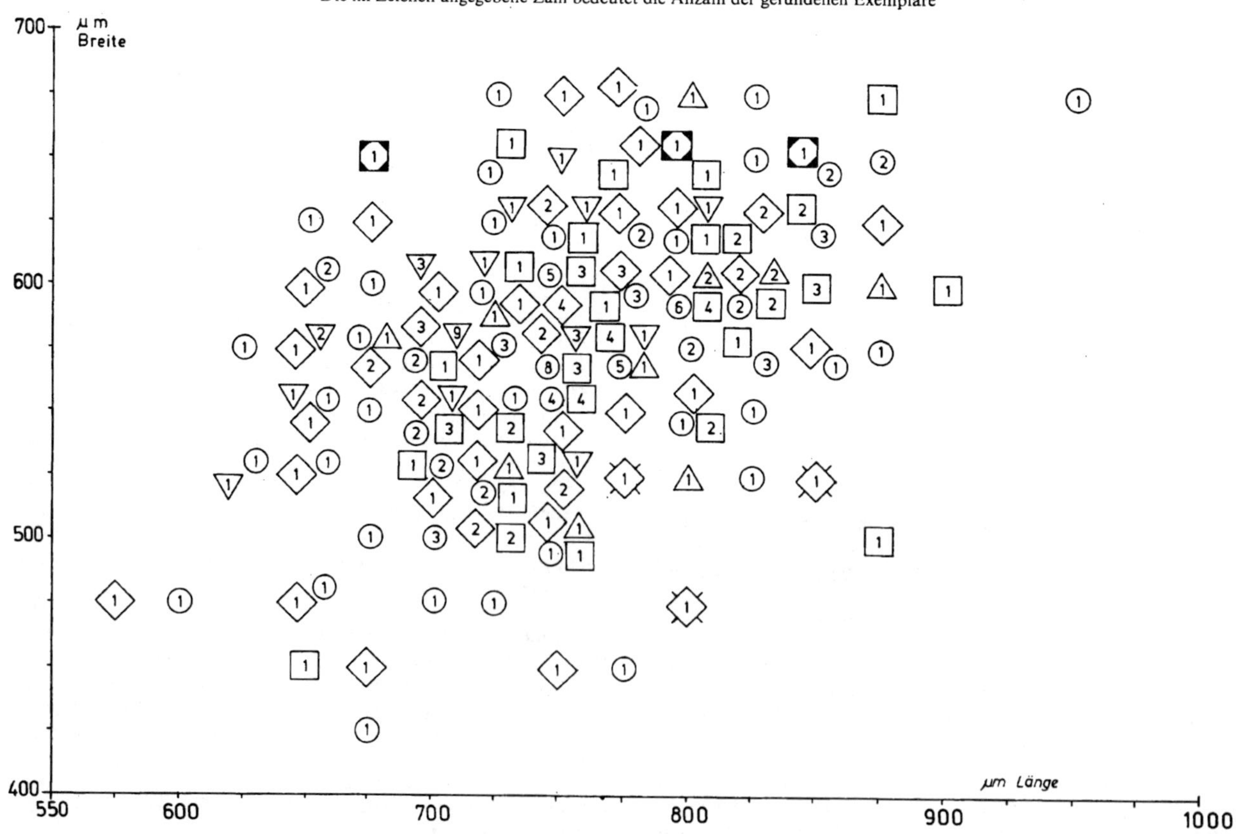

Tab. 2. Größenbereiche und Windungszahlen des Materials von Bilzingsleben 2

Legende s. Tab. 1, S. 132

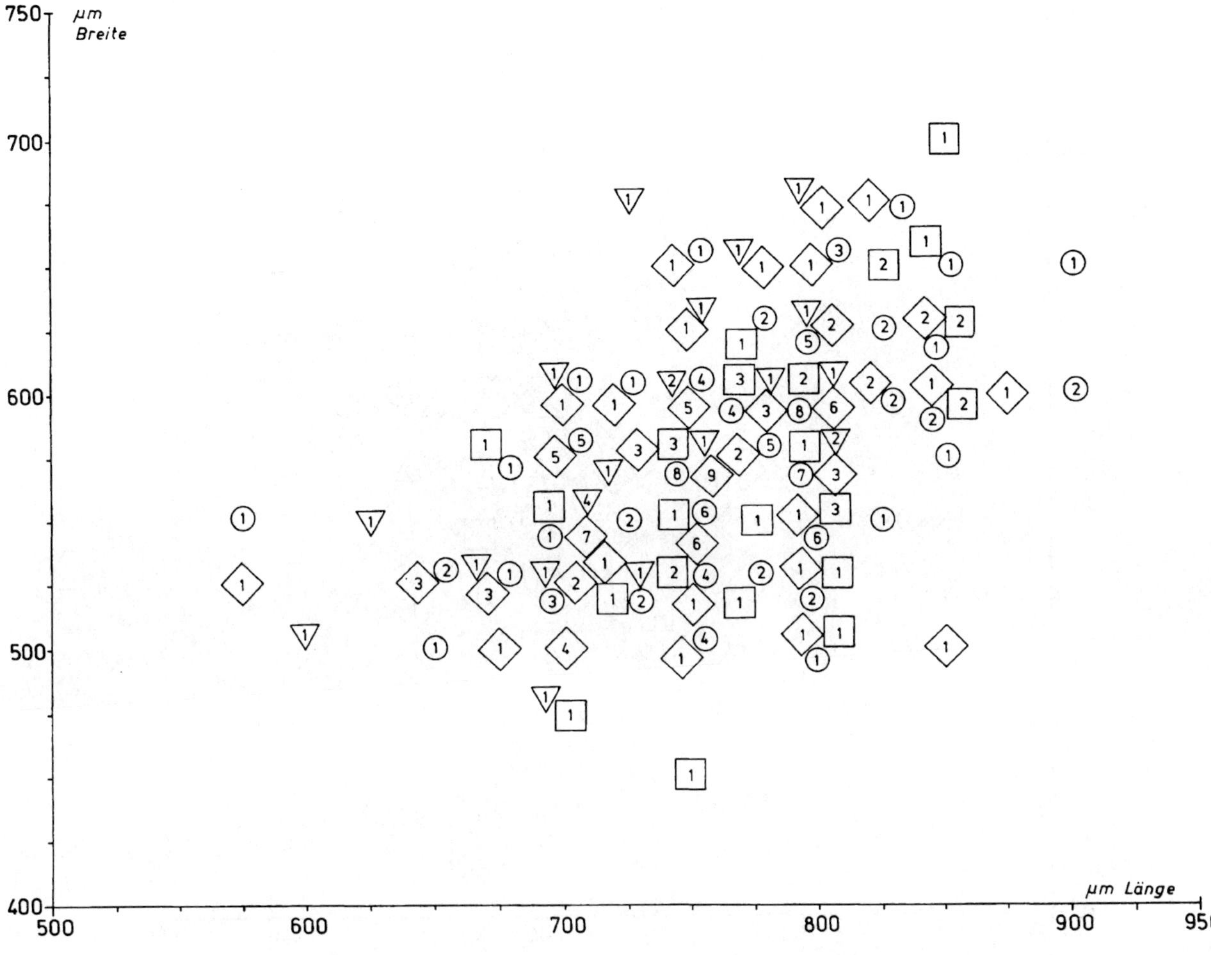

2.3. Burgtonna bei Bad Langensalza/Bezirk Erfurt

Tab. 6—8, 15—21; Taf. I, Fig. 5—7

Aus Burgtonna wurden drei Proben untersucht, die MANIA alle in das Riß/Würm(Eem)-Interglazial einstuft.

1. Burgtonna Bt. 6. Travertinbruch am Ort, etwa am Mittelteil des Nordostprofils.
2. Burgtonna Bt.-N. Nördlicher Bruch, Charasande mit festem Charophytentravertin, über *Helix banatica*-Horizont.
3. Burgtonna — Ort Bt.-O. 1,5 m über der Basis im Travertinbruch im Ort. Aus Travertinsanden mit vereinzelten Lagen der Strukturfraktion mit Blattabdrücken des Eichenmischwalds entnommen.

Auch die Kalkhüllen von Burgtonna gehören einem einheitlichen Taxon an. Aus diesen Funden wurde der Holotypus der *Charites thuringica* n. sp. ausgewählt.

2.4. Weimar/Bezirk Erfurt

Tab. 15—21; Taf. II, Fig. 1—3

Aus dem Ilmtravertin der Parkhöhlen bei Weimar überließ STEINER zwei Proben: Weimar W AI und Weimar W C6. Er stellt die Fundschicht in des Eem-Interglazial. Es zeigte sich, daß die Fossilien der Probe W AI nur ganz geringfügige Unterschiede zu dem Material, daß der Autor bereits 1968 bearbeitete und das aus redaktionellen Gründen erst 1975 publiziert wurde, aufweisen. Die Unterschiede dürften sich wohl daraus ergeben, daß damals für die variationsstatistische Analyse in der Darstellung der Kurven die absoluten Zahlen angegeben wurden, in vorliegender Arbeit jedoch die ausgleichende Berechnung erfolgte. Sowohl die 1975 veröffentlichten Reste wie auch die Fossilien der Probe W AI gehören zu *Charites thuringica* n. sp. Als Besonderheit muß jedoch erwähnt werden, daß, von wenigen Einzelexemplaren abgesehen, sämtliche Fossilien konkave Hüllzellenwände besitzen. Es handelt sich folglich um junge (Lebensalter) Fruktifikationen. Da die Kalkinkrustation sofort nach der Befruchtung der Oogonien einsetzt, wurden die Zygoten offensichtlich kurz nach der Befruchtung von den Mutterpflanzen getrennt und eingebettet. Die Fossilien sind Kalkhüllen, die Fragmente der Oosporenmembranen enthalten. Es war nicht möglich, ganze Oosporenmembranen aus dem Material zu präparieren.

Fünf Exemplare sind noch im organischen Verband mit Thallusfragmenten. Leider sind die Thallusreste zu kurz und zu schlecht erhalten, um die Berindungsverhältnisse exakt erkennen zu können. Da die rezenten Charophyten nach der Thallusmorphologie diagnostiziert werden, wären hier even-

Tab. 3. Größenbereiche und Windungszahlen des Materials von Bilzingsleben 3

Legende s. Tab. 1, S. 132

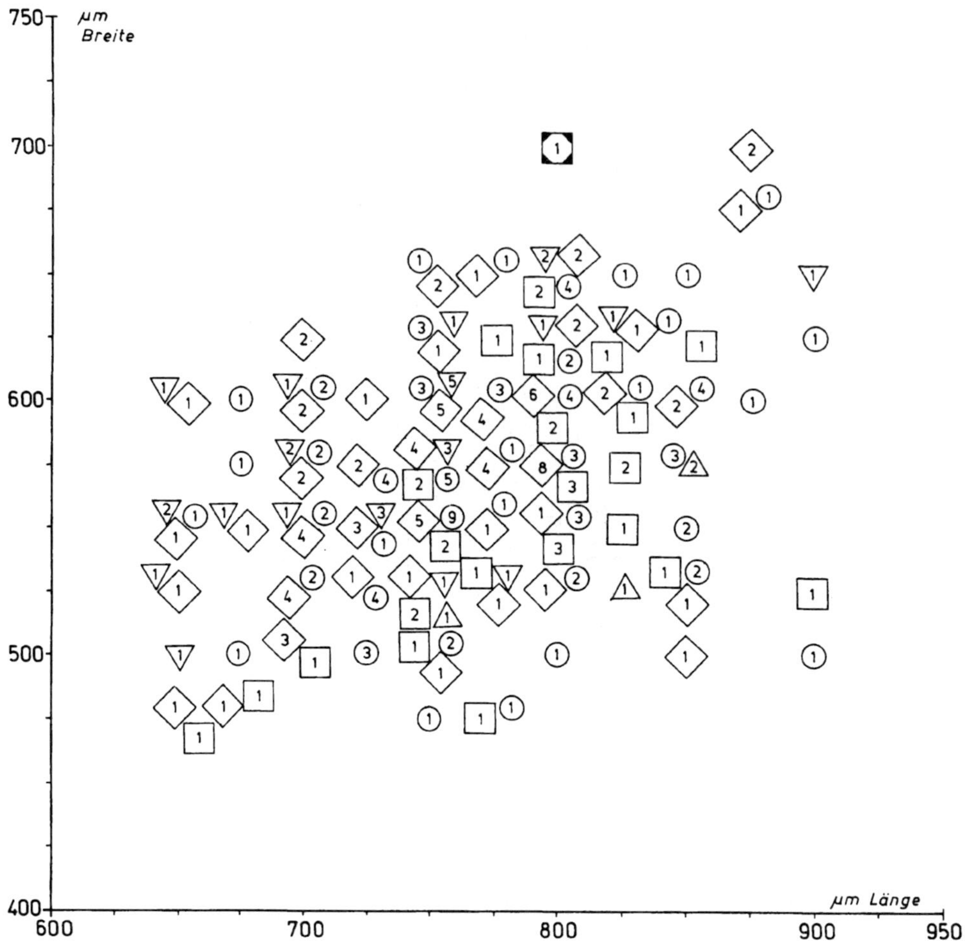

tuell taxonomische und phylogenetische Fragen zu lösen gewesen. Deshalb sollte an weiterem Material, das aus den Parkhöhlen zu bergen wäre, bevor diese nicht mehr zugänglich sind, nachgeforscht werden.

Die Fossilien der Probe W C6 bilden ebenfalls ein einheitliches, taxonomisch jedoch anderes Material. Sie fallen aus dem Rahmen aller bisher bekannten fossilen Charophytentaxa einschließlich der ehemaligen „Formenkreise". Sie werden daher als neues Taxon betrachtet und *Charites ilmica* n. sp. benannt.

2.5. Mücheln im Geiseltal/Bezirk Halle

Tab. 9, 15—21

Aus einer gut gegliederten limnischen Serie des Spätglazials bis Holozäns bei Mücheln-Möckerling entnahm MANIA eine Probe hellgelben Travertinsandes mit größeren Travertinknollen (Stengeltravertin). Da die Begleitfauna eine typische Allerød-Fauna ist, sind die Fossilien eindeutig eingestuft. Es konnten 250 Oogonien, die als homogenes Material zu *Charites thuringica* n. sp. gehören, untersucht werden. Da es sich um vollständig erhaltene Oogonien handelt, konnten diese Fossilien zusammen mit denen von Herbsleben das Material für die diagnostischen Eigenschaften der Oosporenmembranen der neuen Species liefern. Dieses ist besonders wichtig, weil sie die Verwandtschaft zur rezenten *Chara vulgaris* L. s. l. untermauern.

2.6. Bad Langensalza/Bezirk Erfurt

Tab. 10, 11, 15—21; Taf. III, Fig. 6; Taf. IV. Fig. 1, 2

Die Probe Bad Langensalza La 66/16 wurde den Charophytensanden zwischen festen Travertinbänken am westlichen Stadtrand von Bad Langensalza im Steinbruch mit holozänem Travertin entnommen. Trotz des holozänen Alters handelt es sich bei den Fossilien nur um Kalkhüllen. Bereits der visuelle Eindruck ergab vor der Variationsanalyse, daß zwei verschiedene Charophytentaxa vorliegen müssen. Es wurden einmal 250 Reste der *Charites thuringica* n. sp. ausgelesen und zum anderen fanden sich 125 Exemplare von nahezu kugeliger Gestalt, die *Charites sphaerica* n. sp. benannt wurden.

Tab. 4. Größenbereiche und Windungszahlen des Materials von Bilzingsleben 4

Legende s. Tab. 1, S. 132

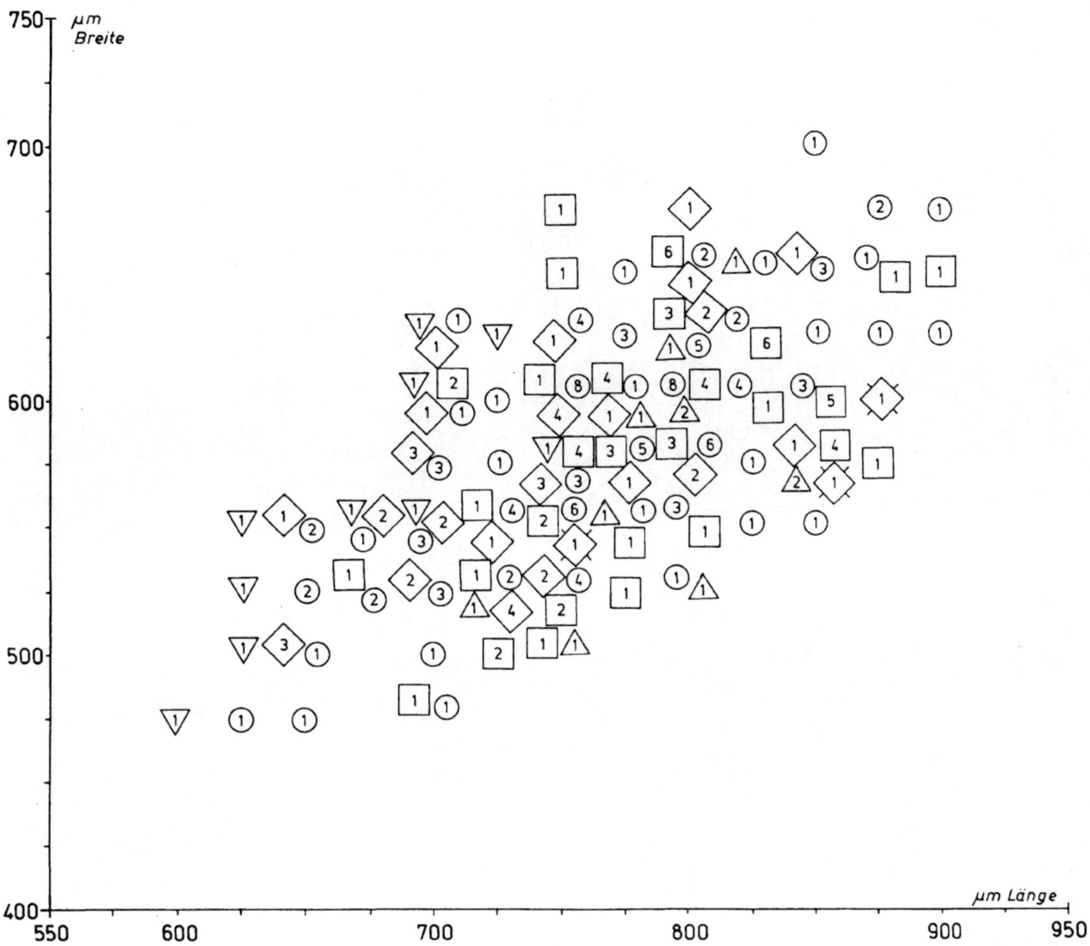

2.7. Greußen bei Sömmerda/Bezirk Erfurt

Tab. 12, 15—21; Taf. IV, Fig. 3—6

Aus den Charasanden mit Travertinknollen von Greußen überließ MANIA eine Probe. Sie entstammt dem ca. 8 m mächtigen, mittelholozänen Travertinkomplex nördlich des Ortes. Das Fossilmaterial scheint „angelöst" zu sein. Dabei läßt sich nicht entscheiden, ob die Kalkhüllen mit Fragmenten von Oosporenmembranen bereits bei der Einbettung einem lösenden Einfluß unterlagen. Bei der Aufbereitung wurde das Material nicht anders als die oben beschriebenen Proben behandelt. Auf jeden Fall besitzen die Fossilien eine rauhe Oberfläche. Die Analyse ergab ein homogenes Material, das von keinem bisher bekannten Taxon erfaßt wird. Vollständig erhaltene Oosporenmembranen konnten nicht gefunden werden. Die Fossilien wurden *Charites lata* n. sp. benannt.

2.8. Herbsleben östlich Bad Langensalza/Bezirk Erfurt

Tab. 13, 14, 15—21; Taf. IV, Fig. 3—6

Die Probe Herbsleben B3 wurde dem mittelholozänen Travertinkomplex im Ried von Herbsleben entnommen. Es ist ein Charophytensand mit versinterten Schilfstengeln. Die meisten Charophyten-Fruktifikationen sind vollständig erhaltene Oogonien, wenn auch die Oosporenmembranen im allgemeinen nur sehr zart erhalten sind. Deshalb wurden lediglich einzelne Membranen präpariert.

Nach der Morphologie, nicht nach der Größe der Reste, liegen zwei verschiedene Organtaxa vor. Es wurden 250 Oogonien, die eindeutig zu *Charites thuringica* n. sp. gehören, ausgewertet. Daneben fanden sich aber noch 100 anomal gestaltete Fossilien. Diese haben neben den normalen, kräftig ausgebildeten, üblichen Linkswindungen der Hüllzellen noch

Tab. 5. Größenbereiche und Windungszahlen des Materials von Mühlhausen

Legende s. Tab. 1, S. 132

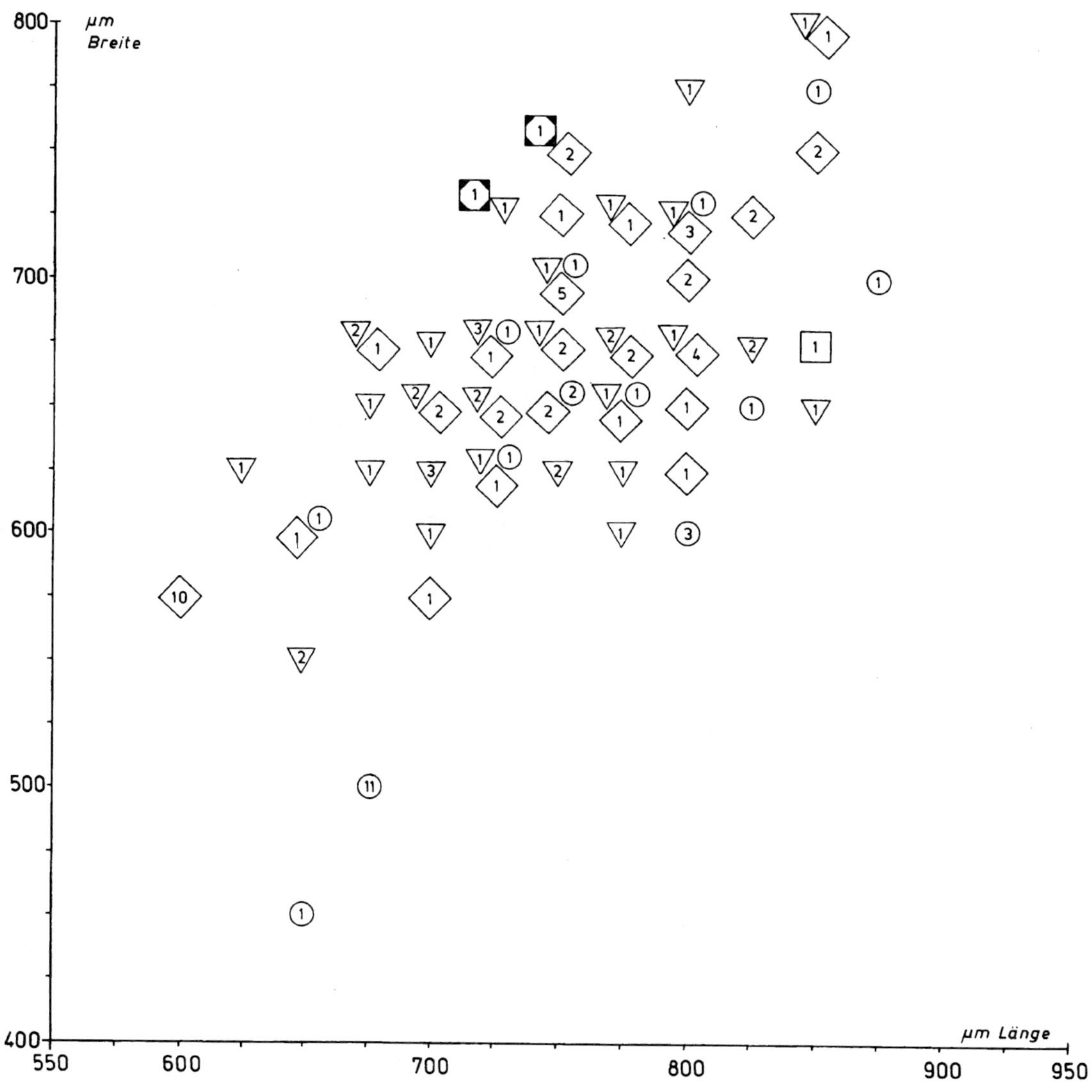

etwas schwächere Gegenwindungen. Bereits 1975 (S. 37) hat der Autor auf einige Fossilien von Strausberg bei Berlin hingewiesen und dort auf Taf. 4, Fig. 9—12 abgebildet. Damals lagen nur wenige Exemplare vor, sie wurden deshalb als Krüppel betrachtet. Nun fanden sich in Herbsleben genügend Fossilien, um eine variationsstatistische Analyse durchführen zu können. Danach ist ihnen ein eigenständiges Taxon zuzuerkennen, sie werden *Charites contraspiralis* n. sp. benannt.

3. Taxonomie der quartären Charophytenreste

Charites prima n. sp.

Tab. 21

Synonymie: Charophytenformenkreis Mary Kocher (NÖTZOLD 1975, Taf. 1, Fig. 4—8; Taf. 15, 42, 47, 52, 60).

Diagnose: Fossile weibliche Fruktifikationsorgane der Gattung *Charites* HORN AF RANTZIEN 1959, 600—850 µm lang, 400—600 µm breit, mit 11—12 seitlich angeordneten sichtbaren Spiralwindungen, L/B = Q 1,20—1,70[1]).

[1]) Bei allen Angaben über Längen, Breiten, Verhältnis $L/B = Q$ und Windungszahlen dieser Arbeit wurden extreme Einzelpaare und Werte ausgelassen, da sonst die Diagnosen zu weit gefaßt worden wären und diese Exemplare ohnehin nur vereinzelte Abnormitäten sind.

Tab. 6. Größenbereiche und Windungszahlen des Materials von Burgtonna-Ort, Bt-O

Legende s. Tab. 1, S. 132

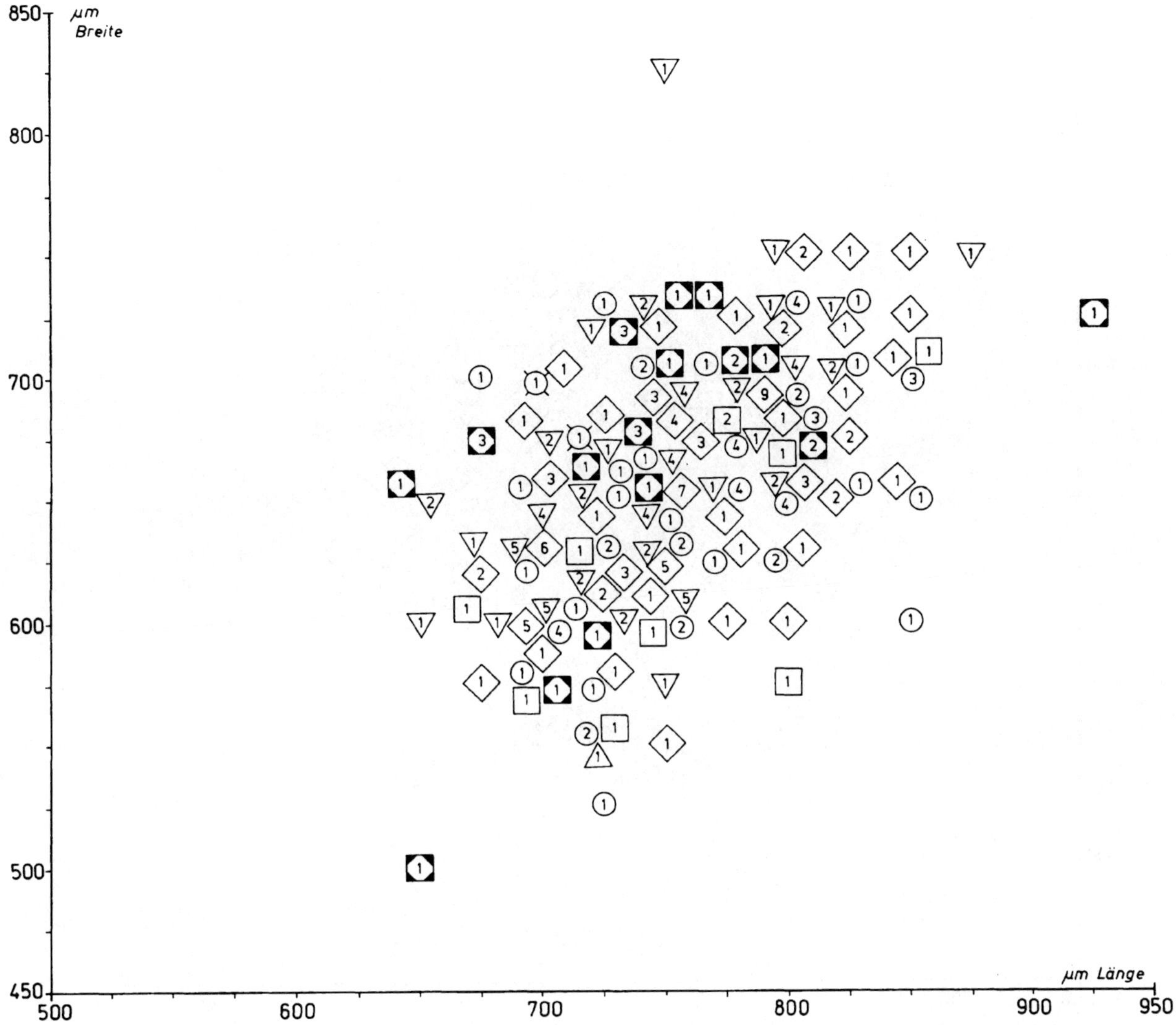

Locus typicus: Kiesgrube „Mary Kocher" bei Sessenheim/Elsaß/Frankreich. (Die Art wurde zur Vollständigkeit hier mit aufgenommen.)

Stratum typicum: Plio-pleistozäne Grenzschichten.

Holotypus: Präparat MK 4. Abbildungen in Nötzold 1975, Taf. 1, Fig. 4, 4a, 4b.

Derivatio nominis: „prima" als erster Vertreter der Gattung im Quartär. Die Maxima der betrachteten Fossilien liegen bei: L = 650 µm; B = 550 µm; Q = 1,30.

Charites cava n. sp.

Tab. 1—4, 15—21; Taf. I, Fig. 1—4

Synonymie: Charophytenformenkreis Eem (Nötzold 1975; Taf. 2, Fig. 18—32; Taf. 3, Fig. 1—12; Taf. 20, 21, 42, 47, 52, 60).

Diagnose: Fossile, weibliche Fruktifikationsorgane der Gattung *Charites* Horn af Rantzien 1959, 575—950 µm lang, 425—700 µm breit, mit 9—13 seitlich sichtbaren Spiralwindungen, L/B = Q 1,14—1,53.

Locus typicus: Bilzingsleben bei Heldrungen (Unstrut), Bezirk Halle/DDR.

Stratum typicum: Mindel/Riß (Holstein)-Interglazial.

Holotypus: Präparat Bilzingsleben Nr. 1, Taf. I, Fig. 1—4.

Derivatio nominis: „cava" die „hohle" Fruktifikation.

Weitere Fundstellen der Species: Strausberg, Bezirk Frankfurt (Oder)/DDR; Kittlitz bei Lübbenau, Bezirk Cottbus/DDR.

Maxima: L = 750 µm; B = 575 µ m; Q = 1,31.

Charites angusta n. sp.

Synonymie: Charophytenformenkreis Kummerow (Nötzold 1975; Taf. 2, Fig. 11—17; Taf. 17, 42, 47, 52, 60).

Diagnose: Fossile, weibliche Fruktifikationsorgane der Gattung *Charites* Horn af Rantzien 1959, 650—800 µm lang, 475—675 µm breit mit 8—12 seitlich sichtbaren Spiralwindungen, L/B = Q 1,0—1,50.

Tab. 7. Größenbereiche und Windungszahlen des Materials von Burgtonna Bt 6

Legende s. Tab. 1, S. 132

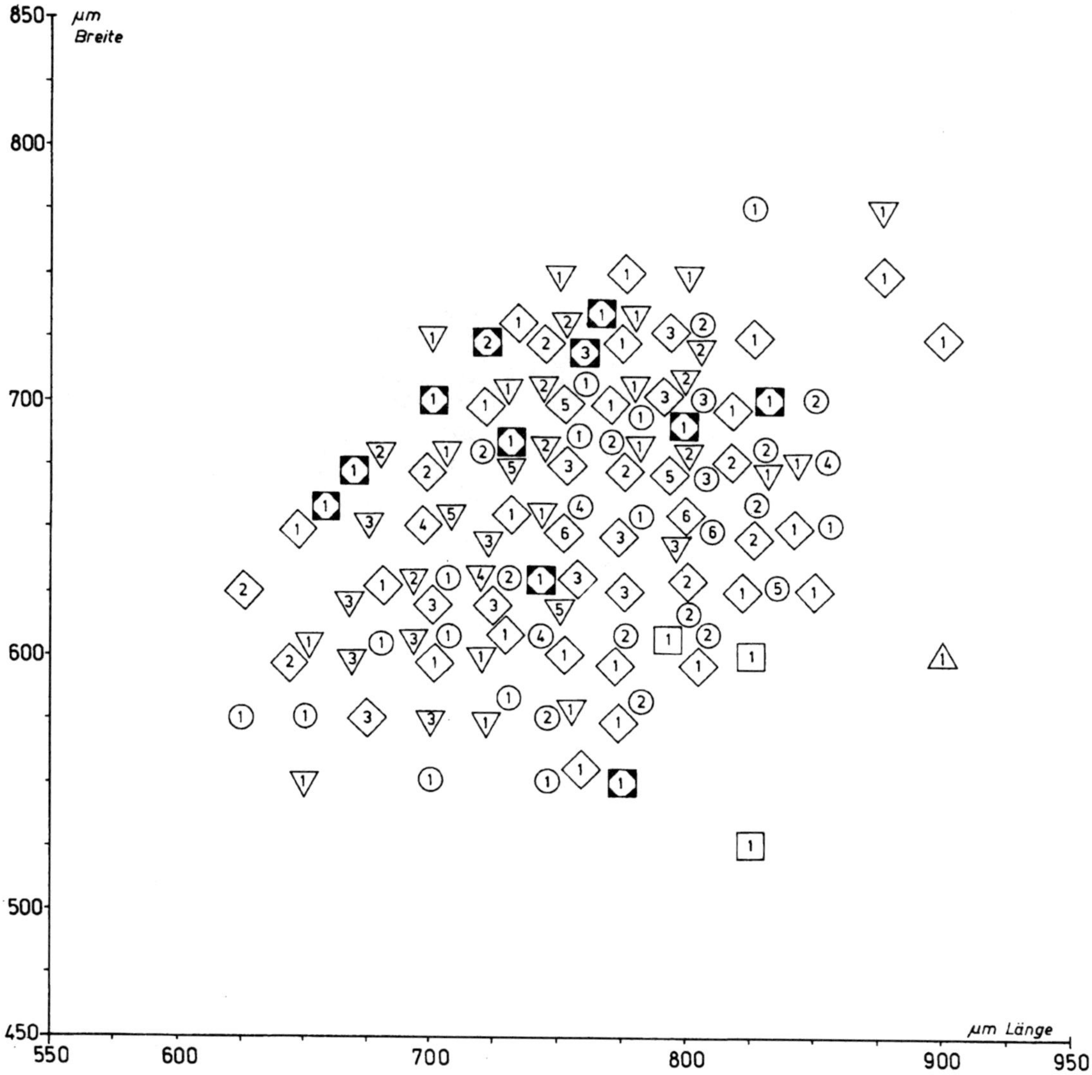

Locus typicus: Kummerow, Bezirk Frankfurt (Oder)/DDR.
Stratum typicum: Mindel/Riß (Holstein)-Interglazial.
Holotypus: Präparat Kummerow Ku. 6, Abb. in NÖTZOLD 1975, Taf. 2, Fig. 16.
Derivatio nominis: „angusta" die „schmale" Fruktifikation. Von der Art ist keine weitere Verbreitung bekannt.
Maxima: L = 700 µm; B = 575 µm; Q = 1,25.

Charites thuringica n. sp.

Tab. 5—10, 13, 14, 15—21; Taf. I, Fig. 5—7

Synonymie: Charophytenformenkreis Museumsinsel (NÖTZOLD 1975; Taf. 7, 8, Taf. 12, Fig. 21—26, Taf. 24, 42, 47, 52, 60).

Diagnose: Fossile, weibliche Fruktifikationsorgane der Gattung *Charites* HORN AF RANTZIEN 1959, Hüllzellen 550 bis 950 µm lang, 425—825 µm breit, mit 8—12 seitlich sichtbaren Spiralwindungen, L/B = Q 1,0—1,50; Oosporenmembranen 450—850 µm lang, 350—700 µm breit, mit 7—12 Spiralwindungen, hellbraun bis schwarz.

Locus typicus: Burgtonna bei Bad Langensalza, Bezirk Erfurt/DDR.
Stratum typicum: Riß/Würm (Eem)-Interglazial.
Stratigraphische Verbreitung: Mindel/Riß bis Mittelholozän.
Holotypus: Präparat Burgtonna Bt. 1, Taf. I, Fig. 5—7.
Derivatio nominis: „thuringica" wegen der weiten Verbreitung in Thüringen.
Weitere Fundstellen: Museumsinsel Berlin; Weimar, Parkhöhlen; Mühlhausen, Bezirk Erfurt; Mücheln/Geiseltal, Bezirk Halle; Herbsleben und Bad Langensalza, beide Bezirk Erfurt/alle DDR.
Maxima: L = 750 µm; B = 650 µm; Q = 1,10—1,20.
Die Oosporenmembranen des Taxon entstammen den Fundstellen Berlin (Museumsinsel), Mücheln und Herbsleben.

Charites ilmica n. sp.

Tab. 15—21; Taf. II, Fig. 1—3

Diagnose: Fossile, weibliche Fruktifikationsorgane der Gattung *Charites* HORN AF RANTZIEN 1959, 475—725 µm lang, 375—600 µm breit, mit 8—12 seitlich sichtbaren Spiralwindungen, L/B = Q 1,0—1,35.

Tab. 8. Größenbereiche und Windungszahlen des Materials von Burgtonna, nördlicher Bruch

Legende s. Tab. 1, S. 132

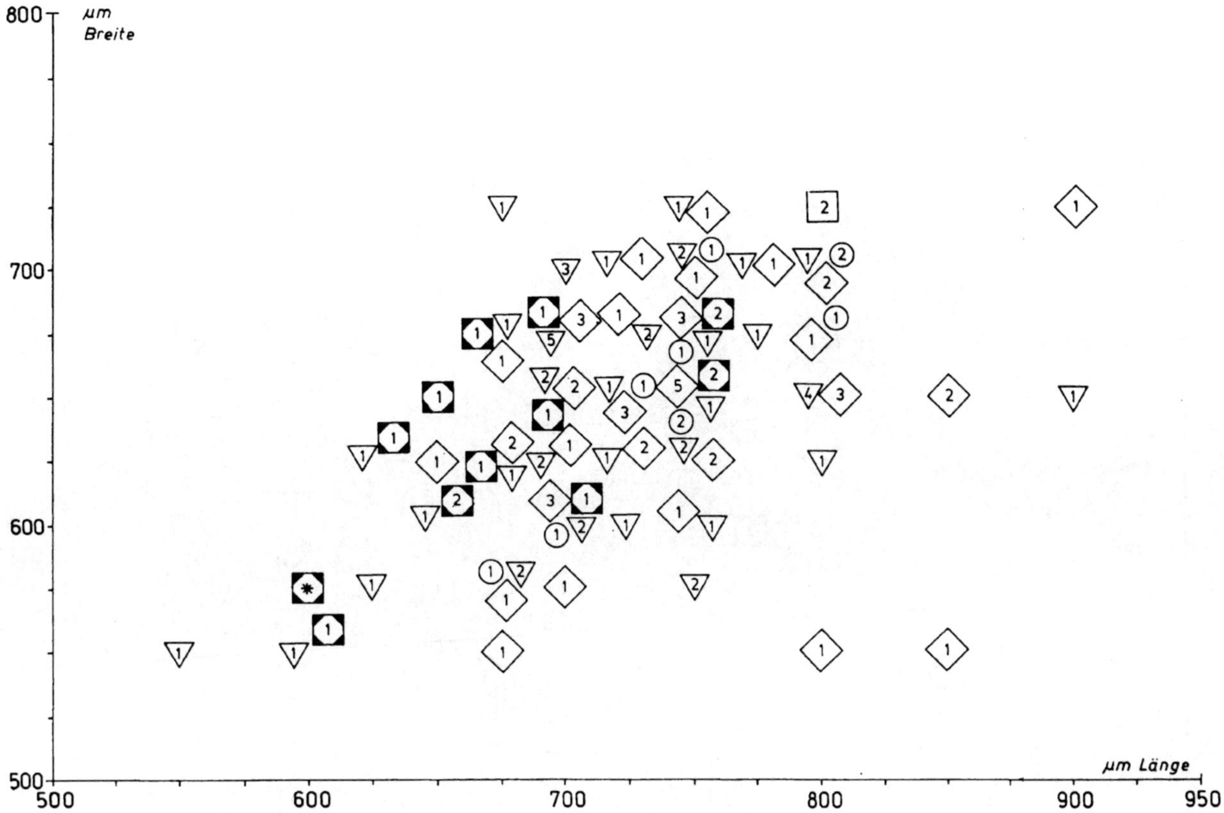

Locus typicus: Weimar, Parkhöhlen, Bezirk Erfurt/DDR.
Stratum typicum: Riß/Würm (Eem)-Interglazial.
Holotypus: Präparat Weimar C6 Nr. 1, Taf. II, Fig. 1—3.
Derivatio nominis: „ilmica" wegen des Vorkommens in den Ilmtravertinen.
Von der Species ist keine weitere Verbreitung bekannt.
Maxima: L = 575 µm; B = 500 µm; Q = 1,12—1,22.

Charites eemica n. sp.

Tab. 21; Taf. II, Fig. 4—6

Synonymie: Charophytenkreis *Molassica* (NÖTZOLD 1975; Taf. 26, 43, 48, 53, 60).

Diagnose: Fossile, weibliche Fruktifikationsorgane der Gattung *Charites* HORN AF RANTZIEN 1959, 375—650 µm lang, 275—450 µm breit, mit 9—12 seitlich sichtbaren Spiralwindungen, L/B = Q 1,15—1,70. Oosporenmembranen 300—550 µm lang, 200—350 µm breit, mit 8—11 Spiralwindungen, dunkelbraun bis schwarz.

Locus typicus: Königsaue bei Gatersleben, Bezirk Halle/DDR.
Stratum typicum: Riß/Würm (Eem)-Interglazial.
Holotypus: Präparat Königsaue Kö. 1, Taf. II, Fig. 4—6.
Derivatio nominis: „eemica" nach der Fundstrate Eem.
Eine weitere Verbreitung der Art ist bisher nicht bekannt.
Maxima: L = 450 µm; B = 325 µm; Q = 1,40.

Charites elongata n. sp.

Tab. 21; Taf. II, Fig. 7—9

Synonymie: Charophytenformenkreis Rhythmus III (NÖTZOLD 1975, Taf. 27, 43, 48, 53, 60)!

Diagnose: Fossile, weibliche Fruktifikationsorgane der Gattung *Charites* HORN AF RANTZIEN 1959, 500—800 µm lang, 275—525 µm breit, mit 9—13 seitlich sichtbaren Spiralwindungen, L/B = Q 1,35—2,0. Oosporenmembranen 400 bis 700 µm lang, 200—450 µm breit mit 8—12 Spiralwindungen, dunkelbraun bis schwarz.

Locus typicus: Königsaue bei Gatersleben, Bezirk Halle/DDR.
Stratum typicum: Interstadial III (nach MANIA)/Würm.
Holotypus: Präparat Königsaue Kö. Nr. 2, Taf. II, Fig. 7—9.
Derivatio nominis: „elongata" nach der langgestreckten Gestalt.
Von der Species ist keine weitere Verbreitung bekannt.
Maxima: L = 650 µm; B = 400 µm; Q = 1,60.

Charites seroglacialis n. sp.

Tab. 21; Taf. III, Fig. 1—3

Synonymie: Charophytenformenkreise Eem — Basistorf, Bølling und Allerød (NÖTZOLD 1975; Taf. 23, 25, 28, 42, 43, 47, 48, 52, 53, 60).

Tab. 9. Größenbereiche und Windungszahlen des Materials von Mücheln

Legende s. Tab. 1, S. 132

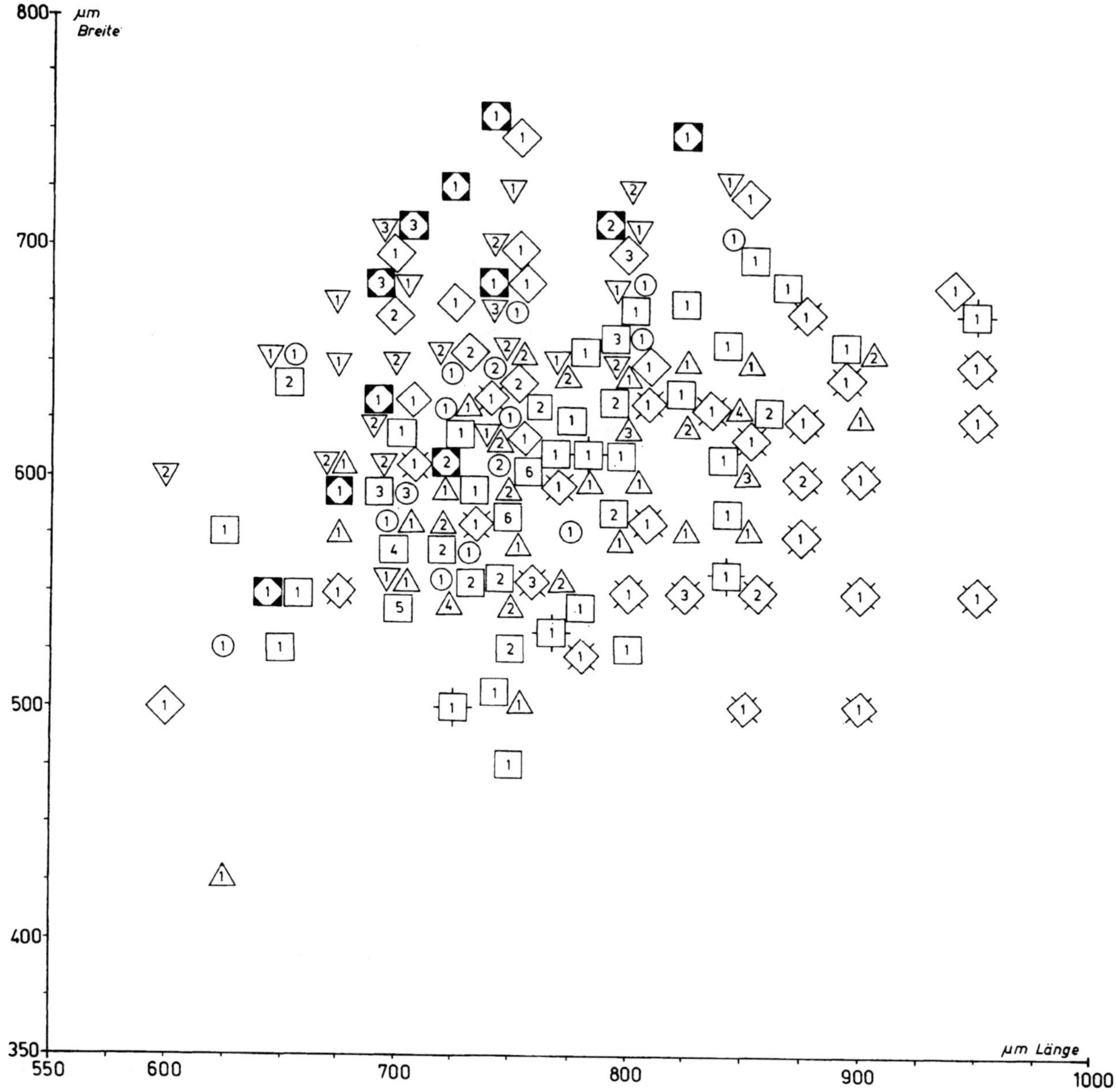

Diagnose: Fossile, weibliche Fruktifikationsorgane der Gattung *Charites* HORN AF RANTZIEN 1959, 500—850 µm lang, 300—725 µm breit, mit 8—12 seitlich sichtbaren Spiralwindungen, L/B = Q 1,05—1,45. Oosporenmembranen 400 bis 750 µm lang, 275—625 µm breit, hellbraun bis schwarz.

Locus typicus: Königsaue bei Gatersleben, Bezirk Halle/DDR.

Stratum typicum: Bølling-Interstadial.

Stratigraphische Verbreitung: Riß/Würm (Eem)-Interglazial bis Allerød-Warmzeit.

Derivatio nominis: „seroglacialis" = zum Spätglazial gehörig.

Weitere Fundstelle: Ruds Vedby, Seeland/Dänemark.

Maxima: L = 625—675 µm; B = 500—600 µm; Q = 1,20.

Charites lata n. sp.

Tab. 12, 15—20; Taf. III, Fig. 4, 5

Diagnose: Fossile, weibliche Fruktifikationsorgane der Gattung *Charites* HORN AF RANTZIEN 1959, 625—900 µm lang, 550—800 µm breit, mit 7—10 seitlich sichtbaren Spiralwindungen, L/B = Q 0,98—1,28.

Locus typicus: Greußen bei Sömmerda, Bezirk Erfurt/DDR.

Stratum typicum: Mittelholozän.

Holotypus: Präparat Greußen G.1, Taf. III, Fig. 4, 5.

Derivatio nominis: „lata" = breite Fruktifikation.

Von der Species keine weitere Verbreitung bekannt.

Maxima: L = 800 µm; B = 700 µm; Q = 1,15.

Tab. 10. Größenbereiche und Windungszahlen des Materials von Bad Langensalza La 66/16

Legende s. Tab. 1, S. 132

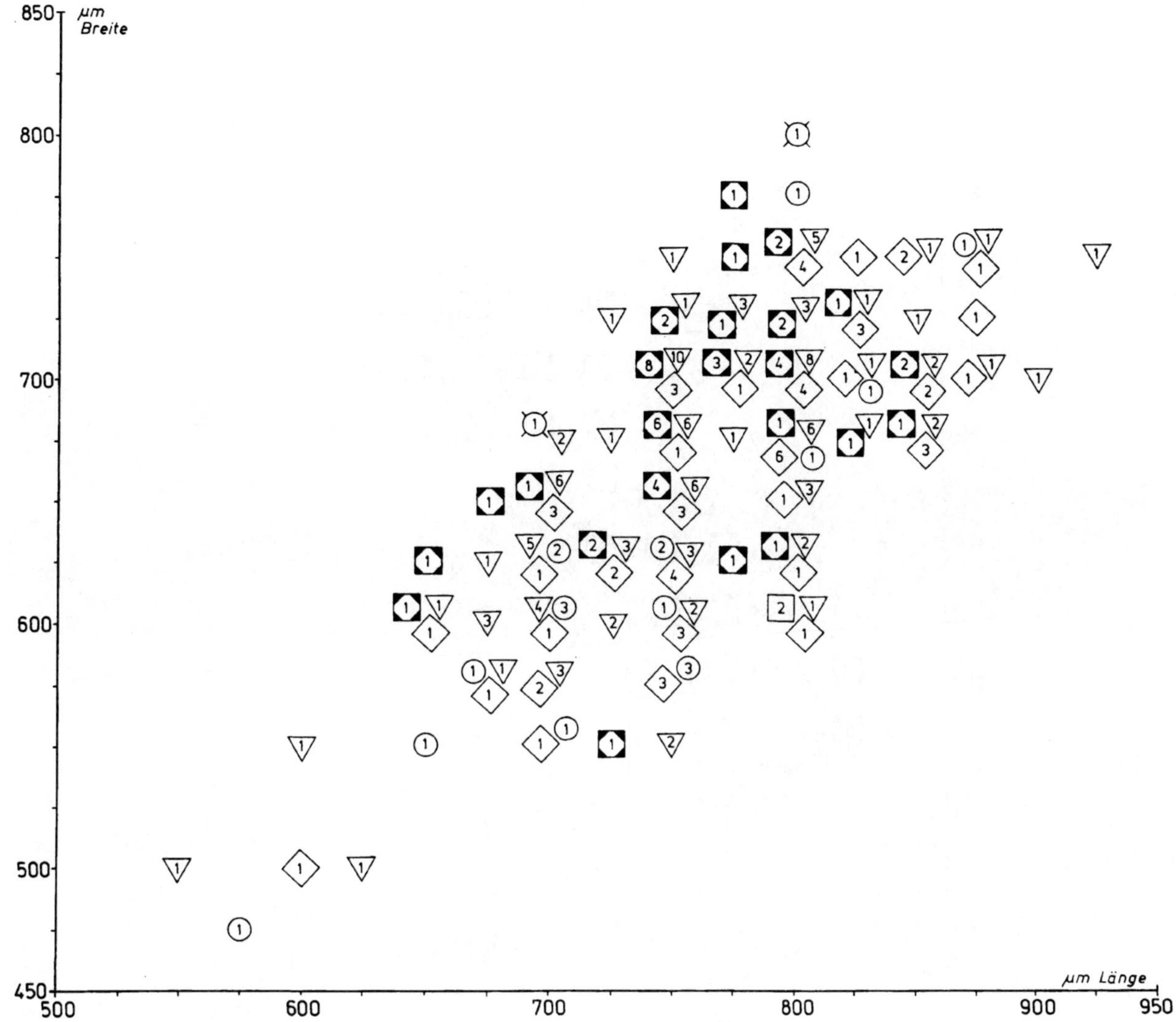

Charites sphaerica n. sp.

Tab. 11, 15—21; Taf. III, Fig. 6; Taf. IV, Fig. 1, 2

Diagnose: Fossile, weibliche Fruktifikationsorgane der Gattung *Charites* HORN AF RANTZIEN 1959, 600—750 µm lang, 525—775 µm breit, mit 7—10 seitlich sichtbaren Spiralwindungen, L/B = Q 0,90—1,16.

Locus typicus: Bad Langensalza, Bezirk Erfurt/DDR.
Stratum typicum: Mittelholozän.
Holotypus: Präparat Langensalza La. 1, Taf. III, Fig. 6.
Derivatio nominis: „sphaerica" wegen der fast kugeligen Gestalt.
Die Art fand sich bisher nur am locus typicus.
Maxima: L = 675—700 µm; B = 675 µm; Q = 1,02.

Charites contraspiralis n. sp.

Tab. 14, 15—21; Taf. IV, Fig. 3—6

Diagnose: Fossile, weibliche Fruktifikationsorgane der Gattung *Charites* HORN AF RANTZIEN 1959, 650—950 µm lang, 475—775 µm breit, mit 8—10 sich in Seitensicht kräftig abzeichnenden Spiralwindungsumgängen der linksgewundenen fünf Hüllzellen sowie bis zu gleich vielen, sich schwächer abzeichnenden gegenläufigen Windungen, so daß auf den Hüllzellen rhombische Felder entstehen. Die Gegenwindungen zeichnen sich auf den Oosporenmembranen nicht ab.

Locus typicus: Herbsleben bei Bad Langensalza, Bezirk Erfurt/DDR.
Stratum typicum: Mittelholozän.
Holotypus: Präparat Herbsleben H. 1, Taf. IV, Fig. 3—6.
Derivatio nominis: „contraspiralis" wegen der zusätzlichen, gegenläufigen Spiralwindungen.
Wegen der Zartheit der Oosporenmembranen wurde von der Präparation einer größeren Anzahl zur Bestimmung ihrer Größenvariation abgesehen.
Maxima: L = 800 µm; B = 675 µm; Q = 1,02.

Einige gleichartige Fossilien wurden von Strausberg bei Berlin, Bezirk Frankfurt (Oder), als „Krüppelexemplare" beschrieben (NÖTZOLD 1975, S. 37, Taf. 4, Fig. 9—12). Nunmehr liegen über 100 Exemplare vor, die der variationsstatistischen Analyse unterzogen werden konnten. Danach muß den Fossi-

Tab. 11. Größenbereiche und Windungszahlen des Materials von Bad Langensalza, Fossilien, die zu *Sphaerochara* tendieren

Legende s. Tab. 1, S. 132

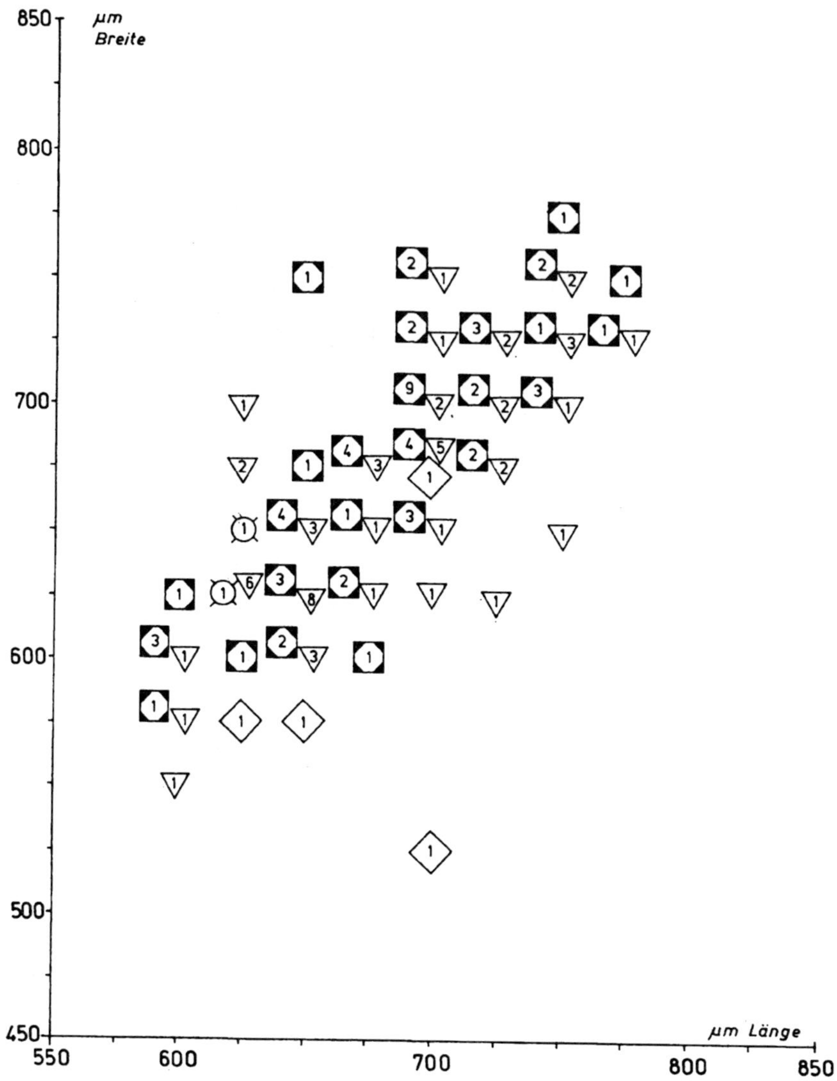

lien der Charakter eines eigenständigen Taxons zugestanden werden. Dem Autor bringt das aber ein taxonomisches Problem. Alle neophytischen Charophyten haben, ob sie verkalken oder nicht, fünf „linksgewundene" (nach der botanischen Definition, die zoologische betrachtet Windungen entgegengesetzt) Hüllzellen ihrer weiblichen Fruktifikationsorgane. Diese fünf linksgewundenen Hüllzellen werden bei Diagnosen stillschweigend vorausgesetzt bzw. gehören bereits zur Gattungsdiagnose. Für die neuen Taxa dieser Arbeit ist das mit der Zugehörigkeit zur Gattung *Charites* HORN AF RANTZIEN ausgedrückt. Die „gegenläufigen" Windungen stellen die *Charites contraspiralis* eigentlich außerhalb aller Diagnosen, nicht nur der Genera, sondern sogar der Familie und Klasse der Charophyceae. Für einen derartigen Fall schweigt sich der ICBN aus. Der Autor erachtet es aber als nicht gerechtfertigt, auf Grund eines einzelnen Organtaxons, noch dazu bei derartig ausgefallenen Fossilien, eine Cäsur in der Charophytentaxonomie durchzuführen und auch noch ein neues Genus für die Fossilien aufzustellen. Da die Stücke nach allen anderen Eigenschaften der Gattung *Charites* entsprechen, werden sie vorerst unter diesem Taxon geführt. Sollten später weitere entsprechende Fossilien gefunden werden, kann immer noch eine Revision erfolgen.

4. Die Beziehungen der neuen Fossilarten zu rezenten und fossilen Charophyten

Die rezenten Charophyten werden nach ihren vegetativen Organen diagnostiziert. Diese liegen dem Paläobotaniker bisher nur in ganz wenigen, vereinzelten Fällen, im organischen Verband mit ihren Fortpflanzungsorganen vor. Aus dem europäischen Neophytikum ist dem Autor bisher kein derartiges Material bekannt geworden. Es kann also nur ein Organvergleich der weiblichen Fortpflanzungsorgane erfolgen. Dabei ergeben sich zwei Problematika. Erstens haben die früheren Autoren die Oogonien in der Länge von der Basis bis zum Gipfel, einschließlich der Gipfelknötchen gemessen,

Tab. 12. Größenbereiche und Windungszahlen des Materials von Greußen G

Legende s. Tab. 1, S. 132

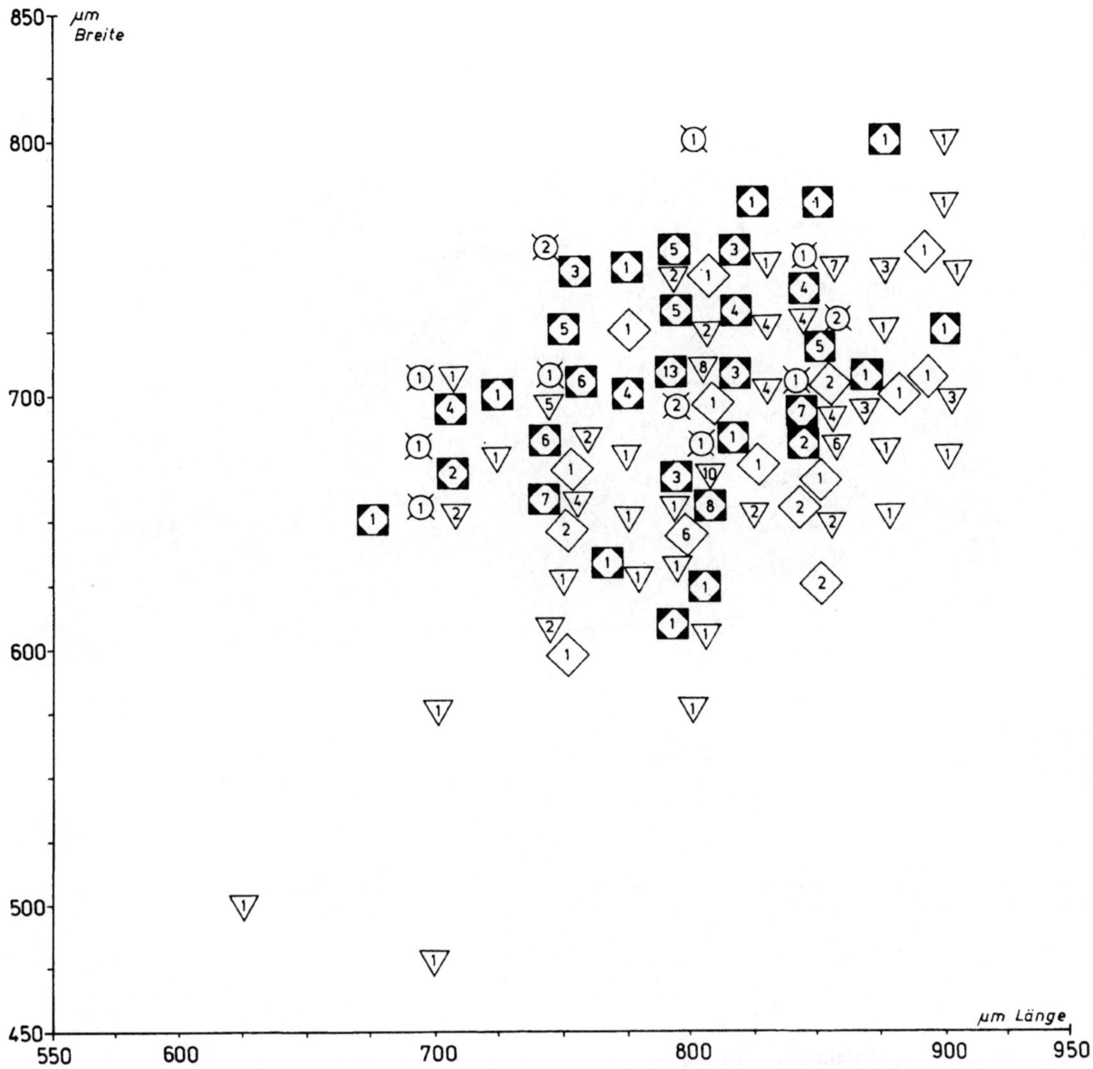

die Fossilien werden aber, weil sie oft ausgekeimt vorliegen, von der Basis bis zur Dehiszenzfurche gemessen. Man muß daher einen Unterschied von ±150 µm dabei in Rechnung stellen. Denn die absolute Größe ist nun einmal ein wichtiges diagnostisches Merkmal. Die Breiten sind dagegen direkt zu vergleichen. Da die früheren Autoren die Windungen einschließlich der Gipfelknötchen zählten, ist die Windungsanzahl um 1—2 zu reduzieren.

Das zweite Problem betrifft die Taxonomie der rezenten Charophyten. Wie bereits ausgeführt (NÖTZOLD 1975) haben CORILLION und WOOD einen Teil der rezenten Charophytentaxa revidiert, dabei aber die Fortpflanzungsorgane nicht berücksichtigt. Dadurch ergaben sich taxonomische Änderungen, denen sich der Autor nur in einigen, wenigen Fällen anschließen kann. In anderen Beispielen scheinen die Revisionen durch CORILLION und WOOD, nach den weiblichen Fortpflanzungsorganen zu urteilen, ungerechtfertigt.

Ohne Zweifel ist die Erweiterung der Variationsbreite der Oogonien durch die Vereinigung von *Chara vulgaris* L. (*Ch. foetida* A. BR.) mit den Arten *Ch. rabenhorsti* A. BR., *Ch. contraria* KÜTZING, *Ch. jubata* A. BR., *Ch. imperfecta* A. BR. und *Ch. dissoluta* A. BR. realer als die frühere enge Fassung der am weitesten verbreiteten rezenten *Chara*-Art. Daher ist in diesem Falle der Neukombination zuzustimmen. Die Arten *Chara braunii* GMELIN, *Ch. baueri* A. BR. (= *Ch. scoparia* BAUER), *Ch. canescens* LOISELEUR (= *Ch. crinita* WALLROTH) werden als eigenständige Taxa anerkannt. Nun könnte es sich der Autor leicht machen und sagen WOOD hat unter *Chara globularis* THUILL. die Arten *Chara fragilis* DESVAUX, *Ch. connivens* SALZMANN, *Ch. aspera* (DETH.) WILLDENOW, *Ch. galioides* DC., *Ch. strigosa* A. BR., *Ch. tenuispina* A. BR., *Ch. kokeilii* A. BR. und *Ch. delicatula* AGARDH vereinigt und er schließt sich dem an. Dabei ist jedoch zu betonen, daß die Stammform der *Chara globularis* var. *globularis* bei uns nicht vorkommt! Gleichermaßen wäre dann der Vereinigung von *Chara hispida* L. mit *Ch. aculeolata* KÜTZING zuzustimmen. Dann gäbe es in Europa noch sieben rezente Arten der Gattung *Chara*. Betrachtet man das vorliegende Fossil-

Tab. 13. Größenbereiche und Windungszahlen des Materials von Herbsleben H/B 3

Legende s. Tab. 1, S. 132

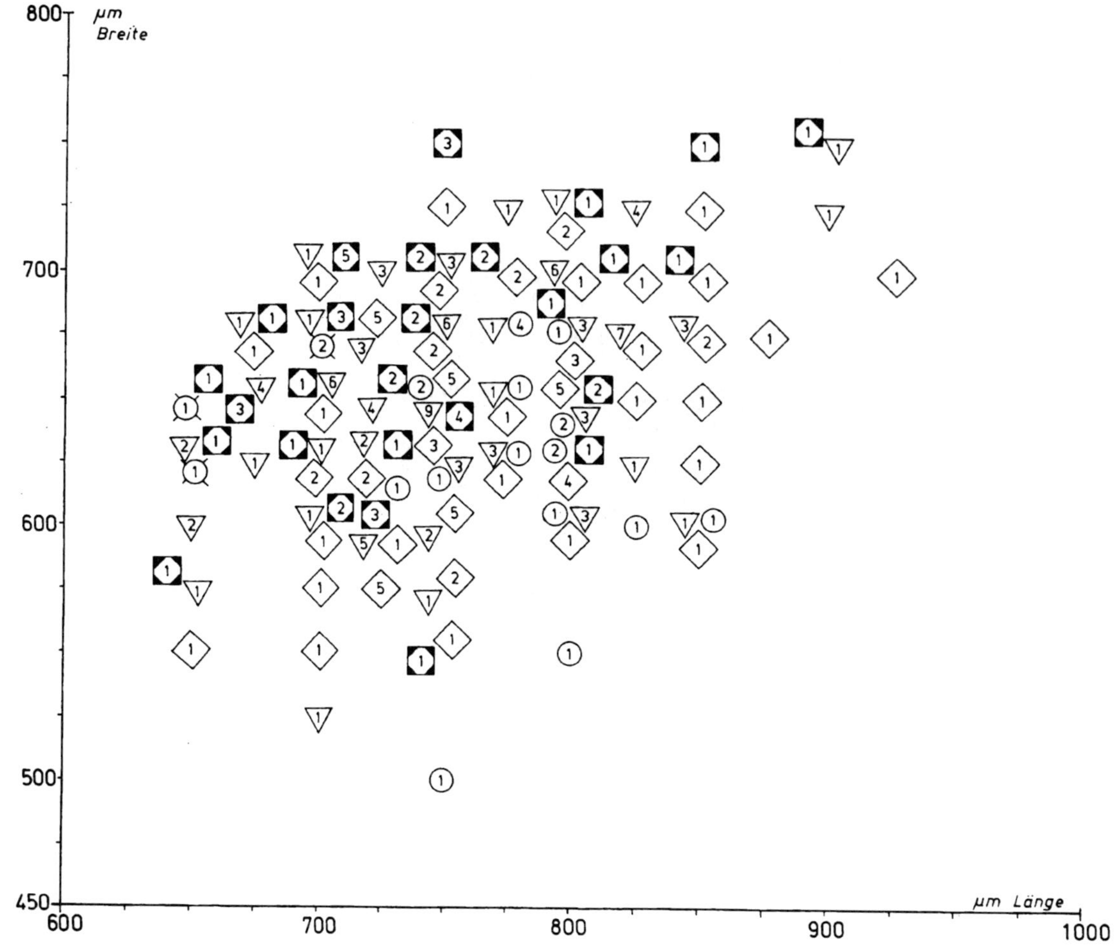

material ist es möglich, ja sogar wahrscheinlich, daß einige weitere rezente Taxa zu vereinigen sind. Aber der generellen Neukombination WOODS muß nach den variationsstatistischen Untersuchungen von über 20000 Fruktifikationsorganen aus dem Neophytikum Mitteleuropas widersprochen werden.

Wenn die Materialien der verschiedensten Fundstellen und Proben in der graphischen Darstellung eindeutige, unterschiedliche Gaußsche Glockenkurven zeigen, so ist ihnen auch ihr eigenständiges Taxon zuzuerkennen. Es ist ebensowenig einzusehen, daß rezente taxonomische Kombinationen, die ohne Variationsanalyse der generativen Organe erfolgten, übernommen werden sollten. Bei den Charophyten besteht Oogamie, die höchste Form sexueller Fortpflanzung und die Rezentbotaniker, die sonst im Pflanzenreich das System nach den Fruktifikationsorganen aufstellen, negieren hier die generativen Organe. Aber auch hier gilt der Grundsatz: „A fructibus eorum cognoscetis eos". Beim Rezentvergleich werden daher die oben genannten Arten noch als selbst- und eigenständig betrachtet.

Charites prima n. sp. steht den rezenten Arten *Chara canescens* LOISELEUR und *Ch. aspera* (DETH.) WILLDENOW nahe. Letztere Art vereinigt WOOD (1962) mit einer Unzahl anderer Taxa unter *Ch. globularis* THUILL. Von den tertiären Gyrogoniten steht *Tectochara rochettiana* (HEER) MÄDLER 1955 am nächsten.

Charites cava n. sp. Selbst wenn man die unterschiedlichen Gipfel beachtet, gibt es keine rezente Art mit einem nur annähernden Variationsbereich. Am nächsten steht noch *Chara strigosa* A. BR., die jedoch, und da gibt es keine Unterschiede im Meßverfahren, breitere Oogonien hat. An tertiären Taxa steht *Charites polonica* (UNGER) m. 1975 am nächsten, hat aber etwas kleinere Gyrogonite.

Charites angusta n. sp. Wie bereits 1975 ausgeführt, liegen die Oogoniten der rezenten „*Chara contraria* und *Ch. dissoluta*", nach WOOD, dem in diesem Falle zuzustimmen ist, völlig im Variationsbereich der *Chara vulgaris* L. Nach Abzug der Spiralwindungsumgänge am Gipfel stimmen die Fossilien in allen diagnostischen Merkmalen mit den Oogonien der rezenten Art überein. Die Fossilien müßten also zu dieser Art gestellt werden, wenn diese nicht nach ihren vegetativen Organen diagnostiziert würde. Da letztere dem Paläobotaniker nicht vorliegen, ist es, wie bereits früher mehrfach ausgeführt (u. a. NÖTZOLD 1977, S. 28) nach dem ICBN nicht zulässig, den Namen des rezenten Taxons für das Organtaxon der Fossilien zu übernehmen. Der Autor bedauert es, immer wieder Unterricht in den Nomenklaturregeln geben zu

Tab. 14. Größenbereiche und Windungszahlen des Materials von Herbsleben, Exemplare mit Gegenwindungen

Legende s. Tab. 1, S. 132

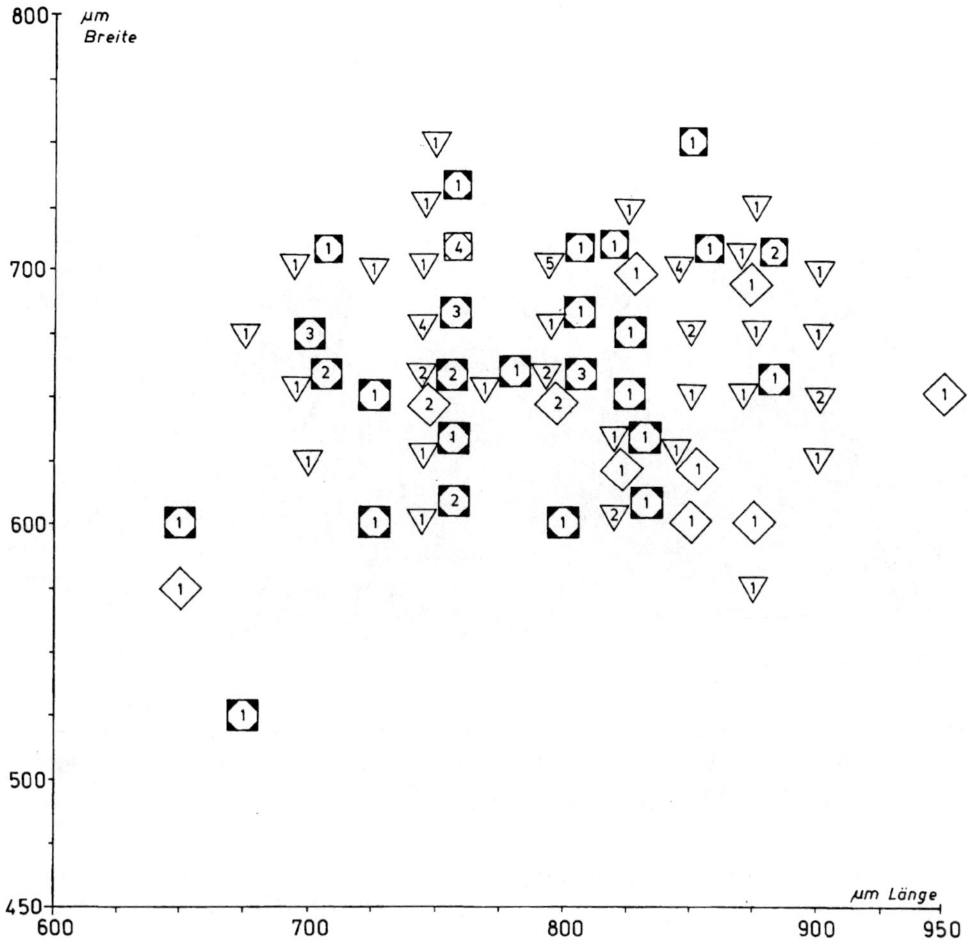

müssen, aber selbst in jüngerer Zeit wurden wieder Verstöße gegen den ICBN begangen, der aber selbst, nicht zuletzt durch gewisse Unklarheiten, einen großen Teil der Schuld für diese Verstöße zu verantworten hat. An tertiären Arten stehen den Fossilien·*Tectochara rochettiana* und *T. tasnadii* (RASKY) MÄDLER 1955 (wie bereits 1975, S. 48 beschrieben) den vorliegenden Fossilien am nächsten.

Charites lata n. sp. Unter den rezenten Charophyten mit kalkigen Hüllzellen gibt es keine Art, die dieser neuen Spezies mit ihren breiten Organen gleicht. Am nächsten kommt ihr wohl *Chara aculeolata* KÜTZING, deren Oogonien bezüglich der Breite noch eine größere Variation zeigen, in der Länge besteht sogar Übereinstimmung. Jedoch die Anzahl der seitlich sichtbaren Spiralwindungen von 7—10 ist bei den Fossilien deutlich geringer als bei den rezenten Oogonien. An tertiären Gyrogoniten hat *Tectochara lineata* (L. GRAMBAST m. 1975) fast die ganze Variationsbreite, sowohl der Größe wie auch der Windungsanzahl nach. Außerdem steht *Tectochara tasnadii* nahe.

Charites ilmica n. sp. Die Oogonien der *Chara canescens* haben fast den gleichen Variationsbereich wie *Charites ilmica*, jedoch die Windungsanzahl der rezenten Art ist weit höher. Letzteres gilt auch für den Vergleich mit *Chara vulgaris* sensu WOOD, die in ihrem weiten Variationsbereich denjenigen der fossilen Art einschließt. Von den tertiären Taxa kommt zum Vergleich nur *Charites voltzii* (UNGER) m. 1975 in Frage, allerdings sind deren Gyrogonite etwas kleiner und besitzen weniger Spiralwindungsumgänge.

Charites elongata n. sp. Wie bereits 1975 betont, besteht für 90% der Fossilien des ehemaligen „Formenkreises Rhythmus III", der in *Charites elongata* aufgeht, völlige Übereinstimmung zu *Chara canescens*. Die Oogonien der *Chara vulgaris* sensu WOOD sind größer als die vorliegenden Fossilien. Die Übereinstimmung von 80% der Fossilien mit den Gyrogoniten der *Charites molassica* (STRAUB) HORN AF RANTZIEN und von 40% mit denen der *Tectochara tornata* REID & GROVES) MÄDLER 1955 wurde bereits 1975 erwähnt.

Charites eemica n. sp. Hier kann auf die Ausführung von 1975 über den ehemaligen „Formenkreis Eem-Mitte" verwiesen werden.

Charites sphaerica n. sp. Unter den rezenten Charophyten gibt es keine Art, die derartige, fast kugelige Organe im Größenbereich der *S. sphaerica* hat. Nach der Gestalt stehen *Nitellopsis obtusa* (DESVAUX) J. GROVES und *Lychnothamnus barbatus* (MEYEN) V. LEONHARDI nahe. Diese haben jedoch weit größere Oogonien. Von den tertiären Taxa kommt die

Tab. 15. Kurven der relativen Längen der Fossilien

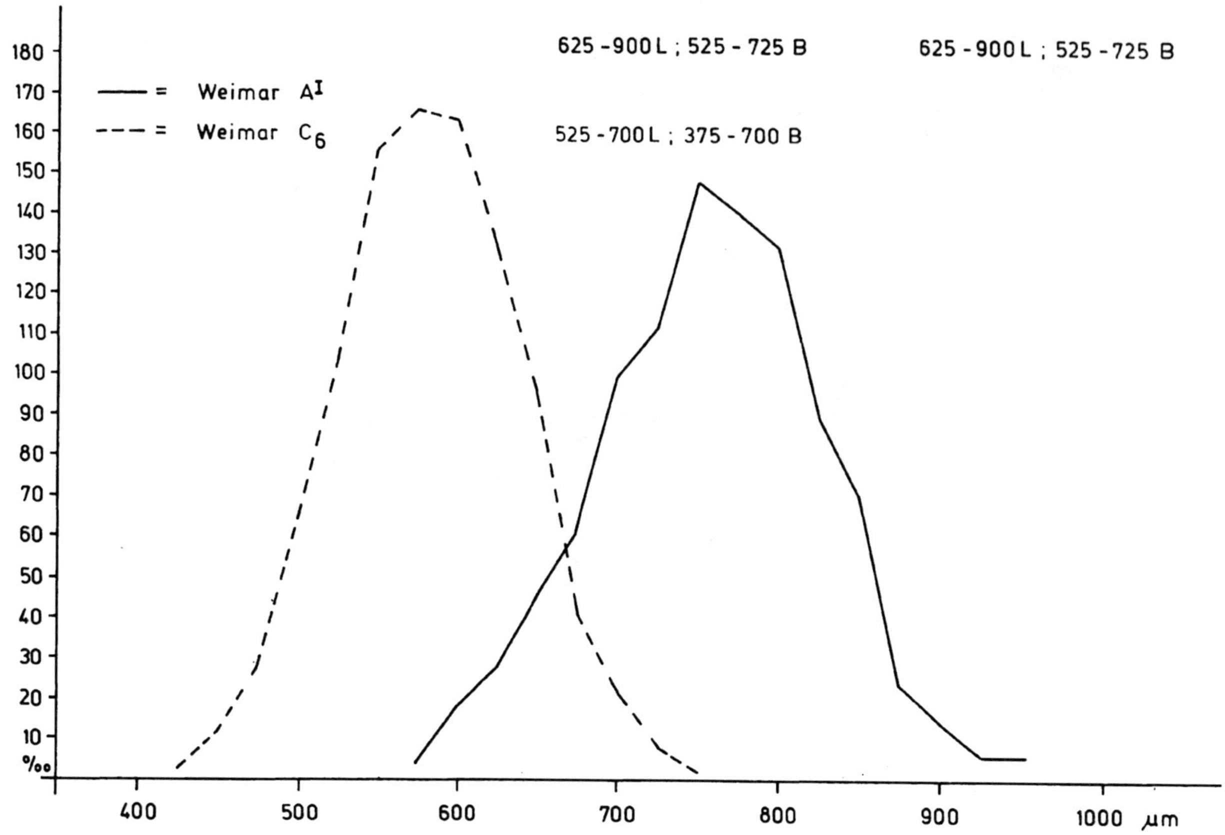

Tab. 16. Kurven der relativen Breiten der Fossilien

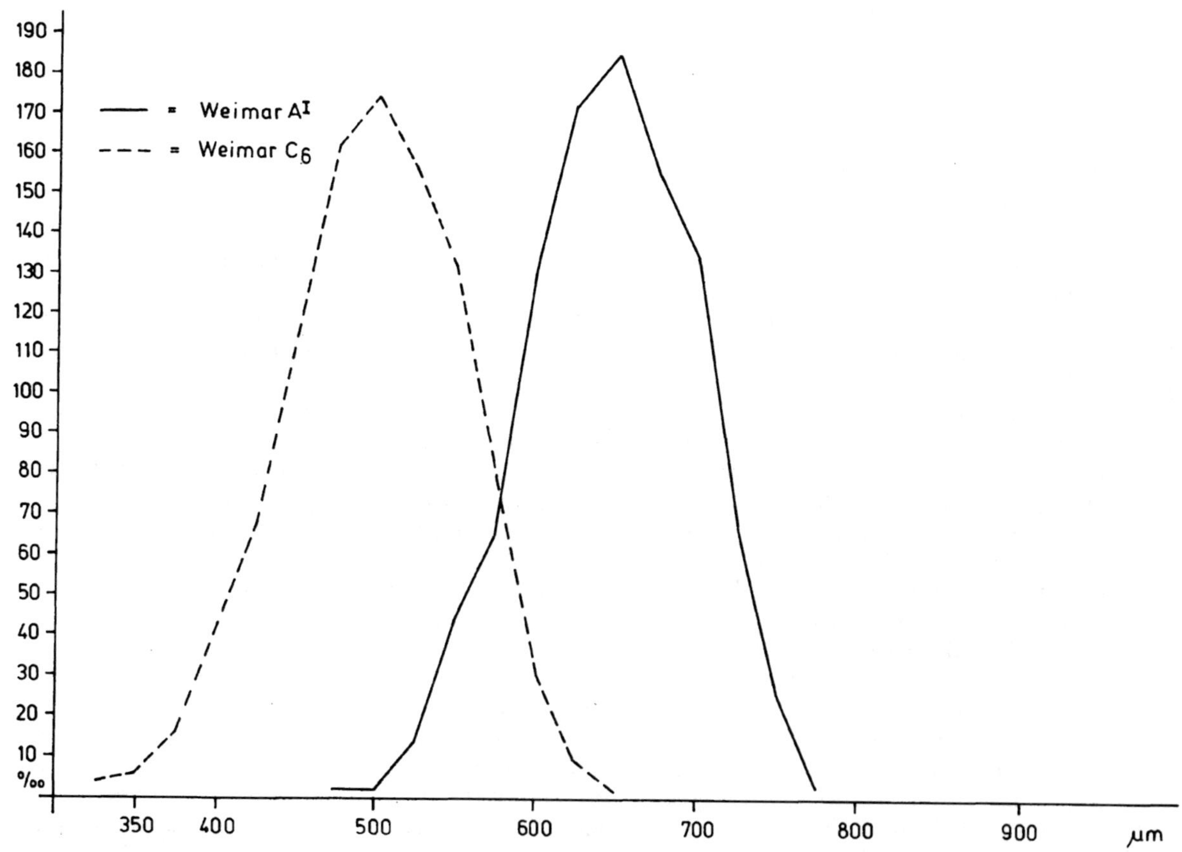

Tab. 17. Häufigkeitskurven der Quotienten Q = L/B der Fossilien

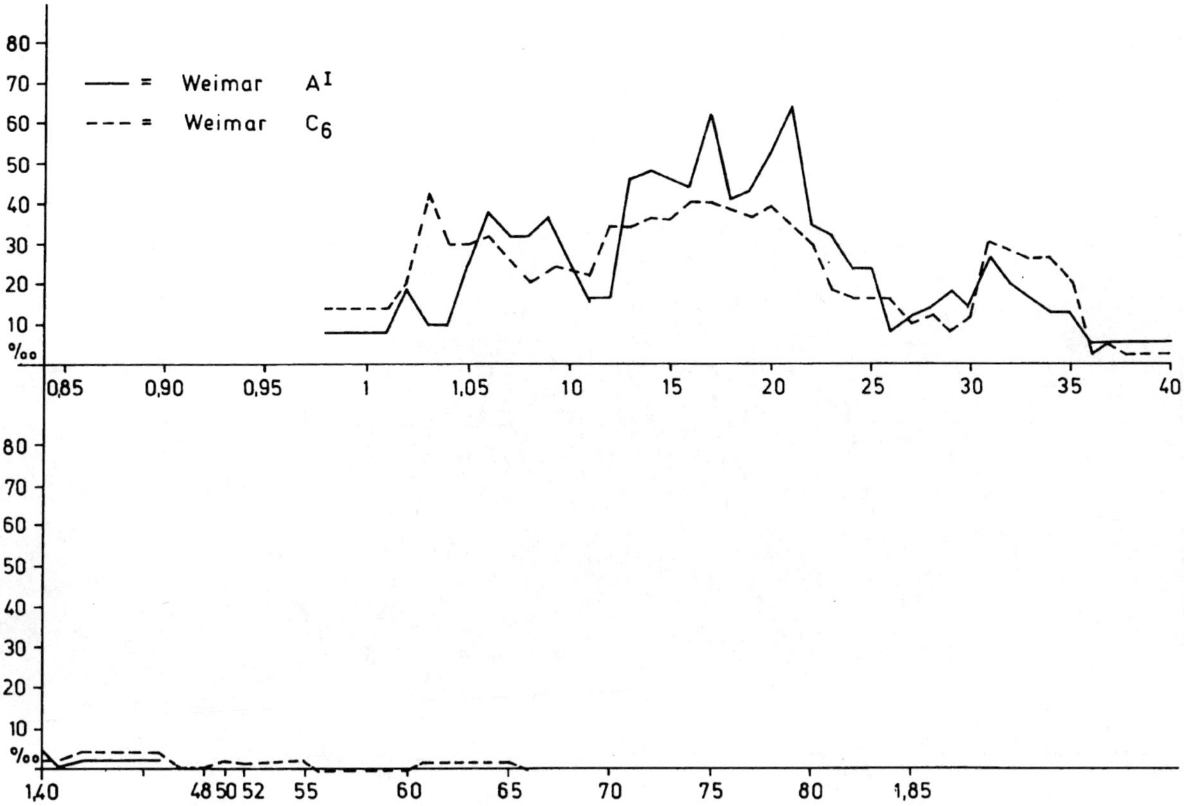

Tab. 18. Häufigkeitskurven der relativen Längen der Fossilien

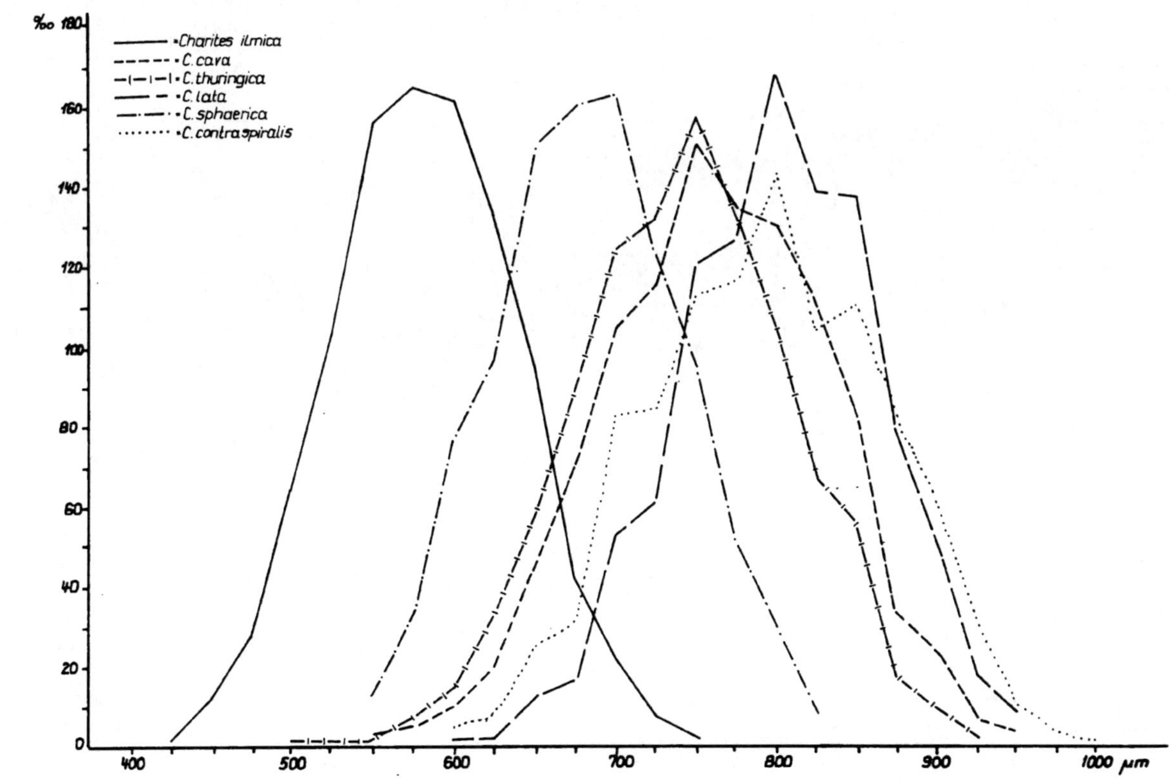

Tab. 19. Häufigkeitskurven der relativen Breiten der Fossilien

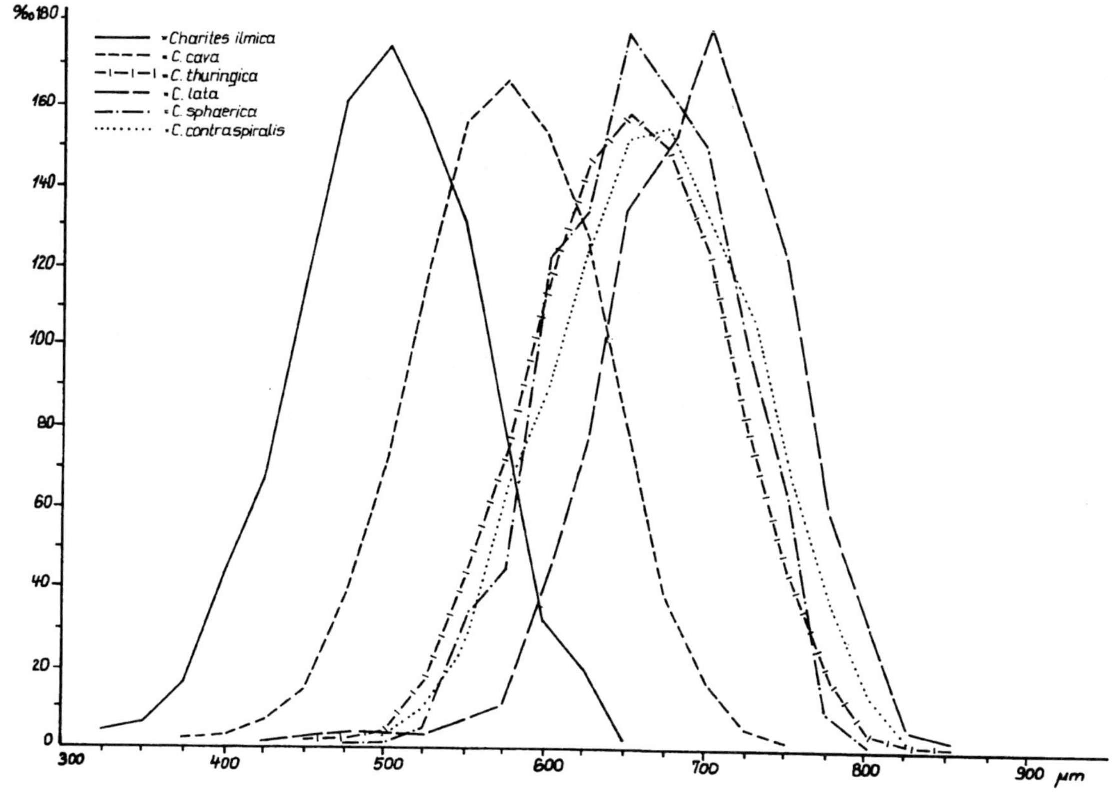

Tab. 20. Häufigkeitskurven der Quotienten Q = L/B der Fossilien

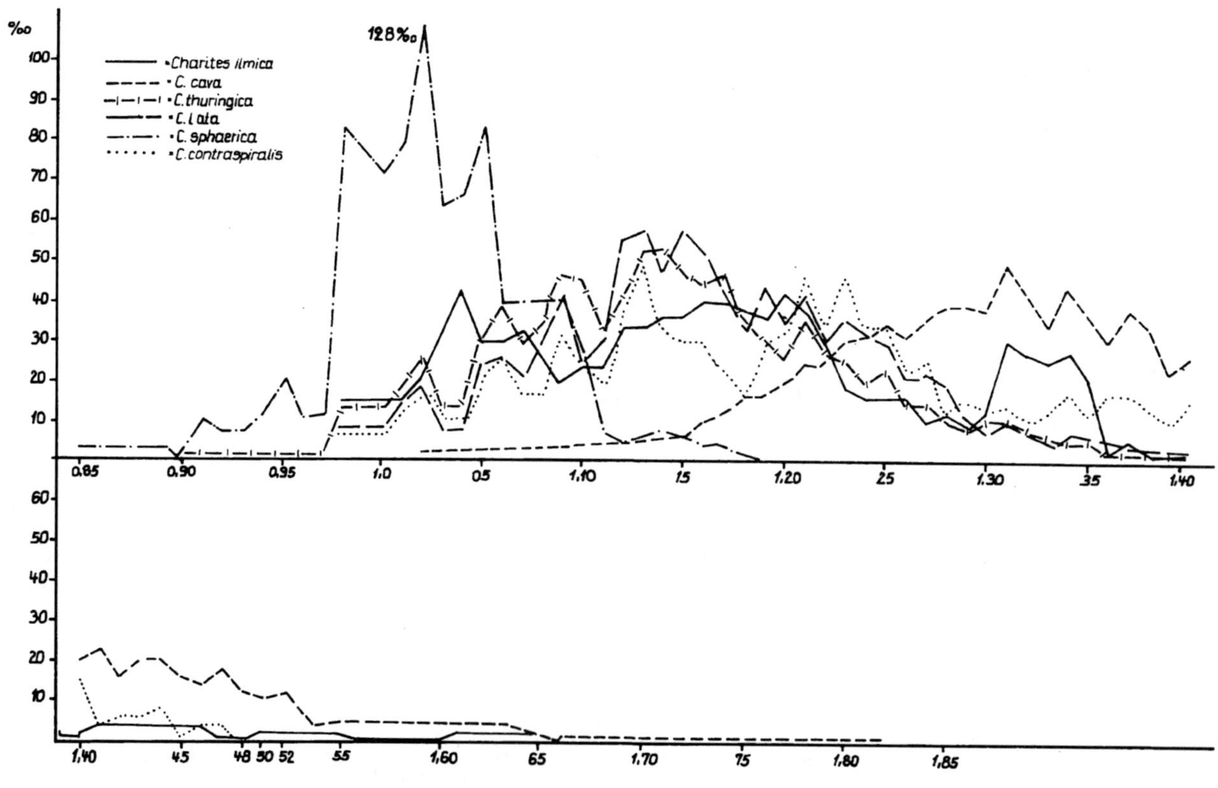

Tab. 21. Größenbereiche der neuen quartären Charophytenarten

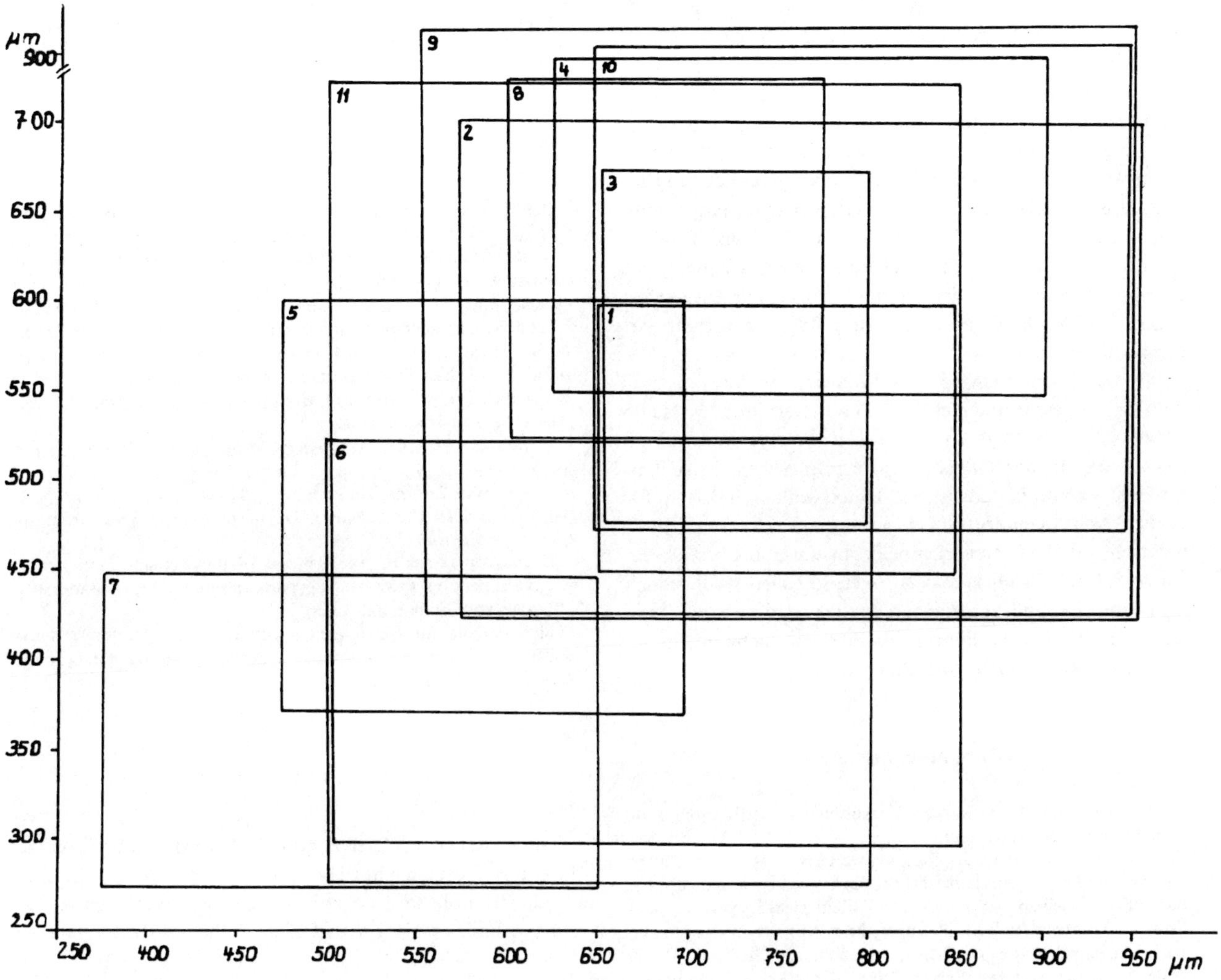

1 — *Charites prima*; 2 — *Charites cava*; 3 — *Charites angusta*; 4 — *Charites lata*; 5 — *Charites ilmica*; 6 — *Charites elongata*; 7 — *Charites eemica*; 8 — *Charites sphaerica*; 9 — *Charites thuringica*; 10 — *Charites contraspiralis*; 11 — *Charites seroglacialis*

Gattung *Tectochara* nicht zum Vergleich in Betracht. Am nächsten stehen noch die Gyrogonite der Gattung *Sphaerochara* STRAUB. Aber alle Sphaerocharen haben weit kleinere Gyrogonite. *Charites sphaerica* nimmt praktisch eine Mittel- bzw. Übergangsstellung zwischen den tertiären Sphaerocharen und den rezenten Gattungen *Lychnothamnus* und *Nitellopsis* ein.

Charites thuringica n. sp. Die am weitesten verbreitete *Ch. thuringica* kommt mit ihren fossilen Formen der *Chara vulgaris* L. sensu WOOD am nächsten. Die Fossilien sind aber ca. 100 µm größer, das ist jedoch, bedenkt man die sich deckenden Gaußschen Glockenkurven der Graphiken über Fossilien von sieben Fundstellen ein recht beachtlicher Unterschied. Selbst wenn man der Neukombination der *Chara globularis* durch WOOD zustimmen könnte, läge *Ch. thuringica* nicht in ihrem Variationsbereich. Es gibt keine tertiäre Gyrogonitenart, die eine derartige Größe besitzt. Normalerweise haben die weiblichen Fruktifikationsorgane der Charophyten eine Variationsbreite bis zu 300 µm in der Länge und 200 µm in der Breite. Lediglich die tertiäre *Tectochara meriani* (UNGER) L. & N. GRAMBAST und *Charites thuringica* sprengen diesen Rahmen. Innerhalb des weiten Variationsbereichs der Fossilien liegen die tertiären Taxa *Charites polonica* (UNGER) m. 1975, *Tectochara straubi* MÄDLER 1955, *T. tasnadii* (RASKY) MÄDLER 1955 und *T. tornata* (REID & GROVES) MÄDLER 1955, ohne daß sie jeweils auch nur annähernd den gesamten Variationsbereich der neuen Art ausfüllen.

Charites contraspiralis n. sp. Es bedarf wohl keiner weiteren Erörterung, daß weder eine rezente noch eine fossile Charophytenart bekannt ist, die zu den abnorm gestalteten Fossilien in Beziehung gesetzt werden kann.

Charites seroglacialis n. sp. Als 1975 die „Charophytenformenkreise Eem — Basistorf, Bølling und Allerød", die in diese neue Spezies eingehen, postuliert wurde, erfolgte der Hinweis, daß es natürlich gewagt ist, überhaupt einen Vergleich zwischen spät- und postglazialen Oogonien und tertiären Gyrogoniten zu führen. Diese Meinung vertritt der Autor auch heute noch. Anders ist es mit dem Vergleich zu rezenten Taxa. Es wurde bereits damals darauf hingewiesen, daß es fast ein Ding der Unmöglichkeit sei, daß *Chara vulgaris* L., die mit ihren infraspezifischen Taxa heute etwa 80% der Charophytenflora der Welt ausmacht, so wenig Beziehung zu quartären Charophyten haben sollte. Dabei wurde der offensichtlich zu klein angesetzte Variationsbereich der Oogonien dieser rezenten Art angezweifelt. Diese schwache Stelle in der Charophytensystematik wurde durch WOOD (1962) beseitigt. 1975 wurde vom Autor betont, daß die „Formenkreise", die hier vereinigt werden, weitgehende Übereinstimmung zur „Chara contraria KÜTZING" zeigen. Damit ist die Deckung mit dem Variationsbereich der *Chara vulgaris* L. sensu WOOD gegeben. Wenn nicht die oben geschilderten Einschränkungen, die der ICBN erfordert, gemacht werden müßten, würde der Autor nicht scheuen, diese Fossilien zur rezenten *Ch. vulgaris* zu stellen. Aber noch wissen wir nicht, wie die vegetativen Organe der *Charites seroglacialis* ausgesehen haben, und nach diesen wird nun einmal die rezente Art diagnostiziert.

Bei der Beschreibung der *Charites cava* wurde praktisch das Gleiche gesagt und hier liegt ein weiterer Grund, weshalb weder *Ch. cava* noch *Ch. seroglacialis* zur rezenten *Chara vulgaris* gestellt werden können. Die graphischen Darstellungen der variationsstatistischen Untersuchung lieferten für beide Fossiltaxa eindeutige Gaußsche Glockenkurven, jedoch unterschiedliche! Wenngleich beide im Variationsbereich der *Chara vulgaris* L. sensu WOOD liegen so füllen sie ihn doch nicht aus. Es muß zugestanden werden, daß sich die Taxa nahe stehen, aber sie können als Fossilien nicht mit dieser rezenten Form vereinigt werden.

Zusammenfassung

Von NÖTZOLD (1975) wurden Charophytenfruktifikationen aus dem mitteleuropäischen Raum in offener Nomenklatur beschrieben. Diese konnten nun nach der Bergung weiterer Fossilien, vor allem aus Thüringen, nomenklatorisch definiert werden. Dazu lagen in der DDR Fossilien der Fundstellen Bilzingsleben, Bezirk Halle, Thomas-Müntzer-Stadt Mühlhausen, Bezirk Erfurt, Burgtonna bei Bad Langensalza, Bezirk Erfurt, Mücheln im Geiseltal, Bezirk Halle, Bad Langensalza, Bezirk Erfurt, Greußen bei Sömmerda, Bezirk Erfurt, und Herbsleben östlich Bad Langensalza, Bezirk Erfurt, vor.

Die Taxa: *Charites prima* n. sp., *Charites cava* n. sp., *Charites angusta* n. sp., *Charites thuringica* n. sp., *Charites ilmica* n. sp., *Charites eemia* n. sp., *Charites elongata* n. sp., *Charites seroglacialis* n. sp., *Charites lata* n. sp., *Charites sphaerica* n. sp., *Charites contraspiralis* n. sp. wurden aufgestellt.

Einige Fundstellen führten eine neue Erhaltungsform fossiler, weiblicher Fruktifikationsorgane, die „Leeren Hüllzellen".

Im Vergleich mit tertiären und rezenten Organen zeichneten sich einige phylogenetische Entwicklungen ab.

Summary

Charophyte reproductive structures from Central Europe were described by Nötzold (1975) using a preliminary informal nomenclature. Following the discovery of additional fossil material, mainly from Thuringia, it has been possible to describe these taxa according to international nomenclatural usage. The fossils in question have come from the following localities in the GDR: Bilzingsleben (Bezirk Halle), Mühlhausen (Bezirk Erfurt), Mücheln im Geiseltal (Bezirk Halle), Bad Langensalza (Bezirk Erfurt), Greussen (Bezirk Erfurt) and Herbsleben (Bezirk Erfurt).

The following new taxa are described: *Charites prima* n. sp., *Charites cava* n. sp., *Charites angusta* n. sp., *Charites thuringica* n. sp., *Charites ilmica* n. sp., *Charites eemica* n. sp., *Charites elongata* n. sp., *Charites seroglacialis* n. sp., *Charites lata* n. sp., *Charites sphaerica* n. sp., *Charites contraspiralis* n. sp.

From several localities preservation of a feature previously unrecorded in fossil material, the coronulae of the female reproductive structures was observed.

From comparisons with modern and fossil Tertiary charophyte material it was possible to postulate certain phylogenetic lines of evolution.

Резюме

Репродуктивные органы харофитов были описаны Нотцольдом (1975) из среднеевропейской области по открытой номенклатуре. Эти репродуктивные органы могли быть таксономически определены только после получения новых ископаемых прежде всего из Тюрингии. Для этого в ГДР имелись ископаемые из местонахождений Бильцингслебен (округ Галле), Томас-Мюнтцер-штадт-Мюльгаузен (округ Эрфурт), Бургтонна около Бад Лангензальца (округ Эрфурт), Греусен около Зоммерда (округ Эрфурт) и Гербслебен к востоку от Бад Лангензальца (округ Эрфурт).

Установлены таксоны: *Charites prima* n. sp., *Charites cava* n. sp., *Charites angusta* n. sp., *Charites thuringica* n. sp., *Charites ilmica* n. sp., *Charites eemica* n. sp., *Charites elongata* n. sp., *Charites seroglacialis* n. sp., *Charites lata* n. sp., *Charites sphaerica* n. sp., *Charites contraspiralis* n. sp.

В некоторых местонахождениях обнаружены новые остатки женских репродуктивных органов хорошей сохранности — „пустые коробочки для спор".

При сравнении третичных и современных репродуктивных органов харофитов выявлены некоторые черты филогенетического развития.

Literatur

In der Arbeit von NÖTZOLD (1975) wurde ein umfassendes Literaturverzeichnis mit über 300 Titeln gegeben. Daher werden hier nur die neueren und wichtigen Publikationen berücksichtigt und im übrigen auf diese frühere Arbeit verwiesen.

BASZYNSKI, T.; KARCZMARZ, K., 1977: Investigations on the production of inorganic matter by Charophyta associations. — In: Acta Hydrobiol. — Kraków **19** (1). — S. 1—7.

DAILY, F. K., 1969: *Lamprothamnium* in Amerika. — In: J. Phycol. — **3**. — S. 201—207.

—, 1969: A *Lamprothamnium succinctum* (Charac.) with imperfect Cortex. — In: Bull. Torr. bot. Club. — **96** (6). — S. 650—660.

—, 1970: Fossil Charophytes from South Dakota and Iowa with a review of modern species. — In: Amer. Midl. Natural. — **84** (2). — S. 365—375.

—, 1973: Lime-shells of *Nitellopsis obtusa* — fossil and extent affinities. — In: Bull Torr. bot. Club. — **100** (2). — S. 75—78.

DAILY, F. K.; DURHAM, J. W., 1966: Miocene Charophytes from Ixtapa, Chiapas, Mexico. — In: J. Palaeont. — **40** (5). — S. 1191 to 1199.

FEIST-CASTEL, M., 1975: Répartition des Charophytes dans le Paléocène et l'Éocène du basin d'Aix-en-Provence. — In: Bull. Soc. geol. Fr. 7. ser. — Paris **17**. — S. 88—97.

—, 1976: Les Charophytes dans le Paléogène du Sud de la France (Provence, Languedoc, Aquitaine). — In: Étude syst. et biostrat. — 82 S.: 10 Taf.

—, 1977: Evolution of the Charophyte floras in the Upper Eocene and Lower Oligocene of the Isle of Wight. — In: Palaeontology. — London **20** (1). — S. 143—157: Taf. 21, 22, (1977a).

—, 1977: Étude floristique et biostratigraphique des Charophytes dans les séries du Paléogène de Provence. — In: Géol. Médit. — **4**. — S. 109—138, (1977b).

—, 1977: Description du nouveau genre *Krassavinella* (Charophytes, Characeae) et répartition de *K. lagenalis* (STRAUB) dans l'Oligocène supérieur de la molasse suisse. — In: Eclog. geol. Helvetiae. — Basel **70** (3). — S. 771—775, (1977c).

FEIST, M.; RINGEADE, 1977: Étude bistratigraphique et paléobotanique (Charophytes) des formations continentales d'Aquitaine, de l'Éocène supérieur au Miocène inférieur. — In: Bull. Soc. geol. Fr. 7. ser. — Paris **19** (2). — S. 341—354.

GEVIN, P.; FEIST, M.; MONGEREAU, N., 1974: Decouverte de Charophytes d'age Éocène au Glib Zegdou (Sahara algérien). — In: Bull. Soc. Hist. natur. Afr. Nord. Alger. — **65** (1/2).

GRAMBAST, L., 1970: Principes de l'utilisation stratigraphique des Charophytes. Applications au Paléogène d'Europe occidentale. — In: Mem. Bur. Rech. Geol. Minères. — Paris **77** (1). — S. 319 bis 328.

—, 1972: Étude sur les Charophytes tertiaires d'Europe sur les occidentale. I. Genre *Tectochara*. — In: Paléobiol. contin. — **3** (2). — S. 1—30.

—, 1977: Étude sur les Charophytes tertiaires d'Europe occidentale. II. Espèces nouvelles de l'Éocène inférieur. — In: Paleobiol. Contr. — **8** (1). — S. 1—27.

GRAMBAST, L.; SOULIE-MÄRSCHE, I., 1972: Sur l'Ansienneté la Diversification des *Nitellopsis* (Charophytes). — In: Paleobiol. Contr. — **3** (3). — S. 1—14.

KARCZMARZ, K., 1963: Przyczynek do badań nad flora ramienic Lubelszczyzny (Polska wschodnia). Część I. — In: Acta Soc. Bot. Poloniae. — **32** (1).

—, 1966: Przyczynek do badân nad flora ramienic Lubelszczyzny (Polska wschodnia). Część III. — In: Acta Soc. Bot. Poloniae. — **35** (2).

—, 1967: Variabilité et distribution de *Lychnothamnus barbatus* (MEYEN) v. LEONHARDI. — In: Acta Soc. Bot. Poloniae. — **36** (3).

—, 1973: On the ecological requirements of *Chara delicatula* AGARDH. — In: Ann. Univ. Lublin. Sect. C. — **28** (11).

KARCZMARZ, K.; ŁUCZYCKA, A., 1974: Materialy do flory ramienic środkowej i poludniowej Polski (A contribution to the flora of Charophyta of Central and Southern Poland). — In: Acta Hydrobiol. — Kraków **16** (1). — S. 113—119.

KARCZMARZ, K.; ŁUCZYCKA, A.; OCHYRA, R., 1976: Materialy do flory ramienic środkowej i poludniowej Polski (A Contribution to the flora of Charophyta of Central and Southern Poland). . Część 2. — In: Acta Hydrobiol. — Kraków **18** (2). — S. 193 bis 200.

KNOBLOCH, E., 1975: Taxonomische und variationsstatistische Untersuchungen an süddeutschen tertiären Charophyten. — In: Acta Univ. Car. Ser. Geol. — Praha (2). — S. 167—186.

KRASSAVINA, L. K., 1971: A comparative study of recent and fossil of the fructification in *Nitellopsis obtusa* and of the gyrogonites in the species *Tectochara*. — In: Bot. Ž. — Moskva **56** (1). — S. 106—117. — In Russisch.

MÄRSCHE, I., 1970: Contribution á l'étude des Charophytes éocènes en Aquitaine orientale. — Toulouse, Thése spécialité.

NÖTZOLD, T., 1975: Charophytenreste aus dem Neophytikum Mitteleuropas. — In: Abh. staatl. Mus. Mineral. u. Geol. Dresden. — Dresden **23**. — 265 S.

—, 1977: *Nitellopsis obtusa* (DESV.) J. GROVES und *Lychnothamnus barbatus* (MEYEN) v. LEONHARDI — lebende Fossilien? — In: Ber. naturforsch. Gesellsch. Freiburg im Breisgau. — Freiburg i. Br. **67**. — S. 217—234.

SOULIE-MÄRSCHE, L., 1973: Donnes generales sur les Charophytes du Plio-Pleistocene. — Bull. Assoc. Fr. l'étude Quatern. — (1973) (2). — S. 69—72.

—, 1974: Nouvelles espèces de Charophytes Eocénes du Bassin d'Aquitaine. — In: C. R. Kong. Nat. Gesellsch. Wiss. Toulouse 1971. — Paris **96** (2). — S. 109—144.

—, 1975: Entwicklungslinien der *Nitellopsis* (Characeen) vom Obermiozän bis zum Quartär. — In: Congr. region. Commit. Medit. Neogene Strat. — Bratislava.

WOOD, R. D., 1962: New combinations and taxa in the revision of Characeae. — In: Taxon. — Utrecht **11**. — S. 7—25.

TAFEL I

Fig. 1—4: *Charites cava* n. sp., Holotypus
Fig. 1: Seitenansicht, Fig. 2: Gipfel, Fig. 3: Basis, alle Vergr. 100:1
Fig. 4: Stielloch, Vergr. 300:1

Fig. 5—7: *Charites thuringica* n. sp., Holotypus
Fig. 5: Seitenansicht, Fig. 6: Gipfel, Fig. 7: Basis, alle Vergr. 100:1

TAFEL II

Fig. 1—3: *Charites ilmica* n. sp., Holotypus
Fig. 1: Seitenansicht, Vergr. 100:1; Fig. 2: Gipfel, Vergr. 450:1; Fig. 3: Stielloch, Vergr. 500:1

Fig. 4—6: *Charites eemica* n. sp., Holotypus
Fig. 4: Seitenansicht, Fig. 5: Stielloch, beide Vergr. 100:1; Fig. 6: Gipfel, Vergr. 150:1

Fig. 7—9: *Charites elongata* n. sp., Holotypus
Fig. 7: Seitenansicht, Fig. 8: Basis, beide Vergr. 100:1; Fig. 9: Gipfelpartie, die das Herausbrechen der einzelnen Gipfelknötchen zeigt

TAFEL III

Fig. 1—3: *Charites seroglacialis* n. sp., Holotypus
Fig. 1: Seitenansicht, Fig. 2: Gipfel, Fig. 3: Basis, alle Vergr. 100:1

Fig. 4, 5: *Charites lata* n. sp., Holotypus
beide Seitenansichten, Vergr. 100:1

Fig. 6: *Charites sphaerica* n. sp., Holotypus
Seitenansicht, Vergr. 100:1

TAFEL IV

Fig. 1, 2: *Charites sphaerica* n. sp.
Fig. 1: Gipfel, Fig. 2: Basis, beide Vergr. 100:1

Fig. 3—6: *Charites contraspiralis* n. sp., Holotypus
Fig. 3: Seitenansicht, Fig. 4, 5: Gipfel bei herausgebrochenem Gipfelknötchen, Fig. 6: Basis, alle Vergr. 100:1

TAFEL I

TAFEL II

TAFEL III

TAFEL IV

Quartärpaläontologie 6 157—171 Berlin 1986

Les mammifères du Pléistocène inférieur d'Izvoru (Département d'Olt, Roumanie)

Par COSTIN RADULESCO et PETRE SAMSON, București

Avec 5 figures et 6 tableaux

Contenu

Abstractum

On donne la description des restes de Mammifères provenant des dépôts stratifiés d'Izvoru, localité située dans l'ouest du Bassin dacique. On discute l'âge géologique du site qui comporte deux niveaux à Mammifères, nommés Izvoru-1 (Elephantidae, Cervidae) et Izvoru-2 (Insectivora, Rodentia, Lagomorpha). Les niveaux fossilifères représentent deux phases successives dans le Pléistocène inférieur (zone de Senèze).

Es werden Säugetierreste beschrieben, die aus Ablagerungen von Izvoru stammen, einer Fundstelle im Dazischen Becken. Das geologische Alter der Fundstelle, die zwei fossilführende Horizonte umfaßt: Izvoru-1 (Elephantidae, Cervidae) und Izvoru-2 (Insectivora, Rodentia, Lagomorpha) wird diskutiert. Die fossilführenden Horizonte repräsentieren zwei aufeinanderfolgende Phasen im Unterpleistozän (Senèze-Zone).

Mammal remains originating from deposits at Izvoru, a fossiliferous locality in the Dacic basin, are described. Also discussed is the geological age of the locality which shows two horizons: Izvoru-1 (Elephantidae, Cervidae) and Izvoru-2 (Insectivora, Rodentia, Lagomorpha). The fossiliferous horizons represent two succeeding phases of the Lower Pleistocene (zone of Senèze).

Описаны остатки млекопитающих из отложений Извору — местонахождения, расположенного в Дакийском бассейне. Обсуждается геологический возраст местонахождения, включающего два костеносных горизонта: Извору-1 (Elephantidae, Cervidae) и Извору-2 (Insectivora, Rodentia, Lagomorpha). Костеносные горизонты представляют собой две последовательные фазы нижнего плейстоцена (зона Сенез).

1. Introduction

Découverte en 1977, la faune de Mammifères d'Izvoru a déjà fait l'objet de quelques considérations d'ordre général (ANDREESCO et al. 1981; FERU et al. 1978, 1979, 1983). Etant donné l'importance de cette localité dans la succession des sites fossilifères plio-pléistocènes du Bassin dacique non seulement par son contenu paléontologique (Mollusques et Mammifères associés), mais encore par la possibilité d'établir des raccords à distance avec les termes équivalents des principaux schémas biostratigraphiques d'Europe, il nous a paru utile de faire une description détaillée de l'ensemble faunique recueilli à Izvoru.

Abréviations

- H_1 = hauteur (THALER 1962)
- H_2 = hauteur (VAN DE WEERD 1976)
- Hai = hauteur de la discontinuité antéro-interne de l'émail prise par rapport à la base de l'angle rentrant lingual postérieur
- Hgm = hauteur de la gouttière mimomyenne (RADULESCO et SAMSON 1983)
- Hpi = hauteur de la discontinuité postéro-interne de l'émail prise par rapport à la base de l'angle rentrant lingual postérieur
- Hri = hauteur du repli insulaire (RADULESCO et SAMSON 1983)
- L = longueur maximale
- Lca = longueur du complexe antérieur (VAN DER MEULEN 1975)
- l = largeur (JÁNOSSY et VAN DER MEULEN 1975)
- lbm = largeur de la boucle mésiale (CHALINE 1980)
- lca = largeur du complexe antérieur (CHALINE 1980)

Les mensurations, dans le texte et dans les tableaux, sont en mm.

Remerciements — Nous avons utilisé à des fins comparatives une petite collection, encore inédite, de Micromammifères provenant de Villány-3 (= Villány-Kalkberg) et de Betfia-2 (= Băile

Episcopiei = Püspökfürdö), offerte par T. KORMOS au Professeur I. SIMIONESCO et conservée au Laboratoire de Paléontologie de l'Université de Bucarest. Nous sommes bien reconnaissants au Dr. TH. NEAGU pour l'aimable permission de consulter la collection susmentionnée.

2. Stratigraphie

Le village d'Izvoru est situé à 11 km à l'ouest de Slatina, chef-lieu du département d'Olt (FERU et al. 1978, carte de la fig. 1). La faune a été découverte dans une exploitation locale de sables à la suite de fouilles pratiquées sur 12 m^2. La succession des couches (fig. 1), de bas en haut, commence par des niveaux de sables en alternance avec des limons sableux qui se développent sur 6 m d'épaisseur (couches 1—8). Le niveau inférieur de sables (couche 1) a livré quelques restes rapportables à *Archidiskodon meridionalis* et *Allocaenelaphus* sp. Ce niveau fossilifère a été distingué sous le nom d'Izvoru-1.

La couche 9, épaisse de 6 à 10 cm, disposée en lentilles, est constituée par un limon sableux olive pâle (5Y 6/3) avec des taches brunes (10Y 4/3); de cette couche, dénommée Izvoru-2, proviennent des restes de Mollusques et de Vertébrés (Poissons, Amphibiens, Chéloniens, Oiseaux, Mammifères).

Les Mollusques sont représentés par: *Bogatschevia sturi* (HÖRN.), *Wenziella wilhelmi* (PEN), *W. pseudosturi* (HAL.).

La liste des Mammifères (y compris le nombre de pièces) est donnée dans le tabl. 1.

La séquence se continue par une alternance d'argiles et de limons sableux (couches 10—15) jusqu'au sol actuel (couche 16).

Tableau 1. Composition et fréquence des Mammifères d'Izvoru-2

	N	%
Ordre Insectivora		
Desmana nehringi KORMOS	3	2,46
Petenyia hungarica KORMOS	2	1,64
Beremendia fissidens (PETÉNYI)	1	0,82
Ordre Lagomorpha		
Hypolagus brachygnathus KORMOS	1	0,82
Ordre Rodentia		
Citellus primigenius KORMOS	4	3,28
Trogontherium boisvilletti boisvilletti (LAUGEL)	1	0,82
Clethrionomys cf. *kretzoii* (KOWALSKI)	1	0,82
Mimomys coelodus cf. *coelodus* KRETZOI	11	9,02
Mimomys reidi HINTON	3	2,46
Mimomys coelodus/reidi	56	45,90
Mimomys pitymyoides JÁNOSSY & VAN DER MEULEN	5	4,10
Kislangia rex (KORMOS)	4	3,28
Borsodia cf. *hungarica* (KORMOS)	11	9,02
Mimomys sp. indet.	19	15,57

3. Description du matériel

3.1. Izvoru-1

3.1.1. Archidiskodon meridionalis (NESTI)

Matériel: axis.

La vertèbre a l'apophyse épineuse et les apophyses transverses brisées; l'épiphyse caudale n'est pas encore soudée. Nous n'avons constaté aucune différence morphologique ou dimensionnelle en comparaison des pièces homologues provenant du riche gisement à Eléphant méridional d'Irimeşti (Fîntîna lui Mitilan).

Mentionnons, en ce qui concerne les dimensions, que la largeur de la surface articulaire pour l'atlas mesure 231; le diamètre transverse du canal neural (à la partie craniale) atteint 75.

3.1.2. Allocaenelaphus sp.

Matériel: fragment de frontal droit avec le départ du bois; fragment basilaire de bois de mue gauche.

Le frontal, provenant d'un exemplaire adulte, est muni d'un pédicule osseux court et massif; en avant du pédicule, le frontal est déprimé. Le fragment de bois laisse reconnaître la présence du premier andouiller, en position typiquement basilaire, brisé à son insertion; le merrain présente, au-dessus du premier andouiller, sur la face antéro-supérieure, une crête mousse, délimitée du côté externe par une dépression superficielle; un bord arrondi marque la séparation entre la face antéro-supérieure et la face externe.

Le fragment de bois de mue possède le départ du premier andouiller situé juste au-dessus de la meule; le merrain présente, ainsi que dans le cas décrit précédemment, une crête et une dépression au-dessus de l'andouiller basilaire; le bord qui marque la limite entre la face antéro-supérieure et la face externe est aussi visible; à quelques 80 mm au-dessus de la première ramification, le merrain a une section caractéristique, déterminée par la formation d'une large dépression sur la face externe, en même temps que la face interne est convexe (fig. 2.1).

Dimensions:

	Iz-1/03	Iz-1/04
diamètre transverse du pédicule osseux	67,0	—
diamètre antéro-postérieur du pédicule osseux	47,5.	—
diamètre transverse de la meule	74,0	56,0
diamètre antéro-postérieur de la meule	—	69,0
circonférence de la meule	225,0	202,0
diamètre vertical du merrain au-dessus de l'andouiller basilaire	63,0	47,0
diamètre horizontal du merrain au-dessus de l'andouiller basilaire	—	36,0

Les deux exemplaires évoquent de près la morphologie des bois de *Allocaenelaphus arambourgi* RADULESCO & SAMSON (fig. 2.2) de Rotbav-Silvestru (localité type) dans la Dépression de Braşov (sud-est de la Transylvanie) (RADULESCO et SAMSON 1967). L'état très incomplet des pièces d'Izvoru ne permet pas une détermination spécifique exacte; il est possible que nous soyons en présence d'une forme de *Allocaenelaphus* plus primitive que *A. arambourgi* de Transylvanie, constatation en accord avec l'âge géologique plus récent de l'association faunique de Rotbav-Silvestru, considérée comme équivalente de l'horizon kaïrien du complexe faunique de Taman (= Waalien B) (FERU et al. 1983).

Mentionnons qu'une forme de *Allocaenelaphus*, proche de celle d'Izvoru, fait son apparition aussi à Irimeşti (Fîntîna lui Mitilan) dans la moyenne vallée de l'Olteţ, également dans le Bassin dacique (phase II du Pléistocène inférieur, zone de Senèze) (FERU et al. 1983; SAMSON et RADULESCO 1973).

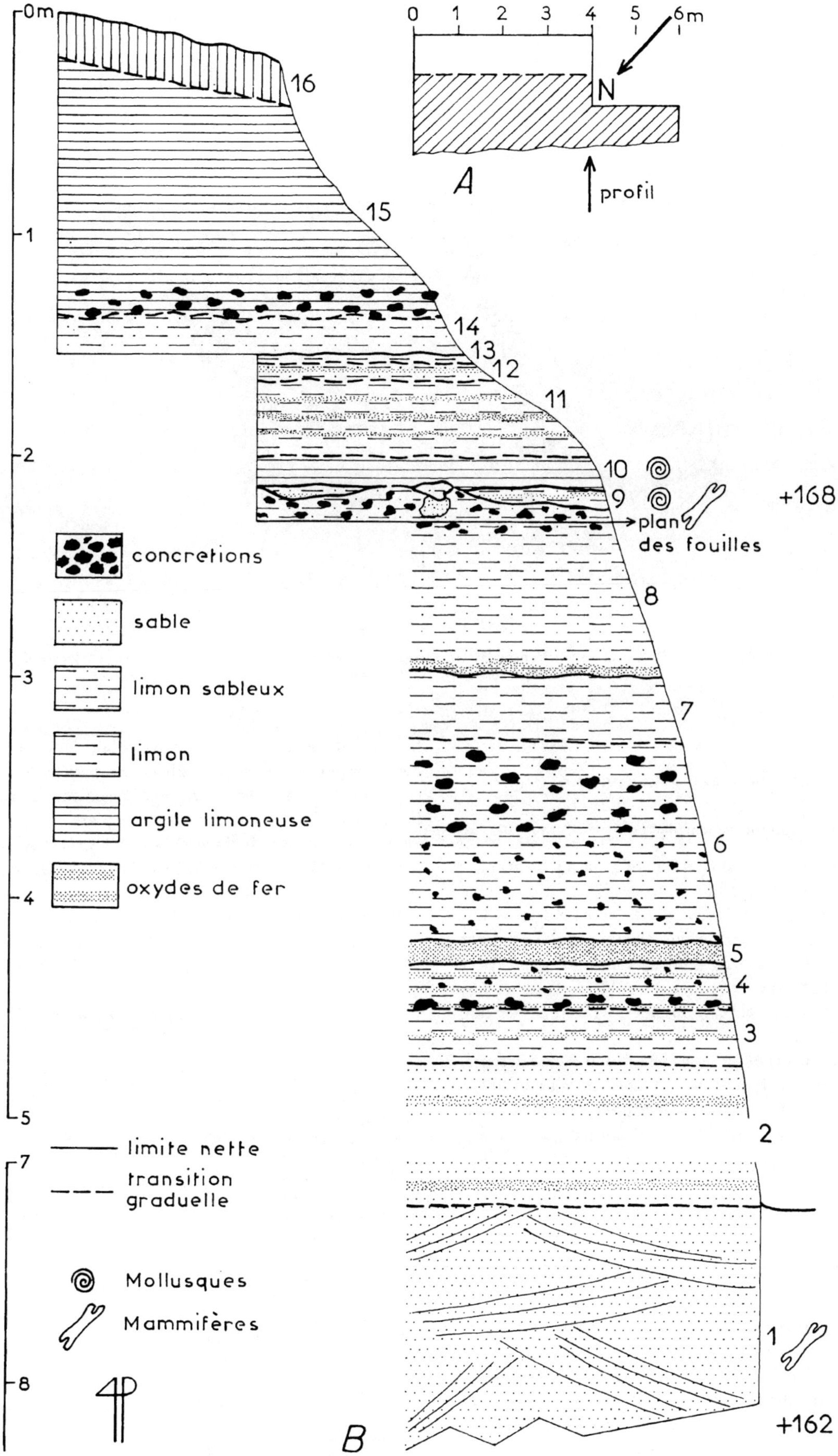

Fig. 1. Izvoru

A — Emplacement des fouilles (en hachures obliques la zone fossilifère); B — Succession stratigraphique (paroi SW): couche 1 = Izvoru-1; couche 9 = Izvoru-2

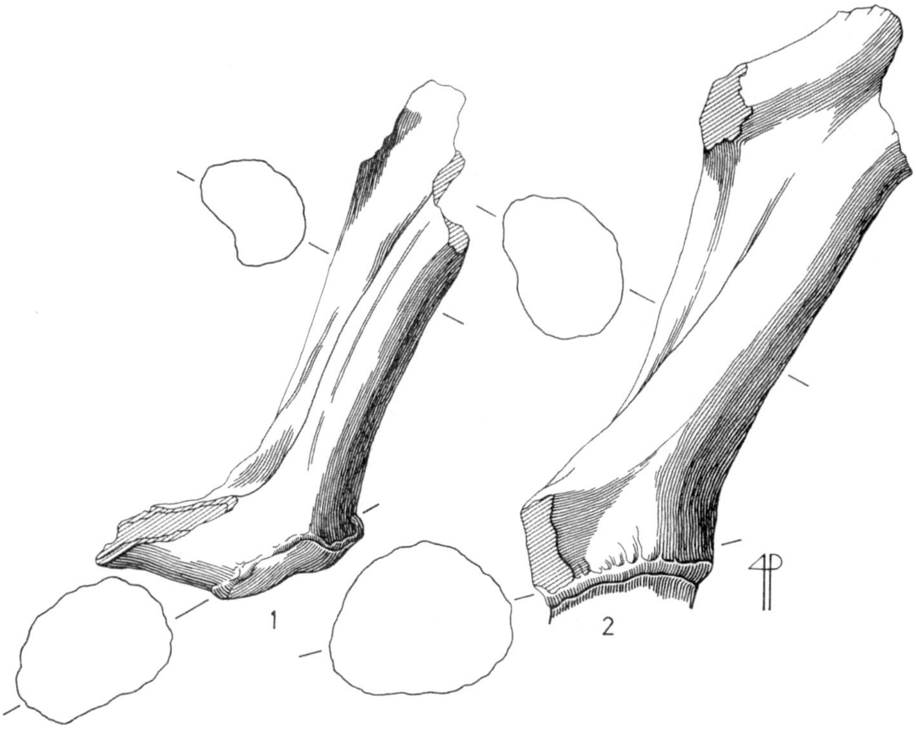

Fig. 2. Izvoru-1 (Iz-1/nos)

1 — *Allocaenelaphus* sp., bois de mue sin/04. Rotbav-Silvestru (RS/nos): 2 — *Allocaenelaphus arambourgi*, bois dext. (inversé), paratype/02 (RADULESCO et SAMSON 1967, fig. 2, pl. II, fig. 1). Vue externe (1/3)

3.2. Izvoru-2

3.2.1. Desmana nehringi KORMOS

Matériel: M_1, 2 fragments de mandibules.

M_1 — La molaire est à un degré d'abrasion relativement avancé (fig. 3.1a); le cingulum est développé sur le côté antérieur de la couronne et se continue du côté externe jusqu'au niveau de l'hypoconide; un cingulum distal est aussi visible; le cingulum lingual, vestigial, est limité à la région du paraconide; en vue interne, l'entoconide présente un fort diamètre antéro-postérieur et la cuspide accessoire postérieure est bien développée.

Une M_1 de Villány-3, qui nous sert de terme de référence, à peine entamée par l'usure, se distingue, comparativement à notre pièce, par son cingulum mieux exprimé et sa taille moins forte, comme il ressort des valeurs indiquées ci-dessous:

	Iz-2/093	Villány-3
longueur	3,08	2,97
largeur antérieure	1,90	1,80
largeur postérieure	2,20	2,15

Nous rapportons à la même espèce deux fragments de mandibules qui se caractérisent, à la différence de la majorité des exemplaires de *D. nehringi*, par la faible hauteur de la branche horizontale; leurs dimensions sont indiquées dans le tabl. 2; nous avons ajouté, à des fins comparatives, les mensurations d'un exemplaire en provenance de Villány-3.

Par la hauteur réduite de la branche horizontale, les deux fragments d'Izvoru évoquent aussi la structure de *D. kormosi* SCHREUDER, espèce mise en évidence dans le Bassin dacique dans des dépôts plus anciens (Pliocène supérieur de Slatina-1) (FERU et al. 1978). Cependant, par l'épaisseur de la branche horizontale, dont les valeurs sont toujours supérieures aux valeurs maximales indiquées pour *D. kormosi* (RZEBIK-KOWALSKA 1971; SCHREUDER 1940), les exemplaires d'Izvoru se rapprochent de *D. nehringi*.

Il est intéressant de relever la présence d'un Desmaniné (*Desmana* sp.) de taille plutôt faible dans la faune de Kamyk en Pologne; par la hauteur de la branche horizontale de la

Tableau 2. Mensurations comparatives de la mandibule de *Desmana nehringi*

	Iz-2/094	Iz-2/095	Villány-3
Hauteur linguale de la branche horizontale en arrière de M_3	3,60	3,45	3,90
Epaisseur de la branche horizontale en arrière de M_3	1,80	1,80	1,90
Distance entre le bord inférieur de la mandibule et le foramen mandibulaire	36,00	—	40,00
Distance entre le bord postérieur de la mandibule et le bord antérieur du foramen mandibulaire	3,10	—	—
Distance entre le bord postérieur de la mandibule et le bord antérieur de la branche ascendante au niveau du foramen mandibulaire	5,50	—	—
Distance entre le bord postérieur de l'alvéole de M_3 et le bord antérieur du foramen mandibulaire	4,10	—	4,50

mandibule, cette forme est intermédiaire entre *D. kormosi* et *D. nehringi* (RZEBIK-KOWALSKA 1971).

Rappelons, enfin, qu'on a signalé en Hongrie, dans la faune de Kisláng (KRETZOI 1954), sous le nom de *Desmagale pannonica*, une forme de la taille de *D. nehringi*, mais à mandibule plus éffilée. Ces deux mentions semblent indiquer la présence d'un Desmaniné différent de *D. nehringi* dans les faunes du Pléistocène inférieur d'Europe Centrale. Malheureusement, les documents d'Izvoru sont trop incomplets pour apporter une solution à ce problème.

3.2.2. Petenyia hungarica KORMOS

Matériel: fragment de mandibule avec M_{1-2}, I supérieure.

Le fragment de mandibule est réduit à la portion de la branche horizontale correspondant aux molaires. Le foramen mentale est situé sous la limite antérieure de l'hypoconide de M_1; il est précédé par une faible dépression au niveau du protoconide de la même dent. La hauteur (interne) et la largeur de la mandibule sous M_2 mesurent 1,78 et 0,95 respectivement.

Les molaires (fig. 3.2a) se distinguent par le développement accentué du cingulum labial qui se continue sur toute la paroi distale des couronnes; le cingulum interne est moins fort; mentionnons la présence caractéristique d'une crête de l'entoconide bien développée en hauteur.

La dernière molaire manque, mais ses alvéoles sont conservés d'où la possibilité d'évaluer la longueur de M_{1-3} qui aurait atteint 4,0 approximativement. Les longueurs de M_1 et de M_2 sont 1,65 et 1,47.

L'exemplaire d'Izvoru présente les particularités dentaires du genre *Petenyia*. Relevons, cependant, sa taille plus forte comparativement aux pièces de la station type de l'espèce (Villány-3) (KORMOS 1934). Précisons de même que les découvertes de *P. hungarica*, en Europe centrale et centre-orientale, se rapportent généralement à des exemplaires proches par leur taille des spécimens de Villány-3. Mentionnons qu'une forme de taille grande, comparable à celle d'Izvoru, est signalée à Kamyk dans un contexte faunique plus tardif à *Allophaiomys pliocaenicus* KORMOS (KOWALSKI 1960).

3.2.3. Beremendia fissidens (PETÉNYI)

Matériel: M^{1-2}.

Les deux pièces d'Izvoru, provenant du même individu, ne diffèrent pas en ce qui concerne la morphologie (fig. 3.3a), de leurs homologues observées sur un maxillaire de Villány-3.

RZEBIK-KOWALSKA (1976), en révisant le matériel de *B. fissidens* de Pologne, a mentionné comme caractéristiques de M^1 le parastyle étroit, l'hypocône assez réduit, le faible développement de la crête qui relie le protocône à la base du métacône, l'échancrure du bord postérieur des couronnes. M^2 se distingue par sa forme rectangulaire. Ces particularités se retrouvent aussi bien sur les pièces d'Izvoru que sur leurs correspondantes de Villány-3. Relevons sur nos pièces un développement un peu plus important de l'hypocône et l'échancrure du bord postérieur des couronnes moins exprimée par rapport à l'exemplaire de Villány-3.

En ce qui concerne les dimensions, la forme d'Izvoru se distingue par sa taille grande qui évoque les exemplaires les plus forts de *B. fissidens*. Nous indiquons ci-dessous les mensurations comparatives de notre matériel et de celui de Villány-3:

	Iz-2/098		Villány-3	
	M^1	M^2	M^1	M^2
Longueur externe (maximale)	—	2,25	2,57	2,14
Longueur médiane (minimale)	2,10	1,88	2,05	1,72
largeur antérieure	2,70	2,70	2,70	2,65
largeur postérieure	—	2,25	2,72	2,15

3.2.4. Hypolagus brachygnathus KORMOS

Matériel: I supérieure, fragment.

L'exemplaire se distingue par sa robustesse, comme il ressort des mensurations indiquées plus bas:

Diamètre mésio-distal	3,45
Diamètre vestibulo-lingual	2,20
Distance entre le sillon antérieur et la face mésiale	1,49

3.2.5. Citellus primigenius KORMOS

Matériel: 2 M_3, d^4, P^3.

M_3 se caractérise par la largeur antérieure accusée et le développement réduit de la partie distale (zone de l'hypoconide); un sillon superficiel, parallèle à l'axe antéro-postérieur de la couronne, marque sur la face occlusale la limite entre l'hypoconide et l'entoconide, sans diviser la paroi distale de la molaire.

Bien que les M_3 proviennent de deux exemplaires d'âge différent, l'un adulte (fig. 3.4a), l'autre très jeune, leurs dimensions sont superposables comportant 3,45 et 3,40 pour la longueur et la largeur antérieure, respectivement.

d^4 se distingue par sa brachyodontie et ses racines divergentes; le parastyle volumineux est séparé du reste de la couronne par une large fossette; le métaconule est soudé par sa partie externe au métacône, sa partie interne, mieux individualisée, présentant une crête qui la relie au protocône; la fossette du cingulum distal est large.

Dimensions: $L \times l = 2{,}65 \times 2{,}65$

P^3 a les mêmes particularités que l'unique pièce homologue de Villány-3, décrite par GROMOV (1965). Notons ainsi, sur notre exemplaire, le développement très faible du cingulum antéro-externe (fig. 3.5a) et la présence d'une seule pointe centrale (malgré l'usure avancée de la dent, la présence d'une seconde pointe en position mésiale est peu vraisemblable). Les deux racines primaires sont intimement soudées, le sillon de coalescence étant visible sur la face postérieure de l'unique racine résultée (fig. 3.5d).

Dimensions: $L \times l = 1{,}68 \times 1{,}95$

La morphologie aussi bien de M_3 que de P^3 correspond aux caractéristiques de *C. primigenius* (GROMOV 1965). L'espèce fait son apparition en Europe centrale à Villány-3 (KORMOS 1934). Dans la succession des faunes de Mammifères plio-pléistocènes du Bassin dacique, la présence de *C. primigenius* est enregistrée, pour la première fois, à Izvoru.

3.2.6. Trogontherium boisvilletti boisvilletti (LAUGEL)

Matériel: M^3.

La pièce provient d'un individu âgé, sur la face occlusale aussi bien le sinus que les synclinaux, au nombre de 5, étant

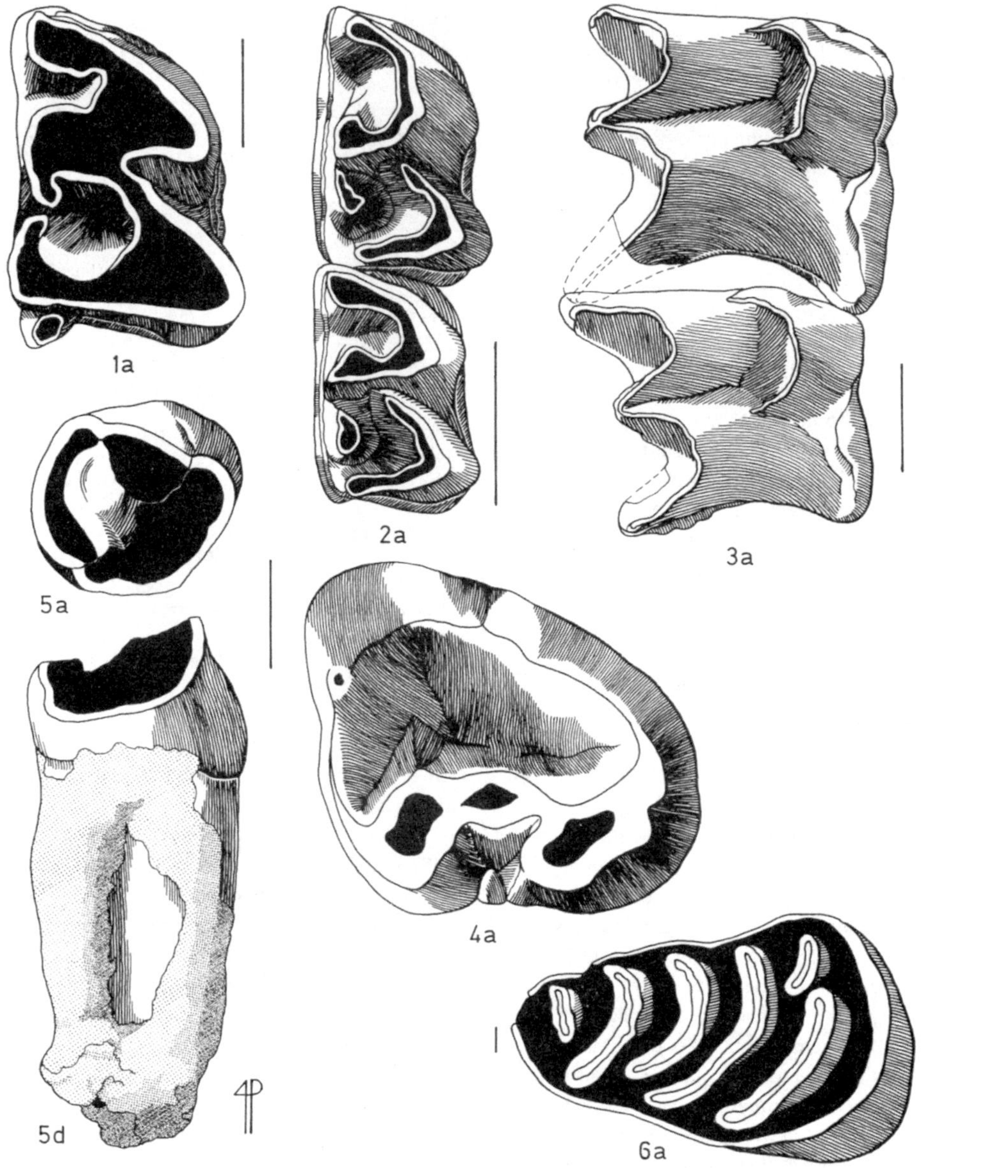

Fig. 3. Izvoru-2 (Iz-2/nos)

1 — *Desmana nehringi*, M_1 dext./093; 2 — *Petenyia hungarica*, M_1-M_2 dext./096; 3 — *Beremendia fissidens*, M^1-M^2 dext./098; 4,5. *Citellus primigenius*: 4 — M_3 sin./087, 5 — P^3 sin./090; 6 — *Trogontherium boisvilletti boisvilletti*, M^3 dext./091

(dans toutes les fig., les vues: a — occlusale, b — latérale externe, c — latérale interne, d — postérieure; les échelles correspondent à 1 mm)

transformés en ilôts (fig. 3.6a). La molaire se remarque par la longueur accusée (12,9), en même temps que la largeur a une valeur plutôt réduite (8,4).

Les grandes formes du genre *Trogontherium* sont encore imparfaitement connues; leur nomenclature n'est pas claire et il n'y a pas un accord sur les noms à utiliser. Dans la succession des sites fossilifères de la vallée de l'Olt, une forme grande de *Trogontherium* fait son apparition au niveau chronologique marqué par la faune d'Izvoru. Rappelons qu'une forme grande est présente aussi, dans le Bassin dacique, à Irimeşti (Fîntîna lui Mitilan) (Samson et Radulesco 1973). Nous avons assimilé cette forme de taille grande du Pléistocène inférieur de Roumanie à *T. boisvilletti boisvilletti*. Elle est précédée dans le Bassin dacique par une forme de moindre taille, considérée comme une sous-espèce chronologique (*T. boisvilletti dacicum* Radulesco), mise en évidence, à la limite Pliocène/Pléistocène, dans les faunes de Slatina-3 et de Tetoiu-1 (= Bugiuleşti-1) (Andreesco et al. 1981; Conovici et Radulesco 1972).

3.2.7. Clethrionomys cf. kretzoii (Kowalski)

Matériel: M^1.

La pièce qui nous occupe se distingue de l'ensemble des M^1 d'Izvoru par ses dimensions plus réduites (tabl. 4). Comme il s'agit d'un exemplaire encore jeune, les champs de dentine présentent de légères communications entre eux; une confluence plus marquée s'observe entre T1—T2 (fig. 4.7a); une quantité réduite de cément colmate les angles rentrants; l'émail, relativement épais, n'est pas differencié. En vues laterales interne (fig. 4.7c) et externe, on constate que les échancrures de l'émail sont modérément hautes; en revanche, l'échancrure distale (au contact avec la molaire suivante) a établi une discontinuité dans l'émail de la face occlusale. Bien que les racines ne soient pas complètement constituées, la molaire était, fort probablement, triradiculée.

On trouve très peu d'informations sur M^1 de l'espèce qui nous intéresse. ALEXANDROVA (1976) mentionne pour une M^1 de Liventsovka (horizon 1) la présence de 3 racines et la confluence visible entre les prismes dentaires; l'espèce est rapportée au genre *Pliomys (P. kretzoii)*.

Dans le Bassin dacique, *C. kretzoii* est mieux documenté dans l'association de Drăgăneşti-Olt (FERU et al. 1979). Une M^1 de cette dernière localité, qui nous sert de terme de référence, présente le dessin de la face occlusale très ressemblant à celui de la pièce d'Izvoru, mais ses dimensions sont un peu plus réduites. Les étroites similitudes entre les deux pièces examinées nous font croire qu'une forme rapportable à *C. kretzoii* est présente aussi dans la faune d'Izvoru. En l'absence de M_1 qui est la pièce la plus caractéristique dans l'identification de l'espèce, la détermination de l'exemplaire d'Izvoru ne peut être faite en toute certitude.

3.2.8. Mimomys coelodus cf. coelodus KRETZOI

Matériel: 9 M_1, fragment de mandibule avec M_{1-2}, M^3.

M_1 — Dans un travail antérieur (RADULESCO et SAMSON 1983), nous avons décrit sous ce nom un lot de M_1 (N = 7) caractérisées par la persistance des structures mimomyennes (puiselet, pli mimomyen) et la présence d'une quantité modérée de cément dans les angles rentrants (fig. 4.1a—2a); le degré d'hypsodontie est accusé (fig. 4.1b—2b). Sur le seul exemplaire suffisamment jeune pour posséder encore les discontinuités de l'émail sous la face occlusale (fig. 4.1b), le paramètre E de CHALINE (1974) a une valeur élevée (4,20); il en est de même des paramètres Ea (4,10) et Eb (3,70) de VAN DE WEERD (1976).

Rappelons que *M. coelodus alutae* RADULESCO & SAMSON, décrit de Drăgăneşti-Olt, qui est moins évolué que la forme d'Izvoru, se caractérise par des valeurs plus basses des paramètres susmentionnés (RADULESCO et SAMSON 1983).

Nous avons attribué l'échantillon d'Izvoru, compte tenu de la taille (tabl. 3) et de la morphologie de M_1, à la sous-espèce nominative. Nous suggérons la possibilité de rapporter à la même forme encore 3 M_1 endommagées par des corrosions latérales dont les longueurs sont les suivantes: 2,85—3,00—3,05.

M_2 — Relevons la différenciation de l'émail et la confluence partielle entre T1—T2 (fig. 4.3a). Il est intéressant de relever la structure de la racine distale de cette molaire et sa position par rapport à l'incisive. A la différence des formes de *Mimomys* plus archaïques, la racine distale de M_2 d'Izvoru ne se situe plus sur l'incisive *(Acrorhiza)*; elle se déplace vers l'extérieur, mais sa partie linguale garde un contact très étroit avec le bord labial de l'incisive; dans cette phase de sa transformation morphologique, la racine distale de M_2 est raccourcie et présente le bord inférieur oblique (fig. 4.3d).

M^3 — La pièce est endommagée par des corrosions latérales. Comparativement à M^3 de *M. reidi*, cet exemplaire conserve le puiselet; le premier synclinal externe est plus profond, isolant mieux le premier triangle par rapport au lobe antérieur. La taille est plus grande (tabl. 6).

3.2.9. Mimomys reidi HINTON

Matériel: fragment de mandibule avec M_1, 2 M^3.

M_1 se caractérise par la réduction des structures mimomyennes; le puiselet est absent (fig. 5.1a); le repli insulaire est peu développé verticalement d'où, en vue latérale externe, une remarquable différence de hauteur par rapport à la base de la gouttière mimomyenne (fig. 5.1b). Les échancrures latérales de l'émail sont élevées; celle du côte interne, moins haute, n'a pas encore déterminé une interruption de l'émail sur la face occlusale du lobe postérieur. La boucle antérieure est relativement étroite, peu déjetée vers le côté lingual. Une quantité modérée de cément colmate les angles rentrants; l'émail est légèrement differencié (fig. 5.1a).

Les dimensions de la pièce sont indiquées dans le tabl. 3.

Il est important de mentionner afin de fixer les particularités de l'espèce que la racine distale de M_2, en jugeant par son alvéole, s'appuyait partiellement sur l'incisive.

M^3 — La pièce la mieux conservée consiste en deux champs de dentine, séparés entre eux au niveau du premier synclinal interne et du deuxième synclinal externe dont les extrémités

Tableau 3. Mensurations des M_1 d'Arvicolidés

	Iz-2/nos	L	l	Lca	lba	lca	H_1	H_2	Hri	Hgm
Mimomys coelodus cf. *coelodus*										
	004	2,85	1,35	1,20	—	1,17	2,90	2,65	—	2,50
	005	2,90	1,25	1,30	—	1,20	4,10	3,80	—	—
	002	2,98	1,28	1,25	0,85	1,21	4,35	4,20	3,05	3,90
	006	3,05	1,38	1,25	0,78	1,22	2,30	2,05	—	1,90
	003	3,05	1,46	1,25	—	1,30	4,15	4,00	—	3,90
	001	3,10	1,28	1,32	0,78	1,15	3,30	3,15	2,45	3,25
	007	—	1,30	1,25	0,80	1,18	—	3,65	—	3,00
Mimomys reidi										
	062	2,70	1,25	1,10	0,60	1,14	3,35	3,35	0,66	3,05
Mimomys pitymyoides										
	066	2,70	1,20	1,12	0,62	1,02	1,35	1,20	0,82	1,05
Kislangia rex										
	072	4,35	2,15	1,85	1,25	1,85	4,55	4,25	—	—
Borsodia cf. *hungarica*										
	076	2,64	1,09	1,20	0,60	1,04	3,70	3,40	2,60	3,15
	077	—	1,05	1,10	0,30	0,95	—	—	1,68	—

Tableau 4. Mensurations des M^1 d'Arvicolidés

	Iz-2/nos	L	l	H_2	Hai	Hpi
Clethrionomys cf. *kretzoii*	065	2,08	1,03	2,75	2,05	2,10
Mimomys coelodus/reidi	012	2,28	1,33	2,45	—	—
	013	2,37	1,45	3,35	2,65	—
	014	2,47	1,40	4,30	3,30	3,55
	015	2,50	1,50	4,10	3,40	3,25
	016	2,50	1,50	2,25	—	—
	017	2,50	1,44	3,95	—	—
	018	2,50	1,45	2,25	—	—
	019	2,52	1,55	2,00	—	—
	020	2,55	1,48	2,20	—	—
	021	2,55	1,50	2,25	—	—
	022	2,55	1,55	2,75	—	—
	023	2,55	1,55	4,00	3,35	—
	024	2,55	1,57	4,00	—	—
	025	2,55	1,58	3,40	—	—
	026	2,57	1,42	3,90	—	3,70
	027	2,58	1,53	2,80	2,60	—
	028	2,58	1,55	4,40	—	—
	029	2,62	1,57	2,30	—	—
	030	2,63	1,55	3,70	3,40	3,45
	031	2,67	1,55	2,50	—	—
Mimomys pitymyoides	067	2,30	1,34	—	—	—
Kislangia rex	073	—	2,50	—	—	—
Borsodia cf. *hungarica*	078	2,20	1,25	3,65	—	—
	079	2,27	1,35	3,30	—	—
	080	2,55	1,45	3,15	—	—

sont à peu près en contact sur la ligne médiane de la face occlusale; le puiselet est absent; l'émail est différencié; la quantité de cément est réduite (fig. 5.2a). Une autre pièce, plus usée, présente une morphologie similaire.

Les dimensions sont indiquées dans le tabl. 6.

Par son degré d'hypsodontie, l'absence de puiselet, la différence importante de hauteur entre le repli insulaire et la gouttière mimomyenne, M_1 d'Izvoru évoque de près la pièce type de *M. reidi* de Weybourne Crag (HINTON 1926).

Les rapports entre la racine distale de M_2 et l'incisive montrent que la forme d'Izvoru est plus archaïque comparativement à *M. pusillus* (MÉHELY). Notons que cette dernière espèce, qui est considérée comme très proche ou parfois même synonyme de *M. reidi*, possède la racine distale de M_2 en position labiale par rapport à l'incisive (MÉHELY 1914 et observations sur 4 mandibules de *M. pusillus* de Betfia-2, localité type de l'espèce).

On fait généralement crédit à l'idée que *M. reidi* possède M^1 à 3 racines, à la différence de *M. pusillus* dont M^1 est biradiculée (VON KOENIGSWALD 1977). Il est nécessaire de montrer que cette caractéristique de *M. reidi* est déduite de la structure de M^1 de *M. petenyi*, espèce considérée à tort comme synonyme de *M. reidi* (HINTON 1926). Or il est démontré à présent que *M. petenyi* constitue un maillon dans la lignée évolutive qui conduit à *M. hungaricus*, cette lignée particulière étant distinguée sous le nom sousgénérique de *Borsodia* (JÁNOSSY & VAN DER MEULEN 1975). Etant donné son degré général d'évolution, *M. reidi* du Pléistocène inférieur pourrait bien posséder M^1 à 2 racines au niveau chronologique d'Izvoru-2.

En ce qui concerne M^3, les 2 pièces d'Izvoru ressemblent à M^3 de *M. pusillus* (Méhely, 1914). Elles ont été rattachées, toutefois, à *M. reidi* en raison de la structure de l'alvéole de M_2 mentionnée plus haut.

3.2.10. Mimomys coelodus/reidi

Matériel: fragment de mandibule avec M_1—M_2, 5 M_1, 10 M_2, 6 M_3, 20 M^1, 14 M^2.

M_1 — Plusieurs exemplaires fortement corrodés peuvent être rapportés à *M. coelodus* ou à *M. reidi*; leurs longueurs mesurent approximativement: 2,7—2,8—2,85—2,9—3,0—3,05.

M_2 — Les pièces se distinguent généralement par la différenciation de l'émail et la présence du cément dans les angles rentrants; une communication variable, jamais trop ample, existe toujours entre T1 et T2. La racine postérieure, à cause du contact avec l'incisive, est plus courte que la racine antérieure et présente le bord interne tronqué (fig. 4.4d).

M_3 — Ainsi que sur M_2, l'émail peut être différencié; le cément est présent; les deux triangles d'émail sont largement confluents.

M^1 — Ces pièces se caractérisent par la tendance manifeste à l'isolement des triangles, bien que des communications variables entre les champs de dentine subsistent; l'émail est différencié; il y a toujours du cément dans les synclinaux (fig. 4.5a). Dans tous les cas où les racines sont conservées, elles sont au nombre de deux ($N = 8$); il y a donc fusion entre la racine antérieure et celle du côté interne.

Les échancrures de l'émail sont bien développées (fig. 4.5c); sur la plupart des pièces, elles établissent des discontinuités dans l'émail de la face occlusale.

L'examen des dimensions (tabl. 4) indique une variation de la longueur comprise entre 2,28—2,67. La fréquence plus grande de *M. coelodus*, en jugeant par le nombre de M_1, nous fait croire que la majorité des M^1, étant donné aussi leur taille relativement forte, appartiennent à cette espèce. Rappelons que MÉHELY (1914) indique pour la longueur de M^1 de *M. pusillus* (espèce de taille similaire à *M. reidi*) une variation comprise entre 2,1—2,3. Il est donc possible que les M^1 de moindre taille d'Izvoru (2,28—2,37—2,47—2,50?) appartiennent à *M. reidi*.

M^2 — En ce qui concerne la morphologie, les exemplaires se distinguent par les mêmes particularités que les M^1 décrites précédemment: tendance à l'isolement des prismes dentaires (fig. 4.6a), hypsodontie accentuée (fig. 4.6b—c), présence de cément.

Les dimensions sont indiquées dans le tabl. 5.

3.2.11. Mimomys pitymyoides JÁNOSSY & VAN DER MEULEN

Matériel: M_1, M^1, 3 M^2.

M_1 — La pièce présente un degré d'abrasion avancé; en vue occlusale (fig. 5.3a), on observe une large confluence entre T2—T3; une communication importante existe entre T1—T2; le complexe antérieur se distingue par la persistance du pli mimomyen et par la relative étroitesse de la boucle mésiale, bien qu'un faible angle saillant interne soit indiqué; la base de la boucle antérieure est marquée par un léger rétrécissement des bandes d'émail; une quantité réduite de cément est visible dans les angles rentrants qui sont assez larges par rapport aux angles saillants; l'émail est différencié; en vue latérale externe, on constate que le repli insulaire est profond.

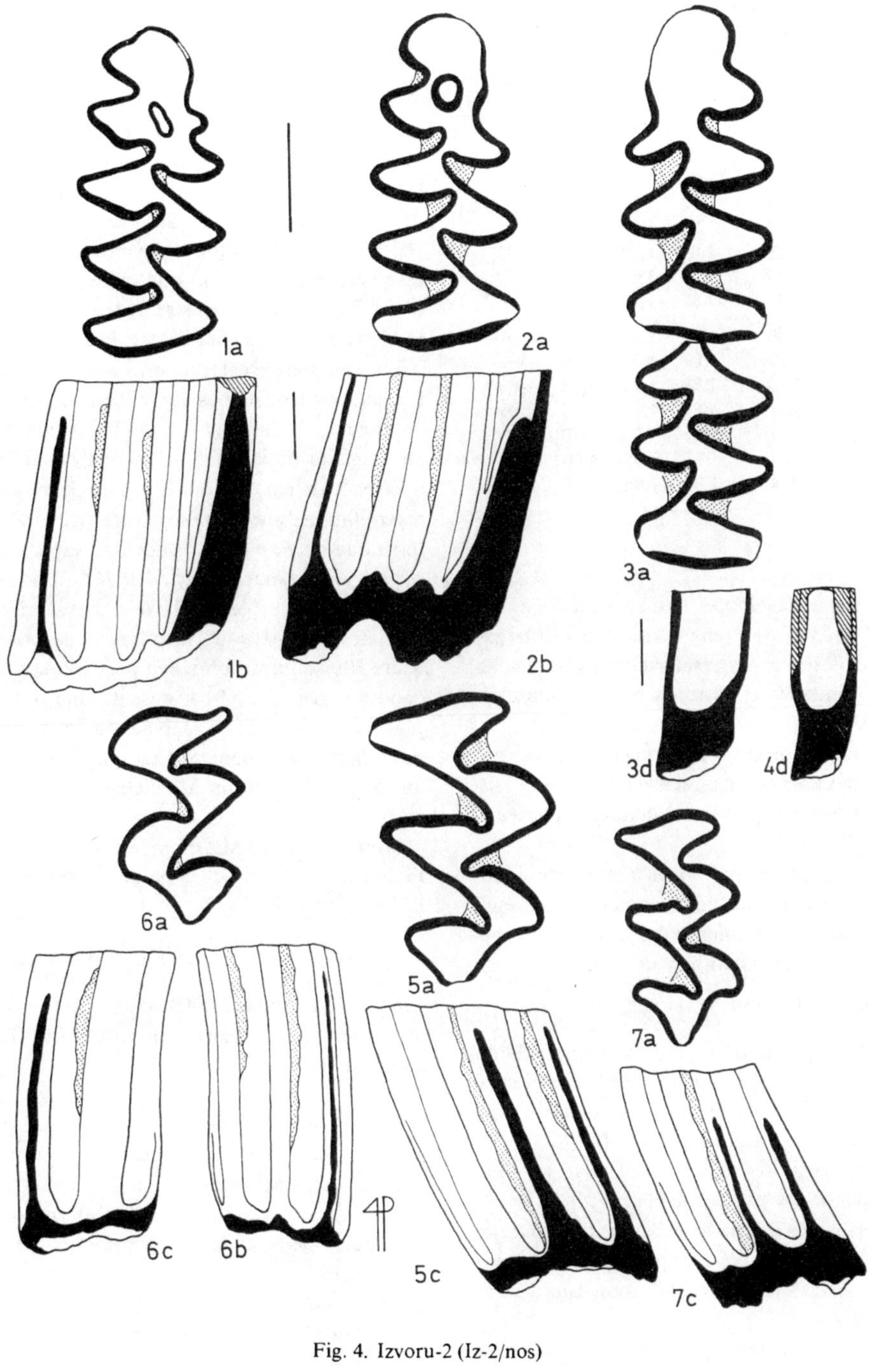

Fig. 4. Izvoru-2 (Iz-2/nos)

1—3. *Mimomys coelodus* cf. *coelodus*: 1 — M_1 dext./002, 2 — M_1 dext./001, 3 — M_1—M_2 sin./006; 4—6. *Mimomys coelodus/reidi*: 4 — M_2 sin./046, 5 — M^1 dext./030, 6 — M^2 sin./038; 7 — *Clethrionomys* cf. *kretzoii*, M^1 dext./065

Les dimensions sont indiquées dans le tabl. 3.

M^1 — Par la communication importante entre les deux triangles anterieurs, cet exemplaire peu usé, se rattache à la même espèce. La longueur de la pièce est modérée (tabl. 4).

M^2 — Les trois exemplaires montrent la confluence très large, caractéristique entre T2—T3 (fig. 5.4a). Leurs dimensions sont indiquées dans le tabl. 5. Mentionnons que le degré d'hypsodontie est remarquable; les échancrures de l'émail sont très hautes (fig. 5.4b) apparaissant sous forme de discontinuités sur la face occlusale.

Par leurs particularités morphologiques, les exemplaires d'Izvoru ressemblent de près aux pièces homologues d'Osztramos-3, localité type de l'espèce (Janossy et van der Meulen 1975). Relevons, toutefois, la persistance du pli mimomyen jusqu' à un degré d'abrasion avancé sur la seule M_1 d'Izvoru. Les M^2 de la localité roumaine ont, comparativement, les champs de dentine mésial et distal, mieux isolés par rapport au rhombe formé de T2—T3.

3.2.12. Kislangia rex (Kormos)

Matériel: M_1, M_2, M^1 fragment, M^3.

M_1 — En vue occlusale, la molaire présente T4 à peu près isolé; la boucle mésiale est aplatie; le quatrième angle

Tableau 5. Mensurations des M^2 d'Arvicolidés

	Iz-2/nos	L	l	H_2	Hpi
Mimomys coelodus/reidi	032	1,75	1,15	3,00	—
	033	1,87	1,15	2,75	—
	034	1,92	1,25	3,70	3,20
	035	1,92	1,27	3,10	—
	036	2,00	1,30	1,80	—
	037	2,00	1,35	2,10	—
	038	2,02	1,22	3,95	3,25
	039	2,07	1,30	2,80	—
	040	2,07	1,37	2,30	—
Mimomys pitymyoides	068	1,80	1,12	3,40	—
	069	1,87	1,09	3,30	—
	070	1,88	1,12	3,15	—
Borsodia cf. *hungarica*	081	1,80	1,10	2,90	2,90
	082	1,81	1,06	3,70	—
	083	1,85	0,95	4,00	2,80
	084	1,87	1,05	3,50	3,10

rentrant lingual est bien développé; les structures mimomyennes semblent absentes; les angles rentrants internes sont profonds, étroits, dirigés transversalement; le chevauchement des synclinides internes et externes est bien marqué; le cément est abondant (fig. 5.5a).

La molaire se distingue par ses fortes dimensions (tabl. 3) (la pièce ayant subi des érosions latérales dans sa partie supérieure, les mensurations se rapportent à la moitié inférieure de la couronne).

M_2 — La dent se remarque par l'isolement accusé des prismes de dentine et leur aplatissement antéro-postérieur; une grande quantité de cément colmate les angles rentrants qui sont étroits et se chevauchent visiblement (fig. 5.6a).

Dimensions: L = 2,65; l = 1,95; H_1 = 2,25; H_2 = 2,40

M^1 — Conservée en partie dans un fragment de maxillaire, la molaire se distingue par l'aplatissement des triangles, le chevauchement puissant des synclinaux et la quantité importante de cément; la taille est grande (tabl. 4).

M^3 — Ainsi que sur les autres pièces, on observe sur cette molaire la nette tendance des prismes dentaires à s'isoler; le puiselet est présent; la partie distale de la dent est raccourcie; le lobe antérieur aplati est bien développé transversalement (fig. 5.7a). La dent était biradiculée; ses dimensions sont indiquées dans le tabl. 6.

La forme d'Izvoru se caractérise par ses dimensions très grandes, la longueur de M_1 étant supérieure aux valeurs (3,9—4,2) indiquées pour les pièces homologues de *K. rex* de Villány-3 (station type) (KORMOS 1934). Par leur morphologie, les molaires de la localité roumaine sont tout à fait comparables aux pièces correspondantes de Hongrie.

Tableau 6. Mensurations des M^3 d'Arvicolidés

	Iz-2/nos	L	l	H_2
Mimomys coelodus cf. *coelodus*	008	1,95	1,00	2,65
Mimomys reidi	063	1,78	0,95	1,80
	064	1,80	0,85	1,37
Kislangia rex	075	2,60	1,67	2,45

3.2.13. Borsodia cf. hungarica (KORMOS)

Matériel: 2 M_1, M_3, 3 M^1, 5 M^2.

M_1 — Une pièce est bien conservée (fig. 5.8), l'autre est incomplète (fig. 5.9). En vue occlusale, les deux exemplaires se caractérisent par des triangles plutôt aigus, à faible développement antéro-postérieur et des angles rentrants larges, dépourvus de cément; le bord externe du complexe antérieur possède une duplication assez importante; l'émail est d'épaisseur uniforme (fig. 5.9a) ou différencié selon le type *Microtus* (fig. 5.8a). Relevons encore la tendance à la formation d'un faible angle saillant lingual sur la boucle antérieure.

En vue latérale externe, on observe que les échancrures de l'émail sont très élevées sur le lobe postérieur et sur la boucle antérieure; la hauteur du repli insulaire est variable; sur l'exemplaire de la fig. 5.8b le repli insulaire est profond, la différence de hauteur qui le sépare de la base du repli mimomyen étant réduite; cette structure est considérée comme caractéristique de *B. petenyi* (MÉHELY), espèce plus primitive que *B. hungarica* (ALEXANDROVA 1976; ZAZHIGIN 1980). Dans le cas précis de la pièce d'Izvoru, nous préférons parler d'un morphotype „*petenyi*", expression de la variabilité dans le cadre d'une forme plus évoluée. Remarquons que la pièce endommagée (fig. 5.9b) présente une différence de hauteur plus importante entre la base du repli insulaire et celle du repli mimomyen, dénotant un type plus progressif, conforme au degré d'évolution de l'ensemble des pièces dentaires d'Izvoru.

Notons sur la M_1 complète que l'extrémité apicale de l'échancrure interne du lobe postérieur se trouve à 0,18 sous la face occlusale.

Les dimensions de M_1 sont indiquées dans le tabl. 3.

M_3 — La structure deltodonte est visible, mais encore imparfaite (communication entre T1—T2). La longueur atteint 1,30.

M^1 — En vue occlusale, les triangles apparaissent aigus et les synclinaux larges; l'émail peut être différencié dans le sens *Microtus* (fig. 5.10a); les interruptions de l'émail sont hautes: sur 2 M^1 elles ont établi des discontinuités sur la face occlusale; sur l'autre M^1 l'interruption antéro-externe est séparée de la surface d'abrasion par un intervalle de 0,58 (fig. 5.10b).

Les dimensions sont indiquées dans le tabl. 4.

M^2 — Les exemplaires présentent les mêmes particularités de structure que M^1. Les interruptions de l'émail sont bien développées; dans un cas, l'interruption linguale est moins haute (fig. 5.11c).

Les mensurations des pièces sont indiquées dans le tabl. 5.

La forme d'Izvoru-2, par la morphologie des molaires et leur degré d'hypsodontie, a les plus grandes affinités avec *B. hungarica*, surtout avec les morphotypes à duplication plus exprimée sur le bord externe du complexe antérieur de M_1 (KORMOS 1938), particularité qui confère à nos pièces une nuance conservative.

Soulignons qu'un terme de comparaison très utile nous à été fourni par le matériel dentaire de Drăgăneşti-Olt, déterminé comme *Borsodia* cf. *lagurodontoides* (CHEVTCHENKO) (FERU et al. 1979), dont le degré d'évolution est moins exprimé qu'à

Fig. 5. Izvoru-2 (Iz-2/nos)

1, 2. *Mimomys reidi*: 1 — M_1 dext./062, 2 — M^3 sin./063; 3, 4. *Mimomys pitymyoides*: 3 — M_1 dext./066, 4 — M^2 dext./068; 5—7. *Kislangia* rex: 5 — M_1 dext./072, 6 — M_2 dext./073, 7 — M^3 dext./075; 8—11: *Borsodia* cf. *hungarica*: 8 — M_1 dext./076, 9 — M_1 dext./077, 10 — M^1 dext./078, 11 — M^2 sin./083

Izvoru-2. En effet, à Drăgănești-Olt les molaires ont les échancrures de l'émail moins élevées et parfois M^1 est à trois racines.

4. Chronologie

4.1. Caractéristiques de la faune

Dans l'ensemble, la faune d'Izvoru représente une étape évoluée dans la succession des faunes de Mammifères du Pléistocène inférieur (phase II du Pléistocène inférieur in: Samson et Radulesco 1973). Elle succède aux faunes de Slatina-3 (Mollusques et Mammifères associés, paléomagnétisme normal = événement d'Olduvai) (Andreesco et al. 1981) et de Tetoiu-1 (Pléistocène inférieur, phase I in: Samson et Radulesco 1973). La limite Pliocène/Pléistocène, en Roumanie, passe sous ce niveau (Andreesco et al. 1981; Feru et al. 1978; Samson et Radulesco 1973).

4.1.1. Izvoru-1

En nous rapportant au Bassin dacique, le niveau fossilifère d'Izvoru-1 correspond, comme nous l'avons montré (Feru et al. 1983), aux faunes de Tetoiu-2/Irimeşti, Milcovu-2, Frăteşti-2 (gros Mammifères) et de Drăgăneşti-Olt (Micromammifères). Nous rappelons, comme caractéristiques de ce niveau à *Archidiskodon meridionalis meridionalis*, un *Trogontherium* de grande taille (*T. boisvilletti boisvilletti*), l'association de deux Equidés sténoniens de taille différente (*Allohippus stenonis mitilanensis* SAMSON, *Allohippus* ex gr. *major* (BOULE)) (SAMSON 1976), une forme primitive de *Allocaenelaphus*, *Praealces gallicus* (AZZAROLI) et *Megalovis*. La plupart de ces formes ainsi que *Paracamelus alutensis* (STEFANESCO), *Allactaga*, *Borsodia* cf. *lagurodontoides*, *Clethrionomys kretzoii*, *Mimomys newtoni* MAJOR et la fréquence plus grande de *M. pliocaenicus* (MAJOR) peuvent indiquer une détérioration climatique, allant dans le sens de l'accentuation du cachet continental. Ajoutons que la faune de Drăgăneşti-Olt, par le degré d'évolution des Rongeurs, précède celle d'Izvoru-2 (RADULESCO et SAMSON 1983), comme nous allons voir ci-dessous.

4.1.2. Izvoru-2

La faune d'Izvoru-2 n'a pas encore un équivalent en Roumanie (FERU et al. 1983); l'apparition de *Citellus primigenius*, *Mimomys pitymyoides* et *Kislangia rex* lui est propre; le Castoridé est le même que dans le niveau précédemment discuté (*T. boisvilletti boisvilletti*). Le phylum de *Mimomys coelodus* se distingue par un haut degré d'évolution d'où son assimilation probable à la sous-espèce nominative (*M. coelodus* cf. *coelodus*). Rappelons que dans l'association de Drăgăneşti-Olt, le même phylum est représenté par une forme moins évoluée (*M. coelodus alutae* RADULESCO & SAMSON). En ce qui concerne le genre *Borsodia* (JÁNOSSY & VAN DER MEULEN 1975), précisons que la forme d'Izvoru-2 (*B.* cf. *hungarica*) est plus progressive que celle de Drăgăneşti-Olt (*B.* cf. *lagurodontoides*).

Les Mollusques (*Bogatschevia sturi*) ainsi que les Mammifères indiquent une amélioration du climat pendant la formation du gisement d'Izvoru-2 (ANDREESCO et al. 1981, TCHÉPALYGA 1980), mais conservant encore une nuance steppique (*Citellus*, *Borsodia*).

Si nous nous rapportons maintenant à l'ensemble des Arvicolidés, il faut souligner que parmi les 110 molaires découvertes, complètes ou fragmentaires (tabl. 1), aucune n'indique l'existence de formes arhizodontes; dans ce sens nous avons en vue les espèces de *Allophaiomys* et de *Lagurodon*[1]).

Cette constatation pose l'alternative suivante: ou

— la faune d'Izvoru-2 précède l'apparition des Campagnols arhizodontes, ou
— la faune d'Izvoru-2 est contemporaine des formes arhizodontes, mais celles-ci n'ont pas encore atteint le sud de la Roumanie (ou sont-elles extrêmement rares?).

Nous allons discuter ce problème plus loin.

4.2. Corrélations

Les corrélations que nous avons établies, au niveau continental (FERU & al. 1983), se rapportent principalement aux schémas stratigraphiques de l'Union Soviétique (NIKIFOROVA et al. 1976, TCHÉPALYGA 1980), des Pays Bas (VAN DER MEULEN 1976; VAN DER MEULEN et ZAGWIJN 1974; ZAGWIJN 1974a, b) et de la France (CHALINE 1973; GUÉRIN 1980; HEINTZ 1968), mais non sans prendre en discussion certaines stations très connues de Hongrie (KRETZOI 1956, VAN DER MEULEN 1976) et de Pologne (KOWALSKI 1975a).

4.2.1. Izvoru-1

La faune d'Izvoru-1 peut être située à la partie supérieure de l'horizon domachkinien du complexe faunique odessan (Psekups et niveau moyen de Saratovskaïa), au début de l'Eburonien, dans la partie moyenne de la zone de Senèze (la localité type) ou de la zone 18. Il faut mentionner que la zone de Brielle de CHALINE (1973) (apparition de *Allophaiomys pliocaenicus*) n'est pas l'équivalent de la zone de Senèze comme l'indique GUÉRIN (1980), parce qu'elle correspond à la partie supérieure de l'Eburonien (VAN DER MEULEN et ZAGWIJN 1974), étant par conséquent contemporaine de la zone de Peyrolles.

4.2.2. Izvoru-2

La faune d'Izvoru-2 pourrait correspondre au début de l'horizon bochernitsien du complexe faunique odessan (avant l'apparition des Campagnols arhizodontes du niveau supérieur de Kryjanovka), à une phase de réchauffement de l'Eburonien (antérieurement à Brielle), à la partie supérieure de la zone de Senèze ou de la zone 18. Dans cette étape, au moins dans les régions situées au sud des Carpates, comme nous l'avons montré, les Arvicolidés arhizodontes semblent absents.

En ce qui concerne l'Europe centrale, les rapports avec les sites fossilifères de Hongrie, placés pour la plupart dans des zones karstiques, ne sont pas faciles à établir. Il s'agit surtout de Villány-3 et Villány-5, dont les associations des Arvicolidés à molaires radiculées, révisées récemment par VAN DER MEULEN (1976), sont presque superposables à celle d'Izvoru-2, à l-exception de *Mimomys pliocaenicus* qui manque dans la localité roumaine; d'ailleurs, les représentants du phylum de *M. pliocaenicus* sont extrêmement rares dans le Bassin dacique pendant le Pliocène supérieur et le Pléistocène inférieur, sauf dans le cas du gisement de Drăgăneşti-Olt. En général, ce phylum est remplacé par celui de *Mimomys minor-M. coelodus* (RADULESCO et SAMSON 1983) dont l'évolution est parallèle au phylum de *Mimomys polonicus-M. pliocaenicus*.

Comme cette différence un'est pas essentielle, il en reste une autre plus importante: l'apparition des Arvicolidés arhizodontes, *Lagurodon arankae* KRETZOI à Villány-3 et *Allophaiomys deucalion* KRETZOI à Villány-5. Ce qui complique encore le problème, c'est que VAN DER MEULEN (1976) fait correspondre Villány-3 au Tiglien C5 (niveau à Micromammifères) et Villány-5 au sommet du Tiglien. Par conséquent, les deux localités hongroises seraient antérieures aussi bien à Izvoru-2 qu'à Drăgăneşti-Olt, bien que, répétons-le, ces derniers sites ne contiennent pas des Campagnols arhizodontes.

---

[1]) Les Lemmings (*Lemmus, Dicrostonyx*) sont exclus de nos propos, parce qu'ils n'ont jamais pénétré dans les zones méridionales de la Roumanie.

On revient à l'alternative que nous avons soulevée dans le paragraphe précédent, mais non sans faire quelques observations sur les gisements discutés. Les voici:

— en dehors des listes fauniques (KRETZOI 1956; VAN DER MEULEN 1976), une description adéquate des espèces, concernant surtout leur degré d'évolution, manque complètement, ce qui empêche toute comparaison utile;

— l'homogénéité des faunes peut être mise en doute, étant donné les conditions de l'exploitation, sans prendre garde à une possible stratification (FERU & al. 1978) dans des dépôts parfois très épais (18 m à Villány-3) (KRETZOI 1956, p. 183); ajoutons que *A. deucalion* provient de la partie sommitale de Villány-5 (KRETZOI 1956, p. 188);

— d'après des recherches récentes, les Arvicolidés arhizodontes, aussi bien les Lagurinés *(Lagurodon)* que les Microtinés *(Allophaiomys pliocaenicus)* font leur apparition, dans la Plaine russe (niveau supérieur de Kryjanovka), dans l'horizon bochernitsien (NIKIFOROVA et al. 1976) lequel, situé vers la fin de l'Eburonien (donc contemporain de Brielle) mais principalement dans le Waalien A, est nettement plus tardif que l'actuel encadrement chronologique de Villány-3 et Villány-5;

— comme *A. deucalion* représente une forme primitive, on doit accepter son apparition avant celle de *A. pliocaenicus*, donc antérieurement à Kryjanovka (niveau supérieur) et à Brielle, mais il n'est peut-être pas nécessaire de le descendre jusqu'à la fin du Tiglien.

Nous reviendrons sur cet encadrement, plus loin, quand nous allons discuter les gisements de Pologne.

En nous rapportant, encore une fois, à la Hongrie, nous retenons le site de Kisláng, qui pourrait être assez rapproché de celui d'Izvoru-2. En effet, aussi bien l'association que le degré évolutif des Arvicolidés[2]) (KRETZOI 1954) présentent des similitudes dans les deux localités; mais, la présence de *L. arankae* à Kisláng paraît indiquer un âge géologique un peu plus récent du site hongrois.

En ce qui concerne la Pologne, il est vraisemblable de situer Izvoru-2 près du niveau chronologique de Kadzielnia (KOWALSKI 1958), bien qu'il y ait peu d'éléments (*Clethrionomys kretzoii, Mimomys pitymyoides*) de part et d'autre qui permettent une mise en parallèle satisfaisante. Dans un travail antérieur (FERU & al. 1979), nous avons considéré, compte tenu de l'absence de formes arhizodontes, que le niveau d'Izvoru-2 peut précéder Kadzielnia où *A. deucalion* est présent (KOWALSKI 1977; VAN DER MEULEN 1976). Il nous paraît nécessaire maintenant d'apporter quelques retouches à cette manière de voir. Comme nous l'avons déjà montré (FERU & al. 1979) il y a beaucoup de ressemblances fauniques entre Drăgăneşti-Olt et Kadzielnia, la localité polonaise étant toutefois un peu plus tardive par l'apparition de *Allophaiomys* qui manque dans le Bassin dacique. Il est possible que les deux localités qui nous occupent s'inscrivent, en succession, dans le même intervalle de dépression climatique (présence de *Lemmus* à Kadzielnia). Or, les Rongeurs d'Izvoru-2 sont plus évolués que ceux de Drăgăneşti-Olt. Il ne serait donc pas exclu que l'association d'Izvoru-2, qui est sûrement ultérieure à celle de Drăgăneşti-Olt et s'inscrit dans une phase de réchauffement, soit aussi un peu plus tardive que la faune de Kadzielnia, succédant ainsi à la dépression climatique reconnue à Drăgăneşti-Olt et dans le site polonais. De toute façon, Izvoru-2 représente dans l'évolution des Rongeurs une étape antérieure au niveau chronologique marqué par la faune de Kamyk où les formes arhizodontes sont non seulement plus fréquentes, mais encore plus évoluées (*A. pliocaenicus*) (KOWALSKI 1960, 1975a).

Nous rencontrons en ce qui concerne l'encadrement chronologique de Kadzielnia les mêmes difficultés que dans le cas des gisements de Villány-3 et Villány-5. KOWALSKI (1975b) situe la localité polonaise, également, dans le Tiglien.

En rappelant, une dernière fois, l'alternative concernant la place chronologique du gisement d'Izvoru-2, même si on accepte le second terme de cette alternative, précisons que sa faune ne pourrait être plus récente que celles avec *A. deucalion* et en aucun cas aussi ancienne que le Tiglien. Par suite il reste, en attendant une description détaillée du stade évolutif des associations fauniques des sites hongrois et polonais, à repenser soit leur encadrement chronologique, soit leur homogénéité.

Sommaire

Le travail présente le contenu paléontologique des deux niveaux fossilifères mis en évidence dans la formation fluvio-lacustre d'Izvoru, localité située dans le Bassin dacique à 11 km ouest de Slatina. Le niveau inférieur, désigné sous le nom d'Izvoru-1, a livré quelques ossements de gros Mammifères ((*Archidiskodon meridionalis, Allocaenelaphus* sp.). Le niveau supérieur, nommé Izvoru-2, a fourni surtout des restes de Micromammifères renfermant entre autres *Citellus primigenius, Mimomys coelodus* cf. *coelodus, M. pitymyoides, Kislangia rex* et *Borsodia* cf. *hungarica*. Soulignons que les Arvicolidés ne sont représentés que par des formes rhizodontes. Notons encore la présence sur ce niveau d'une forme grande de *Trogontherium (T. boisvilletti boisvilletti)*. Les Mollusques associés aux Micromammifères comportent *Bogatschevia sturi, Wenziella wilhelmi* et *W. pseudosturi*.

A l'échelle du Bassin dacique, Izvoru-1 peut être mis en parallèle avec l'horizon fossilifère de Tetoiu-2/Irimeşti (IIe phase du Pléistocène inférieur de Roumanie) (SAMSON et RADULESCO 1973). Dans l'état actuel des recherches, Izvoru-2 n'a pas encore d'équivalent chronologique en Roumanie.

A l'échelle continentale, Izvoru-1 peut être corrélé avec la partie supérieure de l'horizon domachkinien du complexe faunique d'Odessa, le début de l'Eburonien, la zone de Senèze (localité type). Izvoru-2 peut être situé au début de l'horizon bochernitsien du complexe faunique d'Odessa, dans une phase de réchauffement de l'Eburonien (antérieurement à Brielle) et dans la partie supérieure de la zone de Senèze.

Summary

The palaeontological finds of two fossiliferous horizons of fluvio-lacustrine deposits at Izvoru, a village situated 11 km west of Slatina, Dacic basin, are described. In the lower geological horizon — referred to as Izvoru-1 — fossil remains of large mammals have been found (*Archidiskodon meridionalis, Allocaenelaphus* sp.). In the upper horizon — referred to as Izvoru-2 — only small mammals have been excavated including *Citellus primigenius, Mimomys coelodus* cf. *coelodus, M. pitymyoides, Kislangia rex*, and *Borsodia* cf. *hungarica*. It should be noted that the Arvicolidae are only represented by rhizodont species. Furthermore a large form of *Trogontherium (T. boisvilletti boisvilletti)* is known from this horizon. The fossil molluscs of this horizon include *Bogatschevia sturi, Wenziella wilhelmi*, and *W. pseudosturi*.

[2]) Nous ne prenons pas en considération les gros Mammifères qui, de toute évidence, appartiennent à plusieurs niveaux chronologiques distincts.

Izvoru-1 has been correlated within the Dacic basin with the fossil horizon of Tetoiu-2/Irimesti (phase 11 of the Lower Pleistocene of Romania) (SAMSON & RADULESCU 1973). In the present state of research it is not possible to name an exact chronological equivalent within this area for Izvoru-2.

Izvoru-1 is possibly correlated in the continental succession with the upper horizon of the Domashkinian of the Odessa faunistic complex, early Eburonian, zone of Senèze (type locality). Izvoru-2 possibly corresponds to a phase of warming within the Eburonian (prior to horizon of Brielle), upper part of the Senèze zone.

Zusammenfassung

In der vorliegenden Arbeit werden die paläontologischen Funde zwei fossilführender Horizonte der fluviolakustrinen Ablagerungen von Izvoru, einem Ort im Dazischen Becken, der 11 km westlich von Slatina liegt, beschrieben. Im liegenden Horizont I als Izvoru-1 bezeichnet — konnten Reste von Großsäugern (*Archidiskodon meridionalis, Allocaenelaphus* sp.) nachgewiesen werden. Aus dem hängenden Horizont — als Izvoru-2 bezeichnet — liegen nur Kleinsäugerreste vor, darunter *Citellus primigenius, Mimomys coelodus* cf. *coelodus, M. pitymyodes, Kislangia rex* und *Borsodia* cf. *hungarica*. Es ist zu betonen, daß die Arvicoliden nur durch rhizodonte Formen vertreten sind. Weiterhin liegt aus diesem Horizont eine Großform von *Trogontherium (T. boisvilletti boisvilletti)* vor. Die mit den Kleinsäugern assoziierten Mollusken umfassen *Bogatschevia sturi, Wenziella wilhelmi* und *W. pseudosturi*.

Izvoru-1 kann im Raume des Dazischen Beckens mit dem Fundhorizont von Tetoiu-2/Irimesti (II. Phase des Unterpleistozäns in Rumänien) (SAMSON & RADULESCO 1973) korreliert werden. Im gegenwärtigen Untersuchungsstadium kann für Izvoru-2 aus diesem Raum noch kein sicheres chronologisches Äquivalent genannt werden.

Izvoru-1 kann im überregionalen Bereich mit dem obersten Horizont des Domaschkinien des Odessa Faunenkomplexes, frühes Eburonien, Zone von Senèze (Typuslokalität) korreliert werden. Izvoru-2 kann mit dem frühen Bochernitsien des Odessa-Komplexes korreliert werden, das einer Erwärmungsphase des Eburonien entsprechen dürfte (früher als der Horizont von Brielle), oberster Teil der Senèze-Zone.

Резюме

В предлагаемой работе описаны палеонтологические находки из двух костеносных горизонтов озерно-речных отложений из Извору, местности в Дакийском бассейне, которая расположена в 11 км западнее от Слатина. В нижнем горизонте, обозначенном как Извору-1, обнаружены остатки крупных млекопитающих (*Archidiskodon meridionalis, Allocaenelaphus* sp.). В верхнем горизонте, обозначенном как Извору-2, обнаружены остатки только мелких млекопитающих, в том числе *Citellus primigenius, Mimomys coelodus* cf. *coelodus, M. pitymyoides, Kislangia rex* и *Borsodia* cf. *hungarica*. Необходимо отметить, что полевки представлены только корнезубыми формами. Кроме того, в этом горизонте обнаружена крупная форма *Trogontherium (T. boisvilletti boisvilletti)*. В ассоциации с мелкими млекопитающими собраны моллюски, включающие *Bogatschevia sturi, Wenziella wilhelmi* и *W. pseudosturi*.

Извору-1 может быть сопоставлен с самой верхней частью Домашкинского горизонта (Одесский фаунистический комплекс), ранним эбуронием, зоной Сенеза (типовое местонахождение). Извору-2 может быть сопоставлен с нижней частью Бошерницкого горизонта (Одесский фаунистический комплекс), которая может соответствовать фазе потепления в эбуронии (ние горизонта Бриеля); верхняя часть зоны Сенез.

Bibliographie

ALEXANDROVA, L. P., 1976: Rodents of Anthropogene of the European part of the USSR. — In: Acad. Sci. USSR, Tr. geol. Inst. — Moskva **291**. — P. 1—100, En Russe.

ANDREESCO, I.; RADULESCO, C.; SAMSON, P.; TCHÉPALYGA, A. L.; TROUBIKHINE, V. M., 1981: Chronologie (Mollusques, Mammifères, Paléomagnétisme) des formations plio-pléistocènes de la zone de Slatina (Bassin dacique), Roumanie. — In: Trav. Inst. Spéol. „Emile Racovitza". — Bucureşti **20**. — P. 127—137

CHALINE, J., 1973: Biogéographie et fluctuations climatiques au Quaternaire d'après les faunes de rongeurs. — In: Acta zool. Cracov. — Kraków **18** (7). — P. 141—165.

—, 1974: Un nouveau critère d'étude des *Mimomys*, et les rapports de *Mimomys occitanus-Mimomys stehlini* et de *Mimomys polonicus* (Arvicolidae, Rodentia). — In: Acta zool. Cracov. — Kraków **19** (16). — P. 337—355.

—, 1980: *Mimomys salpetrierensis* n. sp., forme relique datée de 14000 B. P. dans la grotte de la Salpétrière (Gard). — In: Geobios. — Lyon **13** (4). — P. 645—651.

CONOVICI, M.; RADULESCO, C., 1972: Sur la découverte d'un nouveau Castoridé (Rodentia, Mammalia) dans le Villafranchien d'Olténie. — In: Trav. Inst. Spéol. „Emile Racovitza". — Bucureşti **11**. — P. 327—333.

FERU, M.; RADULESCO, C.; SAMSON, P., 1978: Biostratigraphie (Micromammifères) des dépôts plio-pléistocènes de la zone de Slatina (dép. d'Olt). — In: Trav. Inst. Spéol. „Emile Racovitza". — Bucureşti **17**. — P. 117—134.

FERU, M.; RADULESCO, C.; SAMSON, P., 1979: Biostratigraphie (Micromammifères) des dépôts plio-pléistocènes du domaine gétique de la Dépression valaque. — In: Trav. Inst. Spéol. „Emile Racovitza". — Bucureşti **18**. — P. 141—169.

FERU, M.; RADULESCO, C.; SAMSON, P., 1983: Succession des Mammifères plio-pléistocènes dans le Bassin dacique (Roumanie). — In: An. Inst. Geol. Geofiz. — Bucureşti **59**. — P. 161—168.

GROMOV, I. M., 1965, In: GROMOV, I. M.; BIBIKOV, D. I.; KALABOUKHOV, N. I.; MEIER, M. N.: La Faune de l'URSS, Mammifères T. 3 (2) (Marmotinae). — Moskva—Leningrad: Nauka. — 468 Pp. — En Russe.

GUÉRIN, C., 1980: Les Rhinocéros (Mammalia, Perissodactyla) du Miocène terminal au Pleistocène supérieur en Europe occidentale. Comparaison avec les espèces actuelles. — In: Docum. Lab. Géol. Lyon. — Lyon **79** (2). — P. 423—783.

HEINTZ, E., 1968: Principaux résultats systématiques et biostratigraphiques de l'étude des Cervidés villafranchiens de France et d'Espagne. — In: C. R. Acad. Sci. — Paris **266**. — P. 2184—2186.

HINTON, M. A., 1926: Monograph on Voles and Lemmings (Microtinae) living and extinct. 1. — London: Brit. Mus. Natur. Hist. — 488 Pp.

JÁNOSSY, D.; MEULEN, A. J. VAN DER, 1975: On *Mimomys* (Rodentia) from Osztramos-3, North Hungary. — In: Kon. nederl. Acad. Wet., Proc. (B). — Amsterdam **78** (5). — P. 381—391.

KOENIGSWALD, W. v., 1977: *Mimomys* cf. *reidi* aus der villafranchischen Spaltenfüllung Schambach bei Treuchtlingen. — In: Mitt. bayer. Staatssamml. Paläont. u. hist. Geol. — München **17**. — P. 197—212.

KORMOS, T., 1934: Neue Insektenfresser, Fledermäuse und Nager aus dem Oberpliozän der Villányer Gegend. — In: Földtani Közlöny. — Budapest **64**. — P. 296—321.

—, 1938: *Mimomys newtoni* F. MAJOR und *Lagurus pannonicus* KORM., zwei gleichzeitige verwandte Wühlmäuse von verschiedener phylogenetischer Entwicklung. — In: Math.-naturwiss. Anz. ung. Akad. Wiss. — Budapest **57**. — P. 356—379.

KOWALSKI, K., 1958: An Early Pleistocene fauna of small mammals from the Kadzielnia Hill in Kielce (Poland). — In: Acta Palaeont. Pol. — Warszawa **3** (1). — P. 1—47.

—, 1960: An Early Pleistocene fauna of small mammals from Kamyk (Poland). — In: Folia Quaternaria. — Kraków **1**. — P. 1—24.

—, 1975: Earliest appearance of lemmings in the Quaternary of Poland. — In: Trabajos sobre Neogeno-Cuaternario, Actas I Coloquio Internacional sobre Biostratigrafia Continental del Neogeno y Cuaternario inferior. — Madrid **4**. — P. 99—104, (1975a).

—, 1975: *Pliomys lenki* in Polen. — In: Acta zool. Cracov. — Kraków **20** (10). — P. 422—427, (1957b).

—, 1977: Fossil Lemmings (Mammalia, Rodentia) from the Pliocene and Early Pleistocene of Poland. — In: Acta zool. Cracov. Kraków **22** (7). — P. 297—317.

KRETZOI, M., 1954: Bericht über die calabrische (villafranchische) Fauna von Kislång, Komitat Fejér. — In: Jahrb. ung. geol. Anst. — Budapest (1953). — P. 239—265.

—, 1956: Die altpleistozänen Wirbeltierfaunen des Villányer Gebirges. — In: Geol. Hung., Ser. Palaeont. — Budapest **27**. — P. 131—264.

MEHELY, L. VON, 1914: Fibrinae Hungariae. Die ternären und quartären Wurzelzähnigen Wühlmäuse Ungarns. — In: Ann. Mus. Nat. Hung. — Budapest **12**. — P. 155—243.

MEULEN, A. J. VAN DER, 1975: Middle Pleistocene small mammals from the Monte Peglia (Orvieto, Italy) with special reference to the phylogeny of *Microtus* (Arvicolidae, Rodentia). — In: Quaternaria. — Roma **17**. — P. 1—144.

—, 1976: The smaller mammal assemblage. — In: Preliminary report on a field campaign in the continental Pleistocene of Tegelen (The Netherlands)/ed. by M. FREUDENTHAL; T. MEIJER; A. J. VAN DER MEULEN. — In: Scripta geol. — Leiden **34**. — P. 14—27.

MEULEN, A. J. VAN DER; ZAGWIJN, W. H., 1974: *Microtus (Allophaiomys) pliocaenicus* from the Lower Pleistocene near Brielle, The Netherlands. — In: Scripta geol. — Leiden **21**. — P. 1—12.

NIKIFOROVA, K. V.; KRASNOV, I. I.; ALEXANDROVA, L. P.; VASSILIEV, I. M.; KONSTANTINOVA, N. A.; TCHÉPALYGA, A. L., 1976: Oscillations climatiques et stratigraphie détaillée des dépôts du Pliocène supérieur-Pléistocène inférieur du sud de l'URSS. — In: XXVe Congr. Internat. Géol., Comm. Géol. Sov. — Moskva. — P. 101—119, En Russe.

RADULESCO, C.; SAMSON, P., 1967: Sur un nouveau Cerf mégacérin du Pléistocène moyen de la Dépression de Braşov (Roumanie). — In: Geol. Romana. — Roma **6**. — P. 317—344.

RADULESCO, C.; SAMSON, P., 1983: La lignée *Mimomys minorcoelodus* (Rodentia, Mammalia) dans le Bassin dacique. — In: Trav. Inst. Spéol. „Emile Racovitza". — Bucureşti **22**. — P. 57—64.

RZEBIK-KOWALSKA, B., 1971: The Pliocene and Pleistocene Insectivores (Mammalia) of Poland, I. Erinaceidae and Desmaninae. — In: Acta zool. Cracov. — Krakôw **16** (9). — P. 435—461.

—, 1976: The Neogene and Pleistocene Insectivores (Mammalia) of Poland, III. Soricidae: *Beremendia* and *Blarinoides*. — In: Acta zool. Cracov. — Kraków **21** (12). — P. 359—385.

SAMSON, P., 1976: Les Equidés fossiles de Roumanie (Pliocène moyen-Pléistocène supérieur). — In: Geol. Romana. — Roma **14** (1975). — P. 165—352.

SAMSON, P.; RADULESCO, C., 1973: Les faunes de Mammifères et la limite Pliocène-Pléistocène en Roumanie. — In: Trav. Inst. Spéol. „Emile Racovitza". — Bucureşti **12**. — P. 191—228.

SCHREUDER, A., 1940: A revision of the fossil water-moles (Desmaninae). — In: Arch. neerl. Zool. — Leiden **6** (2—3). — P. 201—333.

THALER, L., 1962: Campagnols primitifs de l'Ancien et du Nouveau Monde. — In: Coll. Internat. CNRS. — Paris **104**. — P. 387—399.

TCHÉPALYGA, A. L., 1980: Paléogéographie et Paléoécologie des bassins des mers Noire et Caspienne (Région ponto-caspienne) au Plio-Pléistocène. — Moskva, Rés. Thèse Dr. Sc. Géogr., 50 Pp., En Russe.

WEERD, A. VAN DE, 1976: Rodent faunas of the Mio-Pliocene continental sediments of the Teruel-Alfambra region, Spain. — In: Utrecht Micropaleont. Bull., Spec. Publ. — Utrecht **2**. — 217 Pp.

ZAGWIJN, W. H., 1974: The Pliocene-Pleistocene boundary in western and southern Europe. — In: Boreas. — Oslo **3**. — P. 75—97, (1974a).

—, 1974: Les problèmes de la subdivision du Pléistocène moyen des Pays Bas. — In: Bull. AFEQ. — Paris **40—41** (3—4). — P. 105—107, (1974b).

ZAZHIGIN, V. S., 1980: Late Pleistocene and Anthropogene Rodents of the South of Western Siberia. — In: Acad. Sci. USSR, Tr. geol. Inst. — Moskva **339**. — 156 Pp., En Russe.

Quartärpaläontologie 6 173—183 Berlin 1986

A Fossil Bison from the Moscow Region and the Geological Conditions of its Locality

By Valery Yuryevich Reshetov, Ekaterina Leonidovna Dmitrieva and Yuri Maksimovich Vasiliev, Moscow

With 2 Figures, 1 Table, and 4 Plates

Contents

Abstractum

Description, classification, and geological conditions of the fossil site of *Bison priscus mediator* Hilzheimer from the Moscow district are given. A comparative analysis of cranial pecularities of fossil bison has permitted a consideration of the phylogenetic relationship in the genus *Bison*.

In der vorliegenden Arbeit wird die Beschreibung, Klassifikation und Geologie der Fundstelle eines Schädels von *Bison priscus mediator* Hilzheimer aus der Umgebung von Moskau vorgelegt. Eine vergleichende Analyse der kranialen Besonderheiten der Bisonten erlaubt Rückschlüsse auf phylogenetische Verwandtschaftsbeziehungen.

Описание, классификация, геологические условия захоронения ископаемого бизона из Подмосковья — *Bison priscus mediator* Hilzheimer. Сравнительный анализ краниологических признаков позволил высказать соображения о филогенетических взаимоотношениях в роде *Bison*.

1. Geology and Chronology of the Deposits

The remains of a fossil bison were found in 1981 in the west of the Moscow region in deposits of the Moscow Glaciation.

Within the confines of the East European plain, the Moscow Glaciation is otherwise known as the Moscow Second Middle Pleistocene Glaciation (the First Middle Pleistocene, maximum, Glaciation is the Dniepr Glaciation and it corresponds to the Samarov-Bakhtin Glaciation of Siberia, the Saale Glaciation in Central Europe and the Riss Glaciation in the Alps). The counterpart of the Moscow Glaciation in the Asian part of the USSR is the Tazov Glaciation. In Central Europe its analogue is the Drenthe stage of the Saale Glaciation (Riss II in the Alps), and in North America the Sakagava-Ridge Glaciation (Nikiforova et al. 1982).

Near the southern part of the Volokolamsk terminal moraine, 10 km north-west of the town of Mozhaisk near the Dolginino village in the lower Iskona River area (a left-bank tributary of the Moskva River), a group of kames hills is characterizing the relief. Here, in the north-eastern wall of a sand quarry, the bison skull was found.

The structure of the Quaternary deposits in the Moscow region, and the principal geodynamic and other deformations of the Dolginino kames are shown in Fig. 1 and Fig. 2.

As Fig. 2 shows, the body of the kames is made up of sands and boulder-pebble beds which are intensively folded, the separate layers and bands having intricate contacts. These deposits contain lenses of greenish and brown aleurites (interstadial sediments). Up-section, the deposits are overlain by a basal moraine, which is undisturbed, and by later subaerial cover silts. Nearly all layers and bands are inclined, often contacting each other along subvertical surfaces.

In the north-easternwall of the quarry in the slope of the hill, the stripping has removed the uppermost layers of the section and exposed two buried hollows. One of them is composed of a lens of sands and aleurites. The size of the lens is about 20 m, with layers markedly rising westwardly, the angle of dip reaching 30°. The lens is made up of very thin-layered sands, sandy loams, and aleurites more clayey in the middle lens. A grey colour is predominant in the rock, with some prominent ochreous intercalations; quite noteworthy is the complex fine folding, i.e. corrugation of the sand and aleurite interlayers, with frequent fine breaking-up by subvertical low-amplitude microfaults (to fractions of one centimetre). Such rock makes up the uppermost 3 m of the lens. Below is sand — fine-grained, thin-layered, and with scattered pebble, gravel and boulders. This is also broken up by a system of unidirectional fissures. The thickness of the sands in this part of the lens is about 3—4 m. Apart from micro-fault fissures, there are syngenetic pseudomorphs resulting from multiple-veined ice. The sands and aleurites described are conformably underlain by boulder-pebble beds making up the body of the kame. It is in the lowermost part of the lens described, in silted sands, that the bison skull was discovered.

2. Palynological Analysis

The palynological analysis made by Ivanova (Vasiliev et al. 1982) showed the presence of pollen of the Carboniferous, Mesozoic and Cenozoic (the latter two are predomi-

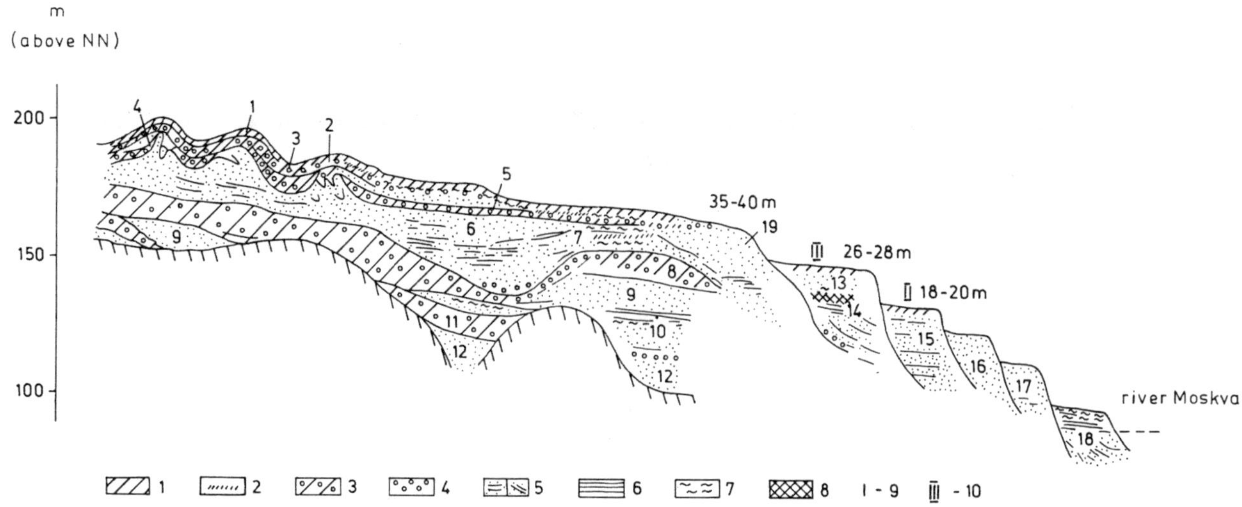

Fig. 1. Scheme of occurrence and relationships of Quaternary deposits in the Moscow area.

Conventional signs: 1 — cover subaerial silts; 2 — fossil soils; 3 — glacial deposits (moraine); 4 — shingle and boulders; 5 — sands (a — horizontally laminar, b — obliquely and diagonally laminar); 6 — clays; 7 — silts and aleurites; 8 — peat.
Digital signs on the figure — stratigraphic horizons and genetic types (for 10 read 20): 1 — Valdai (Würm) deposits; 2 — interglacial (Riss-Würm, Mikulino) and interstadial deposits; 3 — upper Moscow moraine; 4 — intra-Moscow inter-stadial sediments; 5 — lower Moscow moraine (early-Moscow maximum glaciation stage); 6 — Moscow fluvioglacial sediments; 7 — Odintsovo interglacial layers; 8 — Dniepr (Riss) moraine; 9 — Dniepr fluvioglacial sediments; 10 — Likhvin (Mindel-Riss) deposits; 11 — Okskaya (Mindel) moraine; 12 — Prae-Oka, Lower Pleistocene alluvial formations; 13 — upper alluvium formation of above-flood plain terrace III (early Valdaian); 14 — Mikulino (Riss-Würm) alluvium, lower formation of Terrace III; 15, 16, 17 — alluvium of low above-flood plain terraces of Late Valdaian age; 18 — alluvium of the Holocene flood plain; 19 — outwash surface of Moscow Glaciation; 20 — ordinal number of river terrasses

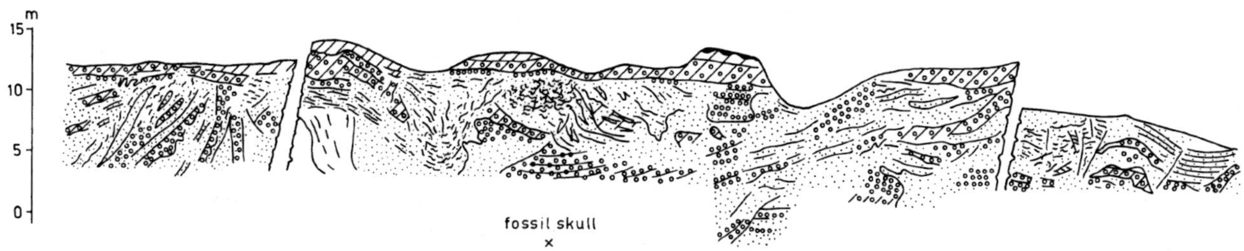

Fig. 2. Occurrence of glacial, water-glacial and interstadial deposits within the limits of the kame near Dolginino village.

× — whereabouts of the bison skull, for the rest of conventional signs see Fig. 1

nant). In addition, there are Pleistocene spores and pollen. In a sample taken from the nasal cavity of the Moscow bison's skull, pollen of the Pleistocene plants was found.

The identified spectrum is characterized by a significant content of the pollen of tree species and shrubs (71%). This is chiefly the pollen of the species with a great ecologic amplitude and vast distributional ranges: *Pinus silvestris* (71%), *Picea* sect. *Eupicea* (7%), *Betula* sect. *Albae* (10%), *Alnus* (2%). Significant quantities (12%) were recovered of *Betula nana* pollen (confined in its present occurrence to the tundra zone, to Alpine and peat bogs in the European USSR and in Western Siberia). The spectrum shows pollen of the species confined today predominantly to the eastern and continental areas [Northern Asia, north-east of the European USSR — *Alnaster fructicosa* (5%), *Larix* (insig.)]. In the group of non-arboral pollen (25% of total spectrum), predominant are *Artemisia* and other herbs (Ranunculaceae, Rosaceae, Caryophyllaceae, etc.). Some herbs are less abundant, Chenopodiaceae being very insignificant. Spores (4%) are represented by ferns and mosses.

The total character of the spectrum and the composition of flora point to a rather cold and continental climate under which conditions, besides forest formations with pine, spruce and birches, treeless spaces occupied by herbaceous associations (wormwood-herbs, grassland-herbs etc.) were also quite wide-spread.

On the basis of obtained palynologic data (we have only the spectrum of a single sample) it is difficult to speak of the palaeogeographic rank of the enclosing deposits. However, taking into account the geographical position of the section concerned, the absence in the spectrum of thermophilic elements and the presence of the pollen of such species as *Betula nana*, *Alnaster fructicosa* and *Larix*, as well as the composition of the spores, may be taken as evidence of conditions similar to the interstadial ones (VASILIEV et al. 1982).

Originally, the Dolginino kame arose as a filling void in the mass of dead ice possibly at the beginning of the degradation of the Moscow Glacier (maximum stage). In the subsequent interstadial, one dead ice had melted, and the fluvioglacial deposits were cut by a hollow. The evidence for interstadial sedimentation conditions in the hollow rests upon the aforementioned results of palynological studies; it is proved also by the fact that moraine silts were observed both in its flanks (and beneath it) and on the top of it. Severe climatic conditions were continuing at the time, proved by the disturbed occurrence of rocks in the slopes of the hollow. The disturbances could have occurred due to solifluction processes in the seasonally thawing ground (gravels, boulders). Evidence to this effect is also found in the fine deformations of aleuritic sediments, in syngenetic and epigenetic fissured frost disturbances in aleurites and sands and, finally in microfaults breaking up these loose rocks.

During the renewed advance of ice in the Moscow Late Glaciation period, all the above-described deposits of the Dolginino kame were subject to far-reaching dislocations as a result of glacio-dynamic processes.

3. Description of the Fossil Bison Skull

The remains of the bison are represented by the skull with the partly damaged facial part, left tibia and right metacarpal belonging to a very large adult bovine.

Cranium: Coll. PIN, USSR Academy of Sciences, No. 4062-1. Very large (see Table 1), massive with very conspicuous rugosities on the bones. Frontal width across posterior orbital edges is 413 mm, between horn-core bases, 360 mm. The facial part is damaged; however, judging from the preserving walls (maxilla), the nasal area was relatively high, tuber molaris being highly elevated above the maxilla surface and showing significant rugosity. The frontal part of the cranium is flattened, with numerous depressions and tubers. Tubular orbits protrude significantly from the cranial surface and are slightly oriented laterally and anteriorly. Orbital margins are covered by rugosities and separate bone growths. Lacrimalia are narrow, recessed at the front and form an isolated narrow ridge wedging in between nasal and maxillary bones. Praeorbital cavity is well preserved in the front part of lacrimal bone. The parietal region and entire occiput descend below the frontal level. Sulcus supraorbitalis is wide and is clearly delineated by a deep and wide furrow with numerous foramina and branchings into orbital part and frontal surface. There is present a certain ossification in the upper part of the channel, which amounts to an indirect evidence of the significant individual age of the specimen described. The length of the horn cores along the maximum curvature is 560 mm (dext.) and 567 mm (sin.), along a straight line — 520 mm (dext.) and 520 mm (sin.). The curvature index is 57%. They are extremely massive (the massivness index being 69.70%) and are steeply bent upwards. Their ends converge and are round in shape. Distance between them is 725 mm. In the upper third of their length the horn-cores are flattened (the flattening index being here 87%). The circumference of the horn-cores at the base is 394 mm (dext.), 392 mm (sin.). The flattening index at the base is 96%, and 97%. In the maximum curvature region they are seen to extend for behind the frontal plane. At the base of the horn-cores there are significant rugosities and protuberances. The upper edge of crista occipitalis is below the base of the horn-cores. The margins of the occipital crest are strongly rugose. Fossae temporales are narrow and deep. At the external and internal edges of the ventral side there occur crests running along jugale and processus zygomaticus. The occiput is very wide. Condyli occipitali in the external edges at the base are elongated to form a ridge bent forward. Auditory bones are rugose, the tympanic bulla being non-inflated. Basioccipitale is a short and relatively massive bone. Tuberes pharingea are large and rugose. Choanae are narrow. The choana walls protrude slightly sideways. The articular surfaces of the mandibular block are wide and long.

Teeth: M^{2-3} dext. Coll. PIN, USSR Academy of Sciences, No. 4062-1; and M^3 sin. Coll. PIN, USSR Academy of Sciences, No. 4062-1. Tooth crowns are moderately worn. Crown height — 30 mm. Tooth length at the ectoloph (M^2) — 36 mm, M^3 dext. — 32 mm, M^3 sin. — 35 mm; tooth width (cheek-lingual) M^2 — 26 mm, M^3 — 26 mm dext. and 25 mm sin. There is an additional crown at M^3. The structure of the teeth points to a significant degree of hypsodontism. Masticatory tooth surface area is insignificant compared to that in the long-horn forms.

Os metacarpale: Coll. PIN, USSR Academy of Sciences, No. 4062-2. Large and massive, with clearly expressed rugosities in the epiphyses region. The bone diaphysis is wide and flat. The index of the ratio of the anterioposterior diameter of metacarpal diaphysis to its width is 67%; the massiveness index of metacarpus (the ratio of the minimal width of metacarpal diaphysis to the length of the bone) is 23%.

Tibia: Coll. PIN, USSR Academy of Sciences, No. 4062-3. Bone length — 530 mm, proximal width-to-length ratio — 32%. Crista tibiae is large and massive, strongly rugose in the upper third. This crest is slightly shorter than half the bone length. Clear-cut relief on the plantar surface designates crests and surfaces for outgoing muscles.

Certain details of the cranium structure: A high nasal cavity (an evidence of breathing in cold air masses), a certain hypsodontism of teeth, etc., indicate that the *Bison priscus mediator* was living under cold forest-tundra conditions in Eurasia in the Pleistocene, but it did not inhabit fully open spaces.

4. Discussion

As seen from the description, the skull of the Mozhaisk bison has certain features which do not permit it to be classified with genuine long-horned *Bison priscus* races. Among such features are: 1. a narrow lacrimal bone, which is peculiar to the majority of recent wisent *Bison bonasus* LINNAEUS (HILZHEIMER 1918; GROMOVA 1935), whereas the majority of North-American *Bison bison bison* LINNAEUS and fossil long-horned forms have a broadened lacrimal bone; 2. a well-expressed praeorbital fossa in the area of the front part of lacrimal bone, a feature mainly peculiar to recent wisent (FLEROV 1965); 3. a flattened frontal part of the skull with numerous fossae and protuberances, which is likewise peculiar to wisent, whilst in bison and in the majority of long-horned forms the frontals are convex (GROMOVA 1935; FLEROV 1965); 4. massive horn-cores steeply bent upwards, their rounded ends being directed inward, so that the maximum distance between them is not at the ends but in the upper third. The circumference of the horn-core at the base exceeds its length measured along a straight line. With respect to these features the Mozhaisk bison is close to bisons of the early Taubach (FLEROV 1979) and those of the *Bison schoetensacki* group (FREUDENBERG

Table 1. Measurements and indices

Measurements and indices (after GROMOVA 1935, with abbreviations)	*B. priscus* „*longicornis*"	*B. priscus* „*deminutus*"	*B. priscus gigas* FLEROV South Siberia, Coll. PIN, USSR Academy of Sciences, No. 4063-1	*B. priscus mediator* HILZHEIMER Mozhaisk, Coll. PIN, USSR Academy of Sciences, No. 4062-1
	Eastern Europe, Siberia GROMOVA (1935)			
Median frontal length	264—337 (11 sp.)	273—304 (10 sp.)	330	335
Postorbital frontal length	237—308 (21 sp.)	226—274 (17 sp.)	243	306
Lateral frontal length	180—224 (18 sp.)	166—209 (17 sp.)	228	210
Maximum length of temporal fossa	192—233 (3 sp.)	204—216.5 (4 sp.)	204(d)	231(d)
Maximum width of temporal fossa	5—16 (7 sp.)	6—22 (4 sp.)	37(d)	37(d)
Depth of temporal fossa	50—62 (7 sp.)	46—62 (5 sp.)	47(d)	64(d)
Maximum frontal width	331—400 (16 sp.)	336—380 (17 sp.)	390	413
Minimum frontal width	297—353.5 (25 sp.)	284—330 (20 sp.)	295	312
Frontal width between horn-cores in the middle of their frontal surface	322—440 (28 sp.)	313—378 (17 sp.)	363	360.5
Parietal length	85—120 (24 sp.)	75—116 (14 sp.)	87	89
Maximum occipital width	264—322 (22 sp.)	265—296 (17 sp.)	302	312
Maximum occipital height	153—195 (30 sp.)	146—170 (18 sp.)	157	167
Minimum occipital height	107—179 (28 sp.)	105—131 (18 sp.)	112	119
Horn-core basal circumference	315—410 (37 sp.)	295—365 (29 sp.)	400	394/392d/s
Frontal diameter of horn-core basal	115—145 (39 sp.)	100—121 (28 sp.)	127	122/123d/s
Diameter perpendicular to the above	98—127 (36 sp.)	87—111 (28 sp.)	97	118/118d/s
Straight length of horn-core	371—535 (25 sp.)	302—423 (22 sp.)	520	321/321d/s
Length of horn-core along maximum curvature	540—715 (33 sp.)	400—540 (25 sp.)	659/656d/s	560/567d/s
Distance between horn-core tips	1026—1360 (15 sp.)	845—1090 (13 sp.)	1345	725
Degree of curvature of horn-cores, %	60.6—89.6 (25 sp.)	70.2—86.9 (21 sp.)	79/79d/s	57/57d/s
Massiveness of horn-cores, %	53.3—75.7 (31 sp.)	60.8—86.5 (25 sp.)	61/61d/s	69/70d/s
Flattening of horn-core basale, %	75.2—94.6 (34 sp.)	79.2—96.9 (27 sp.)	76/76d/s	97/96d/s
Flattening of horn-cores in the upper third, %			98/96d/s	87/87d/s

1914; SOERGEL 1923). Apart from this, large pharyngeal tubercles, rostral growths of occipital condyles, a certain hypsodontism of teeth and very large overall size make the Moscow find similar to bisons of the *Bison priscus* group (GROMOVA 1935; FLEROV 1979).

This relationship points to the fact that what we have is a certain "transitional" stage in the evolution of the European bisons after they had attained large size, already having certain features pointing to their probable direct affinity with the later typically short-horned bisons and recent European wisent.

The ancestor of recent European wisent and American bisons, the fossil *Bison priscus* (BOJANUS) was inhabiting in the Pleistocene a vast territory of the Palaearctic. The genus flourished during the Riss.

Its broad habitation area, the heterogeneity of the landscape and climatic environment and a frequent lack of geological data about fossil materials — all this has accounted and is accounting for significant difficulties in drawing up phylogenetic schemes. There are some summaries which give a sufficiently detailed account of the history of the genus and indicate original forms, systematics and morphology of bisons (GROMOVA 1935; SKINNER and KAISEN 1947; BOHLKEN 1967; GUTHRIE 1970; FLEROV 1979). However, there still exist in the history of bisons certain blank spots and unsolved problems. One of such problems is the origin of the recent European wisent *Bison bonasus* LINNAEUS, and, correspondingly, the history of forest bisons of European Pleistocene. As early as 1918 HILZHEIMER suggested the possibility of the existence of two parallel lineages of bisons in Europe, which were developing independently since the Mindel-Riss Interglacial. One of the lineages inhabited open and semi-open spaces and was represented by the so-called "long-horned" forms, the other was a direct continuation of the Early Pleistocene *Bison schoetensacki* FREUDENBERG and existed in European forest areas until the end of the Würm when it transformed into the recent wisent *Bison bonasus* LINNAEUS. A similar relationship between the two ecological groups is also observable in the case of elephants of the *Mammuthus primigenius* and *Palaeoloxodon antiquus* groups. This hypothesis was supported by SOERGEL (1923). However, GROMOVA (1935) expressed serious doubts about the possibility of co-existence of two sub-species of the primitive bison in Europe.

C. C. FLEROV (1979) in his last review of the history of the genus *Bison* admitted the possibility of co-existence of two lineages of European bisons, proceeding in his assumption from the well-dated material from the Süssenborn, Taubach and Ehringsdorf deposits (FLEROV 1968, 1969). But he subdivided all bisons into three main morphoecological groups: 1. typical inhabitants of the southern Holarctic steppes, feeding solely herbaceous plants. Placed in this group were the largest long-horned forms of primitive bison, such as *B. priscus gigas* (FLEROV), *Bison priscus latifrons* (HARLAN), as well as the living *Bison bison bison* LINNAEUS of the U.S.A.; 2. inhabitants of mixed forests in the Palaeartic temperate zone, mesodendrophages, which were feeding on a mixed diet, predominantly trees and soft forest grasses. Placed in this group were principally ancient bisons of the type *Bison schoetensacki* FREUDENBERG and the recent European wisent *Bison bonasus* LINNAEUS; 3. bisons of the Holarctic northern cold parts, which were living under severe conditions of forest-tundra and were feeding on mixed forest-tundra vegetation.

Placed under this group were bisons *Bison priscus priscus* (BOJANUS), *Bison priscus crassicornis* (RICHARDSON) inhabiting Europe and Asia, North-American *Bison priscus crassicornis* (RICHARDSON), *Bison priscus occidentalis* LUCAS, and *Bison bison athabascae* ROADS (*Bison priscus athabascae* in FLEROV's interpretation) and an intermediate form between the second and third group, which is assumed by different authors to be the ancestor of the recent European wisent — *Bison priscus mediator* HILZHEIMER. The first group of bisons is characterized by the very large size of the horn-cores, the convex frontals associated with this, a hypsodontal tooth system and certain pecularities of the postcranial skeleton. The second group are European forest bisons, appearing in the Cromerian deposits and existing till the Holsteinian, which are characterized by moderate size for the genus as a whole, relatively short, strongly bent and massive horn-cores (basal circumference always superior to straight length), relatively thinner supporting bones (compared to later bisons of the *Bison priscus* group) in the limbs, resembling those of the recent wisent. The latter representative of *Bison schoetensacki* FREUDENBERG were much larger and massive. Bisons of the third group combine different morphological features, both long-horned and short-horned forms being present among them, the latter with respect to structural type showing affinity to the second-group bisons. Teeth in the third-group bisons may have elements of hypsodontism. With respect to time, the distribution of the representatives of the various groups is practically synchronous.

The morphological analysis of features of the Mozhaisk bison skull gives evidence about its belonging to the subspecies *Bison priscus mediator* HILZHEIMER. Speaking of the scope of the subspecies *Bison priscus mediator* HILZHEIMER and its chronological range, one should pay attention to the morpho-ecological and phylogenetic meaning attributable to this group of bisons. GROMOVA identified some of the bisons later classified in the *mediator*-group as the final diminished stage of *Bison priscus longicornis* and placed all the Late Pleistocene bisons in the sub-species *deminutus*. SKINNER and KAISEN (1947) who elaborated their own intricate systematics, taking into account the shape and size of bison horns, considered the GROMOVA sub-species *deminutus* to be a synonym of the group *Bison priscus crassicornis* where they included numerous long-horned forms. C. C. FLEROV revised the scape of the same sub-species and included in it all bisons of the so-called "transitory" group from bisons of the type *Bison priscus priscus* (BOJANUS) to the recent wisent. In this regard we adhere to FLEROV's classification (1979) and place the Mozhaisk bison in the same sub-species. The new find fills the gap between the bisons of the Late Mindel (late Süssenborn) represented by the *schoetensacki*-group and those of the Riss-Würm Interglacial and Würm, thus enabling us to continue the existence of the *mediator* sub-species until the Riss. Certain peculiarities in the cranial structures and tooth hypsodontism of the Mozhaisk bison furnish evidence favouring its classification with bisons of the *priscus* species.

Presumably by the Middle Pleistocene there existed two sub-species of bisons combined into the *priscus* species (which was a direct descendant of *Bison schoetensacki*) forest and forest-tundra *Bison priscus mediator* HILZHEIMER and steppe *Bison priscus priscus* (BOJANUS) (= *longicornis*, in GROMOVA's interpretation). The latter was characterized by a wider habitation area, a significantly larger span of the horns, and certain structural peculiarities of the cranium and skeleton. Both forms underwent stages of dimensional growth in the Middle Pleistocene and thereafter, of diminuting. However, European steppe bisons became totally extinct, whereas the second lineage represented by the *mediator* sub-species has yielded the recent *Bison bonasus* LINNAEUS. The possibility of the occurrence of both forms during the European Pleistocene is indirectly confirmed by the co-existence of the forest (*Bison bison athabascae* ROADS) and the steppe race (*Bison bison bison* LINNAEUS) of bisons in the north of the U.S.A. and Canada. Two sub-species in historical time were living in relatively close-lying areas without intermixing and were easily distinguished by the Indians who gave each sub-species its proper name (SOPER 1941).

The bison find in the Moscow glacial deposits is of great importance for estimation of the process of evolution in the European branch of forest bisons — *schoetensacki-priscus-bonasus* — as well as for the biostratigraphy of Eastern European Quaternary deposits.

Summary

The remains of a fossil bison were found in a sand pit near Dolginino village, north-east of Mozhaisk town (Moscow region) within a kame of the Mozhaisk beds of the Moscow Glaciation. The fossil bones occurred in interstadial deposits of this glaciation. These beds showed traces of the dynamic processes which took place during the second stage of the Moscow Glacial age. The fossil remains — a skull, tibia, and metacarpals — belong to an adult male. The horn cores are flattened antero-posteriorly and are very curved. The horn core spread is 725 mm, the width of the frontal bone between the orbits is 413 mm, and between the horn cores 354 mm.

The above described bison remains belong to *Bison priscus mediator* HILZHEIMER (sensu FLEROV 1979). This form probably represents the early woodland bison stock, distributed in Europe at the end of the Middle Pleistocene (Moscow Glacial age). In a sample for palynological research, taken from the nasal cavity of the skull, pollen of trees and shrubs predominate (71 %). The composition of the spectrum indicates a cold continental character of the climate with boreal woodland as well as unwooded areas.

Zusammenfassung

Reste eines fossilen Bisonten konnten in einer Sandgrube in der Nähe des Dorfes Dolginino, nordwestlich der Stadt Možaisk (Gebiet Moskau) in einem Kames der Možaisk-Ablagerungen der Moskau-Vereisung geborgen werden. Die fossilen Knochen lagen in Interstadial-Sedimenten dieses Kaltzeitkomplexes. Die Ablagerungen zeigen deutlich die Spuren dynamischer Glazial-Prozesse, die während des zweiten Stadials der Moskau-Vereisung abliefen.

Die Fossilfunde — Schädel, Tibia und Metacarpalia — stammen von einem adulten männlichen Tier. Die Hornzapfen sind antero-posterior abgeplattet und stark gebogen. Die größte Breite der Hornzapfen-Auslage beträgt 725 mm, die Breite des Frontal-Schädels zwischen den Orbitae beträgt 413 mm und zwischen den Hornzapfen 354 mm.

Die beschriebenen Bison-Reste gehören zu *Bison priscus mediator* HILZHEIMER (sensu FLEROV 1979). Wahrscheinlich repräsentiert diese Form den frühen Wald-Bison-Stamm, der am Ende des Mittelpleistozäns (Moskau-Glazial) in Europa verbreitet war. In einer Probe für palynologische Untersuchungen, die aus der Nasenhöhle des Schädels entnommen wurde, dominieren Baum- und Busch-pollen (71 %). Die Zusammensetzung des Spektrums läßt auf kaltkontinentalen Klimacharakter mit borealem Wald, durchsetzt von unbewaldeten Gebieten schließen.

Резюме

Остатки ископаемого бизона обнаружены северо-западнее г. Можайска в песчаном карьере у деревни Долгинино (Московская область) в пределах кама на можайской моренной гряде московского оледенения. Остатки бизона приурочены к межморенным отложениям московского ледниковья. Эти слои несут следы гляциодинамических процессов, происходивших при вторичном наступлении московского ледника. Остатки костей — череп, большая берцовая и пястная кости принадлежат взрослому самцу. Роговые стержни сплющены спереди — назад и сильно изогнуты. Размах между их концами — 725 мм, ширина лба между краями глазниц — 413 мм, между основаниями роговых стержней — 354 мм.

Описываемый бизон принадлежит подвиду *Bison priscus mediator* Hilzheimer (в смысле Флерова 1979). Эта форма возможно представляет наиболее ранних лесных бизонов, распространенных в Европе в конце среднего плейстоцена (московское ледниковье). В образце, взятом из носовой полости черепа на споро-пыльцевой анализ, преобладает пыльца кустарников и деревьев (71%). Состав спектра свидетельствует о холодном и континентальном климате, наряду с лесными формациями были распространены безлесные пространства.

References

Bohlken, H., 1967: Beiträge zur Systematik der rezenten Formen der Gattung *Bison*. — In: Z. zool. System. u. Evolutionsforsch. — Frankfurt (Main) **5** (1). — P. 4—110.

Flerov, C. C., 1965: Sravnitel'naja kraniologija sovremennych predstavitelej roda *Bison* (Comparative craniology of recent representatives of the genus *Bison*). — In: Bjull. Mosk. Obşč. Ispyt. Prir., Otd. Biol. — Moskva **20** (1). — P. 40—54.

—, 1969: Die Bison-Reste aus den Kiesen von Süßenborn bei Weimar. — In: Paläont. Abh., A. — Berlin **3** (3/4). — P. 40—48.

—, 1975: Die Bison-Reste aus den Travertinen von Weimar-Ehringsdorf. — In: Abh. zentr. geol. Inst. — Berlin 23. — P. 171—199.

—, 1979: Sistematika i evoljucija. In: Zubr. Morfologija, sistematika, evoljucija, ekologija (Systematics and evolution. In: Wisent. Morphology, systematics, evolution, ecology). Part 1 (1—5). — Moskva: Nauka. — P. 9—127.

Freudenberg, W., 1914: Die Säugetiere des älteren Quartärs von Mitteleuropa mit besonderer Berücksichtigung der Fauna von Hundsheim und Deutschaltenburg in Niederösterreich etc. — In: Geol.-paläont. Abh., N.F. — Jena **12** (4—5). — P. 81—100.

Gromova, V. I., 1935: Pervobytnyj zubr. (*Bison priscus* Boj.) v SSSR [The fossil bison of the U.S.S.R. (*Bison priscus* Boj.)]. — In: Tr. zool. Inst. Akad. Nauk SSSR. — Moskva **2** (2/3). — P. 77—204.

—, 1964: Kratkij obzor četvertičnych mlekopitajuščich Evropy (A brief survey of European Quaternary mammals). — Moskva: Nauka. — 142 Pp.

Guthrie, R. D., 1970: Bison evolution and zoogeography in North America during the Pleistocene. — In: Quart. Rev. Biol. — 45 (1). — P. 1—15.

Krasnov, I. I., 1974: Krivaja solnečnoj radiacii i izmenenie prirodnych uslovij landšaftnoj oboločki v antropogene (Solar radiation curve and the change of natural conditions in the landscape of the Anthropogene.) — In: Kosmos i evoljucii organizmov. Materialy soveščanija „Kosmičeskie faktory i evoljucija organičeskogo mira". — Moskva. — P. 62—83. — (Rotaprint).

Menner, V. V., 1930: Opisanie ostatkov mlekopitajuščich iz mežmorennych suglinkov Odincova (Description of mammalian remains from Odintsovo intermorainic loams). — In: Materialy po prirode Moskovskoj oblasti. — Moskva. — P. 29—35

Nikiforova, K. V.; Kind, N. V.; Ivanova, I. K.; Vasiliev, Yu. M.; Pevzner, M. A.; Tseitlin, S. M., 1982: Problemy geologii i istorija četvertičnogo perioda (antropogena) [Problems of the geology and the history of the Quaternary period (Anthropogene)]. — Moskva: Nauka. — 252 Pp.

Putevoditel' ekskursij A—I, C—I, Podmoskovsko'e (Guide to Excursions A—I, C—I, surroundings from Moscow). — In: XI kongress INQA 1982g. — Moskva. — P. 1—40.

Shantser, E. V., 1947: (On the Early Quaternary [Mindel] glacial deposits in the city of Moscow). — In: Tr. geol. Inst., Ser. Geol. — Moskva 26. — P. 113—125.

Skinner, M. F.; Kaisen, O. X., 1947: The fossil Bison of Alaska and a preliminary revision of the genus. — In: Amer. Mus. Natur. Hist., Bull. — New York **89** (3). — P. 123—256.

Soergel, W., 1923: Die diluvialen Säugetiere Badens. Teil I. Älteres und mittleres Diluvium. — In: Mitt. bad. geol. Landesanst. — Heidelberg 9. — 254 Pp.

Soper, J. D., 1941: History, range and life of the northern bison. — In: Ecolog. Monogr. — **2** (4). — P. 348—412.

Vasiliev, Yu. M.; Dmitrieva, E. L.; Ivanova, N. G.; Reshetov, V. Yu, 1982: Ostatki korotkorogogo bizona v otloženijach Moskovskogo lednikov'ja (The remains of a short-horned bison in the Moscow glacial deposits). — In: XI kongress INQUA, Abstracts. — Moskva. — P. 40—41.

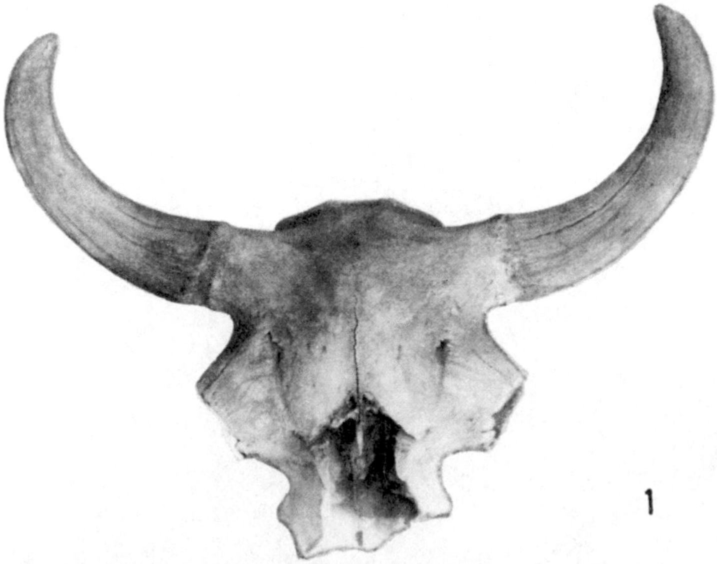

1

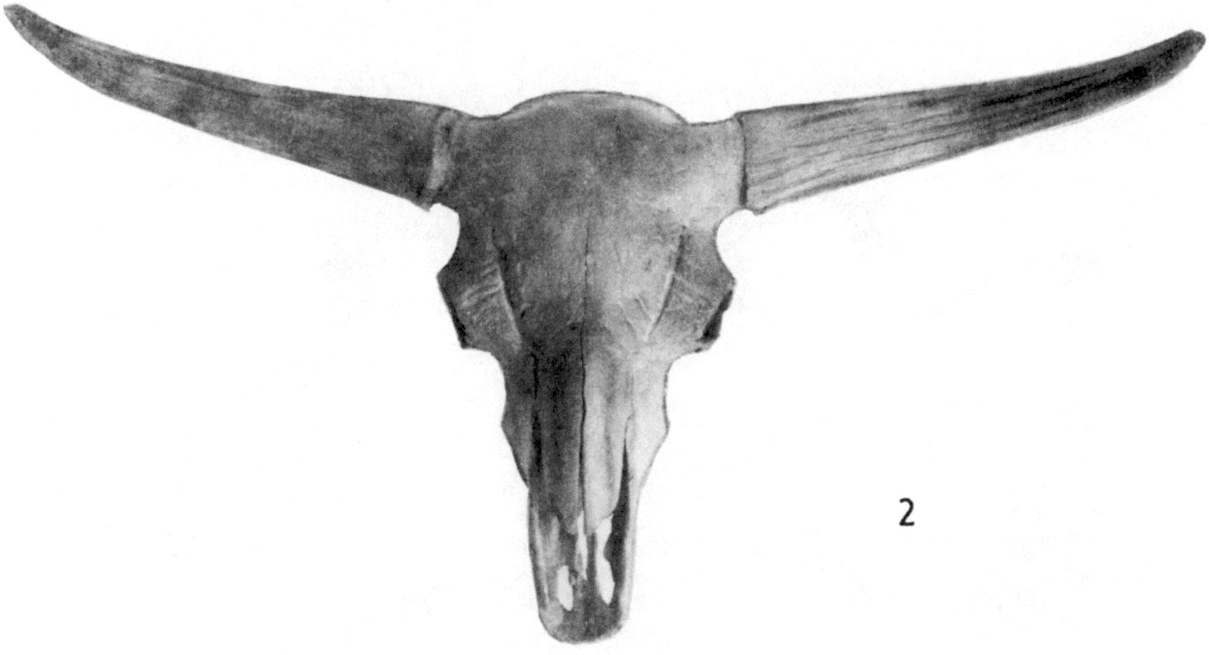

2

PLATE I

Fig. 1: *Bison priscus mediator* HILZHEIMER
Skull, male ad., Moscow district, environs of Mozhaisk, Iskona river, Riss II, Coll. PIN USSR Acad. Sci., No. 4062-1, 1/8 nat. size

Fig. 2: *Bison priscus gigas* FLEROV
Skull, male ad., South Siberia, environs of Pavlodar, Irtysh river, Riss/Würm, Coll. PIN USSR Acad. Aci., No. 4063-1, 1/8 nat. size

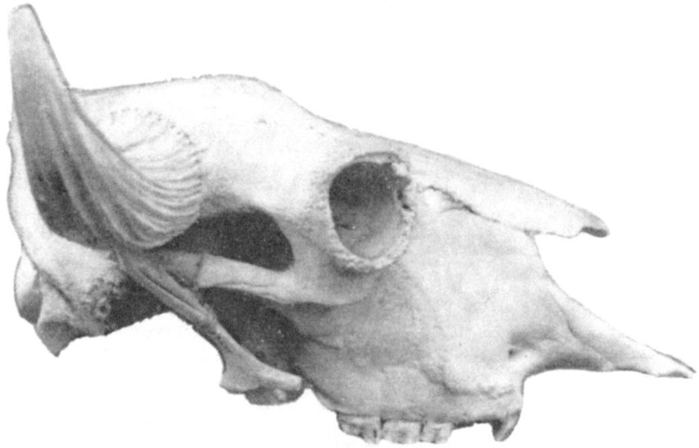

1

2

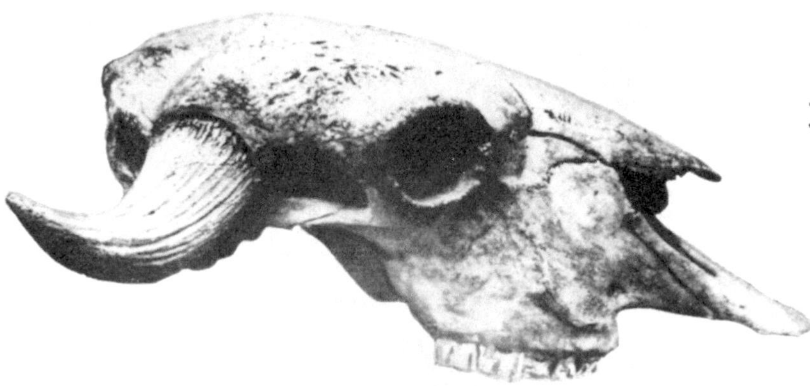

3

4

Fig. 1: *Bison bonasus* LINNAEUS
Skull, male ad., Recent, Coll. PIN USSR Acad. Sci., No. 193, 1/6 nat. size

Fig. 2: *Bison priscus mediator* HILZHEIMER
Skull, male ad., Moscow district, environs of Mozhaisk, Iskona river, Riss II, Coll. PIN USSR Acad. Sci., No. 4062-1, 1/ nat. size

Fig. 3: *Bison bison bison* LINNAEUS
Skull, male ad., Recent, Coll. PIN USSR Acad. Sci., No. 131, 1/6 nat. size

Fig. 4: *Bison priscus gigas* FLEROV
Skull, male ad., South Siberia, environs of Pavlodar, Irtysh river, Riss/Würm, Coll. PIN USSR Acad. Sci., No. 4063-1, 1/6 nat. size

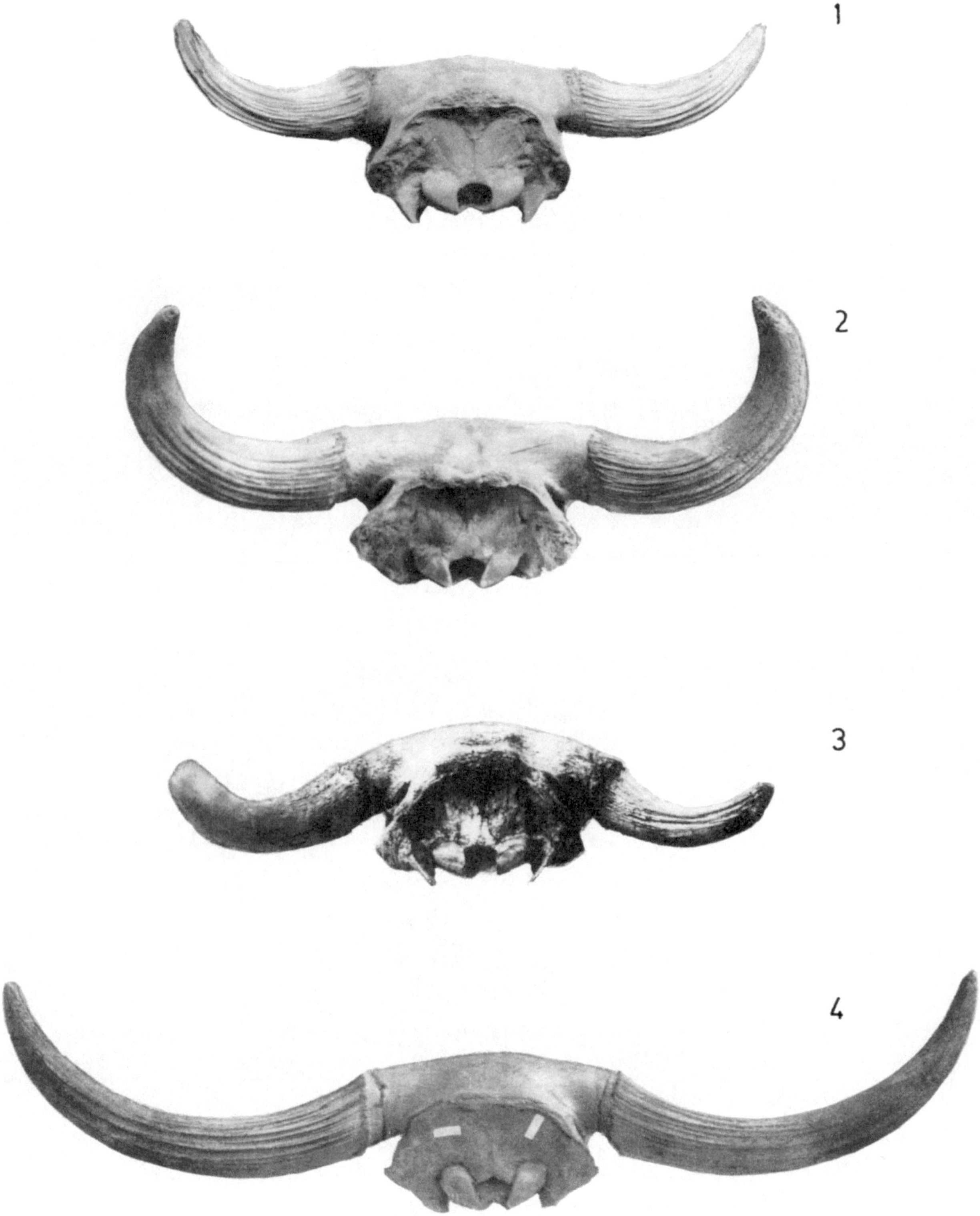

PLATE III

Fig. 1: *Bison bonasus* LINNAEUS
Skull, male ad., Recent, Coll. PIN USSR Acad. Sci., No. 193, 1/7 nat. size

Fig. 2: *Bison priscus mediator* HILZHEIMER
Skull, male ad., Moscow district, environs of Mozhaisk, Iskona river, Riss II, Coll. PIN USSR Acad. Sci., No. 4062-1, 1/7 nat. size

Fig. 3: *Bison bison bison* LINNAEUS
Skull, male ad., Recent, Coll. PIN USSR Acad. Sci., No. 131, 1/7 nat. size

Fig. 4: *Bison priscus gigas* FLEROV
Skull, male ad., South Siberia, environs of Pavlodar, Irtysh river, Riss/Würm, Coll. PIN USSR Acad. Sci., No. 4063-1, 1/7 nat. size

PLATE IV

Fig. 1: *Bison priscus mediator* HILZHEIMER
Tibia sin., male ad., Moscow district, environs of Mozhaisk, Iskona river, Riss II, Coll. PIN USSR Acad. Sci., No. 4062-3, 1/2 nat. size

Fig. 2: *Bison priscus mediator* HILZHEIMER
Metacarpal. dext., male ad., Moscow district, environs of Mozhaisk, Iskona river, Riss II, Coll. PIN USSR Acad. Sci., No. 4062-2, 1/2 nat. size

Fig. 3: *Bison bison athabascae* RHOADS
Metacarpal. dext., male ad., Recent, 1/2 nat. size

Fig. 4: *Bison bonasus* LINNAEUS
Metacarpal. sin., male ad., Recent, 1/2 nat. size

Fig. 5: *Bison bison bison* LINNAEUS
Metacarpal. dext., male ad., Recent, 1/2 nat. size.

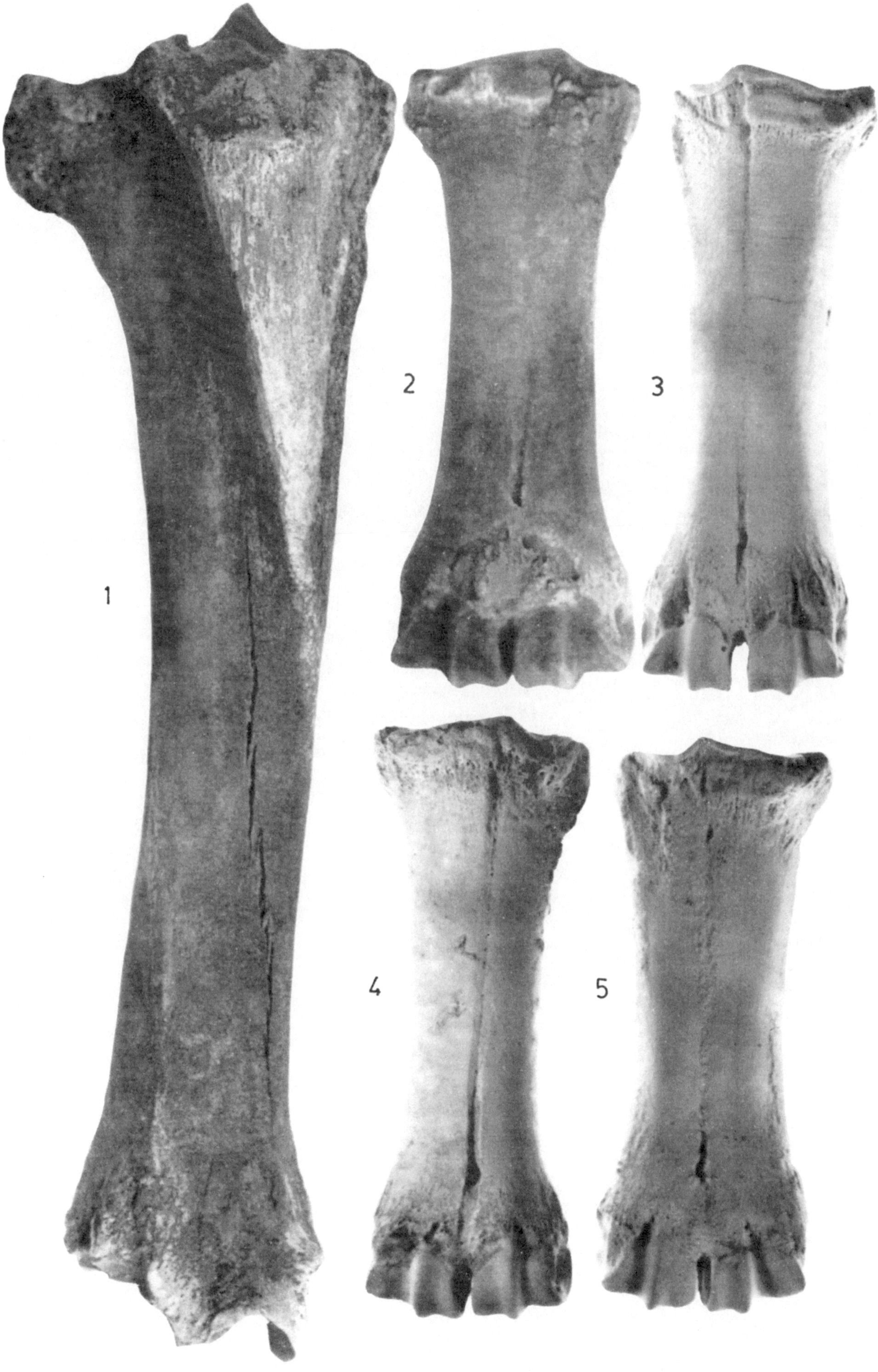
1
2
3
4
5

Quartärpaläontologie 6 185—193 Berlin 1986

On the History of Mammal Fauna of Beringida

By ANDREI VLADIMIROVICH SHER, MOSCOW

With 1 Figure and 2 Tables

Contents

Abstractum

According to the recent palaeontological evidence from NE Asia and NW North America, the ancient land of Beringida was one of the main late Cenozoic speciation centers. The history of the mammal fauna and environment of Beringida during the last 2.5 million years is briefly discussed. The influence of the Beringidan fauna on the faunas of middle latitudes of North America is examined. Dispersal of Beringidan mammals resulted from an early development of cold-climate and cold-dry-climate communities within its territory. Subsequently these communities spread south as climatic cooling progressed in middle latitudes.

Nach dem heutigen paläontologischen Forschungsstand über Nordostasien und im nordwestlichen Nordamerika zu urteilen war das alte Beringida eines der Hauptevolutionszentren des späten Känozoikums. Die Geschichte der Säugetierfauna und der Umweltbedingungen Beringidas während der letzten 2.5 Millionen Jahre wird kurz dargestellt. Der Einfluß der Beringida-Fauna auf die Faunen der mittleren Breiten von Nordamerika wird ebenfalls untersucht. Die Ausbreitung von Beringida-Säugetieren war das Ergebnis der frühen Anpassung an kalte und an kalt-trockene Umweltbedingungen dieser Assoziationen. Mit den in den mittleren Breiten einsetzenden Klimadepressionen geschah dann folgerichtig eine Südausbreitung dieser Assoziationen.

Новейшие палеонтологические данные по северо-востоку Азии и северо-западу Северной Америки показывают, что древняя суша Берингида была одним из основных позднекайнозойских центров формообразования. Кратко рассматривается история фауны млекопитающих и ландшафтов Берингиды за последние 2,5 млн. лет. Анализируется влияние фауны Берингиды на фауну умеренных широт Северной Америки. Радиация млекопитающих Берингиды была обусловлена ранним возникновением криофильных и криоксерофильных сообществ на ее территории и расселением их к югу по мере развития похолодания в умеренных широтах.

1. Introduction

The latest Cenozoic, that is, the end of the Pliocene, the Pleistocene and the Holocene, was a time of radical transformation of the mammalian fauna of the Northern Hemisphere. Important rearrangements occurred in the natural zonation of environments; new communities came into being; there were large invasions by faunistic blocks and a wide dispersal of certain species. Finally, the circumpolar distributions of many species became established.

The late Cenozoic turnover of the biota was the result of the climatic cooling which started in the Pliocene and later resulted in several continental glaciations and the development of vast permafrost areas. Thus the northernmost Holarctic was the source of the most revolutionary transformation of the fauna. The first signs of cooling in high latitudes were as early as the Miocene/Pliocene boundary. The earliest record of forest-tundra environments in the Arctic dates approximately from 5—6 million years ago. The earliest tundra-type communities known in northern Yakutia are about 2.5 million years old.

Cooling in temperate latitudes manifested itself much later. In these regions many characteristic Neogene mammal lineages survived into the Quaternary. Some of these lineages died out slowly, while others flourished. At the same time completely new taxa appeared in Eurasian middle latitudes without obvious immediate ancestors. Lemmini, Dicrostonychini, Alcini, Ovibovini, Rangiferini and Saigini were among them. In other lineages the break between newly-appeared taxa and their probable ancestors is not so large. Nevertheless, distinctly separated species or even groups of subgenus rank are rather common among some ungulates, proboscideans, carnivores and rodents. Thus only some part of mammalian fauna of temperate latitudes was transformed autochthonously. The other component of this fauna consisted of the Quaternary newcomers.

The most essential feature, however, in the late Cenozoic history of Holarctic biota is the origin of two new types of faunal communities. The first is the cryophilous fauna of the open plains that survived into recent times as the fauna of the tundra zone. The second is the cryoxerophilous fauna, or fauna of tundra-steppe type. The latter type of community is not known before the latest Cenozoic. During the Pleistocene it became zonal and virtually disappeared around the Pleistocene/Holocene boundary. Nevertheless, many features of the present day Holarctic animal population are derived from the time of maximum spread of tundra-steppe communities (late Würm/Wisconsin) and their subsequent decay.

When and where did these new types of communities evolve? Where did the Quaternary newcomers come from, whose invasion changed the appearance of the fauna of middle latitudes? Finally, where should we search for the missing links of many late Cenozoic mammalian phyletic lineages?

These major questions of Holarctic biogeography have faced zoologists since the early 1900's. Trying to find adequate answers, such Russian zoogeographers as P. Sushkin, A. Tugarinov, B. Shtegman and others turned to the northern and north-eastern margins of the Eurasian mainland. In so doing they relied upon the studies of the present day fauna. On the other hand, as knowledge of the pecularities of mammals now living in high latitudes deepened, the better was their morphological isolation understood. Such students of northern mammals as C. Flerov, S. Shvartz, R. Hoffmann and others became more and more convinced that the particular arctic adaptations of the Holarctic mammals evolved in relatively early times.

Thus, many zoologists thought, that the northern margins of Eurasia, in particular north-east Siberia, have been an important area of late Cenozoic speciation. They also suggested that the evolution of the Pleistocene and Recent Arctic mammals took place in that area rather early in geological history. As early as the beginning of the 20th century a special name for this area was suggested. W. Kobelt called it "Behringia" and W. Bianki in a Russian edition of Kobelt's monograph (1903) interpreted it as "Beringida".

For a long time Beringida was almost a blank space for palaeontologists. All its known fauna was of the late Pleistocene age and rather similar to the fauna of more southern regions of the same age. So palaeontologists could neither support nor disprove the idea of Beringida as an ancient center of speciation. They used to treat it as a rather abstract land bridge over which faunal interchange between the Old and New World took place from time to time. For neontologists Beringida was *a land* (though hypothetical) rather than *a bridge*. Thus two different concepts of Beringida have come into being.

The discovery and study of the more ancient fauna of Beringida (pre-late Pleistocene) started only about 20 years ago. The main results of these studies are briefly discussed below.

2. Present Knowledge of the History of the Beringidan Mammalian Fauna

2.1. The Fossil Mammal Record of Beringida

As conceived by the author Beringida is a vast palaeogeographic land covering the continental shelf of the Laptev, East Siberian, Chukotsk and Bering Seas, marginal lowlands of Northeast Siberia and Alaska, and the surrounding low mountains with large basins. This forgotten palaeogeographic term has been revived to distinguish *Beringida* from *Beringia* — a narrower and more mobile contact zone between the landmasses of Asia and America (Sher 1976).

The most important palaeontological evidence on the early Beringidan fauna comes from a number of key localities within the Maritime Lowland of Yakutia and its surrounding. Some of these localities include more than one fauna of different ages ranging from late Pliocene to late Pleistocene. One late Pliocene locality — Cape Deceit — has been discovered in Alaska (Guthrie and Matthews 1971). Some ancient faunal elements similar to those of the Yakutian (Olyoran) were discovered in the Yukon Territory but unfortunately they have no exact stratigraphic context (Harington 1978).

As most of the north Yakutian key localities were studied with modern techniques, rather convincing palaeoecological evidence has been obtained. Studies of fossil insects, plant macrofossils, spore- pollen spectra and cryogenic features of bone-bearing deposits are usually in quite close agreement.

The earliest known mammal fauna of Beringida is the Kutuyakh rodent fauna from the lower course of the Kolyma. It comes from the key section of the Krestovka River (Sher et al. 1979). Fossiliferous deposits show clear features of rather low-temperature permafrost including ice-wedge pseudomorphs. The palynological spectra as well as the plant macrofossils and the insect fauna indicate a tundra type of environment. The Kutuyakh fauna certainly lived under a rigorous climate and permafrost conditions. This fauna is quite unique and so far no equivalent faunas have been found in the Holarctic. The fauna is dominated by rooted microtines of the *Mimomys*-group and bog lemmings — *Synaptomys*. The two identified lineages of rooted microtines, *Cromeromys* and *Mimomys*, are represented by rather primitive species characteristic of the Khapryan (= Podpusk-Lebyazhyan) faunal complex (units). These faunal units are recognized in the south of the European part of the USSR and in the south of West Siberia respectively and are usually correlated with the middle Villafranchian faunas of Europe. The evolutionary level of the microtines and the palaeomagnetic evidence (lower part of the Matuyama epoch) allow the Kutuyakh fauna to be referred to a time span of 2.0—2.5 million years ago. At present this is the earliest community of tundra type known in the Holarctic. Unfortunately no fossils of large mammals have been found here, except one fragment which might belong to a reindeer. A small fauna of rooted microtines including a member of the American vole lineage *Phenacomys*, is referred to a later period. This fauna, which has been found in another exposure on the Krestovka River (Sher et al. 1977), is unique because of the first appearance of steppe insects (together with tundra species).

In the Krestovka Key Section the Kutuyakh Suite is separated from the overlying Olyor Suite by an unconformity. Comparison of rodent fauna of the two suites indicates a relatively large chronological gap embracing an important stage of faunal evolution that corresponds to the Odessan (Kizikhan) stage of rodent evolution in middle latitudes. A part of this gap may be represented by the above mentioned fauna with *Phenacomys*. Moreover, some features of this stage may be reconstructed from the preceeding (Kutuyakh) and subsequent (Olyor) faunas as well as from fossils reworked into the lower Olyor deposits from a missing layer younger than the Kutuyakh.

This was probably the final stage of the flowering of rooted microtines in the Far North and also the time of wide dispersal of *Allophaiomys* in the region. This can be assumed to have been the last appearance of *Hypolagus*. During this time *Synaptomys* was replaced by *Lemmus*, and the most primitive

Dicrostonychini (*Predicrostonyx hopkinsi*) probably made their first appearance.

As for the large mammals, we can only guess that during that time span, which can be roughly estimated as 1.8—1.2 million years ago, various groups, whose descendants are known from the following Olyoran age, inhabited Beringida.

The Cape Deceit Fauna in Alaska is of great importance for the understanding of Beringida faunal history. GUTHRIE and MATTHEWS (1971) correlated it with the European Günz and did not indicate its relation to the North American Land Mammal Ages. Present estimates of its age range from middle Irvingtonian (KURTÉN and ANDERSON 1980) to late Blancan (REPENNING 1980). It is clear that the age of the Cape Deceit fauna should be determined on the basis of the Kolymian rather than on the known North American record. This is a typical Beringidan fauna; not without reason REPENNING has called it "a Siberian fauna found in Beringia on the wrong side of the Bering Strait" (1980, p. 43). The Cape Deceit fauna seems to hold an intermediate position between the Kutuyakhan and Olyoran or to correspond to any early phase of the Olyoran mammalian age not yet recognized in West Beringida. Thus, I refer this fauna to the time span roughly estimated as 1.8—1.2 million years ago; which means, if we follow the present estimates of duration of the Irvingtonian Land Mammal Age, to the early Irvingtonian.

The Cape Deceit fauna includes the earliest known species of the collared lemming lineage, *Predicrostonyx hopkinsi*; two archaic voles, one of which ("*Microtus*" *deceitensis*) as V. ZAZHIGIN (pers. comm. 1979) suggests is a member of *Allophaiomys* group, while the other ("*Pliomys*" *deeringensis*) has been referred independently to *Phenacomys* by both J. CHALINE and V. ZAZHIGIN. A lemming related to *Lemmus sibiricus* KERR (= *L. obensis* BRANTS), a ground squirrel (*Citellus* sp.), a marmot and a very large pika were also members of this fauna. Among the large mammals of the Cape Deceit fauna were a large horse, presumably related to the Olyoran *Equus* (*Plesippus*) *verae*, reindeer, elaphoid deer and a wolf. While GUTHRIE and MATTHEWS correlate this fauna with an interval of relatively mild climate, the presence of mammals preferring dry habitats such as ground squirrel, marmot and pika, is rather suggestive.

The Olyor fauna marks a large and important stage in the history of Beringidan mammals. According to our concept, this stage started around 1.2—1.4 million years ago and finished 0.7 million years ago. The fauna of the Olyoran age is known from numerous localities across West Beringida. In all the localities which have been adequately studied, fossils of Olyoran mammals are associated with the upper part of a reversally magnetized zone correlated with the Matuyama epoch. Sometimes this part includes an event of normal polarity interpreted as the Jaramillo Event (VIRINA 1979; VIRINA et al. 1984). Small mammal remains allow the recognition of some stages in the evolution of the Olyoran fauna (ZAZHIGIN in SHER et al. 1979), but its earliest stages remain still unknown.

About 30 mammalian taxa have so far been identified in the Olyor fauna. They include shrews, hares and pikas, ground squirrels and beavers, several voles, brown and collared lemmings, archaic wolverines, wolves, polar foxes and xenocyons, machairodontine and panthera-like cats, bears, archaic mammothes, a large group of horses, reindeer and giant archaic mooses, many musk-oxen, bisons and also probably elaphoid cervids and saiga antelopes. Many mammalian lineages are represented in the Olyoran by rather primitive evolutionary stages characteristic of the latest Villafranchian ("Epivillafranchian"). The whole complex of rich and diverse palaeoecological evidence indicates that the Olyoran fauna existed under conditions of very cold and, at times, fairly dry climate. Its environment alternated repeatedly between forest-tundra and tundra to tundra-steppe (GITERMAN 1980; GITERMAN et al. 1982). It is reasonable to believe that well-established cryoxerophilous communities of tundra-steppe type had already existed in Beringida during Olyoran times as early as about 1.0 million years ago.

A remarkably rich mammalian fauna, including many Olyoran elements, is known from south-western Beringida, in the Verkhoyansk region. Studies in progress of original localities exposed just at the lower-water level of the Adycha River reveal that this fauna includes some more advanced elements alongside those of the Olyoran. The bone-bearing members of the section, which have been examined are supposed to be rather later than the typical Olyoran. It remains obscure whether the Olyoran faunal elements were found in original association with the advanced ones or whether they were reworked from earlier deposits. In any case, the original source of the Olyoran fossils is very near to the river level, because the fossils occur abundantly on the river bars. Moreover, in this natural palaeontological exhibition some fossils have been found which are evidently more archaic than the Olyoran. These include some cervids and carnivores e.g. a very primitive wolverine described by SOTNIKOVA (1982) as *Gulo minor*. A single find of a hipparion tooth (cf. *Neohipparion*) in Northern Siberia made by A. KOTOV in the headwaters of the Adycha may be of the same age. The age of this find is still uncertain but it is evidently pre-Olyoran. I tentatively refer these isolated finds of ancient mammals to a hypothetical Adychan age.

The initial Pleistocene[1]) fauna occurring in the upper part of the Olyor Suite just above the Brunhes/Matuyama boundary has recently been distinguished as the Akanan Fauna. It is still rather similar to the Olyoran, but nevertheless reflects quite noticeable changes. This fauna is marked by the presence of a primitive *Dicrostonyx* (*D. renidens*), instead of *Predicrostonyx* (*P. compitalis*), and various *Microtus* s. str., instead of *Allophaiomys*, as well as by gradual replacement of plesippoid horses by caballoid horses.

In general, the Beringidan fauna of the first half of the Pleistocene is not adequately known. There is still a gap in the fossil record between the immediately post-Olyoran (i.e. Akanan) and late middle Pleistocene ("Rissian") faunas. Represented in the Beringidan record by a few poor localities the existence of this gap conceals the elimination of many of the most primitive Olyoran groups of mammals together with important changes in surviving native lineages leading towards the late Pleistocene and Recent forms. These include the transformation of *D. renidens* into *D. simplicior*, large caballoid horses into small horses and rather archaic species of *Mammuthus* and *Ovibos* into advanced species.

[1]) The lower limit of the Pleistocene officially adopted in the U.S.S.R. is around 0.7 million years ago.

The Beringidan fauna of the second half of the middle Pleistocene (i.e. "Rissian") does not differ essentially from the typical late Pleistocene fauna. Both are usually united in one complex (mammalian age) traditionally referred to as "Upper Palaeolithic" in western Beringida and as Rancholabrean in Alaska. The late Pleistocene fauna of Beringida (= Würm, Wisconsin) is the best understood (GUTHRIE 1968; HARINGTON 1978; SHER 1971). A report on the late Pleistocene key faunal localities in the Kolyma Lowland, for example Duvanny Yar (early Würm, Zyryan) and Alyoshkina Zaimka (late Würm, Sartan) has been published recently (SHER et al. 1979).

Let us consider the broad environmental conditions under which the Beringidan mammal fauna evolved during the second half of the Pleistocene. The alternation of climatic and landscape conditions formed the background to their evolution. In the Far North, however, these changes ranged between north taiga larch-birch forest and essential treeless tundra-steppe. Such alternation is recorded in many of the key sections of western Beringida. Similar changes are reflected in the fossil insect faunas of this region (KISELYOV 1981). It should be noted that during the phases of relatively milder climate, viz. warmer and moister, xerophilous communities were not entirely eliminated, but their expansion became more or less restricted. To judge from this rather small amplitude of environmental change, the mammal fauna of Beringida can be assumed to have also changed rather little during "warm" and "cold" stages. This appears to account for the lack of really "warm", or "interglacial" faunas in western Beringida.

Drastic change of environment at the Pleistocene/Holocene boundary resulted in the degradation of tundra-steppe communities and in the first uncompensated extinction of a group of large mammals. We can infer that the Holocene turnover of the environment on the plains of Beringida was basically different from the earlier, Pleistocene, phases of warmer and milder climate. Thus, in considering the problem of the Great Pleistocene Extinction we believe that environmental factors were of primary importance with anthropogenic factors only secondary. This interpretation has recently found additional strong support from basic difference between the Recent and Pleistocene insect faunas of Beringida (KISELYOV 1981).

2.2. Dispersal of Beringidan Mammals to North America

In recent years many scientists have noticed that in the latest Cenozoic the Beringidan faunal exchange between the Old and the New World was in the form of "one-way traffic" from Asia to North America. In my view, this is accounted for by the important role played by Beringida itself as an area of speciation and subsequent dispersal of mammals into the middle latitudes of North America. An essential component of this dispersal is the autochthones of Beringida. Moreover, even the taxa that can be more reasonably considered as members of "the transit traffic" via Beringida have their own fossil record in this northern land. Evidently, they had to undergo, to some degree, an adaptation to this high-latitude environment. Thus, in the course of the progress of Pleistocene cooling into middle latitudes, the Beringidan faunal elements, which were dominated by derivatives of Asian stocks, were at an advantage to invade the middle latitudes of both continents. With regard to North America, this invasion is registered as the "wave of Old World immigrants".

Let us attempt to consider chronological relations between the known Beringidan fossil record and the first known appearance of "Asiatic" mammals in North America south from the main area of continental glaciations[2]). This attempt is of a preliminary nature only; real correlations becoming possible only after direct comparative study of fossil collections from both regions. Meanwhile, the Beringidan fossil mammals are still only in the initial stage of study, as are the faunas of the extremely important localities in southernmost Canada — Medicine Hat (Units D, E, G) and Wellsh Valley. These are the northernmost of the relatively early faunas of "North America". They are essentially of "North American" composition but represent rather early assemblages which include Beringidan faunal elements.

The left hand part of Table 1 includes an essentially complete taxonomic list of mammals at present known as fossils from Beringida. The stratigraphic range of these taxa is given in terms of stages of faunal evolution in Beringida (mammalian ages).

The right hand column shows stratigraphic range of the same or supposedly related taxa in "North America" in terms of the North American Land Mammal Ages.

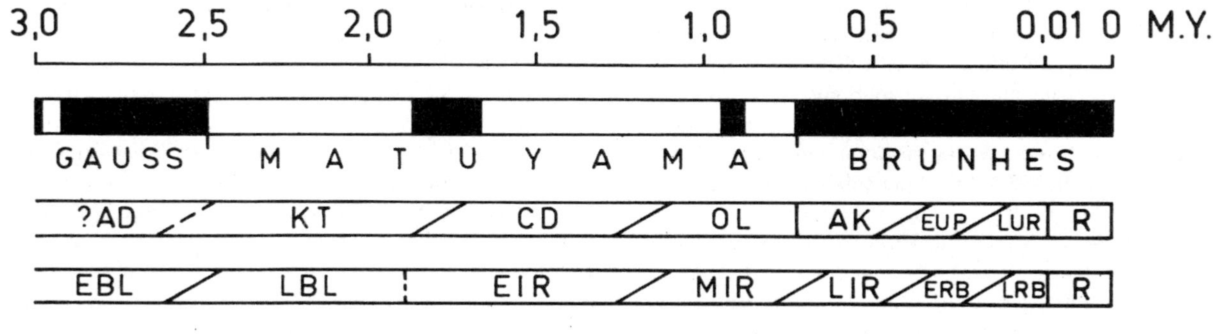

Fig. 1. Probable correlation between the Beringidan and North American land mammal ages, with the time scale in million years B.P. and standard palaeomagnetic scale. See abbreviations at the base of Table 1

[2]) For brevity I shall further refer to this area as "North America".

Table 1. Comparison of late Cenozoic mammalian fauna of Beringida and North America south from the area of continental glaciations

Taxa of fossil mammals known from Beringida	Their known stratigraphic range in Beringida	Stratigraphic range of the same or supposedly related taxa in "North America"	Remarks on the "North American" record
Sorex spp.	CD-R	EBL-R	
Homo sp.	LUP-R	LRB-R	
Megalonyx sp.*	LRB	EBL-LRB	
Hypolagus sp.	?CD	EBL-LBL	(numerous species)
Lepus spp.	OL-R	EIR-R	(group of extant species)
Ochotona sp. (archaic)	KT	—	
Ochotona whartoni	CD-?OL	—	
Ochotona spp.	OL-R	LIR-R	
Tamiasciurus sp.	LUP	LIR-R	
Citellus spp.	CD-R	LIR-R	(group of extant species)
		EBL-EIR	(group of extinct species)
Marmota sp.*	CD-R	LBL-R	
Castoroides sp.*	LRB	EBL-LRB	
Trogontherium cf. *cuvieri*	OL-?AK	—	
Castor spp.	?OL, AK-R	EBL-R	
Erethizon sp.*	LRB	LIR-R	
Ondatra sp.*	RB-R	LBL-R	
Clethrionomys spp.	OL-R	LIR-R	
Phenacomys deeringensis	CD	?EIR	(Wellsh Valley)
		?LIR-R	(*Phenacomys intermedius*)
Predicrostonyx hopkinsi	CD	—	
Predicrostonyx compitalis	OL	—	
Dicrostonyx renidens	AK	—	
Dicrostonyx simplicior	EUP	LRB-R	(*Dicrostonyx hudsonius*)
Dicrostonyx gulielmi-torquatus	LUP-R	LRB-R	
Synaptomys sp.	KT	LBL-R	(numerous species)
Lemmus cf. *sibiricus* (= *obensis*)	CD-R	—	
Mimomys (*Mimomys*) cf. *coelodus*	KT	—	
Cromeromys cf. *irtyshensis*	KT	—	
Cromeromys cf. *intermedius*	?CD-?OL	MIR	(„*Mimomys*" *monahani*)
Allophaiomys cf. *pliocaenicus* (incl. „*Microtus*" *deceitensis*)	CD-OL	EIR-MIR	(„*Microtus*" *pliocaenicus*, „*Microtus*" *deceitensis*)
Microtus (*Microtus*) spp.	AK-R	LIR-R	
Microtus (*Stenocranius*) sp.	?AK, EUP-R	—	
Canis lupus cf. *mosbachensis*	OL-AK	MIR	(small *lupus*-like forms)
Canis lupus (large forms)	UP-R	LIR-R	
Alopex sp.	OL-R	—	
Vulpes sp.	RB-R	RB-R	
Xenocyon sp. (cf. *lycaonoides*)	OL-ERB	?	?*Cuon alpinus* in LRB, Mexico
*Arctodus simus**	LRB	EIR-LRB	
Ursus arctos	OL-R	LRB-R	
*Ursus americanus**	RB	EIR-R	
Thalarctos maritimus	LRB-R	—	
*Mustela erminea**	LRB-R	LIR-R	
*Mustela eversmanni**	LRB	—	
*Mustela nigripes**	LRB	ERB-R	
*Martes nobilis**	LRB	LRB	
Gulo minor	AD	—	
Gulo cf. *schlosseri*	OL	MIR	
Gulo gulo	UP-R	LRB-R	
*Taxidea taxus**	LRB	EBL-R	
Spilogale sp.*	LRB	EBL-R	
Homotherium sp.	OL-AK	MIR-?LRB	
Smilodon sp.*	?LRB	EIR-LRB	
Panthera sp. (cf. *spelaea*)	?OL, AK-LUP	LRB	*Panthera atrox*
Lynx sp.	RB-R	LBL-R	
*Mammut americanum**	RB	EBL-LRB	
Archidiskodon sp.	OL	?EIR-MIR	(„*Mammuthus*" *meridionalis*)
Mammuthus sp. (Olyor type)	OL-AK	LIR-ERB	*Mammuthus columbi*-group
Mammuthus primigenius	UP	ERB-LRB	
Neohipparion sp.	AD	Hemphillian	
Equus (*Plesippus*) spp.	CD-OL, ?AK	EBL-LBL	(*Plesippus*-group)
Equus (*Equus*) cf. *pacificus*-group	OL	EIR-LRB	(*Equus giganteus*/*pacificus*-group)
Equus (*Equus*) spp. (large-sized)	AK-EUP	?	
Equus (*Equus*) spp. (small-sized)	UP	?	The relation of the Beringidan forms to
Equus (subg.?) sp. (very small-sized)	?OL	?	the numerous North American ones is
Equus (*Hemionus*) sp.	?LUP	?	uncertain
Coelodonta spp.	?OL, AK-UP	—	
Camelops sp.*	LRB	EBL-LRB	

Table (Continued)

Taxa of fossil mammals known from Beringida	Their known stratigraphic range in Beringida	Stratigraphic range of the same or supposedly related taxa in "North America"	Remarks on the "North American" record
Cervus spp. (*elaphus*-group)	CD-LUP	?LIR-R	
Libralces latifrons	OL-AK	?LRB	(*Cervalces*)
Alces spp.	EUP-R	LRB-R	
Rangifer sp.	?KT, CD-R	?LIR*, LRB-R	* Medicine Hat
Bos (*Poëphagus*) sp.*	LRB	—	
Bison ex gr. *schoetensacki*	OL-AK	?	
Bison priscus-group	AK-LUP, ?R	?LIR, ERB-R	(several species)
Saiga sp.	?OL, AK-LUP	—	
Soergelia sp.	OL-AK	?LIR	
		EIR-LRB	*Euceratherium*
Praeovibos spp.	?CD, OL-?AK	?	
Bootherium sp.*	LRB	LRB	
Ovibos spp.	?OL, AK-R	ERB-LRB	
Symbos spp.*	RB	?LIR, ERB-LRB	
Ovis sp.	ERB-R	ERB-R	

* — taxon is known as a fossil only from East Beringida (Alaska, Yukon)

AD — Adychan (Pliocene, scattered occurrences without stratigraphic control); KT — Kutuyakhan; CD — Capedeceitan (in addition to the Alaskan type locality, a scanty fauna with *Phenacomys* is known from Krestovka as well as some fossils reworked into the Olyor Suite); OL — Olyoran; AK — Akanan; UP — "Upper Palaeolithic" for Siberia including: EUP — for Early and LUP — for Late "Upper Palaeolithic"; RB — Rancholabrean for Alaska and Yukon including: ERB — for Early and LRB — for Late Rancholabrean; R — Holocene (Recent); EBL — Early Blancan; LBL — Late Blancan; EIR — Early Irvingtonian; MIR — Middle Irvingtonian; LIR — Late Irvingtonian. With the same abbreviations for Rancholabrean and Recent as listed above

The figure 1 gives probable correlations between the Beringidan and "North American" faunal stages. The North American evidence based mainly on the recently published review by KURTÉN and ANDERSON (1980) as well as on the papers by GUTHRIE and MATTHEWS (1971), PEWE (1975) and HARINGTON (1978). The Siberian record given here is mainly according to the author; with the identification and study of small mammals and carnivores by V. ZAZHIGIN and M. SOTNIKOVA respectively. The classification of mammals mainly follows GROMOV and BARANOVA (1981).

Table 1 shows more than 80 taxa of land mammals at present known in the late Cenozoic fauna of Beringida. The real number of recorded species is higher as the presence of more than one Beringidan species is clear in such genera as *Sorex*, *Microtus*, *Clethrionomys* and some others, but we cannot distinguish them reliably at the present level of knowledge.

We consider the mammalian fauna of West and East Beringida as an entity, thus implying that numerous taxa have not yet found in Alaska and Yukon merely because of the small number of pre-Rancholabrean localities. Perhaps, only *Coelodonta*, definitely, and *Trogontherium*, probably, can be regarded as genera of strictly western distribution in Beringida.

Among the taxa known at present only from the Alaskan part of Beringida there are some that will quite probably be found in West Beringida in the future — either the species themselves or their evolutional predecessors (*Marmota*, *Martes*, *Mustela*, *Poephagus*, *Symbos*, *Bootherium*). The remainder are typical American genera that reached Alaska only for some geologically short time during the Rancholabrean or even its latest part: *Megalonyx*, *Erethizon*, *Castoroides*, *Ondatra*, *Arctodus*, *Taxidea*, *Spilogale*, *Mammut*, *Camelops* and probably *Smilodon*. HARINGTON (1980) considers these genera as "survivors of a northern surge or surges of animals from southern North America that occurred before the peak of the Wisconsin glaciation".

Table shows the resemblances in the faunas between Beringida and "North America". Although the Beringidan fauna used to be considered an essentially Asiatic one, the surprising thing is to find 82 percent of 66 genera and subgenera of the land mammals recorded in Beringida, present in the middle latitudes of North America. These common taxa may be divided into several groups.

The largest group comprises the taxa with an earlier 'first appearance' in Beringida than in "North America". Most of them had a rather long evolutionary history in Beringida. This group include. *Ochotona*, *Clethrionomys*, *Phenacomys*, *Dicrostonyx*, *Synaptomys*, *Cromeromys*, *Allophaiomys*, *Microtus* (*Microtus*), *Ursus arctos*, *Gulo*, *Panthera*, *Mammuthus* (Olyor type), *M. primigenius*, *Cervus* (elaphoid group), *Libralces (Cervalces)*, *Alces*, *Rangifer*, *Bison*, *Ovibos*, *Soergelia*, *Ovis* and probably *Bootherium*. Some other taxa, evidently related to Beringida in their origin, can be added to this group. Their fossil record in Beringida starts rather early, but at present the earliest known occurrence is in "North America". These are *Lepus*, *Vulpes*, *Homotherium*, *Archidiskodon*, *Symbos* and (?) *Homo*. With this addition the first group of common taxa amounts to 42 percent of all the known Beringidan taxa.

The second group comprises the Beringidan members of genera that had a fairly long history in the middle latitudes of both the Old and New World. Their history commonly goes far back in the Pliocene. Among these such genera as *Sorex*, *Hypolagus*, *Citellus*, *Marmota*, *Castor*, *Canis*, *Ursus* (*americanus*-group), *Martes*, *Mustela*, *Lynx*, *Equus* and *Neohipparion* can be listed. The present day taxonomy of fossil species of these genera is rather confused; affinities between the Old and New World representatives and their chronological relations being sometimes speculative. Thus we consider that this group accounts for 23 percent of known Beringidan taxa as elements of a Holarctic distribution; their origin in Beringida and direction of dispersal being uncertain.

Evidently some members of group II may be included in group as our knowledge progresses. For instance, some of the extremely varied North American horses could easily be derivatives of some Beringidan caballoid horses. On the other hand, some of the Beringidan horses probably result from a few successful invasions of American forms into the area. But neither the former nor the latter origin can be shown at present, as the solution calls for a thorough revision of the Old and New World horses, which will be, as G. G. SIMPSON puts it, "a truly Herculean or perhaps a Sisyphean task".

The last group of taxa common to Beringida and "North America" comprises the undoubtedly American elements in the Beringidan mammalian fauna. This form, 17 percent of the known Beringidan genera and consist almost entirely of the Rancholabrean immigrants to Alaska and Yukon from the south which are listed above. Perhaps only one member of this group succeeded in crossing the Bering Strait: *Tamiasciurus* (AGADZHANYAN 1980). It is noteworthy that of the remaining 10 taxa six became extinct (a very significant proportion), two retreated to the far south and only two survived in Alaska.

18 percent of known Beringidan mammalian taxa are distinct from the "North American" fauna. Of these some earlier taxa may have later given rise to descendants which spread south [*Praeovibos*, *Xenocyon*, *Mimomys* (*Mimomys*)], while *Predicrostonyx* certainly did so. If definitely established, these taxa should enlarge group I. The reasons for the non-migration of two genera unknown even in Alaska — *Coelodonta* and *Trogontherium* — are disputable. In the rest of the group of Beringidan taxa not known in "North America" we can easily recognize the most typical members of the cryoxerotic (tundra-steppe) fauna of Beringida, such as *Alopex*, *Microtus* (*Stenocranius*), *Lemmus*, *Saiga* and *Bos* (*Poëphagus*).

A suggested grouping of the known Beringidan mammalian taxa is summarized in Table 2.

Table 2. Comparison of the late Cenozoic land mammal fauna of Beringida with the "North American" fossil record, [in %]
(The 66 genera and subgenera recorded in Beringida are taken as 100%)

A. Taxa common to Beringida and "North America"	82
I. Of Beringidan or Palaearctic origin including:	42
a) recorded in Beringida earlier than in "North America"	33
b) recorded at present in "North America" some earlier than in Beringida	9
II. Of Holarctic (uncertain) origin	23
III. Of Nearctic origin	17
including American endemics recorded in Beringida only from the Alaskan Rancholabrean	16
B. Beringidan taxa not known in "North America"	18

The preliminary character of these estimations is evident. Because of the lack of a fossil record this category does not include some genera which certainly inhabited Beringida during the Pleistocene (e.g. *Sciurus* and *Lutra*) or are supposed to have entered North America from Eurasia via Beringida (e.g. *Oreamnos*). The relationship of many Old and New World taxa are far from clear. But it should be noted that 10—15 years ago such estimations were simply impossible and they do seem to be rather suggestive.

Despite these shortcomings, the lists given convincingly demonstrate the powerful dispersal flow of Beringidan mammals into the middle latitudes of North America. Almost half of a rather large list of Beringidan mammals spread into what is now southernmost Canada and the continental United States. Most of these taxa were formerly known as "immigrants from Asia". Now it is possible to regard them more certainly as newcomers from the northern land, Beringida, where they had to undergo adaptation to its rigorous environment to a greater or lesser extent. It has become possible to reach this conclusion due to (1) the appearance of a fossil record of these mammals in Beringida, and (2) the availability of evidence of climatic conditions in Beringida during all the discussed period of the late Cenozoic history, i.e. the last 2.5 million years.

The influx of Beringidan immigrants may be tentatively subdivided into several waves. During the late Blancan and early Irvingtonian the dispersal of elements of the Kutuyakh and Cape Deceit faunas of Beringida occurred. The arrival and fast expansion of bog lemmings, *Synaptomys* in North America is perhaps the most remarkable feature of this stage.

On the basis of the Olyoran and early Akanan faunas of Beringida the middle and late Irvingtonian dispersal appears to have been more active. This wave of dispersal can be roughly correlated with the appearance of the earliest faunas indicating a cooler climate in the middle latitudes of North America, such as Medicine Hat (D, E, G) in southern Alberta, Cudahy in Kansas, Cumberland Cave in Maryland, Java in South Dakota, and others. Some of these faunas are referred to the Kansan Glacial Stage. Significant elements involved in this dispersal include *Clethrionomys*, *Ochotona*, *Microtus* (*Microtus*) and probably *Soergelia* and *Rangifer* (if the latter is reliably recorded in the Unit G of Medicine Hat). Dispersal of *Bison* occurred probably near the end of this stage or some time later.

The middle/late Irvingtonian dispersal, which can be dated very roughly to 1.2—0.6 million years ago, involved many characteristic Beringidan lineages. Nevertheless, dispersal of many rather ancient lineages of Beringidan mammals proved possible only during Rancholabrean times, mainly in the Wisconsin. This is probably true of *Dicrostonyx*, *Panthera*, *Ursus arctos*, *Mammuthus primigenius*, *Alces* and some others. Nevertheless some of the most typical members of the Beringidan cryoxerotic fauna failed to overcome the environmental barriers separating them from temperate North America (*Alopex*, etc., see above).

3. Conclusion

The hypothesis of an ancient northern land of Beringida as a significant center of late Cenozoic speciation is born out by the recent palaeontological discoveries in northeastern Yakutia, Alaska and the Yukon. Permafrost and an environment of tundra type existed there as far back as 2.5 million years ago. Around 1.0 million years ago cryoxerotic (tundra-steppe) communities are recorded in the Kolyma Lowland. Comparison of fossil mammal faunas of Beringida and temperate North America shows that virtually all the late Pliocene and Pleistocene "Old World immigrants to North America" were among the members of a Beringidan fauna. Thus, they had to undergo some adaptations to the rigorous climate. Progressive climatic cooling resulted in dispersal of Beringidan mammals to the middle latitudes of North America. This dispersal occurred in several waves. Generally, the more marked the "cold" and "cold-dry" adaptations of a given taxon, the later is its recorded appearance in temperate latitudes; the most specialized members of Beringidan tundra-steppe fauna never spreading far south.

The mass radiation of Beringidan mammals into temperate North America virtually ruled out a dispersal in the opposite direction. Thus, typical American mammals having reached Alaska for a relatively short time had little chance to move further west. It should be noted that apart from these several, not very successful attempts to colonize Beringida from North America, since the early Pleistocene (Akanan) the Beringidan faunas almost did not accept new members either from the Old or New World temperate latitudes. On the contrary, there was a prevailing radiation from Beringida in both directions. This provides additional evidence of a very ancient composition of the Beringidan fauna, which had probably been established as a whole as far back as the late Pliocene.

Acknowledgements:

I am indebted to those of my colleagues who are involved in the study of geologic and palaeontologic materials from Beringida: V. S. ZAZHIGIN, R. E. GITERMAN, and M. V. SOTNIKOVA (Geological Institute, USSR Academy of Sciences, Moscow), E. I. VIRINA, S. V. KISELYOV and A. A. ARKHANGELOV (Moscow State University). I wish also to thank Drs. E. A. VANGENGEIM (Moscow), B. A. YURTSEV (Leningrad), H. D. KAHLKE (Weimar), J. V. MATTHEWS, Jr. (Ottawa) and R. D. GUTHRIE (Fairbanks) for very useful discussions on the subjects considered in this paper.

Summary

The hypothesis of the ancient northern land of Beringida as a significant center of late Cenozoic speciation is born out by the recent palaeontologic discoveries in northeastern Yakutia, Alaska and the Yukon. Permafrost and an environment of tundra type existed there as far back as 2.5 million years ago. Around 1.0 million years ago cryoxerotic (tundra-steppe) communities are recorded in the Kolyma Lowland.

More than 80 taxa of mammals are already known in the late Cenozoic fossil record of Beringida, including 66 genera and subgenera. It is shown that 82% of these genera and subgenera are also known as fossils in the middle latitudes of North America south of the area of continental glaciations. Most numerous among these taxa common to Beringida and temperate North America are those mammals that had existed in Beringida earlier than their first appearance in the middle latitudes of North America. They amount to almost half of all the known taxa of the late Cenozoic mammals of Beringida and can be considered as the Beringidan element of the temperate North America fauna. As rigorous climates are recorded from Beringida since at least the late Pliocene, we must suppose that the Beringidan mammals were to a greater or lesser extent adapted to a Subarctic environment.

Taxa of undoubted American origin amount to 17% of the known fauna of Beringida. It is noteworthy that almost all of them appeared only for a rather short time during the late Pleistocene in Alaska or the Yukon and failed to migrate further west.

Among the Beringidan genera that did not migrate to temperate North America, many rather typical members of cryoxerotic (tundra-steppe) fauna can be recognized.

It is demonstrated that almost all the late Pliocene and Pleistocene "Asiatic immigrants to North America" were among the member of the Beringidan fauna, so they had acquire some adaptation to a rigorous climate. As climatic cooling progressed in the middle latitudes, these mammals spread to temperate North America. Their dispersal occurred in several waves. Mass radiation of the Beringidan mammals to North America virtually ruled out a dispersal in the opposite direction.

The evidence discussed here supports the idea of great antiquity for the Beringidan fauna which probably had been established, as a whole, as far back as the late Pliocene.

Zusammenfassung

Die Hypothese des alten Nordlands Beringida als bedeutendes Zentrum spätkänozoischer Entwicklung ist das Ergebnis jüngster paläontologischer Entdeckungen in Nordostjakutien, Alaska und Yukon. Permafrost und Tundra-Bedingungen sind aus diesem Raume seit etwa 2,5 Mill. Jahren bekannt und seit etwa 1,0 Mill. Jahren sind cryoxerotische (Tundra-Steppe) Assoziationen aus dem Kolyma-Tiefland nachgewiesen.

Mehr als 80 Säugetier-Taxa, die 66 Gattungen und Subgattungen angehören, sind heute bereits aus dem späten Känozoikum Beringidas bekannt. Es konnte gezeigt werden, daß 82% dieser Gattungen und Untergattungen ebenfalls fossil aus den mittleren Breiten Nordamerikas südlich der Kontinental-Vereisungsgrenze bekannt sind. Am zahlreichsten unter diesen Taxa, die wiederum am häufigsten in Beringida und im temperierten Nordamerika angetroffen werden, sind solche, die in Beringida bereits viel früher existierten als in den mittleren Breiten Nordamerikas. Diese Taxa umfassen fast die Hälfte aller bisher bekannten spätkänozoischen Säugetierarten Beringidas und sie können gleichzeitig als Beringida-Elemente der temperierten Säugetierfauna Nordamerikas angesehen werden. Da in Beringida mindestens seit dem späten Pliozän Klimadepressionen nachzuweisen sind, können wir annehmen, daß die Beringida-Säugetiere bereits in größerem oder geringerem Umfange an subarktische Klimabedingungen angepaßt waren.

Taxa zweifelsfrei amerikanischen Ursprungs erreichen etwa 17% der von Beringida bekannten Säugetierfauna. Dabei erscheint es notwendig aufzuzeigen, daß im Prinzip alle diese Formen erst im späten Pleistozän und dazu nur für eine ziemlich kurze Zeit in Alaska und Yukon erschienen und daß diese Formen nicht weiter nach Westen vordringen konnten.

Unter den Beringida-Gattungen, die nicht das temperierte Nordamerika erreichten, befinden sich eine relativ große Anzahl typisch cryoxerotischer (Tundra-Steppe) Elemente.

Darüber hinaus konnte gezeigt werden, daß fast alle spätpliozänen und pleistozänen „asiatischen Einwanderer nach Nordamerika" als Elemente der Beringida-Fauna zu bezeichnen sind, die einen bedeutenden Anpassungsgrad an ein extremes Klima erreicht hatten. Mit den fortschreitenden Klimadepressionen in den mittleren Breiten konnten sich die hier charakterisierten Faunenelemente in die temperierten Gebiete Nordamerikas ausbreiten, wobei diese Ausbreitung in verschiedenen Wellen erfolgte, und die Radiation der Beringida-Elemente eine Ausbreitung in der entgegengesetzten Richtung ausschloß.

Die hier aufgezeigten Verhältnisse aber unterstützen die Annahme eines hohen Alters der Beringida-Fauna als Komplex, die möglicherweise bis in das späte Pliozän zurückreicht.

Резюме

Новые палеонтологические открытия на территории Северной Якутии, Аляски и Юкона подтверждают гипотезу о древней северной суше — Берингиде — как важнейшем центре формообразования в позднем кайнозое. Не позднее 2,5 млн лет назад здесь формируется многолетняя мерзлота и биоценозы тундрового типа, а около 1,0 млн лет назад — криоксеротические (тундростепные) биоценозы.

В настоящее время в позднем кайнозое Берингиды установлено присутствие более 80 таксонов ископаемых млекопитающих, в том числе 66 родов и подродов. Показано, что 82% этих родов и подродов известны в ископаемом состоянии и в средних широтах Северной Америки к югу от области материковых оледенений. Среди этих таксонов, общих для Берингиды и умеренной Северной Америки, наиболее многочисленны те млекопитающие, которые существовали в Берингиде раньше, чем они впервые появляются в ископаемой летописи умеренных широт Северной Америки. Они составляют почти половину всех таксонов млекопитающих, известных из позднего кайнозоя Берингиды, и их можно рассматривать как Берингидский элемент в фауне умеренных широт Северной Америки. Существование в Берингиде суровых климатических условий по крайней мере с позднего плиоцена заставляет считать их в той или иной степени адаптировавшимися к субарктической обстановке.

Таксоны несомненно американского происхождения составляют 17% известной фауны Берингиды. Примечательно, что почти все они лишь на короткое время в позднем плейстоцене появлялись на Аляске или в провинции Юкон и не смогли расселиться дальше к западу.

Среди берингидских родов, не расселявшихся в умеренные широты Северной Америки — многие наиболее типичные представители криоксеротической (тундростепной) фауны.

Анализ показывает, что почти все позднеплиоценовые и плейстоценовые „азиатские иммигранты в Америку" входили в состав фауны Берингиды и должны были, следовательно, пройти там адаптации к суровым климатическим условиям. По мере распространения похолодания на умеренные широты млекопитающие Берингиды несколькими волнами расселялись к югу. Массовая радиация млекопитающих Берингиды в Северную Америку практически исключала расселение в противоположном направлении.

Проведенный анализ свидетельствует в пользу того, что фауна Берингиды имеет весьма древний облик и ещё, вероятно, в конце плиоцена сложилась как устойчивое фаунистическое целое.

References

Agadzhanyan, A. K., 1980: Fauna mlekopitajuščich plejstocena Čukotki i osnovnye ėtapy ee formirovanija (The Pleistocene mammalian fauna of Chukotka and the main stages of its development). — In: Novejšie otloženija i paleogeografija plejstocena Čukotki. — Moskva. — P. 256—259.

Giterman, R. E., 1980: The history of Pliocene and Pleistocene vegetation in the western part of Beringia. — In: 4th Internat. Palynol. Conf., Proc. — Lucknow 3. — P. 134—139.

Giterman, R. E.; Sher, A. V.; Matthews, J. V., Jr., 1982: Comparison of the development of tundra-steppe environments in West and East Beringia: pollen and macrofossils evidence from key sections. — In: D. M. Hopkins; J. V. Matthews, Jr. (Eds.): Paleoecology of Beringia. — New York: Academic Press. — P. 43—73.

Gromov, I. M.; Baranova, G. I. (Eds.), 1981: Katalog mlekopitajuščich SSSR (pliocen—sovremennost') [Catalogue of mammals of the USSR (Pliocene—Recent)]. — Leningrad: Nauka. — 456 Pp.

Guthrie, R. D., 1968: Paleoecology of the large-mammal community in Interior Alaska during the Late Pleistocene. — Amer. Midl. Naturalist. — **79**. — P. 346—363.

Guthrie, R. D.; Matthews, J. V. Jr., 1971: The Cape Deceit fauna. — Early Pleistocene mammalian assemblage from the Alaskan Arctic. — In: Quaternary Res. — London (1). — P. 474—510.

Harington, C. R., 1978: Quaternary vertebrate faunas of Canada and Alaska and their suggested chronological sequence. — In: Syllogeus Ser. — Ottawa **15**. — 106 Pp.

—, 1980: Faunal exchanges between Siberia and North America: evidence from Quaternary land mammal remains in Siberia, Alaska and the Yukon Territory. — In: Can. J. Anthropol. — **1** (1). — P. 45—49.

Kiselyov, S. V., 1981: Pozdnekajnozojskie žestkokrylye Severo-Vostoka Sibiri (Late Cenozoic Coleoptera of Northeastern Siberia). — Moskva: Nauka. — 116 pp.

Kurtén, B.; Anderson, E., 1980: Pleistocene mammals of North America. — New York: Columbia Univ. Press. — 442 Pp.

Pewe, T. L., 1975: Quaternary geology of Alaska. — In: U.S. Geol. Surv., Prof. Pap. — Washington **835**. — P. 1—145.

Repenning, C. A., 1980: Faunal exchanges between Siberia and North America. — In: Can. J. Anthropol. — **1** (1). — P. 37—44.

Sher, A. V., 1974: Pleistocene mammals and stratigraphy of the Far Northeast USSR and North America. — In: Internat. geol. Rev. — **16**. — P. 1—284.

—, 1976: Rol' Beringijskoj suši v formirovanii fauny mlekopitajuščich Golarktiki v pozdnem kajnozoe (The role of Beringian Land in the development of Holarctic mammalian fauna in the Late Cenozoic). — In: V. L. Kontrimavičus (otv. red.): Beringija v kajnozoe. — Vladivostok: P. 227 241.

Sher, A. V.; Giterman, R. E. . . . , 1980: New data on late Cenozoic deposits of the Kolyma Lowland. — Internat. geol. Rev. — **22** (6). — P. 643—655.

Sher, A. V.; Kaplina, T. N. . . ., 1979: Late Cenozoic of the Kolyma Lowland. — In: 14th Pacific Sci. Congr., Guidebook, Tour XI. — Moscow: P. 1—116.

Sotnikova, M. V., 1978: Novye dannye o chiščnych mlekopitajuščich verchnego pliocena i nižnego plejstocena na territorii Severo-Vostoka SSSR (New evidence on the upper Pliocene and lower Pleistocene carnovorous mammals of Northeast USSR). — In: Bjull. kom. izuč. četv. perioda. — Moskva **48**. — P. 22—30.

—, 1982: K istorii roda *Gulo* v Evrazii (On the history of *Gulo* in Eurasia). — In: Mamontovaja fauna Aziatskoj časti SSSR. — In: Tr. zool. Inst. Akad. Nauk. SSSR. — Leningrad 111. — P. 65—73.

Virina, E. I., 1979: Paleomagnetism of the Kolyma Lowland's Late Cenozoic sediments. — In: 14th Pacific Sci. Congr., Committee 'B'. Sect. B III, Abstr. Pap., — Moscow 2. — P. 185—186.

Virina, E. I.; Zazhigin, V. S., Sher, A. V., 1984: Paleomagnitnaj charakteristika tipovych mestonachoždenij orelskogo faunističeskogo kompleksa (Kolymskaja nižmennost') [Paleomagnetic (Kolyma Lowland)]. — In: Izvest. Akad. Nauk SSSR, Ser. Geol. — Moskva.

Zazhigin, V. S., 1976: Rannie ėtapy ėvoljucii kopytnych lemmingov (Dicrostonychini, Microtinae, Rodentia) — charakternych predstavitelej subarktičeskoj fauny Beringii [Early stages of evolution of collared lemmings (Dicrostonychini, Microtinae, Rodentia) as characteristic representatives of Beringian Subarctic fauna]. — In: V. L. Kontrimavičus (otv. red.): Beringija v kajnozoe. — Vladivostok: P. 280—288.

Quartärpaläontologie 6 195—208 Berlin 1986

Histologie und Histochemie der Haut des Magadan-Mammuts

Von Vladimir Evgenevič Sokolov und Elena Borissovna Sumina, Moskau

Mit 10 Abbildungen, einer Tabelle und einer Tafel

Inhalt

Abstractum

Histologische und histochemische Untersuchungen der Haut eines vor kurzem entdeckten juvenilen Mammut-Kadavers, der nahe Magadan, UdSSR, gefunden wurde, werden diskutiert. Gewisse Teile der Mammut-Haut zeigen eine mehr oder weniger vollständige Erhaltung der Haut-Strukturen, die detaillierte Untersuchungen ermöglichten.

The histochemical analysis of lipid and non-lipid compounds finally led to a model of close complex-interrelationships and correlations between the different classes of chemical substances found in the structures of the fossil mammoth skin.

Обсуждаются гистологические и гистохимические исследования кожи недавно найденного трупа мамонтенка. Определенная часть кожи мамонтенка более или менее полно сохранила морфологические структуры, что делает возможным детальные исследования.

1. Einleitung

Zu Beginn unseres Jahrhunderts erschienen mehrere Arbeiten über sibirische Mammut-Funde, die besonders durch das dreibändige Werk „Wissenschaftliche Resultate einer Expedition, ausgerüstet durch die Kaiserliche Akademie der Wissenschaften zur Ausgrabung eines Mammuts, das 1901 an der Berezovka gefunden wurde" (1903, 1909, 1914) Aufsehen erregten. Die Haut des Berezovka-Mammuts wurde damals von Zalenskij (1909) histologisch untersucht. Er konnte aber anhand der gewonnenen histologischen Schnitte weder Epidermis noch Drüsen erkennen; Bündel von Kollagen-Fasern waren indessen in ihrem morphologischen Bau noch erhalten geblieben. Er untersuchte den äußeren und inneren Bau der Mammuthaare sowie die Unterscheidungsmerkmale der Haargruppen. Die biochemische Untersuchung der Unterhaut-Fettschicht dieses Mammut-Kadavers wurde von Šestakov (1914) durchgeführt, der ebenfalls die Unterhaut-Fettlagen des Ljachovsker Mammut-Kadavers untersuchte. Das Fett dieses Mammut-Kadavers roch nicht faulig, es erinnerte an gewöhnlichen rohen Speck und wurde von Tieren gefressen. Das ausgelassene reine Fett besaß eine hellgelbe Farbe, wurde bei 24,4 °C fest, enthielt 25,92 % Triglyzeride, 73,60 % freie Fettsäuren und 0,48 % Cholesterin sowie harzartige Bestandteile. Nach Angaben des Autors war das Fett in Farbe und Konsistenz reiner Kuhbutter sehr ähnlich. Weiterhin ermittelte Šestakov den qualitativen und quantitativen Bestand an isolierten Fettsäuren. Er stellte fest, daß im Fett viele ungesättigte Fettsäuren, teils freier, teils glyzeridartiger Natur enthalten waren. Um sie von den gesättigten Säuren zu trennen, unterzog er das Fett der Verseifung. Der isolierte Teil der Fettsäuren stellte fast völlig reine Oleinsäure dar. Der andere Teil des Fetts enthielt reine Palmitinsäure sowie 1—2 % niedermolekulare Säuren, wahrscheinlich Miristin- und Oxysäuren. Es wurden 0,5 % nichtverseifbare Bestandteile, 63,3 % Oleinsäuren, 34,2 % Palmitinsäure und 2 % niedermolekulare gesättigte Säuren bestimmt. Der Autor merkte weiterhin an, daß in den untersuchten Fetten die gesättigte Stearinsäure, die in den Fetten der rezenten Grasfresser vorkommt, fehlte. Bei einem von ihm untersuchten indischen Elefanten wurden 45—50 % Olein und 40—45 % Stearin ermittelt. Als Ergebnis seiner Arbeit zeigte der Autor auf, daß sich das Ausgangsfett der untersuchten Mammut-Kadaver nur wenig veränderte, weil der hydrolytische Prozeß der Zerlegung der Glyzeride in Glyzerin und Fettsäuren unter dem Einfluß des feuchten Einbettungssediments verlief.

Unsere Arbeiten widmeten sich besonders dem Studium der Hautdecke des jugendlichen Mammut-Kadavers, der während Erdarbeiten im Raum Magadan, Rayon Kirgiljach, im Dauerfrostboden Anfang Juli 1977 gefunden wurde.

Sechs Hautproben von Rücken, Bauch und der Leistengegend des gefrorenen Tieres wurden am 17. Juli 1977 in 10 % neutralem Formalin fixiert. Die Obduktion zeigte, daß die Unterhautschicht fehlte und die Muskelschicht zerfasert und schlaff war. Im Januar 1978 wurde von diesem Mammut-Kadaver, der zu dieser Zeit in das Zoologische Institut der Akademie der Wissenschaften der UdSSR überführt worden war, eine weitere Probe und zwar von der Sohle des rechten Hinterfußes, entnommen. Bei dieser Probe waren die Haare unzerstört, weshalb wir besser erhaltene Strukturen erwarteten. Die vorher entnommenen sechs Hautproben vom Rumpfe des jugendlichen Mammut-Kadavers zeigten keine Haare mehr.

Für eine vollständige Erforschung der Morphologie, Histologie und Histochemie der Haut des jugendlichen Mammut-Kadavers wendeten wir im Arbeitsprozeß verschiedene Hauteinbettungs-Methoden an, die wir ausführlicher beschreiben wollen, weil sie, wie sich im Verlauf der Untersuchungen zeigte, nicht nur den Bau, sondern auch den chemischen Hautbestand besser erkennen halfen. Außerdem erlaubten sie die Veränderung festzustellen, denen die Haut des mumifizierten Mammut-Kadavers unterworfen worden war.

Sieben Tage nach der Fixierung wurde die Hautprobe ausgewaschen und vor dem Einschluß in Paraffin und Gelatine einer entsprechenden Bearbeitung unterzogen. Aus reinem Äthanol legte man sie anschließend für drei Tage in Anilin. Anilin — $C_6H_5NH_2$ — aromatisches Amin, ein Nitrobenzol-Derivat, ist chemisch dem Xylol, Toluol und Benzol sehr ähnlich. Diese Verbindung wird in unserem Laboratorium seit Jahren verwendet und stellt zur Zeit das beste Zwischenmedium für die Haut vor der Paraffineinbettung dar. Als Kriterium hinreichender Bearbeitung der Haut rezenter Tiere dient gewöhnlich die Transparenz: die Probe gewinnt ein gelatinöses Aussehen. Ähnliches stellte man bei der Haut des Mammut-Kadavers nicht fest. Danach spülte man nach dem Anillin die Proben in Chloroform und brachte sie in ein Gemisch (1:1) von Chloroform[1]) und Paraffin bei 37 °C. In diesem Falle trocknet das Chloroform die Haut weniger aus als andere Reagenzien, welche bei der Paraffineinbettung Anwendung finden. Aus dem Chloroform-Paraffin-Gemisch überführte man die Hautproben in Paraffin. Die Aufenthaltsdauer in den Gemischen und im Paraffin wird durch den Gewebecharakter der Haut und durch die Dicke der Hautproben bestimmt. Die Mammuthaut beließen wir 24 Stunden im Gemisch, im ersten Paraffin 48 Stunden, ebenfalls 48 Stunden im zweiten Paraffin. Da die Hautproben mumifiziert waren, wurde für eine bessere Durchtränkung mit Paraffin eine längere Zeit notwendig.

Zur Erkennung der lipidartigen Komponenten der Strukturen bettet man die Haut zumeist nach der Formalinfixierung in Gelatine ein. Primär waren unsere Arbeiten in diesem Einbettungsmedium auf Lipiduntersuchungen begrenzt worden. Bei den Arbeiten aber zeigte es sich, daß das Entwässern und Entfetten der Mammuthaut vor der Paraffineinbettung die Epithel-Strukturen sowie die des Unterhautzellgewebes verändert hatten und die Gefäße vollständig verschwunden waren. Bei den weiteren Untersuchungen zur Morphologie der Mammuthaut wurden von uns nur Gelatine-Schnitte verwendet. Mit Hilfe dieser gewebeschonenden Einbettung, welche alle Strukturen der Mammut-Haut konservierte, erhielten wir tadellose Resultate nicht nur im Hinblick auf die Lipide, sondern auch für solche Komplexfärbungen wie Resorzin-Fuchsin nach Hart auf elastische Fasern und die Van Gieson-Methode.

Für das Studium der Histochemie der Haut des Mammut-Kadavers wurde ein Komplex sich gegenseitig ergänzender Methoden in Verbindung mit einem Kontrollsystem, welches hinreichend begründete Schlußfolgerungen über die Natur der nachzuweisenden Verbindungen erlaubte, erstellt. Wir konnten natürlich nicht alle Reaktionen zur Identifizierung der Lipide durchführen, da diese durch die Besonderheit des Materials, die Formalinfixierung und andere Gründe begrenzt waren. Die histochemischen Methoden zur Färbung der Schnitte und ihrer Interpretation wurden aus der Monographie „Histochemie" von PIRS (1962), aus „Prinzipien und Methoden der histochemischen Analyse in der Pathologie" (1971, Red. AVCYN et al.) und aus „Histochemie" von KONONSKIJ (1976) übernommen.

Bei den Untersuchungen kamen folgende histologische und histochemische Reaktionen zur Anwendung: Hämatoxylin Karacci, Eosin, Hämatoxylin nach Heidenhain, Van-Gieson-Methode, Resorzin-Fuchsin nach Hart, Mallory-Methode, Feulgen-Reaktion, Sudan-III-Färbung für Lipide, Sudan Schwarz „B", Nilblausulfat, Fischler-Methode für Fettsäuren, Reaktionen mit Schiffscher Lösung für Fett-Aldehyde, Ultraviolett-Reaktion mit Schiffscher Lösung lokalisiert an den Doppelbindungen, Ameisensäure-Reaktion mit Schiffscher Lösung für ungesättigte Fettsäuren, „Pseudoplasma"-Reaktion, Plasmareaktion, beständiges Grün (pH — 2,2) für Gesamteiweiß, Prozion Blau 2 B (pH = 10,5) für aminosaure Lysin-Reste, Histidin, Tyrosin, Threonin und Serin, PAS-Reaktion mit Diastasekontrolle für Glykogen und neutrale Mukopolysaccharide, Azur I (pH = 1,5; pH = 5,8) für sulfatisierte und nichtsulfatisierte Mukopolysaccharide, fünfprozentige Schwefelsäure für Calcium. Außerdem wurde die Imprägnierung gefrorener Hautschnitte von einer Dicke von 90 mk mit salpetersaurem Silber nach BIELSCHOWSKY-GROSS und mit Vergoldung nach LAVRENTEV durchgeführt.

Zur Beurteilung der chemischen Hautveränderungen, die während des Permafrosts eintraten, wurden parallel laufende Reaktionen an mit Formalin fixierten Hautschnitten des Indischen Elefanten (Rüssel), des Sika-Hirschs (Widerrist) und der nichtfixierten Haut des Moschustiers (Schwanz) durchgeführt (vgl. hierzu die Arbeit von SOKOLOV 1973). Da aus der Literatur die normale Histochemie der Haut rezenter Tiere bekannt ist, konnte bei den beschriebenen Methoden der Gewebefärbung die Anwendungsmöglichkeit der Reagenzien kontrolliert werden.

2. Forschungsmethode und Ergebnisse

Unsere Vorarbeiten zeigten, daß man jede der oben genannten Untersuchungsmethoden nur an der Probe des jugendlichen Mammut-Kadavers durchführen konnte, die von der Fußsohle entnommen worden war, da hier fast alle Strukturen erhalten geblieben waren. Alle übrigen Hautproben vom Rumpf des Tieres hatten an der Oberfläche keine Epithelschicht. Der verbliebene Teil der Hautdecke wird anscheinend durch einen tieferen Teil der dermalen Schicht, die durch dicht verflochtene Bündel kollagener Fasern gebildet wird, vertreten. In einigen Abschnitten erhielten sich in der Tiefe der Haut Haarwurzeln.

Im Bindegewebe der Rumpfderma finden sich kleine Gefäße und Fetttröpfchen, was gut in den Gelatineschnitten sichtbar wurde. Die Paraffinschnitte der Rumpfhaut (Dicke 15—20 mk) brachten wir nur mit Mühe auf Objektträger, wo sie sofort in kleinste Teile zerfielen. Dennoch gelang es, einen Teil der Schnitte zu fixieren, nach dem Trocknen jedoch lösten sie sich von den Objektträgern und die weitere Arbeit mit diesen Schnitten wurde eingestellt. Ein anderer Teil der Schnitte löste sich nach der Deparaffinierung im Spiritus.

Wir begannen unsere Arbeiten mit der Darstellung der Zellen und Zellkerne in der Mammuthaut durch verschiedene hämatoxyline Farbstoffe. Dabei wurden sowohl Gelatine- als auch Paraffinschnitte gefärbt. Dieses führte zu der Erkenntnis, daß die Zellen der Epidermis, der Haarfollikel und des Bindegewebes zerstört waren. Die Feulgen-Methode auf DNK brachte, wie zu erwarten, kein positives Resultat. Demgegenüber blieben Zellbündel der quergestreiften Muskeln erhalten, die in der Vergrößerung als Reihen kleiner Knötchen erscheinen. Wie ZENEVSKAJA (1977), die die Muskeln fossiler Mammute und des Bisons untersuchte, annimmt, stellen sie sich als zerfallende Disken, welche aus Acitin und Myosin bestehen, dar. Die Grenze der Muskelfasern wird besonders gut sichtbar in mit Sudan Schwarz „B" gefärbten Präparaten nach der Fettextraktion mit Pyridin. Dieser Färber erlaubt es, die Phospholipide in einer gegen Zerstörung widerstandsfähigen basalen Membran darzustellen. Die relativ gute Erhaltung verdankt die basale Membran der bindegewebsartigen Entstehung und dem kollagenen Eiweiß, welches im Bindegewebe mumifizierter fossiler Tiere sehr gut erhalten ist (ZENEVSKAJA 1977, SUMINA 1977, SUBBOTIN und TUM 1978). In der Mehrzahl waren auch die Fettzellen der Unterhautschicht erhalten.

Die Fußhaut des jugendlichen Mammut-Kadavers ist mit einer dicken Epidermis bedeckt (Abb. 1). Gut erhalten in der Haut sind ebenfalls die Haarfollikel und die Haare selbst, die faserigen Strukturen (Kollagene, retikuläre und elastische Fasern), Gefäße und Fettzellen. Die Unterhautschicht ist von der Derma durch eine Schicht quergestreifter Muskeln abgeteilt. Die Unterhautschicht schließt bindegewebsartige Fasern, welche parallel zur Hautoberfläche angeordnet sind, und Fettlappen ein (Abb. 2). Hier trifft man auch Hauptvenen und -arterien, Vater-Pacinische Körperchen und Sehnen an[2]).

Die mittlere Dicke der Epidermis beträgt 0.200 mm. Da die Zellen der Epidermis zerstört sind, kann man ihre Schichtung schwer unterscheiden. Das erhaltene Pigment in den verhornten Zellen der Epidermis erlaubt es jedoch, die Mal-

---

[1]) An Stelle des Chloroform-Paraffin-Gemischs kann man auch Anilin mit Paraffin im gleichen Mischverhältnis benutzen. Wegen der starken Toxizität des Anilins taten wir dieses nicht.

[2]) Alle mittleren Maße und ihre Ableitungen sind in Tab. 1 zusammengestellt.

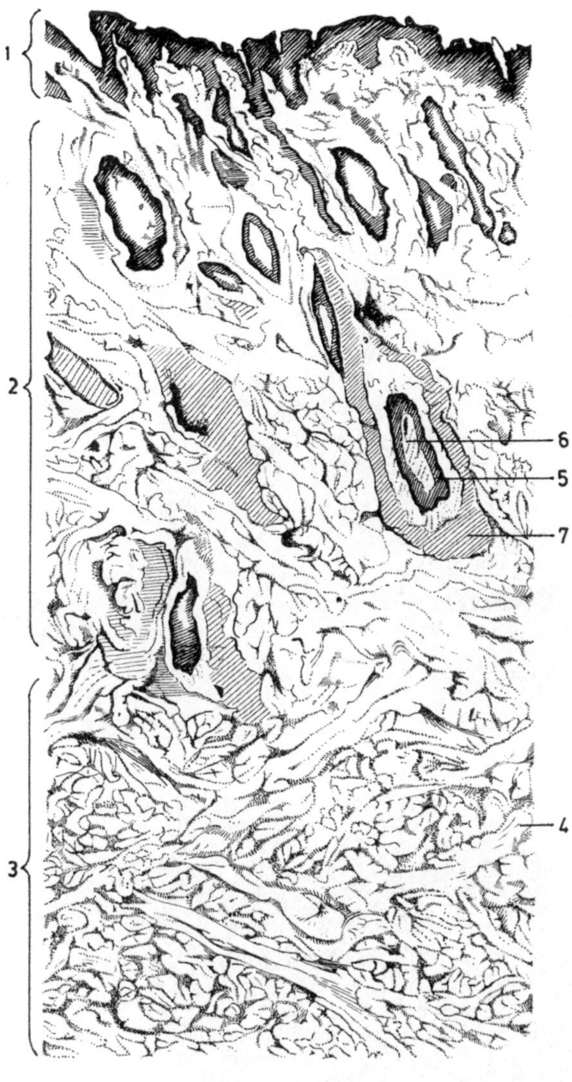

Abb. 1. Vertikalschnitt der Fußhaut des Magadan-Mammuts (Zeichnung nach Hautpräparat, mit 1% Nilblau gefärbt, ×20)

1 — Epidermis; 2 — warzenartige Schicht; 3 — netzartige Schicht; 4 — kollagene Fasern; 5 — Epithel der Haarfollikel; 6 — Haare; 7 — Gelatine

pighi-Schicht und die Hornschicht zu bestimmen. Äußere und innere Grenze der Epidermis sind gewunden.

Die Derma ist durch warzen- und netzartige Schichten charakterisiert. Die Trennung zwischen den Schichten verläuft an den Wurzeln der selten vorhandenen dicken und langen Grannenhaare der ersten Reihe. Die warzenartige Schicht ist dicker als die netzartige (Tab. 1). Die relativ dünnen kollagenen Bündel (12—15 mk) unter der Epidermis bilden eine lockere Verflechtung parallel zur Hautoberfläche. Im mittleren Teil der warzenartigen Schicht verdicken sich die Bündel allmählich auf 32—44 mk. Die dünnen primären Bündel vereinigen sich zu sekundären und tertiären und gehen langsam in große 75—150 mk dicke Bündel über. Diese bilden dann dichte rhombenartige Verflechtungen (Abb. 1). Neben den Haarwurzeln sind dünnere Bündel von Bindegewebe zu dichten Bindegewebstaschen vereinigt. Die warzenartige Schicht wird nur von relativ wenigen Blutgefäßen versorgt. In der Derma finden sich ihre Querschnitte. Alle Gefäße sind von großen Fettzellen umgeben. Spezialfärbungen auf elastische und retikuläre Fasern brachten positive Resultate. Mit den Methoden nach Van-Gieson und Mallory, Resorzin-Fuchsin nach Hart sowie der Silberimprägnierung kann man die retikulären wie die elastischen Fasern neben den kollagenen Bündeln und Haaren darstellen. Besonders reich an elastischen Fasern sind die Gefäße (Abb. 3), Muskeln und Sehnen (Abb. 4). Muskeln, welche die Haare heben, konnten in der warzenähnlichen Schicht nicht entdeckt werden.

Die großen verflochtenen kollagenen Bündel der warzenähnlichen Schicht gehen ohne scharfe Veränderungen in das Bindegewebe der netzartigen Hautschicht über. Im oberen Teil sieht man die größten kollagenen Bündel, welche eine rhombenähnliche Ligatur bilden. In der Regel nehmen bei zahlreichen Säugetieren die verflochtenen Skleroproteide beliebiger Zeichnung in der netzartigen Schicht eine horizontale Lage ein, was auch auf die elastischen Fasern zutrifft. Auf die untersuchte Haut des jugendlichen Mammut-Kadavers trifft das jedoch nicht zu. Die rhombenartige Verflechtung der kollagenen Fasern bricht abrupt ab, es beginnt eine ziemlich dicke Schicht quergestreifter Muskeln, welche die Derma von der Unterhautschicht trennt. In dieser Schicht wechseln dünne kollagene Bündel, die parallel zur Hautoberfläche liegen, mit Lagen von Unterhautfettzellen, Hauptgefäßen und Venen, in denen sogar Klappen sichtbar sind, mit der Unterhautfaszies und Bündeln quergestreifter Muskeln.

An den Stellen des Unterhautfettgewebes sind die Strukturen der Unterhautschicht des Bindegewebes durch große Kristalldrusen auseinandergezogen. Diese erwiesen sich als Fettbildungen und verschwanden im Prozeß der Entfettung und Entwässerung der Haut, der der Paraffinierung voranging, spurlos.

Von der Epidermis in die Tiefe der warzenähnlichen Schicht verlaufen vielfach gewundene Epithelbildungen ohne Haare, die man für Schweißdrüsen halten könnte. Nach Auswertung einer Serie von Hautschnitten aber kamen wir zu dem Schluß, daß es sich um gewundene Haarfollikel dünner Daunenhaare handelt. Solch eine Wurzelform dieser Haare in der Haut war entweder für die Mammute charakteristisch oder aber

Tabelle 1. Meßwerte der Haut Magadan-Mammuts [in mm]

Probe	Haut	Epidermis		Derma		warzenartige Schicht		netzartige Schicht	
Nahe der Sohle der rechten Hinterextremität			in % zur Dicke der Haut		in % zur Dicke der Haut		in % zur Dicke der Derma		in % zur Dicke der Derma
	8,2	0,2	2	8,0	98	5,5	67	2,5	31

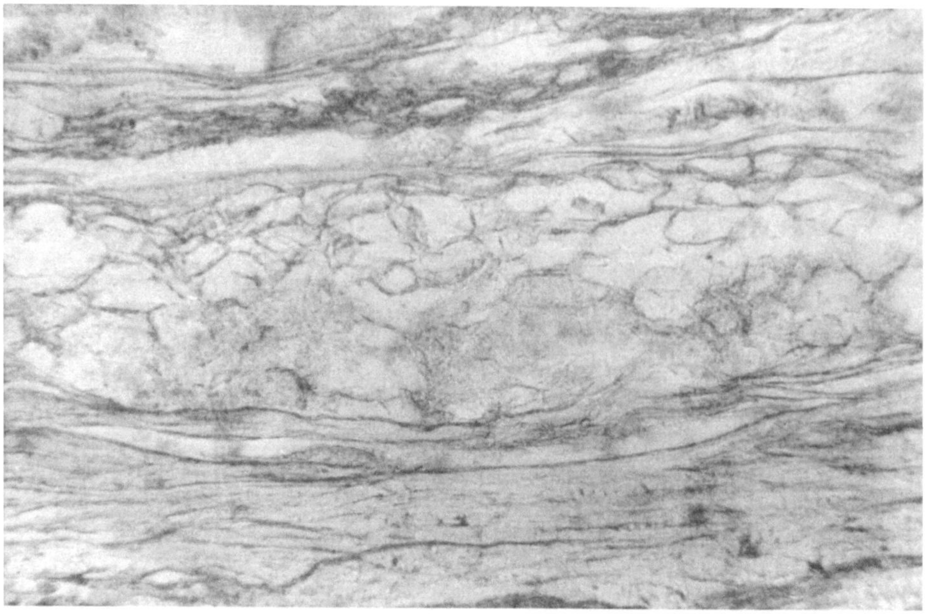

Abb. 2. Fetteil der bindegewebsartigen Unterhautschicht. Färbung mit Resorzin-Fuchsin nach Hart. × 100

Abb. 3. Arteriengefäß der Unterhautschicht. Färbung mit Resorzin-Fuchsin nach Hart. × 100

wurde durch Deformation und Mumifizierung der Haarfollikel hervorgerufen.

Es konnten aber weder Schweiß- noch Fettdrüsen in der Haut des jugendlichen Mammut-Kadavers entdeckt werden.

Die Identifizierung der Lipide begannen wir mit für sie spezifischen Färbern Sudan III und Sudan Schwarz „B". Danach führten wir eine Färbung der Präparate mit Nilblau verschiedener Konzentrationen (1% und 0,02%) und nach der Fischler-Methode durch. Die Beobachtungen zeigten, daß bei allen angewandten Methoden der Färbung auf Fett und fettähnliche Substanzen Epidermis erscheint (Abb. 5; Taf. I, Fig. 1), darunter auch die Hornschicht, Haarscheiden, Gefäße (Abb. 6) und die Stellen in der Haut, wo sie waren, das Unterhautfettgewebe (Abb. 7), die Räume um die Gefäße und Fettzellen in der warzenartigen Schicht und in den Zwischenräumen zwischen allen Bündeln kollagener Fasern und in den Bündeln selbst. Das Fett, welches alle Zwischenräume zwischen den Skleroproteiden ausfüllt, nannten wir „freies Fett". Haare und Muskeln wurden mit Sudan III schwach gefärbt.

Sudan Schwarz „B" färbt wie Sudan III alle Lipide insgesamt kontrastreich und gibt ein annehmbares morpholo-

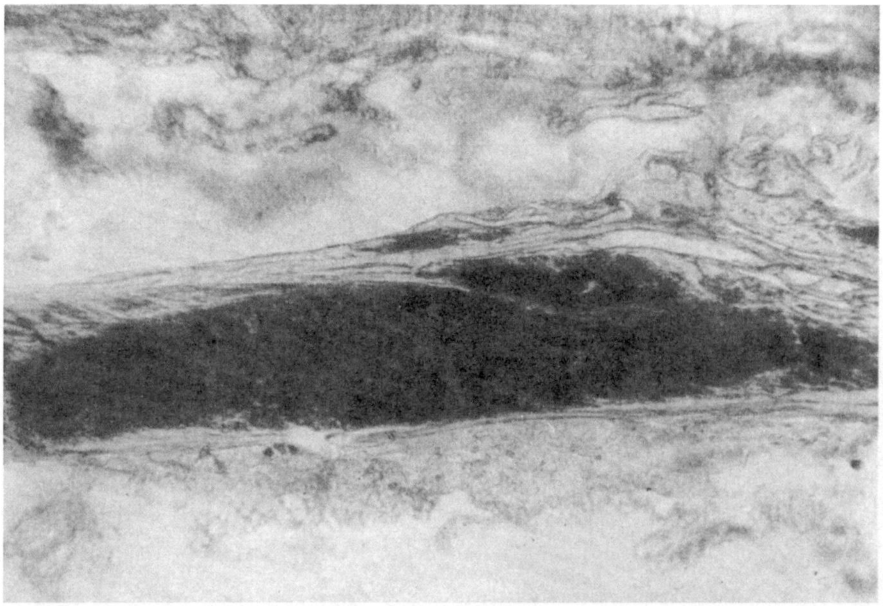

Abb. 4. Sehne der Unterhautschicht. Färbung mit Resorzin-Fuchsin nach Hart auf elastische Fasern. × 100

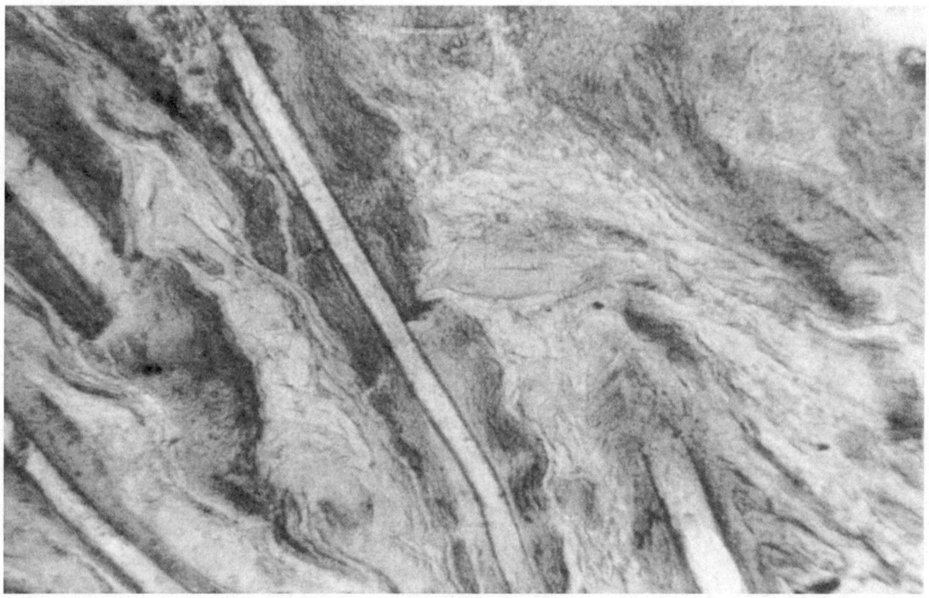

Abb. 5. Epidermis und Haarfollikel der Haut. Färbung mit Sudan III. × 100

gisches Bild der Fettbildungen. Um die Natur der Fettsubstanzen, welche so gut durch die Sudan-Färbung dargestellt werden, zu klären, wurde eine 24stündige Extraktion der freien Fette durch Piridin im Thermostat bei einer Temperatur von 58 °C durchgeführt. Dann färbten wir die extrahierten Gelatine- und Paraffinschnitte mit Sudan Schwarz „B“. Im Ergebnis dieser Färbung unterzogen wir in den Gelatineschnitten der Mammuthaut die neutralen Fette, die freien Fettsäuren und das „freie Fett“ einer Extraktion. Aus den Fettsubstanzen verblieben nur die Phospholipide, an denen Hüllen und Organellen der Zellen so reich sind (Abb. 8). In den Paraffinschnitten färbten wir ebenfalls nur die Phospholipide. Diese Kontrolle erlaubte uns, den Charakter der Fette der Epidermis und der Epithelscheiden aufzuklären. Die Extraktion veränderte nicht nur die Dicke der Epithelschicht sondern auch die Intensität der Färbung. Auf diese Weise konnten wir feststellen, daß in der Epithelschicht nicht nur freie Fettsäuren, welche im ersten Falle durch Piridin und im zweiten Falle durch Äthanol gelöst wurden, sondern auch Phospholipide existieren.

Nach den Orientierungsfärbungen mit Sudan setzten wir die Untersuchung des durch uns entdeckten Fetts durch eine Färbung mit Nilblau fort. Diese Färbung wurde in zwei Farbkonzentrationen durchgeführt: 1 % bei Zimmertempera-

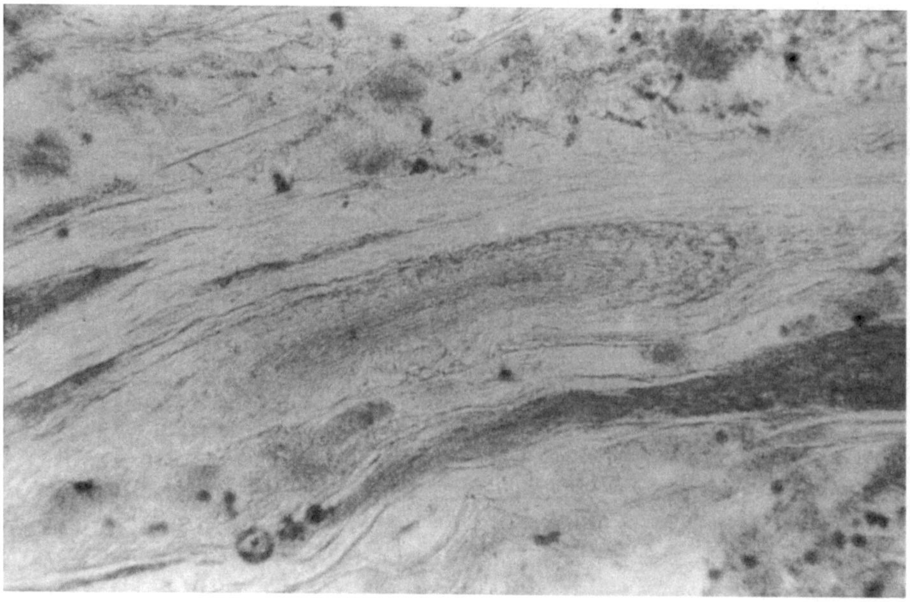

Abb. 6. Arteriengefäß der Unterhautschicht. Färbung mit Sudan III. × 100

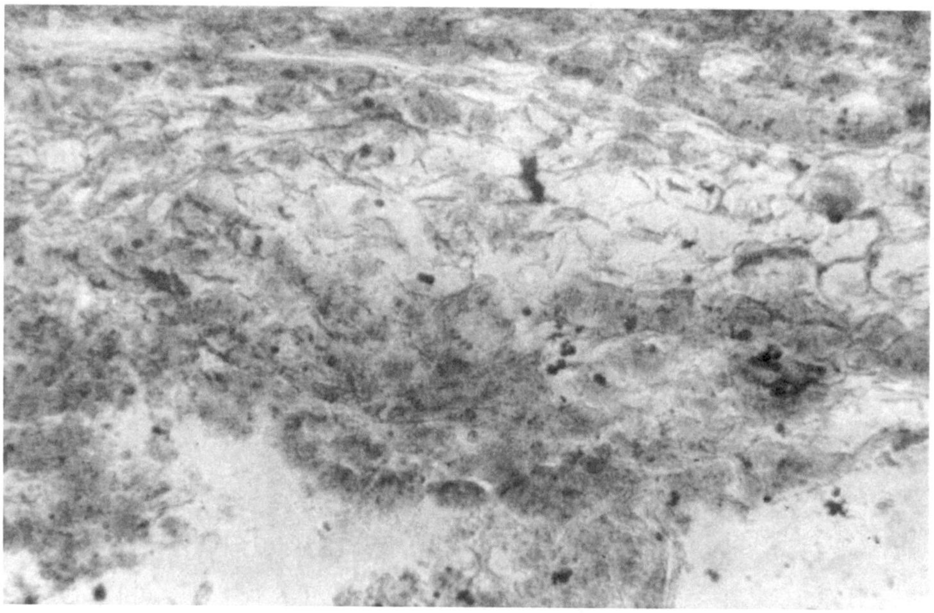

Abb. 7. Unterhautfettgewebe. Färbung mit Sudan III. × 100

tur und 0,02% im Thermostaten (+58 °C), da man den physischen Zustand der zu untersuchenden Lipide berücksichtigen mußte. Es ist anzunehmen, daß diese Methode die Phosphatide blau, die Fettsäuren und Seifen dunkelblau, die neutralen Fette, Cholesterin und seine Äther und Zerebroside rosa-violett färbt und die Möglichkeit bietet, die Fettsäuren zu differenzieren (ŠIBAEVA 1956). Es ist jedoch bekannt, daß Nilblaulösung außer den oben genannten chemischen Substanzen Kerne, Eiweiß und elastische Fasern blau und hellblau färbt. Deshalb interpretieren wir, gleich anderen Autoren, welche nach dieser Methode Gewebelipide untersuchen, die Resultate mit blauer und hellblauer Färbung mit großer Vorsicht. Um das Nilblau von den Nichtfetten zu entfernen verwendeten wir 2%ige Essigsäure.

Nilblau in verschiedenen Konzentrationen und unter verschiedenen Temperaturbedingungen färbt epidermale Bildungen (Taf. I, Fig. 2), „freie" Lipide, Zellen des Unterhautfettgewebes (0,02%, +58 °C), (Taf. I, Fig. 3), Nervenenden, teilweise von uns entdeckte Vater-Pacinische Körperchen, Gefäßöffnungen und innere sowie mittlere Gefäßhüllen (Taf. I, Fig. 4) blau. Eine schwache Reaktion (grünliche Färbung) erfolgt mit dieser Färbung auch in den Muskeln, wobei deren Querstreifung mittels dieser Methode besonders gut sichtbar wird.

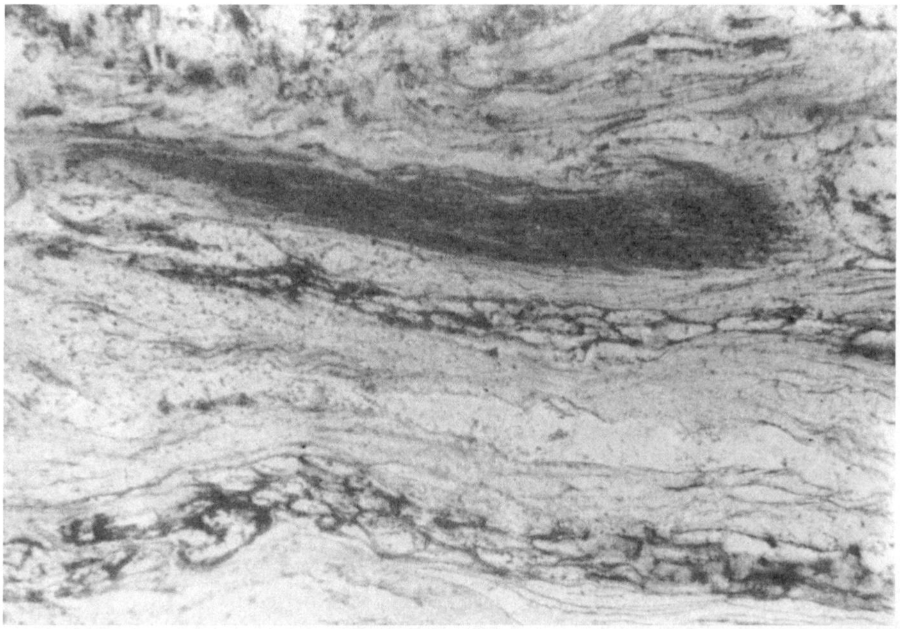

Abb. 8. Fetteile und quergestreifte Muskeln aus der Unterhautschicht. Färbung mit Sudan Schwarz „B" nach Piridin-Extraktion. ×100

Die rosa-violette Färbung der Derma-Fettzellen und des Unterhautfettgewebes erklären wir mit der Existenz freier Fettsäuren und neutralen Fetts (Triglyzeride) in diesen Strukturen. Als Beweis kann die Blaufärbung dieser Bildungen durch Nilblau in einer Konzentration von 0,02% (+58 °C) dienen. Wahrscheinlich verläuft deshalb die Färbung in 1%iger Farblösung mit einer Vermischung des roten Oxasin, das in neutralen Fetten gelöst ist, und des Oxasinsulfats, das in Reaktion mit sauren Lipiden (Fettsäuren) tritt, welche sich tiefblau verfärben. In diesem Falle können wir annehmen, daß die Reaktion auch mit hohen Fettsäuren abläuft. Bestätigt wurde diese Vermutung durch die histochemische Fischler-Methode, welche diese Fettsäuren erkennbar macht.

Neben einigen Haaren vermag man kleine Verdickungen der Epithelschicht, welche mit Sudan III orange gefärbt wurden, zu sehen. Diese kann man für Reste von Fettdrüsen halten, wenn man nur nach dieser Reaktion urteilt und keine weitere Untersuchung der Lipidverbindungen, die durch Sudan-Färbung sichtbar gemacht werden, durchführt. Die Untersuchung mit Nilblau nach Kain bestätigte nicht das Vorhandensein von Fettdrüsen bei den Haaren, da Epidermis und die Schicht der Epithelscheide tiefblau gefärbt wurden. Bei allen rezenten Säugetieren enthält das Zellensekret der Fettdrüsen neutrales Fett in großer Menge und färbt sich dementsprechend rosa-violett. Dennoch können wir nicht endgültig bestätigen, daß die Mammute in der Haut keine Fettdrüsen besaßen. Möglich wäre nämlich, daß das Fett der Haut und der Unterhautschicht sowie der Fettdrüsen unter dem Einfluß verschiedener Faktoren, welche die Zerlegung der Lipide hervorrufen und beschleunigen, Veränderungen unterlag. Dazu gehören der Sauerstoff der Luft, das Licht, Feuchtigkeit, Bakterien, Mikroben und vieles andere. Beim Zusammenwirken aller dieser Faktoren zersetzt sich Fett ungewöhnlich schnell; dabei verläuft der Zersetzungsprozeß gleichzeitig in verschiedene Richtungen und bewirkt ein sehr kompliziertes Zerfallsbild. In den zersetzten Fetten entdeckt man die verschiedenartigsten Verbindungen, die im Ausgangsmaterial fehlen (Šestakov 1914). Die Triglyzeride in den Fettdrüsen des jungen Mammuts konnten sich nach der Zellzerstörung in Glyzeride und freie Fettsäuren, die durch Nilblau blau gefärbt wurden, spalten.

Arterien und Venen der Unterhautschicht liegen in einer bindegewebsartigen Zwischenschicht zwischen Fetteilen. Ihrem Bau und ihrer Lage entsprechend kann man sie zum Muskeltyp rechnen. Sie besitzen eine dicke Wand. Nach Färbung auf elastische Fasern sind leicht drei Hüllen zu unterscheiden: innere, mittlere und äußere. Alle Hüllen der Arteriengefäße sind topographisch und morphologisch vollständig, die Endothelzellen der inneren Hülle aber nicht zu unterscheiden. Sehr gut erhalten und darstellbar sind dagegen die innere elastische Membran (Abb. 9) und die kollagenen und elastischen Fasern in der mittleren und äußeren Hülle. In der Außenhülle sind die elastischen Fasern größer als die kollagenen. Die elastischen Fasern in der mittleren Gefäßhülle sind entlang des Gefäßes orientiert, und die elastischen Fasern in der Außenschicht der Gefäßwand weisen eine vorwiegend schräge oder Längsrichtung auf.

Die mit Sudan III gefärbten Lipide entdeckt man in verschiedener Menge in den Gefäßen und in den Gefäßschichten. Eine ganz klare Orangefärbung zeigen Gefäßinhalt sowie die mittlere und innere Schicht. Die Nilblaufärbung und die Fischler-Methode erlauben es, den chemischen Bestand des Fettes zu präzisieren. Das Gefäßfett, die inneren und mittleren Hüllen stellen sich durch den ersten Färber als saures Lipid dar, durch den zweiten als freie Fettsäuren (Abb. 10). Nach der Fettextraktion mit Piridin und der Präparatfärbung mit Sudan Schwarz „B" verbleibt die Restverfärbung hauptsächlich in der Gefäßöffnung und in den Gefäßwänden. Das Auftreten von Phospholipiden in den Gefäßwänden und im Blut ist als Folge zerstörter Blutzellen, Endothelzellen, Zellen der glatten Muskulatur und modifizierter Bindegewebszellen anzusehen. Möglicherweise zeigen deshalb innere und mittlere Hülle, welche so reich an diesen Zellen sind, eine so gute Reaktion auch auf allgemeines Fett, auf Phospholipide (nach

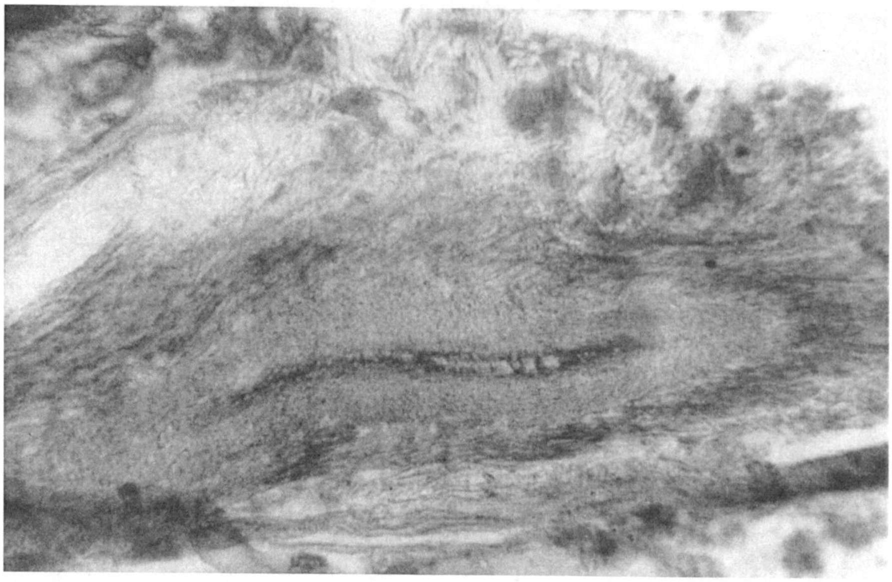

Abb. 9. Hauptateriengefäß in der Unterhautschicht. Darstellung der elastischen Fasern mit Resorzin-Fuchsin nach Hart. Vergr. × 100

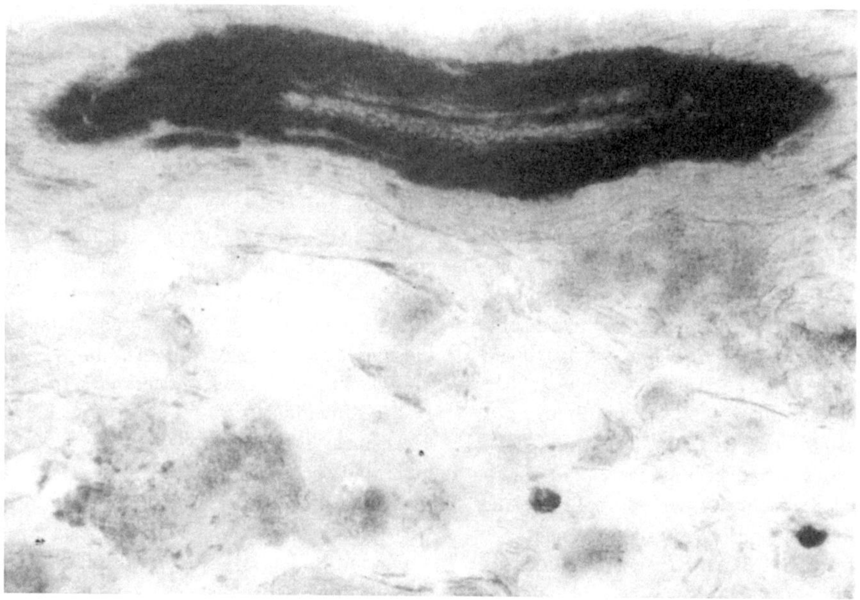

Abb. 10. Arteriengefäß der bindegewebsartigen Unterhautschicht. Gefärbt nach Fischler-Methode. Vergr. × 100

der Extraktion) und auf freie Fettsäuren. Die Färbung der Haut des jugendlichen Mammut-Kadavers nach der Fischler-Methode wurde mit Geweben rezenter Tiere kontrolliert, die Zeit der Differenzierung der Mammuthautschnitte nach der Entfärbung konnte durch die Entfärbung der Erythrozyten in den Hautgefäßen eines Sika-Hirschs kontrolliert werden.

Die Durchsicht der durch Nilblau gefärbten Hautschnitte zeigte, daß durch diese Färbung die Gefäße von den Vater-Pacinischen Körperchen in der Unterhautschicht nicht zu unterscheiden sind, da beide eine Färbung wie saure Lipide zeigen. Die Existenz von Fett und einer ziemlich dicken elastischen Membrane in der Gefäßmitte verstärkt die Färbung dieses Gefäßteils in ein dunkles Blau; das macht sie den Vater-Pacinischen Körperchen sehr ähnlich (nach Angaben von PIRS 1962 färbt sich durch Nilblau auch das elastische Gewebe blau). Die von uns durchgeführte Färbung auf elastische Fasern mit Fuchsin-Resorzin nach Hart half in solchen Fällen bei der Unterscheidung von Gefäßen und Vater-Pacinischen Körperchen. Andere Beweismittel, welche von der Existenz der Körperchen zeugen könnten, haben wir nicht. Die Imprägnierung der Hautschnitte mit Silber zeigte weder Nervenfasern noch deren Enden.

Wir konnten bereits aufzeigen, daß die positiven Reaktionen mit Fett-Färbern eine Lokalisierung der Lipide in der Zone zwischenfaseriger Grenzen in der Derma der Haut des jugendlichen Mammut-Kadavers ermöglichen. In der Haut von Indischem Elefanten und Sika-Hirsch stellte sich Fett in diesem Hautteil durch keine der angewandten Färbungen dar. In den Präparaten der Mammuthaut dagegen, die durch Nilblau gefärbt wurden, erwirbt es eine blaue, manchmal dunkelviolette Farbe, welche auf die mögliche Existenz saurer Lipide hinweist. In diesem Falle handelt es sich um Fettsäuren, Azetalphosphatide und neutrale Fette. Diese Lipide in den Querschnitten der kollagenen Bündel sehen wie Zellkerne aus. Die Entstehung dieses Fetts in der Mammuthaut kann verschiedene Ursachen haben. Einerseits können Fettsäuren, Plasmalogene und Triglyzeride im Bindegewebe nach der Zersetzung der Blut- und Lymphgefäße entstanden sein, andererseits können die Bindegewebszellen als Entstehungsursache in Frage kommen. In diesen Zellen zerfielen nach der Zerstörung die biologischen Komplexverbindungen, und die reaktiven Gruppen der Lipide konnten mit Nilblau in Reaktion treten. Und schließlich konnten zu diesen Lipiden die freiwerdenden freien Lipidekomponenten der kollagenen und retikulären Fasern hinzutreten. Diese sind im allgemeinen an einfache Eiweißverbindungen gebunden. In den retikulären Fasern ist mit Eiweiß eine Lipidekomponente verbunden: die ungesättigte Miristin-Fettsäure; in den kollagenen Fasern zeigten die chemischen Methoden — außer bei den Plasmalogenen — die Existenz noch anderer Lipide an. Diese sind aber histochemisch nicht darstellbar (Pirs 1962). Weiter kann man annehmen, daß das Fett des Bindegewebes auch als Produkt der Tätigkeit von Bakterien und Mikroben entstehen kann.

Die vertikalen Gelatineschnitte der Mammuthaut zeigten eine intensive positive PAS-Reaktion in der blauen Epidermis-Schicht, in den epithelen Schichten des Haarfollikels, in den fibrillären Fasern des Bindegewebes und in den Gefäßen. Besonders klar färbte sich der Inhalt der Gefäße mit Fuchsin-Schwefelsäure. Nach der Intensität der Färbung kann man den Inhalt der Gefäße in der PAS-Reaktion mit dem Glykogen in den Haarfollikeln mit der Haut des Sika-Hirschs vergleichen. Der Schnitt dieses Gefäßes erwies sich als einziger, wobei es nicht gelang mit Hilfe eines Gemischs von Alkohol und Äther die Existenz von Lipiden nachzuweisen. Daher müssen wir annehmen, daß sich die neutralen oder spezifischen Polysaccharide, die in den Erythrozyten oder anderen Blutzellen enthalten sind, zu färben vermochten. Die Blutzellen selbst wurden zerstört, die Polysaccharide konnten jedoch in biologischen Komplexverbindungen erhalten bleiben. Ebenso verblieben sie in der Epidermis und in der Epithelschicht der Haarfollikel. Möglich ist ebenfalls, daß es sich um Bluteiweißverbindungen handelt: Serumalbumine und Globuline, die ebenso wie die Polysaccharide eine positive Schiffsche Reaktion zeigen. Nach der Diastasekontrolle verblieb die Färbung aller Strukturen der Mammuthaut genauso wie bei den unkontrollierten Schnitten.

Es ist bekannt, daß man bei der PAS-Reaktion die reaktionsfähigen Aldehydgruppen, die bei der Schifferschen Reaktion eine rote Färbung zeigen, auf dem Oxydationswege aus α-Glykole- und α-Ketole-Gruppen, α-Aminoalkohol, α-Alkylaminoalkohol, einigen α-Ketoalkoholen, aber auch aus Lipiden erhält. Dabei zeigt sich eine große Gruppe von Verbindungen: Polysaccharide, neutrale und möglicherweise saure Mukopolysaccharide[3]), Mukoproteide und Glykoproteide (Serumalbumin, Serumglobulin, elastische, retikuläre und kollagene Fasern), Glykolipide, Phospholipide (Azetalphosphatide, Lipofuscin, Kardiolipin) (Roskin und Levinson 1957; Pirs 1962). Die positive PAS-Reaktion kann auch durch das Auftreten von Sulfosäuren bei der Oxydation von Jodsäure in Gewebe- oder disulfiden Gruppen bedingt sein (Pirs 1962).

Für die Identifizierung der Verbindungsklassen in der Hautstruktur, welche eine positive PAS-Reaktion geben könnten, setzten wir mehrere Kontrollen und histochemische Methoden ein.

Nach der Diastasekontrolle verblieb die Färbung der von uns bereits beschriebenen Strukturen der Mammuthaut ohne Veränderung. Dies scheint zu beweisen, daß es in den PAS-positiven Strukturen neutrale Mukopolysaccharide gibt, während Glykogen nicht erhalten ist. In den nach dieser Methode gefärbten Paraffinschnitten der Haut gaben die genannten Strukturen eine positive Reaktion, wenn auch eine schwächere. Möglicherweise ist dieses auf den Verlust verschiedener Kohlenwasserstoffe und Phospholipide während der Entwässerung und Entfettung der Gewebe zurückzuführen, ebenso während ihrer Färbung. Eine positive Reaktion wurde ferner mit den Phospholipiden hervorgerufen: die Kontroll-Paraffinschnitte wurden mit Sudan Schwarz „B“ intensiv gefärbt.

Wie unsere Untersuchungen gezeigt haben, war die Mammuthaut sehr reich an verschiedenen Lipiden. Die Anwendung des Schifferschen Reagens' auf freie Aldehydgruppen wurde wegen der Möglichkeit ihrer Bildung im ranzigen Fett vorgenommen. Dabei färbten sich Epidermis, Haarfollikel, Muskeln, Skleroproteide, Gefäße (purpurrot), einige Fettzellen um die Gefäße in der Derma und im oberen Teil der Unterhautschicht rosa. Fettzellen und ihre Teile färbten sich rosa-violett. Alle Hautstrukturen des Sika-Hirschs zeigten diesbezüglich ein negatives Resultat. Nach einer 24stündigen Fettextraktion bei +37 °C durch ein Gemisch von Chloroform und Spiritus (1:1) färbten sich die Fettzellen und Gefäße nicht so, wie das zuvor von uns beobachtet worden war. Das Fett mußte also in einigen Fettzellen der Derma und des Unterhautgewebes teilweise oxydiert worden sein. Bekannt ist, daß sich die Lipide, die freie Aldehydgruppen enthalten, durch die Schiffersche Reaktion purpurrot färben. Die schwache Färbung in den Fettzellen ist wahrscheinlich durch die Existenz einer relativ geringen Anzahl freier Aldehyde im ranzigen (oxydierten) Fett bedingt. In den Gefäßen war dieser Prozeß schon ziemlich weit fortgeschritten; eine purpurrote Färbung beobachteten wir bei dieser Reaktion wie auch bei der PAS-Reaktion. Bekannt ist, daß das Fett der normalen Haut keine Aldehyde enthält und während der histologischen Bearbeitung des Materials diese auch nicht bildet. Langes Aufbewahren des Materials an der Luft oder in Formalin führt zur Bildung sowohl von Karboxyl- (blaue Färbung mit Nilblausulfat) als auch von Aldehydgruppen (positive Schiffersche Reaktion) (Šibaeva 1961). In den Gefäßen der Mammuthaut existierten ebenfalls Karboxylgruppen, die sich mit Nilblau blau färbten. Ein Teil des Fetts der

[3]) Die Darstellung saurer Mukopolysaccharide durch PAS-Reaktion wird derzeit diskutiert.

Unterhautschicht zerfällt beim Einwirken von Fuchsinschwefelsäure in kleine Tröpfchen und tritt zumeist aus den Zellen von Fettschichtteilen aus. Wie unsere weiteren Untersuchungen zeigten, führten alle Reaktionen, bei denen Säuren und Schiffersches Reagens benutzt wurden, zur Hydrolyse der neutralen Fette. Diese zeigten keine Reaktionen.

Zur Darstellung der ungesättigten Lipide in der Mammuthaut und in der Haut des Sika-Hirschs verwendeten wir Ameisensäure-Schiffersches Reagens und Ultraviolett-Schiffersches Reagens. In beiden Fällen wurde eine Rotfärbung der Strukturen, welche ungesättigte Lipide enthalten, nicht erreicht. Nach der ersten Reaktion färbten sich alle Hautstrukturen schwachrosa. Nur die Haare zeigten eine klarere Färbung. In Übereinstimmung mit PIRS (1962) zeigt das Keratin des Haarstiels eine positive rosa Färbung mit Ameisensäure und Schifferschem Reagens. Der Autor erklärte diese Reaktion mit zwei Möglichkeiten: erstens — die Kreatinfärbung erscheint wegen der Existenz eines ungesättigten Komplexes, wahrscheinlich lipider Natur (teilweise oxydiert), in den Haaren, zweitens — diese Reaktion ist durch unbekannte Gruppen bedingt, von denen wir nur wissen, daß es keine Aldehyde sind.

Da wir über Material verfügten, das bereits in neutralem Formalin fixiert war, wurde in der Kette von Reaktionen zur Bestimmung der Verbindungen, die durch die Schiffersche Reaktion dargestellt werden, nur die „pseudoplasmatische" Reaktion durchgeführt. Vor der Färbung eines Teils der Mammuthautschnitte mit Schifferschem Reagens führten wir eine Blockierung der reaktionsfähigen Aldehydgruppen der Azetalphosphate mit Hydroxylamin durch, andere Schnitte bearbeiteten wir mit Sulema (weiche Oxydierung). Parallel zu den Hauptpräparaten führten wir die Färbung der Kontrollschnitte nur mit Schifferscher Reaktion ohne vorbereitende Einwirkung mit Sulema und Hydroxylamin durch. Im Ergebnis dieser Reaktionen erhielten wir bei allen Präparaten eine gleichmäßige Rosafärbung der Strukturen: weder ein Freimachen der potentiellen Aldehydgruppen der Azetalphosphatide (keine violette Färbung) noch eine Blokkierung dieser Gruppen (die Schnitte entfärbten sich nicht) wurde durchgeführt. Bei den Kontrollschnitten gaben alle Hautstrukturen eine positive Reaktion mit Schifferschem Reagens bei verschieden intensiver Färbung. Diese Ergebnisse eindeutig zu erklären ist schwierig, weil in normalen Geweben rezenter Tiere die Hautschnitte nach der Bearbeitung mit Hydroxylamin und nach der Färbung mit Fuchsinschwefelsäure farblos bleiben müssen. Die Wirkung von Sulema bis zur Durchführung der Schifferschen Reaktion muß zu einer Verstärkung der Strukturenfärbung als Ergebnis einer Umwandlung von Verbindungen lipoiden Charakters des Plasmalogen (nach der neuen Nomenklatur Azetalphosphatid) im Plasmal, welches mit Fuchsinschwefelsäure eine violette Färbung zeigt, führen.

Verwendet man die Interpretation der Plasmalereaktion und der „Pseudoplasmalereaktion" verschiedener Forscher, so kann man annehmen, daß als Resultat unserer Forschungen die Verbindungen oder Verbindungsgruppen, welche durch die „Pseudoplasmalereaktion" beim Mammut dargestellt wurden, keine Azetalphosphatide enthalten. Sicherlich sind auch bei allen anderen Reaktionen mit Schifferschem Reagens keine Azetalphosphatide nachzuweisen. Möglicherweise führt die Fixierung der Haut in Formalin zur Oxydation aller Azetalgruppen und schaltet so eine positive Schiffersche Reaktion aus. Daher zeigt auch Sulema bei ihnen keine Wirkung. Andererseits kann man erwarten, daß die Azetalphosphatide in der Mammuthaut längst zerstört waren. Weiter erscheint es möglich, daß an der „Pseudoplasmalereaktion" noch einige andere Gruppen chemischer Verbindungen (vielleicht die Superoxyde), die bei der Autoxydation der Fette gebildet werden, teilnehmen.

Eine andere Reaktion (Plasmale-) auf Azetalphosphatide und eine Kontrolle der Reagenzeigenschaften wurden auf nichtfixierten Hautschnitten des Moschustiers durchgeführt. Eine intensive Plasmalereaktion wurde in der Moschustierhaut nur in den Schweißdrüsen und ihren Kanälen, in den Blutgefäßwänden, in den Nervenfasern und in den elastischen Fasern bei der für Azetalphosphatide typischen purpurroten Färbung entdeckt. Eine schwache positive Reaktion (rosa) in diesen Hautschnitten gab es im Zytoplasma der Zellen mit minimalen fettigen Entartungen der Fettdrüsen, in den Epidermiszellen und in der Epithelschicht der Haarfollikel. Die mit Hydroxylamin behandelten Kontrollschnitte zeigten keine Färbung. Eine andere Reaktion mit Schifferschem Reagens, welches zur Darstellung der Fettaldehyde angewandt wurde, zeigte in den Zellen und in den Kanälen der Fettdrüsen der Moschustierhaut ein negatives Resultat.

Aus allen durchgeführten Reaktionen auf Hautschnitten des Moschustiers kann man schlußfolgern, daß sich eine intensive Plasmalereaktion in den an Azetalphosphatiden reichsten Strukturen entwickelt. Die reaktionsfähigen Gruppen der Azetalphosphatide wurden auf den Kontrollschnitten mit Hydroxylamin blockiert; Fettsekret, abgesondert durch Haut-Fettdrüsen, und Aldehydgruppen, d. h. Produkte der Oxydationsumwandlung des Fetts, sind nicht vorhanden. Die letzte Schlußfolgerung stimmt gut mit den Ergebnissen der Arbeit von ŠIBAEVA (1961) überein. Diese Autorin nimmt an, daß das Fett der normalen Haut keine Aldehyde enthält und im Prozeß der histologischen Bearbeitung des Materials auch keine bildet.

Für das Färben der Paraffinschnitte der Mammuthaut mit Azur I auf sulfatisierte und nichtsulfatisierte saure Mukopolysaccharide wurden Puffergemische von pH = 1,5 und pH = 5,8 ausgewählt. Die Auswahl fiel bei unseren Untersuchungen auf „freie" Sulfate, verschiedene Chondroitinschwefelsäuren und Hyaluronsäure, die man bei hohem pH-Wert nach der sich entwickelnden metachromatischen Färbung darstellen kann (vgl. hierzu Methode nach ŠUBIČ u. a. 1966 und unsere Angaben, die bei den Experimenten mit der Haut des Sika-Hirschs gewonnen wurden). Eine Kontrolle mit Methylisierung und Demethylisierung der von uns dargestellten Verbindungen konnten wir nicht durchführen, da die Mammuthaut diese Reaktionen nicht aushielt.

Bei der Färbung mit Azur I (pH = 1,5) bemerkte man eine intensive Metachromasie in der Epidermis und zwar in der Hauptsache dort, wo sich Zellen der basalen Schicht auch im Epithel der Haarfollikel befinden. Die Skleroproteide der Haut wurden verschieden gefärbt: die kollagenen Fasern im wesentlichen orthochromatisch, die elastischen und retikulären metachromatisch. Die Hornschicht der Epidermis wurde intensiv blau. Beim pH-Wert 5,8 (Färbung mit Azur I) wächst der Grad der Metachromasie in allen von uns genannten Hautstrukturen. Die Auslösung der Metachromasie können während der Färbung mit Azur I außer den sauren Mukopolysacchariden auch Nukleinsäuren, Lipide und schwefelsaurer Äther höherer Alkohole bewirken (KONONSKIJ 1976).

In der Mammuthaut fehlen die Nukleinsäuren (negatives Resultat der Feulgen-Reaktion), alle Lipide verschwinden wahrscheinlich mit Ausnahme der Phospholipide bei der Paraffinierung. Man kann folgerichtig annehmen, daß die metachromatisch gefärbten Komponenten der Hautstrukturen saure Mukopolysaccharide und Phospholipide werden.

Wenn man alle Färbungen der Hautschnitte mit Schifferschem Reagens vergleicht, kann man beispielhaft das Verteilungsbild chemischer Verbindungen in der Haut des jungen Mammuts charakterisieren.

In der Epidermis, den Epithelschichten des Haarfollikel und in den Skleroproteiden färben sich das Keratin in der Hornschicht und in den Haaren, die Phospholipide, neutrale und saure Mukopolysaccharide. Das Fehlen von Glykogen und die Existenz neutraler Mukopolysaccharide bewies die Diastasekontrolle vor der PAS-Reaktion, saure Mukopolysaccharide mit Azur I-Färber, Phospholipide mit Sudan Schwarz „B"-Färbung der Paraffinschnitte. Die Blutgefäße wurden mit PAS-Reaktion und anderen histologischen sowie histochemischen Methoden gefärbt. Sie zeigten nur in Gelatineschnitten Reaktionen auf saure Lipide, Fettsäuren und Phospholipide, da sich die Entwässerung und Entfettung der Haut sowie die Bearbeitung in Anilin und Chloroform tiefgreifend auf die Struktur der Blutgefäße auswirkte. Sie auf Paraffinschnitten mit verschiedenen Färbern darzustellen gelang nicht.

Das negative Resultat von Ameisensäure-Schliff und Ultraviolett-Schliff können wir nicht als Anzeichen für das Fehlen von doppelbindigen Lipiden werten, da zahlreiche histochemische Methoden, einschließlich der genannten, nicht spezifisch sind und biochemische Kontrollen erfordern.

Das Summeneiweiß stellt sich in allen Strukturen der Mammuthaut ebenso dar wie in der Haut des Sika-Hirschs. Einige aminosaure Eiweißreste der Mammuthaut wurden mit Dichlortriasin (blaues Prozion 2B, vgl. Ivanov 1961) dargestellt. Mit dieser Methode im alkalischen Medium kann man im allgemeinen aminosaure Lisinreste, Histidin, Tyrosin und Serin, aber auch neutrale Mukopolysacchare und Glykogen darstellen.

Die intensivste Färbung mit Prozion 2B, pH = 10,5, beobachtet man in der Epidermis und in den Epithelien der Haarfollikel, da hier außer den Aminosäuren auch neutrale Mukopolysaccharide sichtbar werden. An Mukopolysacchariden sind diese Strukturen reicher als die übrigen Komponenten der Mammuthaut. Die Skleroproteide färben sich blau, die verhornten Epidermisteile und die Haarfollikel bleiben ungefärbt (gelb).

Auf den Paraffinhautschnitten, die mit Hilfe von Azeton entwässert und dann an der Luft getrocknet wurden, stellte man Calcium mit 5%iger Schwefelsäure dar. Eine Bildung von Calciumkristallen erfolgte nicht.

Für Konsultationen und ihr Interesse an den gegenwärtigen Forschungen danken die Autoren S. M. Šibaeva und S. A. Sabadas sowie V. M. Gudkova für das Zeichnen des Hautpräparates des Magadan-Mammuts.

Zusammenfassung

Literaturdaten sowie die von uns gewonnenen Ergebnisse zur Erforschung der Mammuthaut zeigen, daß in den Teilen des Mammutkörpers, die nach oder während der Einbettung mehrmals auftauten und wieder froren, die Strukturen nicht erhalten blieben. Die Haut der hinteren Extremität des jugendlichen Mammut-Kadavers von Magadan unterlag diesem Prozeß des wiederholten Auftauens und einer starken Mumifizierung nicht. Daraus resultiert der relativ gute Erhaltungszustand von Haut- und Haarstrukturen. Nur in der Haut der hinteren Extremität waren die Haare nicht ausgefallen.

Alle Hautstrukturen des Mammuts zeigten die verschiedensten postumen Veränderungen, zumeist war ihr Bau jedoch erhalten geblieben. Bei der Anwendung verschiedener histologischer und histochemischer Methoden zeigten sich die für die normale Haut charakteristischen Färbungen. Dieses erlaubte, nicht nur den morphologischen Bau der Haut des fossilen Mammuts zu erforschen sondern auch den chemischen Bestand der histologischen Strukturen. Wie unsere Erfahrungen zeigen, ist eine Einbettung fossiler Mammuthaut in Gelatine vorzuziehen, da sie den morphologischen Bau der Haut nicht zerstört. Paraffinschnitte sollten bei solchen Untersuchungen nur als Ergänzung zur Gelatine-Einbettung verwendet werden.

In der Hautdecke und Unterhautschicht des jugendlichen Mammut-Kadavers von Magadan waren Epidermis, Follikel mit Haaren, Fasern des Bindegewebes, quergestreifte Muskeln, Blutgefäße, Sehnen und Unterhautfettgewebe erhalten. Morphologisch sind diese Strukturen gut darzustellen. Alle Hautzellen — mit Ausnahme der Fett- und Muskelzellen — sind zerstört. Entdeckt wurden Summeneiweiß, einige aminosaure Reste, verschiedene Kohlenwasserstoffe (neutrale und saure Mukopolysaccharide) und Lipide.

Wenn man die Veränderungen berücksichtigt, welche im Ergebnis der Zerstörung der Zellen, ihrer Bestandsstrukturen und ihrer biologischen Komplexverbindungen vor sich gingen, so kann man bei der Interpretation der Daten der histochemischen Untersuchung der Lipidekomponenten entweder von erhaltenen oder von postum entstandenen Verbindungen in der Haut des jugendlichen Magadan-Mammuts sprechen. Was richtig ist, ist schwer zu sagen. Man muß wohl annehmen, daß sich die Lipide histochemisch in der Haut des Magadan-Mammuts auf Grund dekompositorischer Prozesse und Abbau chemischer Verbindungen im Verlauf der letzten 44000 Jahre, in denen sich die Gewebe im Permafrostboden befanden, darstellen. Fäulnisanzeichen konnten nicht festgestellt werden, ein völliger Oxydationszerfall der Gewebestrukturen fand nicht statt.

So zeigen die histochemischen Untersuchungen des Fetts in der Haut des jungen Mammuts, daß Epithel- und Bindegewebe sowie innere und mittlere Gefäßhüllen verschiedene Phospholipide, freie Fettsäuren, darunter die Oleinsäure, erhalten sind. Ohne biochemische Analyse kann man von der Existenz von Palmitin, Stearin und anderen Fettsäuren nicht sprechen. Von den Fettzellen der Derma und des Unterhautfettgewebes waren der äußere wie auch der innere Bau erhalten geblieben. Der chemische Bestand des Fetts in diesen veränderte sich wahrscheinlich nur unbedeutend: außer neutralem Fett (Triglyceride) ist in den Zellen eine geringe Menge von Fettsäuren enthalten. Die Produkte des unvollständigen Zerfalls des Gewebelipids und der Phospholipide konnte in einzelnen Fettzellen und Gefäßen entdeckt werden. Die Azetalphosphatide zeigten keine Plasmalereaktion; möglicherweise wurden sie völlig zersetzt. Auch Nervenfasern konnten nicht entdeckt werden.

Die histochemische Analyse lipider und nichtlipider Verbindungen führt zu Vorstellungen über die engen Wechselbeziehungen und Komplexbildungen zwischen den Klassen von Verbindungen, die in verschiedenen Strukturen der Mammuthaut dargestellt wurden.

Summary

As was already known by literature and as was stated again by investigating the skin of a recently discovered mammoth carcass: those parts of the mammoth body which have been frozen and thawed again for several times when embedded into the permafrost deposits do not preserve cellular structures of the skin. The skin of the hind extremity of the juvenile mammoth carcass of Magadan, however, did not suffer from a process of repeated thawing and freezing and strong mumification resulting in a relative good state of preservation of the skin and hair. Only in the skin of the hind leg the hair was not shed.

The mammoth skin was showing different degrees of posthumous changing, but in general the structures have been preserved. By using a number of histological and histochemical methods the characteristic dying of the skin structures was observed. The state of preservation allowed to investigate not only the morphological structure of the fossil mammoth skin but also the chemical consistence of the histological structures. According to our experience the embedding of fossil mammoth skin in gelatine is preferable because this medium is not destroying the morphological structures of the skin. Paraffin slices should be used by similar investigations as a supplementary method only aside gelatine embedding.

In the skin cover and in the hypodermic cellular tissue of the juvenile mammoth carcass of Magadan: epidermis, follicles with hair, fibres of the connective tissue, cross striped muscles, blood vessels, sinews, and hypodermic sebasic cellular tissue have been prooved. Morphologically all these structures have been well demonstrated. The cover cells of the skin — with exception of the sebacic and muscle cells — have been found destroyed. However, we could proove sum-albumen, remains of amino-acid substances, different hydrocarbonates (neutral and acid mukopolysaccharides) as well as lipids.

Taking the modifications into consideration which took place in the progress of destroying the cells, their structures and biological complex substances, we may interpret the results of our histochemical investigations of the lipid components either as chemical compounds which have been preserved in the cellular tissue or which have been originated posthumous in the skin of the mammoth carcass.

What is the right interpretation is difficult to say. Perhaps we have to assume the lipids originating histochemically in the skin of the fossil mammoth as results of decompository processes and splitting up of chemical compounds in the course of the pas 44000 years in which the organic materials has been lying in the permafrost deposits. Signs or real putrefraction, however, have been not observed; a complete oxydation decay of the tissue structures did not occur.

Thus the histochemical investigations of the fat substances in the skin of the mammoth carcass prooved that epithel and connective tissue as well as inner and middle cover of the vessels contain different phospholipids, free sebacic acids including olein-acid. Without a biochemical analysis, however, it is not possible to speak about the existence of palmitin, stearin and other sebacic acids. The inner and outer structures of the sebacic cells of the derma and of the hypodermic cellular tissue have been found to be preserved. The chemical fat substances in those structures have been altered only to a small extent; aside neutral fat substances (triglyceride) there are small quantitaties of sebacic acids in the cells.

Products of an uncomplete decay of tissue lipids and phospholipids have been found in single sebaceous cells as well as in vessels. The azetalphosphatide were showing no plasmal reaction due perhaps to a complete destruction.

The histochemical analysis of lipid as well as of non-lipid compounds, finally, was leading to a concept of close complex-interrelationships and correlations between the different classes of chemical substances having been found in the structures of the fossil mammoth skin.

Резюме

Литературные и полученые нами данные исследования показывают, что те части тела мамонтов, которые многократно оттаивали и замерзали, не сохраняют свои структуры. Кожа задней конечности изученого молодого мамонта не подвергалась многократному оттаиванию и сильной мумификации. Этим объясняется сравнительно хорошая сохранность структур кожи и волос. Только в коже задних конечностей сохранились невыпавшие волосы. Все структуры кожи мамонта претерпели различные посмертные изменения, но в большинстве случаев, они сохранили свое строение и при применении ряда гистологических и гистохимических методов окрашиваются в цвета характерные для нормальной кожи. Это позволило изучить не только морфологическое строение кожи ископаемого мамонта, но и химический состав гистологических структур. Для этого, как показал наш опыт, кожу лучше всего заливать в желатину, которая не нарушает общего морфологического строения. Парафиновые срезы можно использовать при изучении ископаемых животных только как дополнение к желатиновой заливке.

В кожном покрове и подкожном слое ноги мамонтенка сохранился эпидермис, фолликулы с волосами, соединительнотканные волокна, поперечнополосатые мышцы, кровеносные сосуды, сухожилия и жировая подкожная клетчатка. Морфологически эти структуры хорошо выражены. Все клетки кожи разрушились, за исключением жировых и мышечных. Были обнаружены суммарный белок, некоторые аминокислотные остатки, различные виды углеводов (нейтральные и кислые мукополисахариды) и липидов.

Учитывая изменения, которые произошли в результате разрушения клеток, составляющих их структур, и биологических комплексных соединений, при интерпретации данных гистохимического исследования липидных компонентов мы можем говорить о веществах или сохранившихся, или появившихся посмертно в коже мамонтенка. Что является более правильным, сказать трудно. По-видимому, следует предположить, что липиды гистохимически выявляются в коже мамонтенка благодаря декомпозиционным процессам и ослаблению химических связей (возможно и того и другого), которые произошли за 44000 лет, сохранения тканей в вечной мерзлоте. Никаких признаков гниения нами не обнаружено. Полного окислительного распада тканевых структур не произошло.

Таким образом, полученные результаты по гистохимическому исследованию жира в коже мамонтенка показывают, что эпителиальная и соединительная ткань, внутренняя и средняя оболочки сосудов содержат различные фосфолипиды, свободные жирные кислоты, в том числе олеиновую кислоту. О присутствии пальмитиновой, стеариновой и других жирных кислот без биохимического анализа решить невозможно. Жировые клетки дермы и подкожной жировой клетчатки сохранили как внешнее, так и внутреннее свое строение. Химический состав жира в них, очевидно, изменился незначительно: кроме нейтрального жира (триглицеридов), в клетках содержится еще и небольшое количество жирных кислот. Продукты неполного распада тканевого липида и фосфолипидов были обнаружены в отдельных жировых клетках и сосудах. Ацетальфосфатиды не дали плазмалевой реакции; возможно, что они подверглись полному разложению. Не обнаружены также нервные волокна.

Гистохимический анализ веществ как липидной, так и нелипидной природы приводит к представлению о тесной взаимосвязи и комплексообразовании между классами соединений, выявляемых в различных структурах кожи мамонта.

Literatur

AVCYN, A. P.; u. a. 1971: Principipy i metody gistochimičeskogo analiza patologii (Prinzipien und Methoden der histochemischen Analyse in der Pathologie). — Leningrad: Medicina.

IVANOV, V. B., 1961: Dokl. Akad. Nauk SSSR. — Moskva 2. — In Russisch.

KONONSKIJ, A. I., 1976: Gistochimija (Histochemie). — Kiev: Višča škola.

NAUČNYE rezul'taty ekspedicii snarjažennoj imperatorskoj Akademij nauk dlja raskopki mamonta, najdennogo na reke Berezovke v 1901 godu (Wissenschaftliche Ergebnisse einer Expedition, ausgerüstet durch die kaiserliche Akademie der Wissenschaften, zur Ausgrabung eines Mammuts, das 1901 am Fluß Berezovka gefunden wurde). Bd. 1. — Petrograd, 1903.

PIRS, E., 1962: Gistochimija (Histochemie). — Moskva: Inostrannaja Literatura.

ROSKIN, G. I.; LEVINSON, L. B., 1957: Mikroskopičeskaja Technika (Mikroskopische Technik). — Moskva.

SESTAKOV, P. I., 1914: Naučnye rezul'taty ekspedicii, snarjazennoj imperatorskoj Akademij nauk, dlja raskopki mamonta, najdennogo na reke Berezovke v 1901 godu (Wissenschaftliche Ergebnisse einer Expedition, ausgerüstet durch die kaiserliche Akademie der Wissenschaften, zur Ausgrabung eines Mammuts, das 1901 am Fluß Berezovka gefunden wurde). Bd. 3. — Petrograd.

ŠIBAEVA, S. M., 1961: Archiv patologii. — Moskva **19** (3), (1961a).

—, 1961: Bjull. eksperiment. Biol. i Mediciny. — Moskva 2.

SOKOLOV, V. E., 1973: Kožnyj pokrov mlekopitajuščich (Die Haut der Säugetiere). — Moskva: Nauka.

ŠUBIĆ, M. G.; u. a., 1966: Archiv anatomii, gistologii i embriologii. — Leningrad 1.

SUBBOTIN, V. M.; TUM, JU. V., 1978: Archiv anatomii, gistologii i embriologii. — Moskva 74 (2).

SUMINA, Mamontovaja faunai sreda eë obitanaija v Antropogene SSSR (Die Mammutfauna und ihr Lebensmilieu im Anthropogen der UdSSR). — In: Tr. zool. Inst. — Leningrad 73.

VASIL'EV, JU. M.; DMITRIEVA, E. L.; IVANOVA, N. G.; REŽETOV, V. JU., 1982: Ostatki korotkorogogo bizona v otloženijach Moskovskogo lednikov'ja (The remains of a short-horned bison in the Moscow glacial deposits). — In: XI kongress INKVA, Abstracts. — Moskva. — P. 40—41.

ZALENSKII, V., N., 1909: Naučnye rezul'taty ekspedicii, snarjazennoj imperatorskoj Akademii nauk, dlja raskopki mamonta, najdennogo na reke Berezovke v 1901 godu (Wissenschaftliche Ergebnisse einer Expedition, ausgerüstet von der kaiserlichen Akademie der Wissenschaften, zur Ausgrabung eines Mammuts, das 1901 am Fluß Berezovka gefunden wurde). Bd. 2. — Petrograd.

ZENEVSKAJA, R. P., 1978: Dokl. Mosk. Obšč. ispyt. prir., obščaja Biol. 1975. — In Russisch.

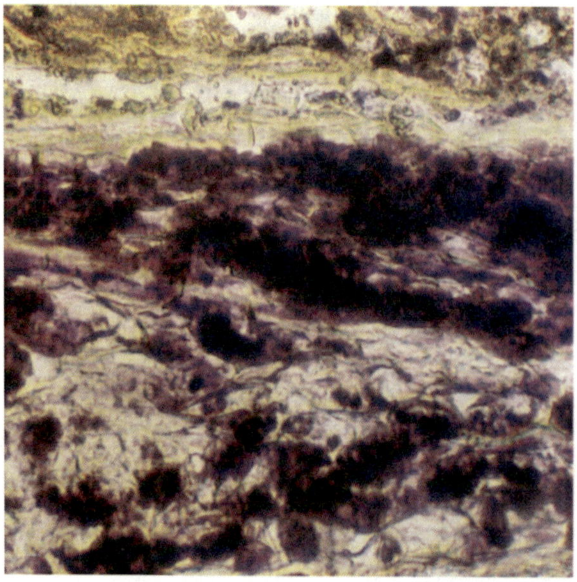

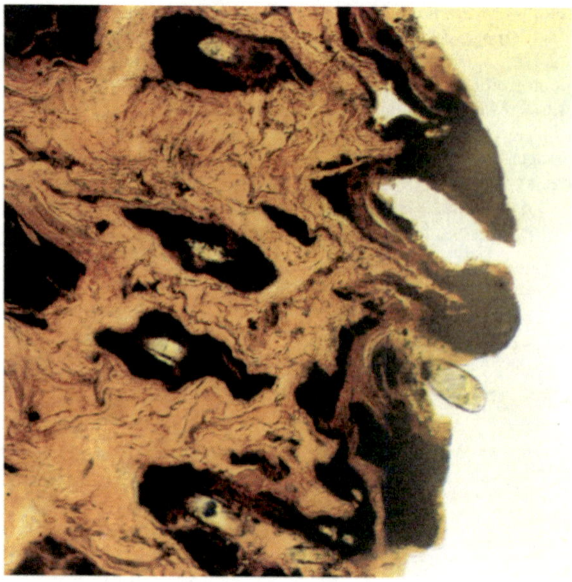

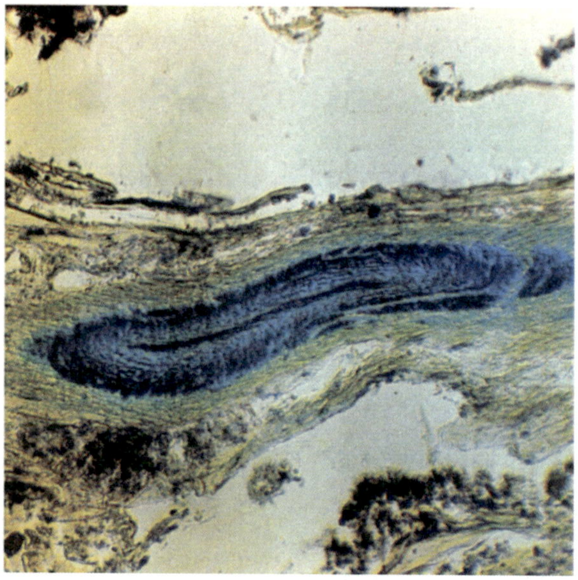

TAFEL I

Hautschnitte des Magadan-Mammuts

Fig. 1. Epidermis und Haarfollikel der Haut. Gefärbt nach Fischler-Methode (oben links)

Fig. 2. Epithelbildungen der Haut. Färbung mit 1 %igem Nilblau (oben rechts)

Fig. 3. Fettgewebe der bindegewebsartigen Unterhautschicht. Färbung mit 0,02 %igem Nilblau (unten links)

Fig. 4. Hauptarteriengefäß der Unterhautschicht. Färbung mit 1 %igem Nilblau (unten rechts)

Fig. 1—4. Vergr. × 100

Quartärpaläontologie 6 209—218 Berlin 1986

Pleistocene Occurrence of Hippopotamus in Britain

By ANTHONY JOHN STUART, Cambridge

With 2 Figures, 4 Tables and one Plate

With Appendix: Pollen Analyses from Hippopotamus Bones from the Cromer Forest Bed Series

By PHILIP LEONARD GIBBARD, Cambridge

With one Table

Contents

Abstractum

Fossil *Hippopotamus* occurrences from Britain are discussed both with respect to their chronological position (Ipswichian, Cromerian) and their mode of deposition (open sites or cave sites).

In addition to previously-known pollen analyses of Ipswichian age, the paper includes pollen analyses from *Hippopotamus* bones from the Cromer Forest Bed Series.

Fossile *Hippopotamus*-Vorkommen in Großbritannien werden im Hinblick auf ihre chronologische Stellung (Ipswichian, Cromer Forest Bed-Serie) wie auch im Hinblick auf die Art ihrer Ablagerung (Freilandvorkommen, Vorkommen in Höhlen) untersucht.

Neben die bereits bekannten Pollenanalysen aus Ablagerungen mit *Hippopotamus*-Resten des Ipswichian wird eine Pollenanalyse von *Hippopotamus*-Knochen aus der Cromer Forest Bed-Serie gestellt.

Данная работа отражает усследования к ископаемым находкам бегемота (*Hippopotamus*) в Великобритании ввиду их хронологического положения (Ipswichian, серия — Cromer Forest Bed) как и ввиду образа их отложения (находки под открытым небом, находки в пещерах).

Сравнивались пыльцевой анализ из отложений с костями *Hippopotamus* из серии Cromer Forest Bed с уже известными пыльцевыми анализами из отложений с остатками

1. Introduction

Fossil hippopotamus remains are fairly common in Pleistocene deposits in England and Wales, and have attracted attention since the beginning of nineteenth century (OWEN 1846). Because the living species are entirely confined to Africa, these occurrences have led inevitably to much speculation as to past climatic conditions.

The present paper is a review of fossil *Hippopotamus* in Britain, concentrating mainly on stratigraphical, palaeoecological and zoogeographical aspects. The taxonomy is only briefly considered here and would benefit from a detailed study.

2. Taxonomy

According to BLANDAMURA and AZZAROLI (1977) the Pleistocene hippopotamuses of Europe belong to two quite distinct species: a) the extinct *Hippopotamus major* (OWEN) found in the upper Villafranchian in southern Europe and the Cromerian of England; and b) *H. amphibius* L. which was probably present in England and elsewhere in Europe during the Last Interglacial (Eemian, Ipswichian), and which survives in Africa at the present-day.

The skull of *H. major* shows several differences to that of *H. amphibius*, e.g. in the far more elongated snout in the former (BLANDAMURA and AZZAROLI 1977). The mandibles can also be distinguished — in particular the lower border of the ramus is concave in *H. major*, convex in *H. amphibius*, and in *H. major* there is a small diastema between P_2 and P_3, and P_2 slopes forwards (ibid.). The features of the mandible described above as characteristic of *H. major* do however also occur to a variable degree in modern *H. amphibius*.

The rich material from Barrington, England, dated to the Ipswichian (Eemean) interglacial (see below) includes several skulls and mandibles, all clearly referable to *H. amphibius* (Fig. 1B and Plate I). The material was extensively monographed by REYNOLDS (1922). Mandibles from Beetley, Norfolk (Ipswichian), Honiton, Devon (? Ipswichian) and a few other sites of probably Ipswichian age are also *H. amphibius* and it is probably safe to assume that all Upper Pleistocene hippopotamuses from Britain are of these species, although most fossils are fragmentary.

The taxonomic position of most of the numerous Cromer Forest Bed Series *Hippopotamus* remains is however less certain. A fine mandible of *H. major* collected at Cromer, Norfolk, and described and figured by OWEN (1846) (Fig. 1A) unfortunately cannot be definitely assigned to a particular

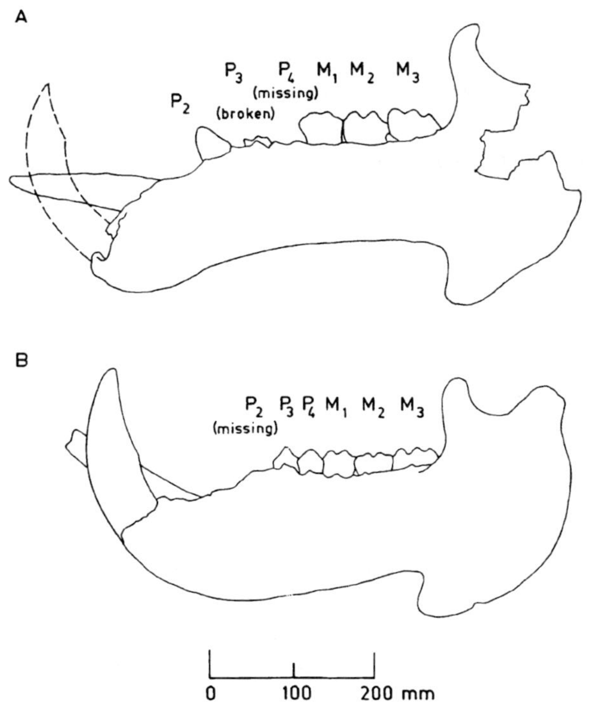

Fig. 1. A — Left mandible of *Hippopotamus major* from the CFBS (exact horizon unknown) of Cromer Norfolk (based on OWEN 1846); B — Left mandible of *Hippopotamus amphibius* of Ipswichian age (zone Ip IIb) from Barrington, Cambridgeshire (based on REYNOLDS 1922). N.B. This is a rather small individual and the cheek teeth are well worn down.

stage (see below). The material pollen dated to the Cromerian stage (sensu stricto) is not sufficiently complete for identification to species level. The CFBS material is here tentatively assigned to *H. major*. A detailed study is necessary to resolve this problem.

3. Stratigraphical Occurrences

Hippopotamus remains have been found at perhaps nearly 100 localities in England and Wales (Fig. 2). As discussed below all of the fossils appear to date from just two stratigraphical horizons within the British Pleistocene sequence viz. the Cromerian and Ipswichian interglacials (Table 1); and are restricted to certain pollen assemblage zones within these stages.

Table 1. Known Stratigraphical Range of *Hippopotamus* in the British Middle and Upper Pleistocene

Stage	Climate	
Flandrian	t	—
Devensian (Weichselian)	c, p, g	—
Ipswichian (Eemian)	t	*H. amphibius*
Wolstonian (?Saalian)	c, p, g	—
Hoxnian (Holsteinian)	t	—
Anglian (?Elsterian)	c, p, g	—
Cromerian (Voigtstedtian)	t	*H.* cf. *major*
Beestonian	c, p, ?g	—

t — temperate; p — evidence of permafrost; g — glacial deposits known

The pollen zonation scheme used here is that of TURNER and WEST (1968), in which each interglacial stage is divided into four main zones reflecting the interglacial vegetational cycle. The pollen assemblage zones and vegetational sequence for the Cromerian and Ipswichian interglacials are given in Table 2.

3.1. Cromerian

The Cromer Forest Bed Series (CFBS), comprising a complex sequence of freshwater and marine deposits, crops out in the cliffs and foreshore along parts of the coasts of Norfolk and Suffolk. It is everywhere overlain by tills and outwash of the extensive Anglian (= Elsterian) glaciation. The CFBS has long been famous for its wealth of fossil vertebrate remains, most of which however were found loose on the beaches, relatively few finds having been recorded in situ.

By carefully recording sections, and by the use of pollen analyses, Prof. R. G. WEST (1979) has entirely revised REID's (1882) classic stratigraphical scheme for these deposits. WEST recognizes a number of biostratigraphical stages represented by the CFBS, reflecting climatic oscillations (FUNNEL and WEST 1977; WEST 1979).

Stage	Climate
Anglian	cold, permafrost, glaciation
Cromerian	temperate
Beestonian	cold, permafrost
Pastonian	temperate
Pre-Pastonian	cold, permafrost

It must be stressed that the Cromerian Stage (sensu stricto) only represents part of the CFBS sequence. It has also become clear recently that there is a considerable stratigraphical hiatus somewhere between the Pastonian and Cromerian stages, as is for example indicated by the marked differences in the vertebrate faunas.

Hippopotamus remains have been found in the CFBS near Cromer, Overstrand, Sidestrand, Trimingham, Mundesley and Bacton, in Norfolk, and Pakefield in Suffolk. Many of these specimens were found by A. C. SAVIN, a local collector, mainly in the early part of this century. His collections, preserved in the British Museum, Natural History, London, are accompanied by catalogues which record basic data on most of the finds. Of 87 catalogued specimens which can be verified as *Hippopotamus*, 40 are recorded from the base of the cliff, only 13 from the foreshore and 34 have no data. This immediately suggests that most, if not all the material is of Cromerian age, since generally speaking in the Overstrand to Bacton sections and at Pakefield, Cromerian deposits comprise the cliff bases and pre-Cromerian deposits occur on the foreshore. There are however some occurences both of pre-Pastonian sediments in the cliffs and of Cromerian deposits in the foreshore (WEST 1980).

More definite evidence for the age of the finds comes from pollen analyses of sediment contained within CFBS *Hippopotamus* bones in museum collections. These analyses by P. L. GIBBARD on specimens from Trimingham, Mundesley and Bacton (see Appendix) confirm a Cromerian age and more specifically indicate pollen zones Cr IIIb and early Cr IV. *Hippopotamus* has not however been recorded in the rich fauna from the Cromerian Freshwater Bed (zone Cr II) at the type locality of West Runton, Norfolk (STUART 1975).

Table 2. Pollen assemblage zone characters and inferred regional vegetation changes for the Cromerian and Ipswichian interglacials; based on WEST (1977)

Vegetational Aspect	Cromerian				Ipswichian			
	Zone		Characters	Vegetation	Zone		Characters	Vegetation
(Early glacial)	(e Ang)		(High NAP)	(Herb vegetation)	(e De)		(High NAP)	(Herb vegetation)
Post-temperate	Cr IV	c b a	*Pinus, Picea, Betula, Alnus*	Coniferous Forest	Ip IV		*Pinus*, NAP *higher*	Coniferous Forest (rather open)
Late-temperate	Cr III	b — a	*Carpinus, Abies*	Temperate Forest with hornbeam and silver fir	Ip III		*Carpinus*	Temperate Forest with hornbeam
Early-temperate	Cr II	b — a	*Pinus, Quercus, Ulmus, Tilia*	Mixed oak forest	Ip II	b — a	*Quercus, Pinus Acer, Corylus*	Mixed oak forest
Pre-temperate	Cr I	b — a	*Pinus, Betula*	Pine-birch forest	Ip I	b — a	*Betula, Pinus*	Birch-Pine forest
(Late-glacial)	(l Be)		(High NAP)	(Herb vegetation)	(l Wo)		(High NAP)	(Herb vegetation)

Unfortunately the precise stratigraphical position of the only CFBS find definitely referable to *Hippopotamus major* (see above) cannot be fixed, as the sediment matrix did not yield pollen and OWEN (1846) merely states that it came from 'fresh-water clay beds, Cromer, Norfolk', BLANDAMURA and AZZAROLI (1977) are not justified in definitely assigning the specimen to the Cromerian interglacial — it could conceivably be of Pastonian age.

3.2. Ipswichian

Remains of *Hippopotamus* are not uncommon in river deposits in the south and Midlands of England, originating either from terraces only a few metres above the modern floodplain or, at a few sites, from below present river level. There are also many finds from cave deposits in Wales and south-western and northern England. Most of these records do not have accompanying stratigraphical data, but in every case where such data are available, an Ipswichian interglacial age is indicated. Finds have been made in association with pollen spectra of Ipswichian sub zone Ip IIb age at Beetley, Norfolk, Barrington, Cambridgeshire, and Trafalgar Square, London, and from early zone Ip III at Swanton Morley, Norfolk (Table 3). Where the *Hippopotamus* remains were collected in a reasonably good stratigraphical context the associated mammal faunas are remarkably similar, whether from open sites or caves (Table 4). These assemblages are characterized by the association of *Crocuta crocuta, Palaeoloxodon antiquus, Dicerorhinus hemitoechus, Hippopotamus amphibius* and *Dama dama.*

There is no evidence for the occurrence of *Hippopotamus* in Britain during the previous Hoxnian (Holsteinian) interglacial. Rich Hoxnian faunas are known from Clacton, Essex Swanscombe, Kent (SUTCLIFFE 1964) and from the type site of Hoxne, Suffolk. These faunas differ from those of Ipswichian age in the absence of *C. crocuta*, and the occurrence of the extinct beaver *Trogontherium cuvieri* and the pine vole *Pitymys* — both of which are absent from the Ipswichian, as well as by the absence of hippopotamus (STUART 1974, 1982).

It appears however that *H. amphibius* was present in Britain during only the middle part of the Ipswichian stage. Hippopotamus remains have not definitely been found in association with *Mammuthus primigenius* (mammoth) and *Equus caballus* (horse), which species characterise the second half of the Ipswichian (STUART 1976). The pollen spectra from *H. amphibius* bones (Table 3) indicate a range from the beginning of sub zone Ip IIb (Beetley) to the beginning of zone Ip III (Swanton Morley). It appears probable that the animal did not survive into the later part of the interglacial, since several sites with abundant faunas of this age are known. Its date of arrival however is uncertain due to the absence or scarcity of faunas from zones Ip I und Ip IIa.

The occurrence of *Hippopotamus* remains in post Anglian Pleistocene deposits in Britain is regarded by most workers as firmly dating those deposits to the Ipswichian interglacial, and the fossil therefore has far-reaching implications for the Pleistocene stratigraphy of many areas. So far this procedure appears sound, but because of this stratigraphical significance it is highly desirable that any future finds of *Hippopotamus* should be carefully recorded *in situ*, and in particular related to pollen analytical studies.

4. Geographical Distribution

4.1. Ipswichian

During the Eemian (Ipswichian) the geographical range of *H. amphibius* in Europe probably included the Balkans, Italy and Iberia, extending northwards through France into Britain, and including Germany but not regions further east. In Britain it is recorded from many sites in southern and central England, and also from Wales, and northern

Table 3. Pollen spectra from *Hippopotamus* bones dated to the Ipswichian interglacial as % total dry land pollen — approximate only. Based on FRANKS (1960), GIBBARD and STUART (1975), TURNER (1975), PHILLIPS (1976), COXON et al. (1980)

Site:	Beetley	Trafalgar Square		Barrington		Honiton	Swanton Morley		
Sample:		10	3	1	2		2	4	SM1
Pollen Zone:	IpIIb*	IpIIb	IpIIb	IpIIb	IpIIb	?	IpIII*	IpIII*	IpIII*
Trees									
Betula	13.5	4	—	0.5	0.5	1	1.5	3	5.5
Pinus	45	2	3.5	2	2	1	6	2.5	7.5
Ulmus	0.5	—	—	0.5	0.5	—	0.5	0.5	2
Quercus	1	7	11	4	4.5	—	1	4	4.5
Tilia	—	1	—	0.1	0.1	—	—	—	—
Alnus	4	1	—	0.5	0.8	—	19.5	27	18.5
Carpinus	—	—	—	0.1	—	0.5	8	4	6.5
Fraxinus	—	0.5	1.5	—	—	—	—	0.5	—
Acer	—	5	11	0.2	0.1	—	0.5	0.5	—
Picea	0.5	—	—	—	—	—	0.5	0.5	1
Total Tree	64.5	21	27	8	8.5	2.5	37.5	42.5	46
Shrubs									
Corylus	6	53	21	2.5	1.5	—	5.5	4	7.5
Salix	1	—	—	1	1	—	—	0.5	2
Hedera	1	—	—	—	—	—	0.5	0.5	—
Glex	1	1	—	—	—	—	0.5	—	—
Total Shrub	9	54	21	3.5	2.5	0	6.5	5	9.5
Herbs									
Gramineae	16	5.5	28	51	37	36.5	26	34	19.5
Cyperaceae	2	6	12	3	5	1	6	2	17
Compositae	2	<1	<1	13.5	24	5.5	10	2	2
Caryophyllaceae	—	2.5	1.5	0.5	1	—	2	1	—
Plantago	—	2	4	11	14	39.5	6	5	3
Ranunculaceae	4	2	4	1	2	6	3	1	1
Trifolium type	—	—	—	—	—	4.5	—	1	—
Umbelliferae	1	—	—	0.5	1	2	1	1	1
Other Herbs	1.5	8	2.5	5.5	4	2.5	2	5.5	2
Total Herbs	26.5	25	50.5	86	87	97.5	56	52.5	45.5
Aquatics	1	<1	<1	1		—	0.1	3.5	5
Lower Plants	64.5	<1	<1	1.5		—	30	20.5	14

*) Early part of zone or subzone.

England. The most northerly record of *Hippopotamus* anywhere in the world is from Stockton-on-Tees, County Durham (latitude 54° 37′ N). The absence of fossil *H. amphibius* from Ireland, Scotland and much of northern England has been widely interpreted as indicating that these areas were beyond the limits of geographical distribution of the species due to climatic factors. Moreover the higher concentration of finds in south-east England has been thought to reflect higher population densities in this area; the records from northern England being regarded as representing sparse populations near the limit of distribution (SUTCLIFFE 1959). A consideration of the distribution of Pleistocene *Hippopotamus* finds in Britain, however (Fig. 2) shows that the above views are unjustified. Firstly it is significant that nearly all of the records from open sites, mostly from fluviatile deposits, are south of the limit of Devensian (Weichselian) till. The one exception is from the raised beach shingle at Sewerby, Yorkshire, which are covered by Devensian till, and exposed in cliff section (CATT and PENNY 1966). The other finds from north of this limit are all from caves, where the deposits were protected against destruction by the Devensian glaciers. It should be stressed that no interglacial faunas, and hardly any interglacial deposits, have been discovered in Scotland or the extreme north of England, presumably because they have either been buried by glacial deposits or eroded away by glaciers and meltwater. Similarly no deposits of definite Ipswichian age are known from Ireland, and the few interglacial deposits of Gortian (Holsteinian) age have so far not yielded any vertebrate fossils. In fact no vertebrate remains whatsoever are known from Ireland prior to the Devensian (Weichselian) cold stage. Furthermore the abundance of *H. amphibius* finds in southern England is clearly a measure of the abundance of Ipswichian deposits there in comparison with far fewer sites in the north of England and in Wales.

It is therefore quite possible that *H. amphibius* did extend into Scotland and perhaps even Ireland. A last point to be taken into consideration is that at Victoria Cave, near Settle Yorkshire, one of the most northerly sites (latitude 54° 05′ N), *H. amphibius*, together with a typical mid-Ipswichian fauna (see above), is recorded at 440 m O.D., which suggests that here it was far from the extreme limit of its range.

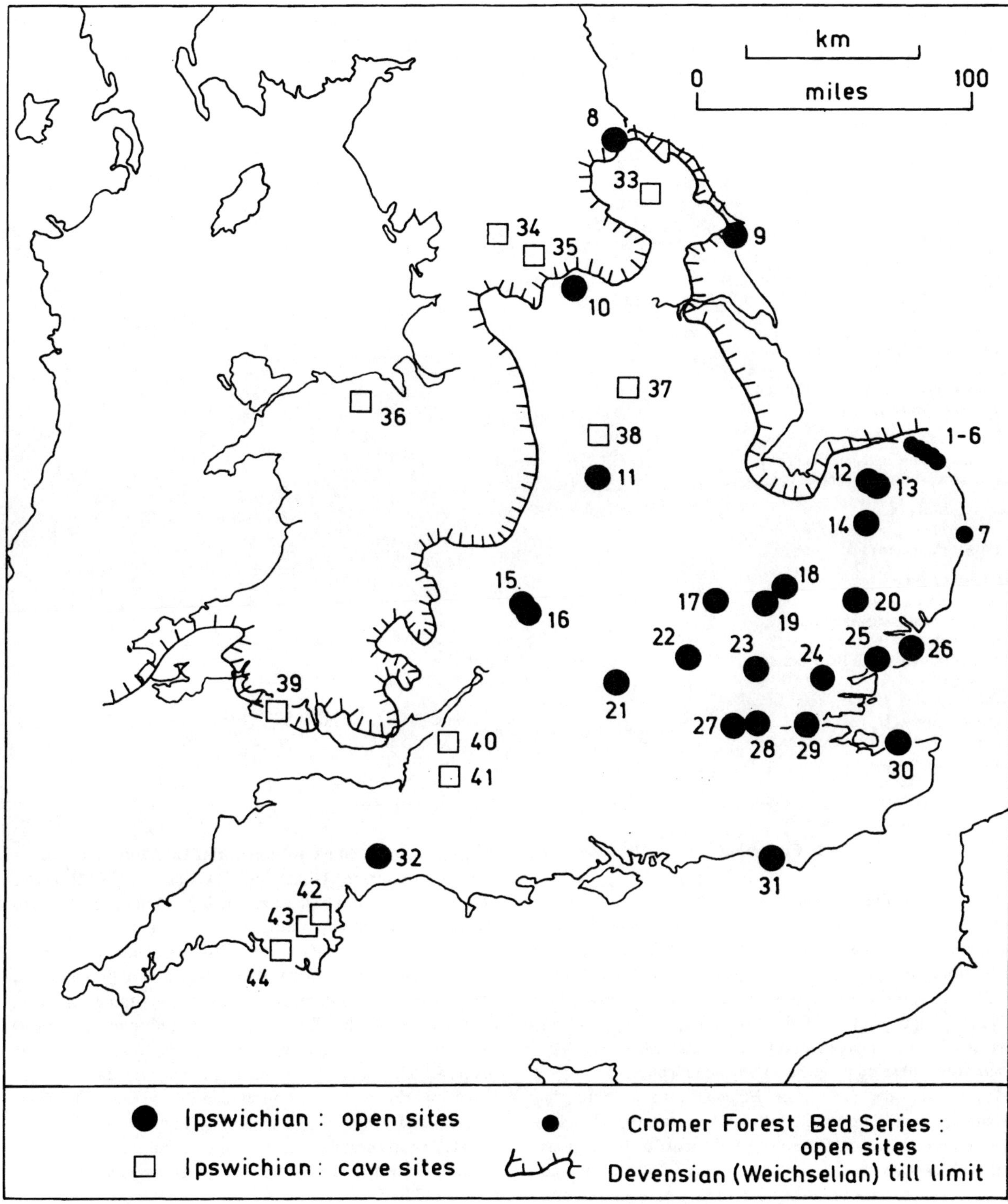

Fig. 2. Distribution of fossil *Hippopotamus* finds in the British Isles, showing all CFBS localities and representative Ipswichian localities

1 — Cromer; 2 — Overstrand; 3 — Sidestrand; 4 — Trimingham; 5 — Mundesley; 6 — Bacton; 7 — Pakefield; 8 — Stockton-on-Tees; 9 — Sewerby; 10 — Leeds; 11 — Boulton Moor; 12 — Beetley; 13 — Swanton Morley; 14 — Shropham; 15 — Wick; 16 — Cropthorn; 17 — Bedford; 18 — Cambridge (Grantchester, Barnwell Abbey etc.); 19 — Barrington; 20 — Lavenham; 21 — Oxford; 22 — Marsworth; 23 — Water Hall Farm; 24 — Chelmsford; 25 — East Mersea; 26 — Walton-on-the-Naze; 27 — Brentford; 28 — Trafalgar Square, London; 29 — Grays; 30 — Tankerton; 31 — Black Rock, Brighton; 32 — Honiton; 33 — Kirkdale Cave; 34 — Victoria Cave; 35 — Raygill Fissure; 36 — Cefn Cave; 37 — Creswell Crags; 38 — Elderbush Cave; 39 — Ravenscliff Cave; 40 — Durdham Down, Bristol; 41 — Milton Hill, Wells; 42 — Tornewton Cave; 43 — Joint Mitnor Cave; 44 — Eastern Torrs Quarry

Table 4. Large mammal faunas, including Hippopotamus, from selectes Ipswichian sites. Data from BOYLAN (1967, 1977), JONES and STANLEY (1975), STUART (1976, 1982), SUTCLIFFE (1960, 1969), SUTCLIFFE and ZEUNER (1962)

	Beetley	Barrington (Ip IIb)	Trafalgar Sq. (Ip IIb)	Swanton Morley (Ip IIb)	Honiton	Sewerby Raised Beach shingle	Boulton Moor	Joint Mitnor Cave	Tornewton Cave Hyaena Stratum	Victoria Cave	Kirkdale Cave
Carnivora											
Canis lupis L., wolf	—	+	—	—	—	—	—	+	+	—	+
Vulpes vulpes (L.), red fox	—	—	—	—	—	—	—	+	+	—	+
Ursus arctos L., brown bear	—	+	—	—	—	+	(+)	+	+	(+)	+
Meles meles (L.) badger	—	+	—	—	—	—	—	+	—	—	—
Crocuta crocuta ERXLEBEN, spotted hyaena	—	+	—	—	—	—	+	+	+	+	+
Felis sylvestris SCHREBER, wild cat	—	—	—	—	—	—	—	+	—	—	—
Panthera leo (L.), lion	—	+	+	—	—	—	—	+	+	—	+
Proboscidea											
Palaeoloxodon antiquus FALCONER & CAUTLEY, straight-tusked elephant	+	+	+	—	+	+	?	+	—	+	+
Perissodactyla											
Dicerorhinus hemitoechus (FALCONER), extinct rhinoceros	—	+	?	—	—	+	?	+	+	+	+
Artiodactyla											
Sus scrofa L., wild pig	—	—	—	—	—	—	—	+	—	—	—
Hippopotamus amphibius L., hippopotamus	+1	+	+	+	+	+	+	+	+	+	+
Megaceros giganteus BLUMENBACH, giant deer	+	+	—	—	—	—	—	+	—	—	+
Dama dama (L.), fallow deer	—	+	+	—	—	—	—	+	+	—	+
Cervus elaphus L., red deer	+	+	+	—	+	—	+	+	+	+	+
Bos primigenius BOJANUS, aurochs	—	+	+	+	+	—	—	—	—	—	—
Bison priscus BOJANUS, extinct bison	—	+	—	—	—	+	—	(+)	—	—	+
Bos sp. or *Bison* sp.	+	—	—	—	—	—	+	—	+	—	—

4.2. Cromerian

It is not possible to infer much from the distribution of *Hippopotamus* in Britain during the Cromerian, since known Cromerian deposits are restricted to a) the coastal exposures of Norfolk, Suffolk (CFBS), b) Sugworth near Oxford, and c) Westbury-sub-Mendip, a site with cave deposits of possible Cromerian age. Neither of the two latter sites has yielded extensive large herbivore remains, so the absence of *Hippopotamus* from them may not be of any significance.

The overall distribution of *Hippopotamus* in both the Cromerian and Eemian interglacials, i.e. confined to the Mediterranean area and Europe west of about 10° E., suggests that both *H. major* and *H. amphibius* were intolerant of cold winters, perhaps because they required ice-free lakes and rivers.

Note added in proof:

Since the above was written FAURE (1981) has summarized fossil occurrences of *Hippopotamus* in W. Europe and v. KOENIGSWALD et al. (1982) has found *Hippopotamus* of Eemian age in Germany (F.R.G.) near Mannheim.

5. Palaeoecology

5.1. Present-day: Hippopotamus amphibius

At the present-day *H. amphibius* occurs over most of equatorial Africa in association with regional vegetation ranging from savanna to equatorial rain forest, up to an altitude of 2.400 m (DORST and DANDELOT 1970). It avoids desert areas, and has been exterminated by man in both North Africa and the extreme south of the continent.

It is always closely associated with water, inhabiting rivers and lakes bordered by permanent grassland. Hippopotamuses are semi-aquatic, spending the day resting and sleeping in the water, which they leave at night to feed in areas 2 km or more from the river. Grasses are the staple food, but aquatic plants are also eaten. The impact of the hippopotamus population on the local vegetation can be spectacular. Their grazing and trampling activities maintain, and probably initiate, short grassland communities, with areas of bare ground, in belts over 1 km wide on either side of a river (LOCK 1972).

5.2. Ipswichian Hippopotamus amphibius

With the exception of material from Honiton and Barrington, the pollen spectra associated with hippopotamus remains of Ipswichian interglacial age (Table 3) are dominated by tree pollen, indicating regional temperate forest vegetation. The Barrington spectra, although dominated by herb pollen, also include significant amounts of thermophilous trees, suggesting that temperate forest was present regionally with extensive herb-dominated vegetation on the river flood-plain

(GIBBARD and STUART 1975). The Honiton spectrum, which consists almost entirely of non-tree pollen (97.5%) may also reflect local vegetational conditions. Perhaps if a much larger number of pollen grains was counted, this would result in the identification of more thermophilous tree taxa, as was the case with the Barrington spectra (ibid.).

With the exception of Honiton, where the paucity of tree pollen precludes a determination of the pollen assemblage zone, all of the Ipswichian hippopotamus spectra can be referred to the middle part of the interglacial. Forests with *Pinus* (pine), *Betula* (birch), *Quercus* (oak), *Alnus* (alder), *Acer* (maple) and *Corylus* (hazel) grew in subzone Ip IIb, joined by *Carpinus* (hornbeam) at the beginning of zone Ip III. Evidence from plants, molluscs, beetles and *Emys orbicularis* (European pond tortoise) (summary in STUART 1976, 1979), indicates that summer temperatures were higher than experienced in the same area today throughout most of the interglacial, and that during zone Ip II, at least, the winters were also mild. The Ipswichian hippopotamuses therefore appear to have lived in England during the warmest parts of the interglacial.

The high levels of herb pollen in these spectra is particularly interesting. None of the spectra shows less than 25% herb pollen, and extraordinarily high herb pollen levels occur at Barrington (86—87%) and Honiton (97.5%). This phenomenon has already been commented upon by TURNER (1975), GIBBARD and STUART (1975), PHILLIPS (1976) and STUART (1976), all of whom stress the importance of large herbivorous mammals in maintaining, and probably initiating, herb-dominated areas on river floodplains. These herb pollen spectra comprise mainly Gramineae (grasses), Compositae (composites), Cyperaceae (sedges) and *Plantago* (plantain). *Plantago*, mostly *P. lanceolata*, although absent from the Beetley early Ip IIb spectrum, is elsewhere unusually well-represented — generally at levels of about 7—16% total herb pollen. It reaches the very high value of about 40% total herb pollen at Honiton. *Plantago* is indicative of trampled and grazed grassland habitats. Judging from their profound influence on the vegetation in present-day Africa (see above), hippopotamuses no doubt played a major role in initiating and maintaining these deforested areas during the mid-Ipswichian in England.

TURNER (1975) has suggested that the habit of hippopotamus of feeding on land and defaecating in the water (LOCK 1972) would lead to over-representation of herb pollen in interglacial sediments, and might partly for the high herb pollen levels recorded at these sites.

Hippopotamus amphibius records, of presumed Ipswichian age, from certain cave sites are of particular interest since the sites are located at some distance from the nearest river. For example, Victoria cave, Yorkshire is more than 2 km from, and about 290 m above, the floodplain of the modern River Ribble, and there are no indications of nearer large bodies of water that might have existed in the Ipswichian. It is probable therefore that hippopotamuses left the river, presumably at night, and climbed the valley slopes to graze on the rich limestone herb flora.

At some cave sites, e.g. Joint Mitnor, hippopotamuses and other mammals were killed by falling into natural pitfall traps. Other hippopotamus records from caves are probably based on remains dragged in by hyaenas *(Crocuta crocuta)*. Significantly the remains from several sites represent immature individuals, some of which may have been killed by hyaenas or lions *(Panthera leo)*.

5.3. Cromerian: Hippopotamus cf. major

The pollen-dating of CFBS hippopotamus finds to the later part of the Cromerian stage (zones Cr IIIb and IV) is rather surprising, in that hippopotamuses are generally thought of as thermophilous animals, and they would therefore be expected to occur in the middle of the interglacial, as they do in the Ipswichian.

The associated vegetation appears to have been conifer-dominated forest, with areas of heathland, and of marsh, with herb-dominated vegetation on the river floodplains (see Appendix). The herb pollen levels (19—33%) are not as high as those recorded in association with Ipswichian hippopotamus finds. Moreover *Plantago* (*P. major/media*, not *P. lanceolata*) is present at only one locality in very low amounts.

The ecology of *Hippopotamus major*, if this was the species involved, may have differed somewhat from that of *H. amphibius*, and this may be related to the differences in associated pollen spectra. Particularly intriguing is the apparent absence of hippopotamus from the middle part of the Cromerian interglacial, zone Cr II. Geographical factors controlling migration from southern Europe however may have been responsible for this phenomenon.

Acknowledgements

I thank Mr. A. CURRANT (British Museum, Natural History), Dr. C. L. FORBES (Sedgwick Museum of Geology, Cambridge) and Mr. P. LAWRENCE (Norwich Castle Museum) for access to specimens in their care. I am also grateful to Dr. C. L. FORBES for permission to use the photograph reproduced in Plate I and to M. RAYMOND GÈZE for information on the distribution of fossil *Hippopotamus*.

The work forms part of a NERC supported research project on mammal faunas of the early Middle Pleistocene (NERC grant to Dr. K. A. JOYSEY), which is gratefully acknowledged.

Summary

Hippopotamus remains are widely recorded from the Pleistocene of England and Wales. They appear to originate from two interglacial stages; the Cromerian and Ipswichian, and to be confined to particular pollen assemblage zones (Cromerian zones Cr IIIb and IV, Ipswichian zones Ip IIb and early III) within these interglacials.

The living species *Hippopotamus amphibius* was probably the only species present during the Ipswichian, and the Cromerian species was probably *H. major*.

Hippopotamuses are known to have reached as far as 54° 35′ N during the Ipswichian, and may have extended considerably further north.

Zusammenfassung

Fossile Flußpferd-Reste sind zahlreich aus dem Pleistozän von England und Wales bekannt. Wahrscheinlich stammen diese Funde aus zwei verschiedenen Interglazial-Stufen, dem Cromerium und dem Ipswichium. Innerhalb dieser Interglaziale sind sie wiederum gewissen Pollen-Zonen zuzuordnen (Cromerium: Zonen Cr III b und IV; Ipswichium: Zonen Ip, IIb und frühe Zone III).

Die rezente Spezies *Hippopotamus amphibius* war wahrscheinlich nur während des Ipswichiums vorhanden, die Spezies des Cromeriums war wahrscheinlich *H. major*.

Während des Ipswichium-Interglazials erreichten Flußpferde in England ein Verbreitung bis 54° 35′ N und es kann angenommen werden, daß sie zu dieser Zeit noch weiter nach Norden vorstoßen konnten.

Резюме

Остатки гиппопотамов широко известны из плейстоцена Англии и Уэльса. Они, по-видимому, происходят из отложений двух межледниковых ярусов, кромерия и ипсвичия, и приурочены к определенным зонам палинологических комплексов в пределах этих межледниковий (кромерские зоны Cr IIIb и IV, ипсвичские зоны Ip IIb и начало зоны III).

Вероятно, единственным видом, существовавшим в течение ипсвичия, был современный вид *Hippopotamus amphibius*, а в кромерии, очевидно, жил вид *H. major*.

Известно, что гиппопотамы в ипсвичии достигали 54° 35′ с.ш., но возможно, что они распространялись значительно дальше к северу.

References

Blandamura, F.; Azzaroll, A., 1977: L'„Ippopotamo Maggiore" di Fillipo Nesti. — In: Atti. Accad. Naz. Linc. — **14**. — P. 169—188.

Boylan, P. J., 1967: The Pleistocene Mammalia of the Sewerby-Hessle buried cliff, east Yorkshire. — In: Proc. Yorks. geol. Soc. — **36**. — P. 115—125.

—, 1977: In: J. A. Catt (Ed.): Yorkshire and Lincolnshire, INQUA X. Congress Excursion Guide. — Norwich: Geo Abstracts.

Catt, J. A.; Penny, L. F., 1966: The Pleistocene deposits of Holderness, East Yorkshire. — In: Proc. Yorks. geol. Soc. — **35**. — P. 375—420.

Coxon, P.; Hall, A. R.; Lister, A.; Stuart, A. J., 1984: New evidence on the stratigraphy, palaeobotany and vertebrate fauna of the interglacial deposits at Swanton Morley. — In: Geol. Mag. — London **117** (6). — P. 525—546.

Dorst, J.; Dandelot, P., 1970: A Field Guide to the Larger Mammals of Africa. — London: Collins.

Faure, M., 1981: Répartition des Hippopotamidae (Mammalia, Artiodactyla) en Europe occidentale. Implications stratigraphiques et paléoécologiques. — In: Géobios. — Lyon **14**. — P. 194—200.

Franks, J. W., 1960: Interglacial deposits at Trafalgar Square, London. — In: New Phytol. — London **59**. — P. 145—152.

Funnel, B. M.; West, R. G., 1977: Preglacial Pleistocene deposits of East Anglia. — In: F. W. Shotton (Ed.): British Quaternary Studies. — Oxford: Clarendon Press.

Gibbard, P. L.; Stuart, A. J., 1975: Flora and vertebrate fauna of the Barrington Beds. — In: Geol. Mag. — London **112**. — P. 493—501.

Jones, P. F.; Stanley, M. F., 1975: Description of *Hippopotamus* and other mammalian remains from the Allenton Terrace of the Lower Derwent valley, S. Derbyshire. — In: Mercian Geol. — **5**(4). — P. 259—271.

Koenigswald, W. v.; Löscher, M., 1982: Jungpleistozäne *Hippopotamus*-Funde aus der Oberrheinebene und ihre biogeographische Bedeutung. — In: N. Jahrb. Geol. u. Paläont., Abh. — Stuttgart 163 (3). — P. 331—348.

Lock, J. M., 1972: The effects of hippopotamus grazing on grasslands. — In: J. Ecol. — **60**. — P. 445—467.

Owen, R., 1846: A History of British Fossil Mammals and Birds. — London: van Voorst.

Phillips, L., 1976: Pleistocene vegetational history and geology in Norfolk. — In: Phil. Trans. Roy. Soc. London, B. — London **275**. — P. 215—286.

Reid, C., 1882: The geology of the country around Cromer. — In: Mem. geol. Surv. U.K.

Reynolds, S. H., 1922: *Hippopotamus*, Vol. 3. Part I. In: A Monograph on the British Pleistocene Mammalia. — London: Palaeontogr. Soc.; Monograph.

Stuart, A. J., 1974: Pleistocene history of the British vertebrate fauna. — In: Biol. Rev. — London **49**. — P. 225—226.

—, 1975: The vertebrate fauna of the type Cromerian. — In: Boreas. — Oslo **4**. — P. 63—76.

—, 1976: The history of the mammal fauna during the Ipswichian Last interglacial in England. — In: Phil. Trans. Roy. Soc. London, B. — London **276**. — P. 221—250.

—, 1979: Pleistocene occurrences of the European pond tortoise (*Emys orbicularis* L.) in Britain. — In: Boreas. — Oslo **8**. — P. 359—371.

—, 1982: Pleistocene Vertebrates in the British Isles. — London: Longman.

Sutcliffe, A. J., 1959: The hippopotamus in Britain. — In: Bull. Mammal. Soc. Brit. Island. — 11. — P. 36—40.

—, 1960: Joint Mitnor Cave, Buckfastleigh. — In: Trans. Proc. Torquay Nat. Hist. Soc. — **13**. — P. 1—26.

—, 1964: The mammalian Fauna. In: C. D. Ovey (Ed.): The Swanscombe Skull. A Survey of Research on a Pleistocene Site. — In: Roy. anthrop. Inst. Great Britain & Ireland, Occasional paper. — 20.

—, 1969: Pleistocene faunas of Devon. — In: F. Barlow (Ed.): Exeter and its Region. — Exeter: University.

Sutcliffe, A. J.; Zeuner, F. E., 1962: Excavations in the Torbryan Caves, Devonshire. 1. Tornewton Cave. — In: Proc. Devon. Archaeol. Exploration Soc. — **5**. — P. 127—145.

Turner, C., 1975: Der Einfluß großer Mammalier auf die interglaziale Vegetation. — In: Quartärpaläontologie. — Berlin **1**. — P. 13—19.

Turner, C.; West, R. G., 1968: The subdivision and zonation of interglacial periods. — In: Eiszeitalter u. Gegenwart. — Öhringen **19**. — P. 93—101.

West, R. G., 1977: Pleistocene Geology and Biology. 2nd Ed. — London: Longman.

—, 1980: The Preglacial Pleistocene of the Norfolk and Suffolk Coasts. — Cambridge: Univ. Press.

Appendix

Pollen Analyses from Hippopotamus Bones from the Cromer Forest Bed Series

Pollen analyses were undertaken from sediment adhering to three hippopotamus bones in the British Museum of Natural History. The bones were originally collected from sections at Trimingham (M6593), Mundesley (M6619) and Bacton (M6951) in Norfolk. Pollen preparation followed standard chemical procedures and the results of the analyses are shown in the Table 1 in which pollen nomenclature follows Andrew (1970).

All three counts show similar pollen spectra, dominated by tree pollen taxa with *Pinus*, *Betula* and *Alnus*, low numbers of *Ulmus* pollen and notably the pollen of the 'Late Temperate' trees *Abies* and *Picea*. The forest vegetation included *Juniperus* and *Corylus* shrubs and forest herbs. Heathland vegetation is represented by the pollen of Ericales, *Calluna* and *Succisa* and the spores of *Lycopodium* and *Pteridium*. Alder carr that grew on wet ground beside the river in each case is represented by *Alnus* pollen and *Polypodium* spores. *Salix* also grew beside the rivers at the Mundesley and Bacton localities. Drier, herb-dominated communities near the rivers are indicated by high pollen frequencies of Gramineae, together with the pollen of Compositae (*Bellis* Type, *Cirsium* Type, Liguliflorae), Chenopodiaceae, *Plantago major/media* Type and *Vicia* Type. A rich marsh community occurred

Table 1. Pollen counts from *Hippopotamus* remains from the Cromer Forest Bed Series. Calculated using the pollen sum: Total pollen + spores — aquatics

Pollen Zone	Triming-ham (M6593) Cr IIIb	Mundes-ley (M6619) Cr IV	Bacton (M6951) Cr IV
Trees			
Betula	23.1	12.0	11.5
Pinus	23.1	31.3	12.9
Ulmus	1.4	0.4	2.4
Quercus	0.4	—	—
Alnus	6.0	10.6	19.5
Carpinus	0.4	0.2	0.5
Abies	7.9	3.5	2.4
Picea	1.8	7.3	4.8
Total trees	64.4	65.4	53.6
Shrubs			
Corylus	0.4	0.6	0.5
Juniperus	2.3	1.2	1.9
Salix	—	0.2	0.5
Total shrubs	2.8	2.0	2.9
Herbs			
Graminea	15.3	9.8	11.5
Cyperaceae	2.8	2.5	3.3
Calluna	0.9	1.6	1.4
Ericales	—	0.6	0.5
Bellis Type	0.9	0.2	1.9
Cirsium Type	—	0.4	—
Artemisia	—	0.2	—
Compositae (Liguliforae)	—	0.2	0.4
Cenopodiaceae	0.5	1.6	11.0
Filipendula	1.8	—	0.4
Vicia Type	—	0.2	—
Plantago major/media Type	—	0.2	—
Polygonum amphibium	0.5	—	—
Ranunculaceae	1.8	1.2	1.9
Succisa	—	—	0.5
Saxifraga	—	0.2	—
Umbelliferae	1.8	—	—
Total herbs	25.5	19.1	33.0
Pteridophytes			
Lycopodium	1.8	0.8	0.5
Osmunda	—	—	0.5
Polypodiaceae undifferentiated	3.2	7.5	7.2
Pteridium	—	0.4	0.5
Sphagnum	4.2	4.9	1.9
Total pteridophytes	8.8	13.6	10.5
Aquatics			
Myriophyllum spicatum	0.4	—	—
Potamogeton	0.9	0.8	—
Sparganium Type	0.9	0.6	1.4
Typha latifolia	0.5	0.8	—
Total aquatics	2.8	2.2	1.4
Total pollen & spores	222	502	212
Pre-Pleistocene pollen + spores	1.8	—	0.5
Dinoflagellate cysts	1.3	2.9	4.2

locally and included *Filipendula*, *Polygonum*, Cyperaceae and *Osmunda*, together with aquatic vegetation indicated by the pollen of *Sparganium* Type, *Typha latifolia*, *Potamogeton* and *Myriophyllum spicatum*.

The presence of the 'Late Temperate' forest elements *Abies* and *Picea* indicate that the spectra represent the second half of a Middle Pleistocene interglacial stage, i.e. pollen assemblage biozones III and IV (after TURNER and WEST 1968). *Abies* is recorded only from the Cromerian and Hoxnian stages in Britain (WEST 1970, 1979). However, all three of the hippopotamus bones come from the Cromer Forest Bed Series and therefore can have only been derived from either the Pastonian or the Cromerian interglacial stages (see WEST 1979). The high frequencies of *Abies* pollen preclude a Pastonian correlation and indicate a Cromerian age for the fossil hippopotamus remains. Based on the mutual frequencies of *Abies* and *Picea* pollen, the Trimingham bone (M6593) was derived from beds of Cr IIIb age, whilst the spectra from both the Mundesley (M6619) and the Bacton (M6591) bones indicate a Cr IV age.

In all three cases it appears that the hippopotamuses lived in rivers that flowed through 'Late Temperate' forest dominated by coniferous trees with herb-dominated areas on the river floodplains. The possibility that these open areas were maintained by animal grazing should be borne in mind (cf. GIBBARD and STUART 1975; TURNER 1975). Wetter areas on the floodplains supported a diverse herb flora and alder carr grew locally.

ANDREW, R., 1970: The Cambridge pollen reference collection. In: D. WALKER; R. G. WEST (Eds.): Studies in the vegetational history of the British Isles. — Cambridge: Univ. Press. — P. 225 to 231.

GIBBARD, P. L.; STUART, A. J., 1975: Flora and vertebrate fauna of the Barrington Beds. — Geol. Mag., London **112**. — P. 493 to 501.

TURNER, C., 1975: Der Einfluß großer Mammalier auf die interglaziale Vegetation. — Quartärpaläontologie, Berlin **1**, p. 13 bis 19.

TURNER, C.; WEST, R. G., 1968: The subdivision and zonation of interglacial periods. — Eiszeitalter u. Gegenwart, Öhringen **19**, p. 93—101.

WEST, R. G., 1970: Pleistocene history of the British flora. In: D. WALKER; R. G. WEST (Eds.): Studies in the vegetational history of the British Isles. — Cambridge: Univ. Press. P. 1—21.

—, 1979: The Preglacial Pleistocene of the Norfolk and Suffolk coasts. — Cambridge: Univ. Press.

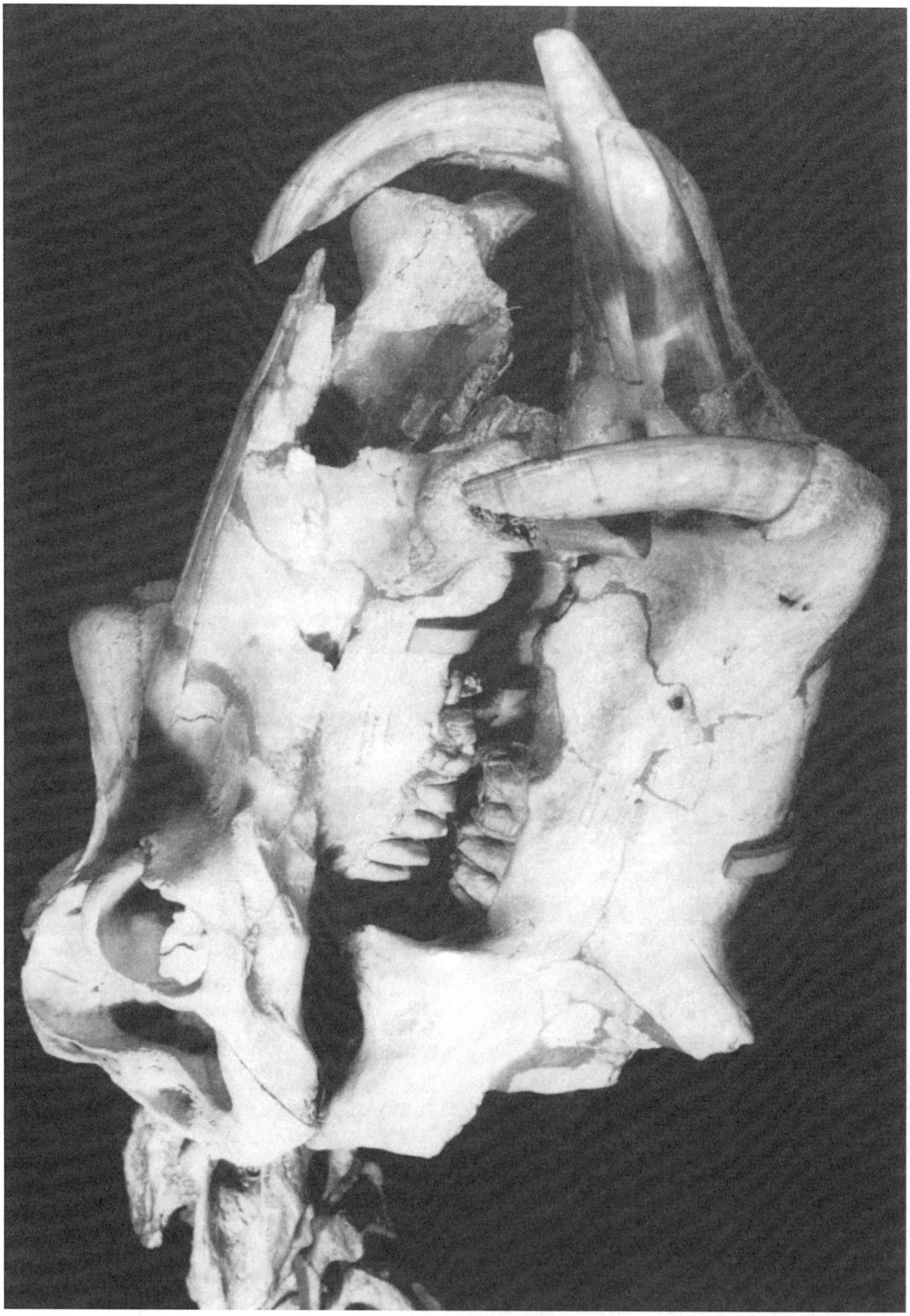

PLATE I

Skull of *Hippopotamus amphibius*, of Ipswichian subzone Ip IIb age, from Barrington, Cambridgeshire. The specimen, part of a composite mounted skeleton, is displayed in the Sedgwick Museum of Geology, University of Cambridge, England

Quartärpaläontologie 6 219—221 Berlin 1986

Die Entwicklungsetappen der Hipparion-Fauna in Kasachstan (UdSSR)

Von FIRUSA ABLAEVNA TLEUBERDINA, Alma-Ata

Inhalt

Abstractum

Eine Analyse der *Hipparion*-Fauna Kasachstans erlaubt es, Hauptetappen ihrer Entwicklung zu bestimmen.

An analysis of the *Hipparion* fauna of Kazakhstan allows us to set out different major stages of evolution.

Анализ изучения гиппарионовой фауны в Казахстане позволил выделить основные этапы в её развитии.

1. Einleitung

Hipparionen-Faunen zogen bereits seit langem die Aufmerksamkeit der Paläontologen und Geologen für die Ziele der Stratigraphie und Paläogeographie aber auch zur Systematik und Phylogenie ihrer Hauptvertreter auf sich. Die allgemeine Veränderungsrichtung des Bestandes von Hipparionen-Faunen im gesamten Zeitablauf ihrer Existenz ist untrennbar mit den bedeutenden Veränderungen der physisch-geographischen Umwelt verbunden, die in der zweiten Hälfte des Neogens vor sich gingen. Die Wandlungen im Artenbestand und in den ökologischen Besonderheiten der Hauptvertreter der *Hipparion*-Fauna erscheinen recht deutlich bei der Gegenüberstellung der Assoziationen unterschiedlichen geologischen Alters. Die vorliegenden Daten zur *Hipparion*-Fauna Kasachstans, die Bestandsanalyse dieses Komplexes und das Evolutionsniveau ihrer Elemente zeigen, daß der Formierungsprozeß der *Hipparion*-Fauna in mehreren Etappen vor sich ging.

2. Die Entwicklungsetappen der Hipparion-Fauna in Kasachstan

Als die früheste Entwicklungsetappe der *Hipparion*-Fauna dieses Raums kann man die fossilen Assoziationen ansehen, die aus grau-grünen Tonen am linken Ufer des Irtyš nahe Pavlodar, vom Kalkaman-See, nordöstlich der Stadt Ermak, und entlang der Trasse des Irtyš-Karaganda-Kanals: nördlich der Siedlung Evgenevka und im Forst Karašigar sowie in Zentralkasachstan bei der Getreidefarm Nurinskij und am Fluß Kulanutpes am Borkutbaj vorliegen.

Die Fauna dieser Fundstellen umfaßt folgende Taxa: Pisces — Siluridae, Esocidae; Amphibia — Ranidae; Reptilia — Emydidae gen. indet.; Aves — Falconidae; Mammalia — *Talpa* sp., *Condilura* sp., *Ochotona* sp., *Agispelagus* sp., *Monosaulax savinovi* LYČEV, *M. pansus* COPE, *Monosaulax* sp., *Stenofiber* sp., *Asiacastor major* LYČEV, *Protalactaga* sp., Dipodidae gen. indet., *Promephites alexeevi* SCHL., *Promephites* sp., *Parataxides crassa* ZD., *Mastodon* cf. *turicensis* SCHINZ., *Anchitherium aurelianense* (CUV.), *Hipparion* sp., *Aceratherium* cf. *incisivum* (KAUP), *Aceratherium* sp., *Aprotodon* (?) sp., *Chilotherium* sp., *Microstonyx major* GERV., Suidae gen. indet. Tragulidae gen. indet., *Prodremotherium* sp., *Cervavitus* sp. (primitive Form), *Gazella deperdita* GERV., Artiodactyla gen. indet.

Die Besonderheit dieses Faunenhorizonts ist das Vorherrschen solcher Formen, die an feuchte Biotope, an bewaldete und versumpfte Gebiete angepaßt waren, wobei die in ihrem Bestand vorhandenen Superstiten, Elemente der *Anchitherium*-Fauna, folgende Formen umfassen: *Agispelagus*, *Prodremotherium*, Tragulidae, *Anchitherium* und primitive Mastodonten. *Anchitherium*-Reste sind hier sehr selten, dafür sind Funde primitiver Mastodonten recht häufig. Gleichzeitig erscheinen und entwickeln sich Bewohner trockenerer und offener Gebiete: *Hipparion*, Gazellen, Antilopen. Ihre Reste sind jedoch noch nicht zahlreich und liegen zumeist als Bruchstücke vor. Allgemein gesagt stellt diese Assoziation eine frühe Entwicklungsetappe der *Hipparion*-Fauna dar, die wahrscheinlich der zweiten Hälfte des Miozäns zuzuweisen ist (LYČEV 1963; AUBEKEROV u. a. 1968, 1970).

Gegen Ende der ersten Hälfte des Mäot verringerte sich der Anteil der an feuchte Biotope angepaßten Formen beträchtlich; die letzten Elemente der *Anchitherium*-Fauna starben aus. Gleichzeitig vergrößerte sich die Anzahl solcher Arten, die an Savannen oder Steppen angepaßt waren. Es liegt also eine ganz bestimmte Gesetzmäßigkeit vor: die Ablösung archaischer miozäner durch evoluierte progressive Formen, die sich den veränderten Umweltbedingungen angepaßt hatten.

Vom Ende des späten Miozäns bis zum Ende des Pont erreichte die typische *Hipparion*-Fauna auf dem Territorium

Kasachstans eine weite Verbreitung. Diese Fauna ist besonders von der klassischen Fundstelle „Gusinyj pereljot" bei der Stadt Pavlodar und vom Hohlweg Tulkisaj südwestlich der Stadt Arkalyk bekannt (SAVINOV 1972). Sie umfaßt mehr als 60 Arten, darunter etwa 35 Säuger. Pisces — Percidae: *Lucioperca* sp., *Perca* sp.; Reptilia — Tryonychidae gen. indet., Lacertidae, Agamidae, Colubroidae; Aves — *Struthio chersonensis* BR., *Sushkinia pliocaena* TUG.; Amphibia — Bombidae, Pelobatidae, Bufonidae, Ranidae; Mammalia — Insectivora — Soricidae: *Similisorex orlovi* STOG. & SAV.; Erinaceidae: *Erinaceus* sp., *Hiroptera* gen. indet.; Lagomorpha: *Proochotona* cf. *eximia* CHOM.; Sciuridae: *Eutamias* sp.; Myoxidae; Dipodidae: *Sicista bagaevi*, *Lophocricetus vinogradovi*; Allactaginae: *Proalactaga varians*; Dipodinae; Cricetidae; Carnivora — *Martes* sp., *Parataxidea crassa* ZDANSK., *Ictitherium hipparionum* GERV., *Ictitherium robustum* NORD. *Crocuta eximia* ROTH & WAGN., *Machairodus irtyschensis* ORLOV, *Promephites alexejevi* SCHLOSS.; Pinnipedia — *Semantor macrurus* ORLOV; Proboscidea — *Mastodon* sp.; Perissodactyla — *Hipparion elegans* V. GROM., *H. longipes* V. GROM., *Chilotherium* sp., *Sinotherium* sp.; Artiodactyla — *Cervavitus orlovi* FLER., *Samotherium irtyschensis* GOD., *Palaeotragus* sp., *Sivatherium* sp., *Gazella deperdita* (GERV.), *Tragocerus frolovi* PAVL., *Parapseudotragus taurica* (BOGATSCHEV), *Paratragocerus caucasicus* SOKOLOV, *Procapreolus* sp., *Spiroceros* sp.

Für den frühen Pavlodar-Komplex ist das Vorhandensein mäotischer Elemente wie *Chilotherium*, *Samotherium*, *Cervavitus*, *Promephites* usw., ferner die geringe Vielfältigkeit der *Hipparion*-Formen charakteristisch. Bei den Kleinsäugern ist das Vorhandensein archaischer Springmäuse und von Hamster typisch. Allgemein gesagt stellen diese Assoziationen eine höhere Entwicklungsetappe der *Hipparion*-Fauna dar und unterscheiden sich von der vorhergehenden durch den größeren Anteil von Steppenelementen.

Gegen Ende des unteren Pliozäns setzte eine verstärkte Aridisierung ein, wodurch sich die Umweltbedingungen ent-entsprechend extrem veränderten. Dieses wiederum löste eine weitere Umgruppierung der *Hipparion*-Fauna aus. Stenotope Formen starben aus und es setzte eine allmähliche Verarmung der *Hipparion*-Formen ein. Diese Verringerung im Artenbestand der *Hipparion*-Fauna ist gut am Beispiel der Fundstelle Karabastuz im Semipalatinsker Irtyš-Gebiet sowie von Kalmakpaj am Saissansee zu verfolgen.

Die Karabastuz-Fauna ist durch folgende Taxa charakterisiert: Reptilia — *Testudo karabastuzica* KUSNETZOV; Aves — *Struthio* sp.; Mammalia — Hyaenidae; *Hipparion* (zwei Arten), *Dicerorhinus* sp.; *Mammut (Zygolophodon)* sp., *Pliocervus karabastuzicus* ABDRACHMANOVA, *Eostylocerus propria* ABDR., *Palaeotragus* sp., *Samotherium irtyschense* GOD., *Tragocerus* sp.

Die Kalmakpaj-Fauna (ŽEGALLO 1966) wird durch Reptilien, Vögel sowie zahlreiche Carnivoren: *Crocuta*, *Ictitherium*, *Martes*, *Vormela* und *Machairodus* bestimmt. Dazu kommen an Herbivoren: *Chilotherium*, *Hipparion houfenense*, *Sinotherium*, Cervidae gen. indet., *Palaeotragus*, *Tragocerus* und *Gazella*.

Die Fauna dieser Fundstellen stellt einen verarmten Bestand dar. Man muß hier allerdings berücksichtigen, daß beide Lokalitäten noch unzureichend erforscht sind. Gleichzeitig kann man aber feststellen, daß in jedem Faunenbestand die Grundelemente, die für den frühen *Hipparion*-Faunenkomplex typisch sind, vorhanden sind. Einige der Formen erlauben es, von einem verhältnismäßig jungen Alter der Faunen von Karabastuz und Kalmakpaj zu sprechen. Viele Raubtiere zeigen einen hohen Spezialisierungsgrad an, ebenfalls sind die Hipparionen (besonders die von Kalmakpaj) hoch spezialisiert; sie zeichnen sich durch Hypsodontie, relativ lange Zahnreihen und reduzierte Seitenzehen aus. Gleichzeitig liegen von hier bereits *Pliocervus*-Hirschreste vor, die eigentlich erst für einen späteren Zeitabschnitt charakteristisch sind. Wir stellen daher die Assoziationen dieser beiden Fundstellen an das Ende des unteren bis in das mittlere Pliozän.

Schärfere Klimaschwankungen und mit ihnen verbundene weitere Veränderungen im Bestand der *Hipparion*-Fauna sind für die folgende Etappe charakteristisch, die Etappe des Verlöschens der *Hipparion*-Fauna. Diesen Vorgang kann man sehr gut am Beispiel der Esekartkansker Fauna aus der Tekes-Senke südöstlich von Alma-Ata verfolgen (AUBEKEROVA 1974, 1977). Zum Bestand dieser Assoziation gehören: Reptilia — Emydidae; Aves — *Struthio* sp.; Mammalia — *Lepus* sp., *Ochotona* sp., *Castor fiber* L., *Felis (Lynx)* sp., *Anancus kazachstanicus* AUBEKEROVA, *Hipparion houfenense* TEILHARD DE CHARDIN & YOUNG, *Hipparion* sp., *Dicerorhinus orientalis* (SCHLOSSER), *Dicerorhinus* sp., *Gigantocamelus longipes* AUB., *Cervavitus flerovi* AUB., *Sinomegaceros* sp., *Palaeotragus* sp., *Samotherium* sp., *Gazella sinensis* (TEILH. & PIVET.), *Gazella* ex gr. *dorcadoides* SCHLOSSER, *Antilospira* sp., *Tragocerus* (?).

Bei der Charakterisierung dieser Assoziation muß man anmerken, daß *Hipparion* in dieser Fauna nicht mehr so häufig erscheint wie in den Faunen der frühen Entwicklungsetappen und daß nur späte, hochevoluierte Formen vorliegen. In dieser Fauna treten wahrscheinlich die letzten Hirsche der *Cervavitus*-Gruppe auf. In geringer Anzahl finden sich Kleinformen von *Palaeotragus*, *Samotherium* und *Tragocerus*. Gleichzeitig mit ihnen sind die ersten *Sinomegaceros*-Hirsche nachgewiesen, die zwar noch primitive Charaktere zeigen, aber bereits ein progressives Element der späteren Faunenentwicklung darstellen. In dieser Zeit erscheinen auch eine große Anzahl verschiedener Antilopen, Gazellen, Kamele der Gattung *Gigantocamelus* und Nager. Das heißt, man kann in der Esekartkansker Fauna in gewissem Sinne eine Erneuerung des Faunenbilds durch evoluierte Elemente erkennen. Daneben zeigt das Vorkommen der oben aufgezählten Reliktformen der Früh-*Hipparion*-Fauna in dieser Assoziation von landschaftlich wie klimatisch wieder für diesen Faunentyp günstigen Bedingungen, wenngleich diese nur in einzelnen Teilen des Areals gegeben waren. Möglicherweise erhielten sich solche Bedingungen lediglich an den äußersten Grenzen des ziemlich umfassenden Lebensraums der *Hipparion*-Fauna. Diese Ansicht wird nicht zuletzt dadurch bestätigt, daß Reste der späten Hipparions (*Hipparion* cf. *parvum*, *H.* cf. *tchicoicum*) (ŽEGALLO 1975) von Beteke, dessen Ablagerungen in das obere Pliozän datiert werden, bekannt sind.

Abschließend kann festgestellt werden, daß in jeder Assoziation die Kontinuität der einzelnen Elemente oder einiger Formen des Bestands erhalten blieb, und daß die Grundlage für die folgenden Assoziationen von solchen Formen gebildet wurde, die die größere ökologische Plastizität, die breitere Adaption und die umfassenderen Areale besaßen.

Zusammenfassung

Die Entwicklung der *Hipparion*-Fauna Kasachstans ging in verschiedenen Etappen vor sich. Die frühesten Assoziationen werden in die zweite Hälfte des Miozäns (Sarmat-Mäot) datiert. Diese Zeitspanne wird als Entstehungsperiode der *Hipparion*-Fauna betrachtet. Vom Ende des Miozäns an und im frühen Pliozän erfuhr sie ihre Blütezeit. Danach können am Ende des frühen Pliozäns allmählich Veränderungen im Bestand der *Hipparion*-Fauna beobachtet werden: das langsame Verschwinden ihrer Hauptvertreter und das Erscheinen höher spezialisierter *Hipparion*-Formen, Raubtiere und anderer, gleichfalls höher entwickelter Arten.

Summary

The evolution of the *Hipparion* fauna of Kazakhstan may be divided into different stages. The earliest associations have been dated to the second half of the Miocene (Sarmat-Mäot). This time interval was the period of evolution of the *Hipparion* fauna. After the end of the Miocene and in the Early Pliocene this fauna reached its zenith. Further on, at the end of the Early Pliocene, certain changes in the composition of the *Hipparion* fauna are observed: the gradual disappearance of its main representatives and the appearance of more highly specialized hipparions and carnivores as well as of other similarly highly evolved species.

Резюме

Развитие гиппарионовой фауны в Казахстане происходило в несколько этапов. Наиболее ранние находки датируются второй половиной миоцена (сармат-меотис). Этот период рассматривается как период становления гиппарионовой фауны. С конца миоцена и в раннем плиоцене происходит её бурный расцвет. Затем в конце раннего плиоцена в составе гиппарионовой фауны наблюдаются изменения: постепенное вымирание её основных представителей и появление более высокоспециализированных форм гиппарионов, хищников и др., а также появляются эволюционно новые виды животных.

Literatur

Aubekerova, P. A., 1974: Pozdneneogenova fauna Tekesskoj vpadiny (Die spätneogene Fauna der Tekes-Senke). — Alma-Ata. Kand. Diss., Autoref.

Aubekerova-Tleuberdina, P. A., 1977: Sopostavlenie esekartkanskoj fauny s gipparionovymi i pozdnegipparionovymi faunami Azii (Gegenüberstellung der Esekartkansker Fauna mit den Hipparionen- und Späthipparionenfaunen Asiens). — In: Materialy po istorii fauny i flory Kazachstana. T. 7. Mezokajnozojskaja fauna i flora Severo-Zapadnogo Kazachstana. — Alma-Ata.

Aubekerov, B. Ž.; Aubekerova, P. A.; Lyčev, G. F., 1968: Stratigrafičeskoe položenie i perspektivy ispolzovanija sero-zelenych glin levoberez'ja r. Irtyš dla narodnogo chozjajstva (Die stratigraphische Position und die Perspektive einer Nutzung der graugrünen Tone am linken Ufer des Irtyš für die Volkswirtschaft). — In: Materialy pervoj respublikanskoj naučno-teoretičeskoj konferencii molodych geologov Kazachskoj SSR. — Alma-Ata.

Aubekerov, B. Ž.; Aubekerova, P. A.; Birjukov, M. D.; Lyčev, G. F., 1970: Stratigrafičeskoe položenie i fauna posvonočnych sero-zelenych glin levoberez'ja Pavlodarskogo Priityš'ja (Stratigraphische Lage und Wirbeltierfauna der grau-grünen Tone am linken Ufer des Irtyš nahe Pavlodar). — In: Resenija i trudy Meževedomstvannych i korreljacionnych stratigrafičeskich schem Zapadno-Sibirskoj nižmennosti. — Tjumen'.

Lyčev, G. F., 1963: Neogenovye mlekopitajuščie Malogo Kalkamana (Pavlodarskoe Priirtyš'e) (Neogene Säugetiere aus Malij Kalkaman in der Nähe des Irtyš, Gebiet Pavlodar). — In: Materialy po istorii fauny i flora Kazachstana. T. 4. — Alma-Ata.

Savinov, P. F., 1972: Obščie rezultaty paleobiologičeskich issledovanii Pavlodarskogo Priirtyš'ja (Allgemeine Ergebnisse paläobiologischer Untersuchungen im Irtyš-Gebiet nahe Pavlodar). — In: Teriofauna. T. 1. — Novosibirsk.

Žegallo, 1966: Novoe mestonachoždenie gipparionovoj fauny v Vostočnom Kazachstana (Eine neue Fundstätte einer Hipparionfauna in Ostkazachstan). — In: Vsesojušnoe soveščanie po paleontologii mlekopitajuščich kajnozoja. — Tezisy dokladov. — Tbilisi.

—, 1975: Gippariony Central'noj Azii (Die Hipparionen Zentralasiens). — Moskva, Kand.-Diss., Autoref.

Quartärpaläontologie 6 223—229 Berlin 1986

Deciduous Teeth of Desmatolagus (Lagomorpha, Mammalia) from the Mongolian Oligocene and of related European Genera

By Heinz Tobien, Mainz

With 11 Figures and one Table

Contents

Abstractum

Unknown or little known deciduous teeth of the lagomorphous genus *Desmatolagus* from the Oligocene of the People's Republic of Mongolia, and of some nearly related genera from the European Oligocene are described and compared.

Unbekannte oder wenig bekannte Milchzähne der Hasenartigen-Gattung *Desmatolagus* aus dem Oligozän der Mongolischen Volksrepublik sowie einiger nahe verwandter Gattungen aus dem europäischen Oligozän werden beschrieben und verglichen.

Описываятся ранее не известные или мало известные молочные зубы зайцеобразных рода *Desmatolagus* из олигоцена Монгольской Народной Республики. Приводится сравнение с некоторыми близко родственными видами из европейского олигоцена.

1. Introduction

Thanks to the courtesy of the late Professor Dr. C. C. Flerov, Director of the Palaeontological Museum of the Academy of Sciences of the USSR, the author has had the possibility to see the excellent collection of *Desmatolagus* remains from the Oligocene site of Tatal Gol (Mongolian People's Republic). I was generously permitted to select some specimens for closer study and publication. The following account is dedicated to the memory of Professor Flerov, outstanding mammalogist and palaeontologist, highly artistic painter of living and fossil mammals in their ambiente and a charming, amiable personality and friend.

Published teeth and lacteal dentitions of Eurasiatic ochotonids with rather primitive structure as *Shamolagus* Burke, 1941, *Amphilagus* Pomel, 1853, and *Piezodus* Viret, 1929 from Europe and *Desmatolagus* Matthew & Granger, 1923 from Mongolia are rare. There are one D^3 and D^4 and a D_3 of *Shamolagus* from the Middle Oligocene of Möhren 20 (Bavaria) (Heissig and Schmidt-Kittler 1976, p. 91); D^2 + D^3 of *Amphilagus antiquus* from the Upper Oligocene of Coderet (France) (Tobien 1974, Fig. 10); and two upper Decidui of *Amphilagus fontannesi* from the Middle Miocene of La Grive (France) (Major 1899, pl. 36, fig. 14, 35; pl. 39, fig. 4). So far *Desmatolagus* is concerned Sych and Sych (1977, Figs. 6, 7) published and discussed stereophotos of two specimens with D_3 and D_4 from Mongolia in the American Museum of Natural History. The following data should contribute to fill somewhat the gap in the knowledge of the milk dentition of archaistic lagomorphs.

Besides the late Prof. Flerov, the late Prof. Dr. J. Viret, Prof. Dr. L. David and Prof. Dr. P. Mein, Museum d'Histoire Naturelle et Département des Sciences de la Terre, both Lyon, Dr. J. Hürzeler et Dr. H. Schäfer, Natural History Museum of Basle (Switzerland) very kindly loaned specimens for description and comparison. Dr. R. N. Tedford and L. R. Flynn, both American Museum of Natural History, helped with measurements of *Desmotolagus* milk teeth of the AMNH-collections.

Abbreviations of the Institutions, mentioned in the text and the captions:

AMNH: American Museum of Natural History, New York (U.S.A.)
FL: Département des Sciences de la Terre, Lyon (France)
MB: Naturhistorisches Museum, Basel (Switzerland)
ML: Musée d'Histoire Naturelle, Lyon (France)
PMM: Palaeontological Museum of the Academy of Sciences of USSR, Moscow.

2. Upper Deciduous Teeth of Desmatolagus

The rich *Desmatolagus* collection from the Oligocene of Tatal Gol (Mongolian People's Republic: Fig. 1) in the Palaeontological Museum of the Academy of Sciences of the USSR (see Gureev 1960, p. 12—16) includes some specimens with upper milk teeth: 475—4203 is a right maxillary fragment with the complete deciduous set D^2—D^4, followed by an unworn M^1. From the length of the M^1 the specimen belongs to the smaller species *D. gobiensis* Matthew & Granger, 1923 (Sych 1975, p. 184, table 1). 475—4209 is a fragmentary skull with M^1 left and D^4 — M^1 right, likewise belonging to *D. gobiensis*.

475—4203 (Fig. 2): D^2 — D^4 are rather worn, the basic structural elements of the crowns are, however, recognizable. As usual in lagomorphous and other milk dentitions to a

[1]) abbreviations of the crown elements (see also Tobien 1974, p. 111, figs. 2, 3 and others):
IE: internal element (Innenelement); ME: middle element (Mittelelement); EE: external element (Außenelement); PC: postero-external cingulum (Hinteraußencingulum); AC: anterior cingulum = anterior arm of the internal element (Vordercingulum = Vorderarm des Innenelements); ms: middle syncline (Mittelsynkline); es: external syncline (Außensynkline); pes: postero-external syncline (Hinteraußensynkline); is: internal syncline (Innensynkline).

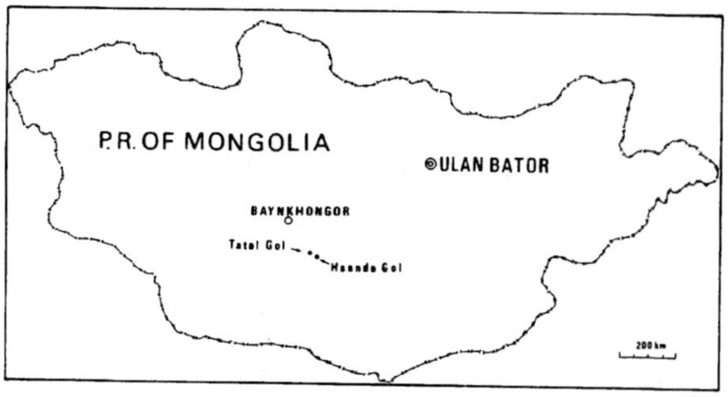

Fig. 1. Topographic situation of the Middle Oligocene locality Tatal Gol (People's Republic of Mongolia) (after KIELAN-JAWOROWSKA and DOVCHIN 1968)

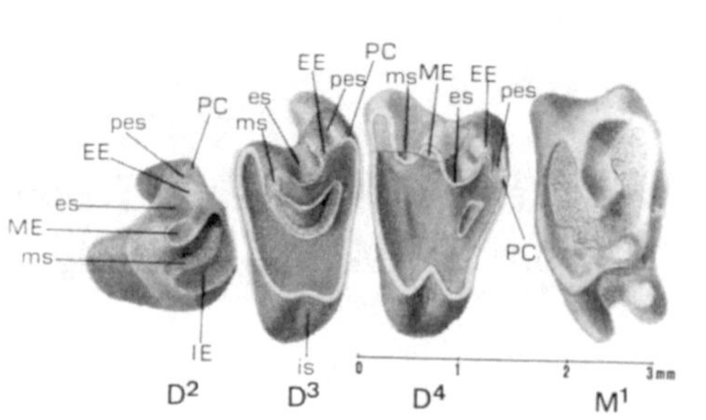

Fig. 2. *Desmatolagus gobiensis* MATTHEW & GRANGER. Middle Oligocene, Tatal Gol (P.R. of Mongolia). D^2—M^1 dext. (inverse). Occlusal view. PMM 475—4203, × 10. D^4 antero-externally reconstructed after PMM 475—4209 (s. Fig. 4)

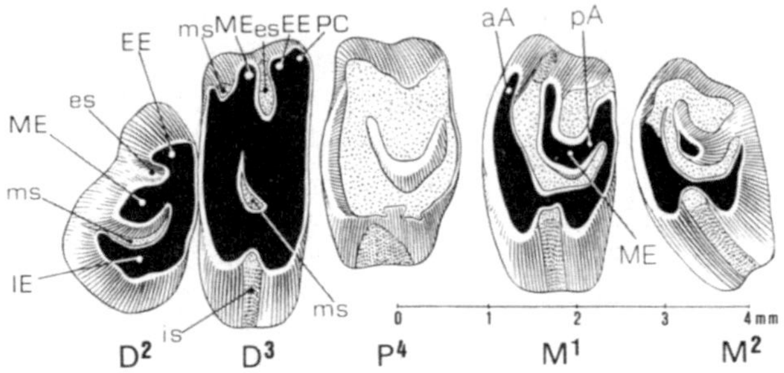

Fig. 3. *Desmatolagus gobiensis* MATTHEW & GRANGER. Middle Oligocene, Hsanda Gol form. Tsagan Nor (P.R. of Mongolia). P^3—M^3 sin. Occlusal view. MB Mo 6. × 10. (see also TOBIEN 1974, Fig. 96)

certain degree the deciduous molars are structurally more advanced than their permanent successors.

The triangular crown of D^2 shows three main elements (IE, ME, EE), separated by two synclines (es, ms)[1]).

At the very external border of the crown a tiny element is visible (PC) that corresponds — after comparison with the following deciduous teeth — unworn premolars and molars, to a postero-external small enamel wall, that very probably corresponds to a cingulum. It is separated by a very small syncline (pes) from the external element (EE).

The internal element is the largest, the middle is smaller and still smaller is the external element. All three elements are connected distally by an enamel crest opened by wear. Compared with the permanent P^2 of *Desmatolagus*, where the middle element is the strongest and the two other decidedly smaller (e.g. MATTHEW and GRANGER 1923, Fig. 10A; BOHLIN 1937, Fig. 11, 1942, Fig. 9:6; SYCH 1975, Pl. 54, Fig. 3), this is a rather advanced structure, e.g. encountered in Miocene *Titanomys*-P^2 (TOBIEN 1963, Fig. 7a) of Western Europe. Compared with the permanent dentition of *Desmatolagus* the structure of D^2 approaches that of P^3 (MATTHEW and GRANGER 1923, Fig. 10A; BOHLIN 1937, Fig. 11:14; GUREEV 1960, Fig. 4A; TOBIEN 1974, Fig. 96; this paper: Fig. 3).

The D^2 crown is supported by two roots: a large lingual root below ME, ms and IE, a smaller buccal root below EE, pes and PC (Fig. 2).

D^3 (Fig. 2) is rather worn, particularly at the lingual side, because of the partial (or unilateral) hypsodonty, a main characteristic of the archaic lagomorphs (TOBIEN 1974, p. 112). The main elements, however, are still visible but in part transformed: the internal element (IE) is elongated and protracted to the antero-external corner, forming the anterior wall of the crown. By homology with the permanent P^3 of *Desmatolagus*, *Amphilagus* and other genera it can be demonstrated that the anterior arm of the internal element (AC) is the enlarged and hypsodont derivative of an anterior basal cingulum (TOBIEN 1974, Figs. 3, 4; 1975, Figs. 2—5).

The middle element (ME) is likewise enlarged, its anterior end is elongated to the buccal border, thereby the shape of the whole element is crescentic. In consequence of the enlargement of IE and ME, the middle syncline (ms) is transformed into a large crescentic valley, that opens on the buccal border of the crown. The structure of the postero-external edge of the crown, however, has not much changed compared with D^2: es, EE, pes, PC are similar to those of D^2.

The outgrowth of IE and AC leads to a trapezoidal contour of the crown, accentuated by a shallow internal syncline (is). The lingual part of the crown, therefore, is shorter in mesiodistal direction than the buccal part. The same holds true for the D^4.

Three roots are present: two outer, smaller ones, not affected by the partial hypsodonty, below the buccal part of the crown (the postero-external one is visible on Fig. 2), and a much larger and longer root below the lingual part of the crown and transformed by hypsodonty.

In general shape and structure the D^3 resembles much more the permanent P^4 than its successor, the P^3 (compare with Fig. 3).

D^4 (Fig. 2) is still more worn than D^3 and D^2, as usual in the complete dentition. The middle syncline (ms) is completely eroded, except the buccal orifice and the postero-external part, the latter as a small islet on the wear surface. Furthermore preserved are: EE, pes and PC (Fig. 2). Measurements see table 1.

The right D^4 in 475—4209 (Fig. 4), a facial skull fragment, with both M^1 is less worn, and the middle syncline therefore is more complete.

The buccal orifice of the middle syncline (ms); the buccal end of the middle element (ME); the external syncline (es), protracted almost to the buccal margin of the crown; the external element (EE), forming a real small conelet; and the anterior labial element (aLE) (see p. 225) are well identifiable. The buccal end of ME is somewhat more elongated than in D^3, but does not yet reach the buccal border of the crown. The internal syncline (is = hypostria) is deeper than in D^3, but the root disposition is identical. As in D^3 in D^4 likewise the lingual part of the crown is narrower than the buccal one, the latter being more elaborated, because the ends of

Table 1. Measurements [mm]
Upper deciduous Molars, length

	D^2	D^3	D^4
Desmatolagus gobiensis PMM 475—4203	1.0	1.3	1.4
Desmatolagus gobiensis PMM 475—4209	—	—	1.5
Shamolagus franconicus (HEISSIG and SCHMIDT-KITTLER 1976, p. 87)	—	2.25	—
Amphilagus antiquus FL 4517b	1.5	1.25	—
Amphilagus fontannesi ML L. Gr. 111	—	1.8	2.0
Piezodus branssatensis MB Cod. 2793	1.8	1.3	1.6

Lower deciduous Molars, length

	D_3	D_4
Desmatolagus gobiensis; ? *Desmatolagus* sp. (SYCH and SYCH 1977, Pl. 3, Fig. 6, Pl. 4, Fig. 7 and pers. communication Dr. R. TEDFORD, AMNH)	1.77 (AMNH 84982)	1.88
	2.78 (AMNH 84977)	2.59
Desmatolagus gobiensis PMM 475—4393	—	1.7
Shamolagus franconicus HEISSIG and SCHMIDT-KITTLER 1976	3.1	—
Amphilagus fontannesi ML L. Gr. 111	2.9	—
Piezodus branssatensis MB Cod. 2796	2.6	2.1

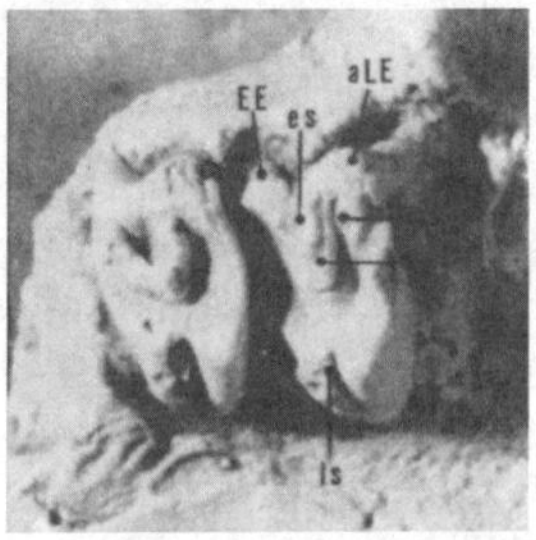

Fig. 4. *Desmatolagus gobiensis* MATTHEW & GRANGER. Middle Oligocene, Tatal Gol (P.R. of Mongolia). Skull fragment with $D^4 + M^1$ dext. Occlusal view. PMM 475—4209. ×6.5

the three main elements are reinforced, and styliform, thus receiving a character as it is known from the molars (Fig. 4).

The synclines are more or less filled with cement. It cannot be excluded, that cement is eroded post mortem by weathering in those synclines, which have presently no cement. The enamel is thin and obviously differs somewhat in thickness from the premolars and molars. Length measurement see table 1.

3. Upper Deciduous Teeth of European Oligocene Lagomorphs

Under *Shamolagus franconicus* from the lower Middle Oligocene of the fissure filling Möhren 20 (Franconia, Bavaria) HEISSIG and SCHMIDT-KITTLER (1976, p. 91, Figs. 12a, b), describe a D^3 and D^4. Together with a lower D_3 (see below) these are in fact the earliest milk teeth of European lagomorphs. Compared with the D^3 of *Desmatolagus* (Fig. 2) the contour of the crown is different (anterior and buccal wall form a right angle, and the posterior IE is larger than the anterior one). The difference in size of the is may be caused by more progressive wear in the Mongolian specimen. For the D^4 (not figured) the authors (1976, p. 91) mention the broader internal syncline compared with the M^1. This is also in accordance with *Desmatolagus* (Figs. 2 and 3). Measurements see HEISSIG and SCHMIDT-KITTLER (1976, p. 87 and table 1).

The D^3 in *Amphilagus antiquus* from Coderet (Fig. 5 see below) has an obtuse angle between anterior and buccal wall, but the posterior IE protrudes lingually more than the anterior one, in a similar way as in *Shamolagus*.

It is the late Oligocene (in the marine scale Chattian, in the continental scale Arvernian: FAHLBUSCH 1976, table 1) site of Coderet (former spelling: Bransat, or Branssat near Coderet) Dep. Allier, France, which has produced a rich collection of the archaistic, partially hypsodont genera *Amphilagus* POMEL, 1853 and *Piezodus* VIRET, 1929 (see TOBIEN 1974, p. 104 and 1975, p. 108 respectively). The collection of the "Dept. des Sciences de la Terre de l'Université de Lyon" preserves a right dentition of *Amphilagus antiquus* with D^2 and D^3 (together with P^4 unerupted) M^1, M^2 and the alveolus of M^3. (FL 4517b; Fig. 5).

D^2: The crown has the three main elements EE, ME, IE in a rather transverse arrangement, and like *Desmatolagus* the internal element is stronger than in the permanent P^2 (Fig. 6).

The three main cusps are also separated by shallower external and deeper middle synclines (es, ms). Compared with *Amphilagus* IE in the *Desmatolagus* D^2 is undeveloped, and the contour of the crown more triangular in the Mongolian species. Obviously because of advanced wear, the tiny postero-external syncline (pes) and the PC behind are already removed in the Coderet specimen.

The root formation is the same as in *Desmatolagus* and the middle syncline is filled with some cement, the external syncline is empty (Fig. 5).

D^3 (Fig. 5): The tooth in FL 4517b is deeper worn than in the Mongolian specimen, therefore the transverse dimension of the wear surface is more elongated. However, the contour of the Coderet D^3 is more rectangular, whereas the *Desmatolagus* crown contour is trapezoidal, which is independent from the state of wear and obviously an archaistic character.

The wear facet preserves the external part of the anterior arm; the lingual orifice of the middle syncline as the central islet of it (ms); and ME, es, EE + PC at the labial border of the crown in a more or less residual condition. The internal syncline is deeper than in *Desmatolagus*, which means an advanced status.

The root formation is as in *Desmatolagus*, already mentioned above. All synclines in *Amphilagus* are filled with cement, in the Mongolian D^3 cement is present only in the middle syncline.

A basic difference concerns the orientation of the external syncline (es). In both D^2 of *Desmatolagus* and *Amphilagus*

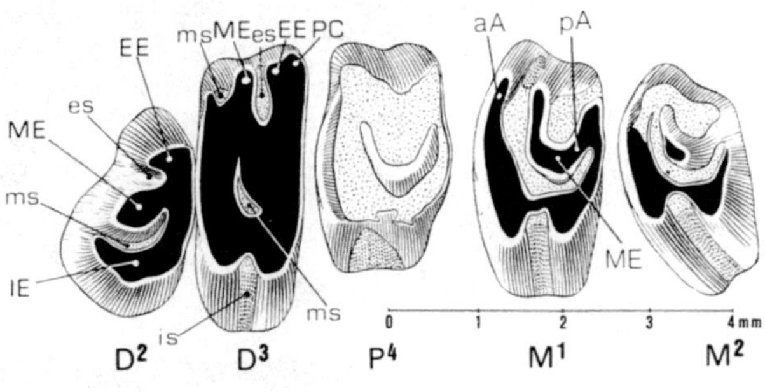

Fig. 5. *Amphilagus antiquues* POMEL. Upper Oligocene (= Upper Arvernian). Coderet (France). D^2, D^3, P^4 (germ), M^1, M^2 dext. (inverse). Occlusal view. FL 4517b. ×10. (see also TOBIEN 1974, Fig. 10)

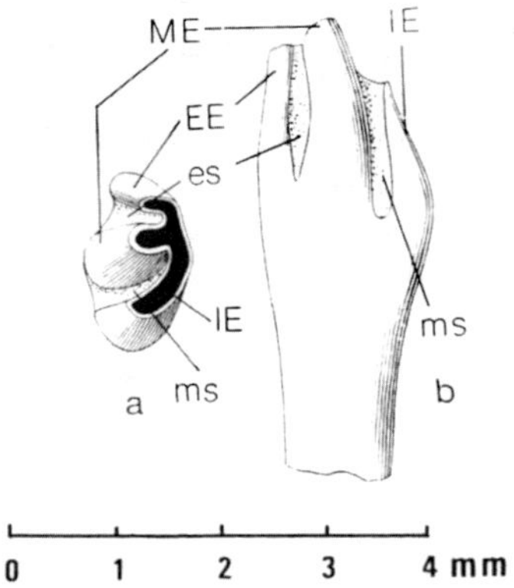

Fig. 6. *Amphilagus antiquus* POMEL. Upper Oligocene (= Upper Arvernian). Coderet (France). P^2 sin. a: occlusal view; b: anterior view. MB Bst. 9447. × 10. (see also TOBIEN 1974, Fig. 2)

the axis has mesiodistal direction (Figs. 2, 5). In the D^3 the axis is oriented oblique, i.e. antero-externally to postero-internally in *Desmatolagus;* transverse, i.e. bucco-lingually in *Amphilagus* (Figs. 2, 5), as in all later ochotonids with remnants of the primary structure. The situation in the *Desmatolagus* D^3 is caused by the shorter anterior arm of ME not yet having reached the buccal border of the crown. This is independent of the state of wear in both decidui. The oblique orientation of es holds also true in the P^4 of a tooth row of *Desmatolagus gobiensis* MATTHEW & GRANGER, 1923 from Hsanda Gol (Fig. 3).

At the present state of knowledge no D^4 of the late Oligocene *Amphilagus antiquus* is known. Measurements see table 1.

A left maxillary fragment of *Amphilagus* (after LOPEZ 1977, p. 8: *Eurolagus* n.g.; the *Amphilagus* and/or *Eurolagus* question will be discussed in a forthcoming paper) *fontannesi* from the Middle Miocene (Astaracian) of La Grive (France) shows the smaller buccal and the larger lingual alveoli of D^2, and the crowns of D^3 and D^4 (Fig. 7).

The buccal border of the D^3 crown obviously differs not much from the same part in *A. antiquus*, especially the far reaching buccal end of ME, both in contrast to the D^3 of *Desmatolagus* (Fig. 2). Nevertheless the oblique, postero-externally to antero-internally directed buccal border of the D^3, very distinct in *Desmatolagus* (Fig. 2), is still somewhat visible in *A. antiquus* and *A. fontannesi* (Figs. 5, 7). The es syncline, oblique in *Desmatolagus* (Fig. 2), is transverse as already in *A. antiquus* (Figs. 5, 7).

In the D^4 of *A. fontannesi* (Fig. 7) the main crown elements are present as in D^3, the anterior arm of ME reaches the buccal border of the crown, and es is transversely oriented. The accessory button-like labial elements (LE: TOBIEN 1974, Fig. 8), anterior (aLE), middle (mLE) and posterior (pLE) one, are stronger developed than in D^3. Traces of them are likewise visible in the *Desmatolagus* (Figs. 2, 4) and *A. antiquus* (Fig. 5) decidui. Measurements see table 1.

MAJOR (1899, Pl. 36, Fig. 35) figured a slightly worn D^4 sin. of *Amphilagus fontannesi* from La Grive with an extremely developed anterior labial element (aLE).

Together with *Amphilagus* appears the related genus *Piezodus* VIRET, 1929 in the late Oligocene of Western Europe. The late Oligocene species *P. branssatensis* VIRET, 1929 from Coderet and other sites is structurally somewhat more advanced than *Amphilagus antiquus* from the same sites (TOBIEN 1975, p. 110) and *Desmatolagus*. This holds also true for the milk dentition, especially for the D^2 (Fig. 8). Here a trace of an internal syncline (is) is already present, which is missing in the *Amphilagus* and *Desmatolagus* D^2 (Figs. 2, 5). The AC element in D^2 is stronger and reaches further buccally than in the *Amphilagus* D^2 (Figs. 5, 8). The labial elements in D^3 and D^4 are well developed, mainly aLE and the degree of unilateral hypsodonty (which generally is less than in the permanent dentitions) is slightly higher than in the milk teeth of *Amphilagus* and *Desmatolagus*. As usual also in *Piezodus* the structures of the anterior lacteal teeth are more advanced than in the succeeding permanent teeth; otherwise

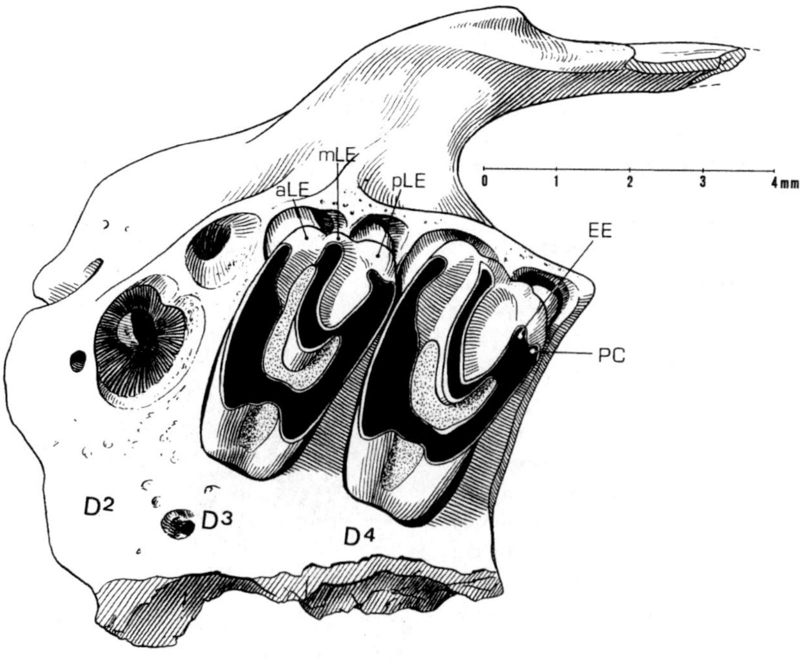

Fig. 7. *Amphilagus fontannesi* (DEPERET). Middle Miocene (= Astaracian), La Grive St. Alban (France). Maxillary fragment sin. with alveoli of D^2, D^3 and D^4. Occlusal view. ML L. Gr. 111. × 10

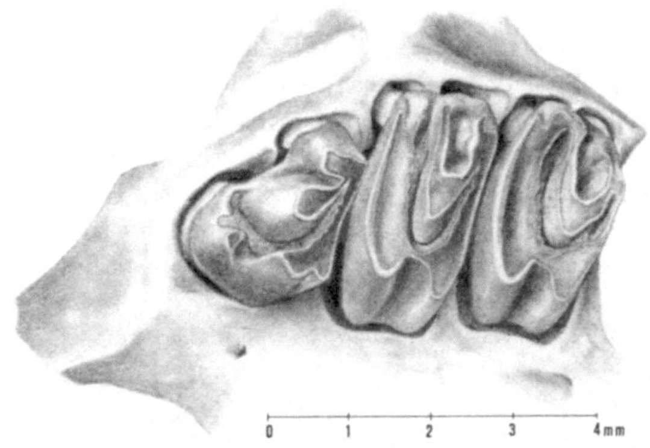

Fig. 8. *Piezodus branssatensis* VIRET. Upper Oligocene (= Upper Arvernian), Coderet (France). Maxillary fragment sin. with D^2 — D^4. Occlusal view. MB Cod. 2793. × 10

said it means that D^2 is more similar to P^3, D^3 resembles more P^4. But D^4 is different from M^1 by the shallow internal syncline, thus more similar to the following P^4, also by the broader crown contour compared with M^1 (Fig. 8: to compare with the permanent dentitions of *Piezodus branssatensis* in TOBIEN 1975, Figs. 2—5). The same holds likewise true for *Amphilagus fontannesi* D^4 (Fig. 7) compared with P^4 and M^1 respectively of the same species (TOBIEN 1974, Figs. 35, 36) and *Desmatolagus* (Figs. 2, 4 versus 3). Measurements see table 1.

4. Remarks on Lower Deciduous Teeth of Desmatolagus and European Oligocene Lagomorphs

SYCH & SYCH (1977, p. 384, Pl. 3, Figs. 6a, 6b, 7a, 7b) described *Desmatolagus* D_3 from the Oligocene of the Tsagan Nor Basin (Mongolian People's Republic). The teeth are — as usual in lagomorphs — trilobate, the anterior lobe is relatively small, compared with the second and third lobe, and with rounded rhomboid, simple contour. Measurements see table 1.

In contrast the D_3 of *Shamolagus franconicus* (HEISSIG and SCHMIDT-KITTLER 1976, Figs. 10a, b) and of *Amphilagus* have a rather broad anterior lobe, as is demonstrated by *A. fontannesi* from La Grive (Fig. 9). No D_3 of the earlier *A. antiquus* have been published so far from late Oligocene sites in Europe. The La Grive and the Middle Oligocene Möhren 20 specimens both show a slight furrow on the antero-external side of the anterior lobe.

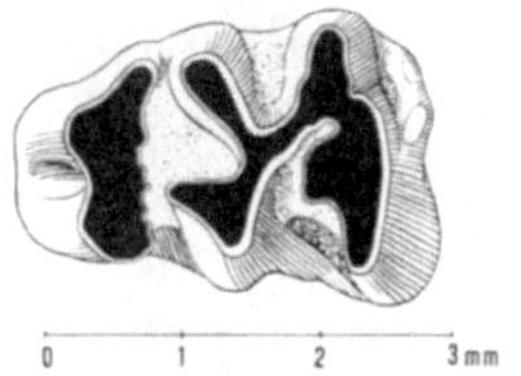

Fig. 9. *Amphilagus fontannesi* (DEPERET). Middle Miocene (= Astaracian) La Grive St. Alban (France). D_3 dext. (inverse). Occlusal view. ML L.Gr. 111. × 10

In a broad anterior lobe of a D_3 of *Piezodus branssatensis* from Coderet this antero-external furrow is enlarged and deepened, and on the lingual wall of the anterior lobe a small accessory groove is visible (Fig. 10). Measurements see table 1.

The former is constantly present in the Neogene European *Prolagus*, the descendant of *Piezodus* (LOPEZ 1974, Pl. 2).

A worn D_4 of *Desmatolagus* sp. is represented together with a M_1 in a mandible fragment from Tatal Gol (Fig. 11). There are two lobes — as usual in lagomorphous D_4 — the anterior being narrower and of rhomboid shape, the posterior larger, more rounded with a posterior fold, which is separated from the lobe by the remnant of a labial synclinid (pes = posterior external synclinid). It is interesting to note, that the situation on the following M_1 is reverted (Fig. 11. Measurement see table 1). Here the lingual synclinid (= pis = postero-internal synclinid; TOBIEN 1974, Fig. 11, 1975, Figs. 7—9) is broader and deeper than the dwindling labial furrow.

The pis element is likewise developed in *Shamolagus franconicus* (HEISSIG and SCHMIDT-KITTLER 1975, Fig. 3a, 1976, Figs. 2, 3a), *Amphilagus antiquus* (TOBIEN 1974, Figs. 11, 14), *A. ulmensis* (TOBIEN 1974, Figs. 29, 31), *Piezodus branssatensis* and *P. tomerdingensis* (TOBIEN 1975, Figs. 7—9, 37, 42, 43, 45). In *Amphilagus fontannesi*, however, it is the posterior external synclinid (= pes), which is larger and deeper and persists longer on the wear surface during attrition (TOBIEN 1974, Figs. 38—40, p. 157): Compared with *A. antiquus*, *Shamolagus*, *Desmatolagus* and *Piezodus* an advanced character, obviously in context with the younger biostratigraphic age of the species *A. fontannesi*. Whether the well developed pes in the Oligocene D_4 indicates a likewise progressive situation needs further investigation.

5. Discussion

Principally, in *Desmatolagus* the basic crown structure and the degree of "molarization", mainly of the D^2 and D^3

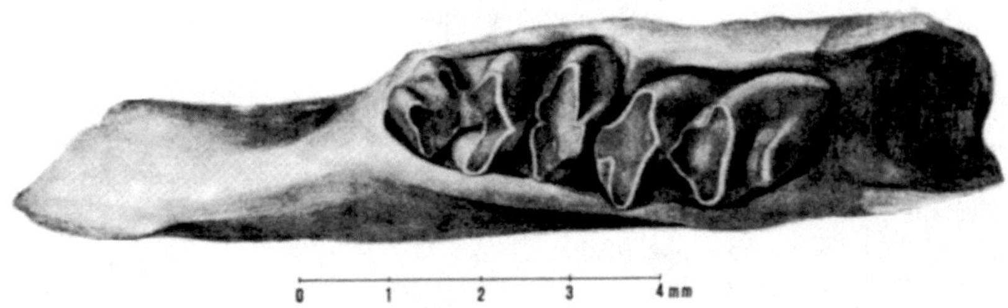

Fig. 10. *Piezodus branssatensis* VIRET. Upper Oligocene (= Upper Arvernian). Coderet (France). Mandibular fragment dext. with D_3 and D_4. Occlusal view. MB Cod. 2796. × 10

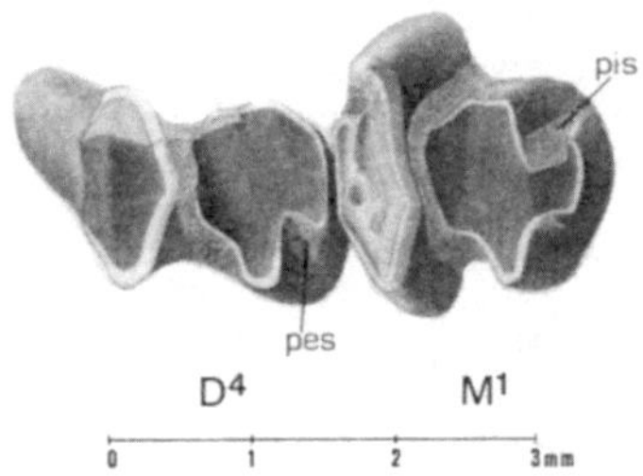

Fig. 11. *Desmatolagus gobiensis* MATTHEW & GRANGER. Middle Oligocene, Tatal Gol (P.R. of Mongolia). D_4 and M_1 dext. (inverse). Occlusal view. PMM 475—4398. × 10

in the upper dentition, is comparable with the European *Shamolagus*, *Amphilagus* and *Piezodus* milk teeth. Different, however, is the general more archaistic structural level of the milk dentition compared with the European genera. Examples are:

a) The more triangular crown contour of the *Desmatolagus* D^2, which might reflect a "pro-lagomorphous" status. In *Amphilagus antiquus* and *Piezodus branssatensis* D^2 the contour is labio-lingually more extended, the three main elements already more in one line. Furthermore the AC element is rather weakly developed than in the mentioned European taxa (Figs. 2, 5, 8). The internal syncline (is) in D^2 is absent in *Desmatolagus* and in *Amphilagus* at least in advanced wear, but present in *Piezodus* (Figs. 2, 5, 8).

b) In the D^3 of *Desmatolagus* the contour of the crown is trapezoidal, i.e. the length on the lingual side is shorter than on the labial one, whereas in the European genera mesial and distal wall of the D^3 crown are more parallel, the contour more rectangular. The direction of the axis of the external syncline is oblique, i.e. antero-externally to postero-internally. In *Shamolagus* (HEISSIG and SCHMIDT-KITTLER 1976, Fig. 12a). *Amphilagus* and *Piezodus* (Figs. 2, 5, 8) it is oriented bucco-lingually, because of the more advanced molarization of the European specimens, in which the anterior arm of the middle element (ME) reaches the buccal border of the crown. This is not the case in *Desmatolagus* (p. 224).

c) In D^4 of *Desmatolagus* the lingual part of the crown is still narrower than the buccal one, and the buccal end of the anterior arm does not yet reach the external border in contrast to *A. fontannesi* and *Piezodus* (Figs. 2, 4, 7, 8).

d) The lower D_3 of *Desmatolagus* has a relatively small anterior lobe, compared with the second and the third (see p. 224). As the small elongated cusp at the anterior base of the P_3 in *Desmatolagus* obviously is a product of the crown base (= cingulum) (SYCH and SYCH 1977, p. 384, Figs. 9a, b) an equal origin can be assumed for the fully developed, anterior lobe of the lagomorphous D_3.

In contrast to *Desmatolagus* the anterior lobes of D_3 of *Shamolagus*, *Amphilagus* and *Piezodus* are broader and with first traces of a secondary differentiation. This means an advanced character (p. 225—226).

The posterior lobe in D_4 has a rounded contour, which reflects to a certain degree the primitive talonid structure, whereas at least in the D_4 of *Piezodus* the first signs of the rhomboid shape of the wear facet occur (Figs. 10, 11).

As usual in lagomorphs and other mammal groups the lacteal dentitions of *Desmatolagus*, *Shamolagus*, *Amphilagus*, *Piezodus* are in parts structurally advanced, compared with their permanent successors. It is well known that the function of the dentition of a juvenile individual has relatively to deliver the same masticatory effects as the dentition of an adult individual: Therefore the higher degree of complication in the lacteal dentitions of many mammals.

On the other hand *Desmatolagus* milk teeth shows some features, which are primitive compared with the European genera. Thus *Shamolagus* is nearer to, if not identical with *Amphilagus* and *Piezodus* more progressive than these two taxa.

This status is likewise valid for the permanent dentitions of these genera.

Summary

As in other lagomorphs and mammals in many characters the lacteal dentition of the Oligocene Asiatic *Desmatolagus* is more advanced than the permanent dentition by reasons of functional morphology. This is likewise true for the European genera *Shamolagus*, *Amphilagus* and *Piezodus* of more or less the same age.

On the other hand the lacteal dentition of *Desmatolagus* is in some respects more primitive than in the mentioned European genera, demonstrating a more progressive habitus of their milk teeth.

At the present state of knowledge the lacteal dentition of *Desmatolagus* shows the most primitive stage of development among the Eurasiatic lagomorph taxa.

Zusammenfassung

Wie bei anderen Lagomorphen und sonstigen Säugetieren ist das Milchgebiß des oligozänen asiatischen *Desmatolagus* in vielen Punkten — aus funktionsmorphologischen Gründen — progressiver als das Dauergebiß. Dies gilt auch für die mehr oder weniger gleichaltrigen europäischen Genera *Shamolagus*, *Amphilagus* und *Piezodus*.

Andererseits ist das Milchgebiß von *Desmatolagus* in mancher Beziehung archaischer als jenes der genannten europäischen Vertreter, die demgegenüber einen progressiveren Eindruck hinterlassen.

Beim gegenwärtigen Kenntnisstand weist das Milchgebiß von *Desmatolagus* unter den eurasiatischen Lagomorphen den ursprünglichsten Zustand auf.

Резюме

Молочные зубы олигоценового азиатского *Desmatolagus*, как и других зайцеобразных и вообще млекопитающих, по многим признакам более прогрессивны, чем постоянные, что связано с функциональной морфологией. Это справедливо также и для европейских родов более или менее близкого геологического возраста — *Shamolagus*, *Amphilagus* и *Piezodus*.

С другой стороны, молочные зубы *Desmatolagus* в некоторых отношениях более примитивны, чем у названных европейских родов, что показывает более прогрессивное строение последних.

На современном уровне знаний молочные зубы *Desmatolagus* обнаруживают наиболее примитивные черты среди всех евразиатских таксонов зайцеобразных.

References

BOHLIN, B., 1937: Oberoligozäne Säugetiere aus dem Shargaltein-Tal (Western Kansu). — In: Palaeont. Sin., N.S. — Nanking C 3. — P. 1—66.

—, 1942: The Fossil Mammals from the Tertiary Deposit of Tabenbuluk, Western Kansu. Part I: Insectivora and Lagomorpha. — In: The Sino-Swedisch Expedition Publ. 20. — In: Palaeont. Sin., N.S. — Stockholm/Nanking C **8**. — P. 1—113.

FAHLBUSCH, V., 1976: Report on the International Symposium on mammalian stratigraphy of the European Tertiary (München, 1975, April 11—14). — In: Newslett. Stratigr. — Berlin [West], Stuttgart **5** (27). — P. 160—167.

GUREEV, A. A., 1960: Lagomorphs of the Oligocene in Mongolia and Kazakhstan. — In: Tr. Paleont. Inst. Akad. Nauk SSSR. — Moskva **77** (4). — P. 5—34. — In Russian.

HEISSIG, K.; SCHMIDT-KITTLER, N., 1976: Neue Lagomorphen-Funde aus dem Mitteloligozän. — In: Mitt. bayer. Staatssamml. Paläont. u. hist. Geol. — München **16**. — P. 83—93.

KIELAN-JAWOROWSKA, Z.; DOVCHIN, N., 1968: Narrative of the Polish-Mongolian Palaeontological Expeditions 1963—1965. — In: Palaeont. Pol. — Warszawa **19**. — P. 7—30.

LOPEZ MARTINEZ, N., 1974: Evolution de la Lignée *Piezodus-Prolagus* (Lagomorpha, Ochotonidae) dans le Cénozoique d'Europe Sud-Occidentale. — Montpellier, Thèse, Univ. Sci. et Techn. du Languedoc. — 165 Pp.

—, 1977: Nuevos Lagomorfos (Mammalia) del Neogeno y Cuaternario Español. — In: Trabajos N/Q. — Madrid **8**. — P. 7—45.

MAJOR, C. I. FORSYTH, 1899: On Fossil and Recent Lagomorpha. — In: Trans. Linn. Soc. — London **7** (2). — P. 433—520.

MATTHEW, W. D.; GRANGER, W., 1923: Nine new rodents from the Oligocene of Mongolia. — In: Amer. Mus. Novitates. — New York **102**. — P. 1—10.

SYCH, L., 1975: Results of the Polish-Mongolian Palaeontological Expeditions — Part VI. Lagomorpha from the Oligocene of Mongolia. — In: Palaeont. Pol. — Warszawa **33**. — P. 183—200.

SYCH, L.; SYCH, B., 1977: Serial Tooth Homology in Ontogeny and Evolution of the Third Lower Premolar in Lagomorphs. — In: Folia Biol. — Kraków **25** (4). — P. 381—389.

TOBIEN, H., 1963: Zur Gebißentwicklung tertiärer Lagomorphen. — In: Notizbl. hess. Landesamt Bodenforsch. — Wiesbaden **91**. — P. 16—35.

—, 1974: Zur Gebißstruktur, Systematik und Evolution der Genera *Amphilagus* und *Titanomys* (Lagomorpha, Mammalia) aus einigen Vorkommen im jüngeren Tertiär Mittel- und Westeuropas. — In: Mainzer geowiss. Mitt. — Mainz **3**. — P. 95 bis 214.

—, 1975: Zur Gebißstruktur, Systematik und Evolution der Genera *Piezodus*, *Prolagus* und *Ptychoprolagus* (Lagomorpha, Mammalia) aus einigen Vorkommen im jüngeren Tertiär Mittel- und Westeuropas. — In: Notizbl. hess. Landesamt Bodenforsch. — Wiesbaden **103**. — P. 103—186.

Quartärpaläontologie 6 231—238 Berlin 1986

On the Composition of the Forage of the Large Herbivorous Mammals of the Mammoth Epoch: Significance for Palaeobiological and Palaeogeographical Reconstruction

By VALENTINA VASIL'EVNA UKRAINTSEVA, Leningrad

With 5 Figures and 1 Table

Contents

Abstractum

In this paper the results of research of the gastrointestinal tracts of a fossil horse, a bison, and four mammothes are discussed, mammals that perished in Siberia in the past 53000—10000 years. The sediments in which the animals have been embedded are described too. The problem of Pleistocene extinction, mainly that of mammoth, is discussed in detail.

In der vorliegenden Arbeit werden die Untersuchungsergebnisse von Pflanzenresten aus dem Magen-Darmtrakt eines Pferdes, eines Bisons und von vier Mammuten, die im Verlaufe der letzten 53000 bis 10000 Jahre in Sibirien verendeten, diskutiert. Ebenso werden die Fundschichten beschrieben, in denen die Kadaver eingebettet waren. Das Problem des Aussterbens einiger Großsäuger im Pleistozän, insbesondere des Mammuts, wird eingehend erörtert.

В статье приведены результаты исследований растительных остатков из желудочно-кишечных трактов лошади, бизона, четырех мамонтов, которые погибли в Сибири в течение последних 53 — 10 тысяч лет, и отложений вмещавших погибших животных. Обсуждается проблема причин вычирания в плейстоцене некоторых животных и, в частности, мамонта.

1. Introduction

As is generally known, a frozen mammoth caracass with the well preserved gastrointestinal tract filled with plant remains was first discovered in Yakutia at the Beresovka river in 1900 (HERZ 1902). This time the Russian Academy of Sciences was sending an expedition to deliver the mammoth. The stuffed body is now on exhibition at the Zoological Museum of Leningrad.

At present, when a fossil caracass or a skeleton is doscovered in any part of the country, the Soviet Academy of Sciences and the Committee for the Investigation of the Mammoth and the Mammoth Fauna will organize integrated expeditions to the respective regions and will superwise the entire research. The Komarov Institute of Botany is in charge of the botanical research. This paper reports the results of the studies of the gastrointestinal contents of large herbivores which died in Siberia at different periods of the Pleistocene and the Holocene. The studies of the Institute also includes investigations of the deposits in which the fossil animals have been embedded.

Six fossil animals, a horse, a bison, and four mammothes (Fig. 1) with well preserved gastrointestinal contents have been found since 1968 in different regions of Siberia.

The vegetation cover is one of the basic factors characterizing habitat conditions of herbivors. There is a close relationship between the latter and the flora according to C. C. FLEROV (1965). A slight change in the patterns of vegetation which serves for as animal forage as well as of the mode of obtaining it has a dramatical effect on their structure. This makes herbivorous animals an ideal object for palaeogeographical reconstructions. The examination of the gastrointestinal contents of fossil animals and the deposits in which they are embedded is therefore of great theoretical and practical importance.

2. Materials and Methods

The gastrointestinal contents of six animals as well as the deposits in which they were embedded have been examined. Studies of this kind require a special technique. Samples for anatomo-morphological, palynological and carpological analyses were taken from the same portions of the gastrointestinal tract, which enabled us to compare the data obtained with different techniques and thus estimate the possibilities of each. Owing to this approach we were able to show that the percentages for the main plant groups, i.e. herbs, shrubs, and low shrubs as established on the basis of macro-remains in the stomach of the Selerikhan fossil horse (Fig. 2, left) approximate those for the same plant groups resulting from the pollen analyses (Fig. 2, right). This illustrates the true percentages of the above plant groups in the feeding of the animal. On the other hand, the palynological method provides a more comprehensive picture of plant taxonomic composition (UKRAINTSEVA 1977), owing to the fact that pollen appears to be least affected by the notion of gastric juice while the tissues of some plants, especially of forb meadow representatives, are almost completely diggested, which makes their identification practically impossible. The identification, if any, can only be achived up to the family rank. The carpological method provides effective results in identifying sedges (EGOROVA 1977), while their pollen cannot be so far identified even down to the genus rank. In addition, the differentiated research of the contents of the gastrointestinal tracts provides possibilities for reconstructing patterns of plant communities which served as pastures shortly before the animal died (Fig. 3).

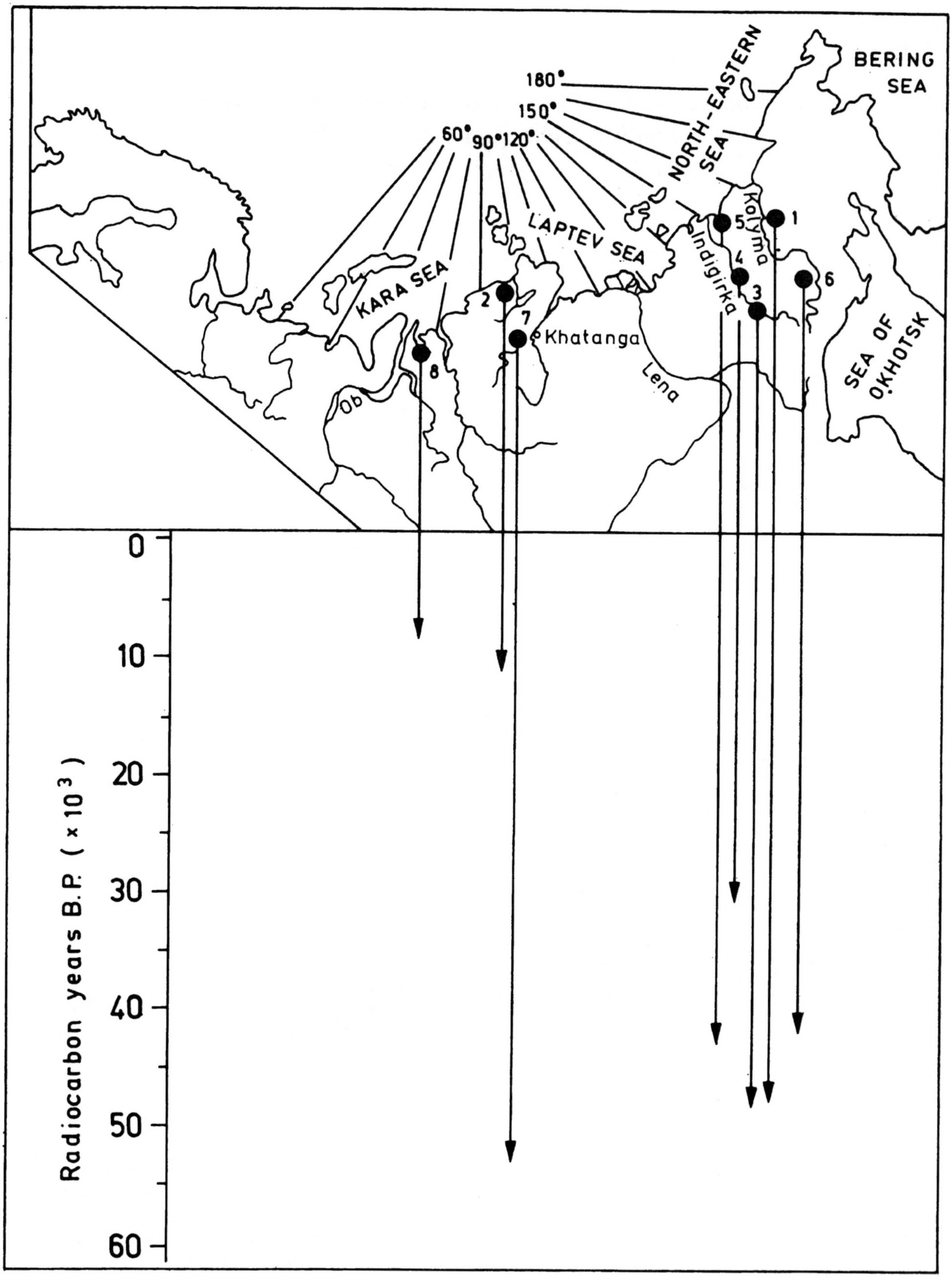

Fig. 1. Sites of frozen caracass and skeleton finds (or of their fragments)

1 — mammoth, Beresovka River, 1900; 2 — mammoth, NW Taimyr, 1948; 3 — horse, Selerikhan River, 1968; 4 — bison, Indigirka River, 1971; 5 — mammoth, Chandrin River, 1971; 6 — mammoth („Dima"), Kirgilyach River, 1977; 7 — mammoth, Large Forested Rassokha River, 1978; 8 — mammoth, Yuribei River, 1979

3. Results and Conclusions

The gastrointestinal contents of all the Siberian finds as well as the fossil plant-remains of the deposits in which the animals were embedded, have by now been fairly well examined. The macro-remains, the pollen, spores, fruits, and seeds have allowed a determination of plants on which the animals grazed shortly before they perished. Percentages in the gastrointestinal forage mass have also been established for the main plant groups — herbs, shrubs, low shrubs, and mooses. The composition of the macro-remains (SUKACZEV 1914; SOLONEVICH et al. 1977; UKRAINTSEVA et al. 1978; SOLONEVICH and VICHIREVA-VASIL'KOVA 1977) as well as of the pollen (TIKHOMIROV and KUPRIYANOVA 1954; KUPRIYANOVA 1957; TIKHOMIROV and KUL'TINA 1973; UKRAINTSEVA [KUL'TINA] 1977, 1982; UKRAINTSEVA et al. 1978) point to a dominant role of herbs in the forage of the representatives of the Mammoth Faunistic Complex. Shrub and low shrub branches and leaves of *Salix* ssp., *Alnus fructicosa*, *Betula exilis*, *Betula nana*, lower branches of some trees *(Larix*, *Betula*, *Chosenia)* also served as forage. It stands for reason, therefore, why the spectra of the gastrointestinal contents show shrub and low shrub pollen accounting for (2) 6—14%; pollen of wood plants making up no more than 4%. Pollen percentages for essential forage plants — grasses, sedges, and forbs — in the gastrointestinal tract spectra are variable in each animal (Fig. 4)[1]), which does not only reflect the role of a particular plant group in the food but also the pattern of plant communities in the area and their peculiarities.

Particularly noteworth in this respect are the results of the studies of the gastrointestinal contents of two animals: the Selerikhan horse (Fig. 1, 3) and the Chandrian mammoth (Fig. 1,5). These results, therefore, require a detailed discussion here. The horse gastrointestinal tract showed predominantly remains of herbaceous plants and their pollen (Fig. 2) — sedges, grasses, and forbs — being represented in practically equal percentage (Fig. 4,3). Along with the herbs the horse ingested moss stems, i.e. *Polytrichum strictum*, *Polytrichum* sp., *Rhutidium rugosum*, *Distichium capillaceum*, *Thuidium abietinum*, also *Bryum* sp., *Turtula ruralis*, *Sphagnum* sp., which occasionally occurred in the forage mass. Shrub branches (*Alnus fructicosa*, *Salix* ssp.), also low shrub branches *(Betula exilis*, *Betula nana)*, lower branches of some trees *(Betula platyphylla*, *Alnus hirsuta)* were also ingested. The variety of ecological plant groups the remains of which have been found in the horse tract (TIKHOMIROV and KUL'TINA 1973, EGOROVA 1977, SOLONEVICH and VIKHIREVA-VASIL'KOVA 1977, UKRAINTSEVA [KUL'TINA] 1977) testify that shortly before the animal died it grassed on a forest dry, shortgrass meadow, alpine-like steppe-meadow with *Carex bigelowii* (GEROD.) EGOR., *Kobresia filifolia* (TURCZ.) CLARKE, *Thalictrum fostidum* L., *Allium strictum* SCHRAD., *Selaginella sibirica* (MILDE) HIERON., *Stellaria jacutica* SCHISCHK., also on moist swamp meadows with *Kobresia* cf. *capilliformis* IVANOVA, *Polygonium scabrum* MOENCH, *Caltha palystris* L., *Juncus* sp., etc. The samples invariably contained pollen of spurce *Piçea obovata*, pine *Pinus sylvestris*, birch *Betula* sp. ex sect. Costatae, alder *Alnus hirsuta*, although in insignificant amounts. Their present areas are over 1000 km south and southwest of the site of the fossil horse. The small pollen amount of the above mentioned trees in the tract content can be explained by the fact that the trees and shrubs had completed their flowering (most trees and shrubs flower in April—May). It was only the tree and shrub pollen which had sunk on other plants and also in the water that reached the horse tract, including a few distant elm and lime pollen grains. However, even this small amount of tree pollen is sufficient to indicate the presence of the above mentioned tree species in the vegetation of the region. These palaeobotanical data also suggests that during the period of 38590 ± 1120 years ago the climate of the adjacent areas was somewhat warmer and more humid than the climate at present, since large woody birch, spruce, and pine were among the forest components of the epoch.

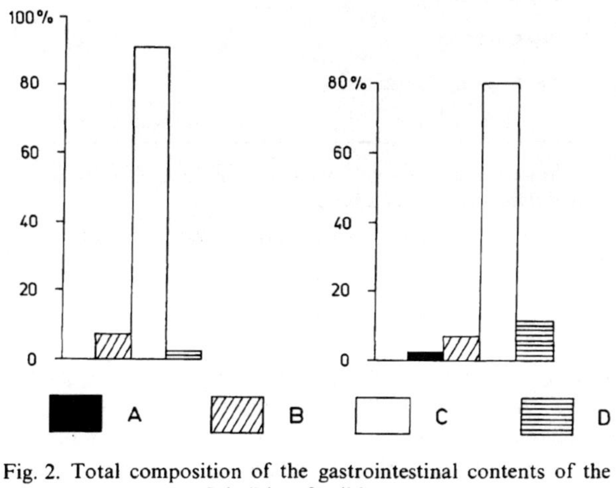

Fig. 2. Total composition of the gastrointestinal contents of the Selerikhan fossil horse

right — pollen and spores; left — macro-remains
A — trees; B — shrubs and low shrubs; C — herbs; D — mosses and ferns

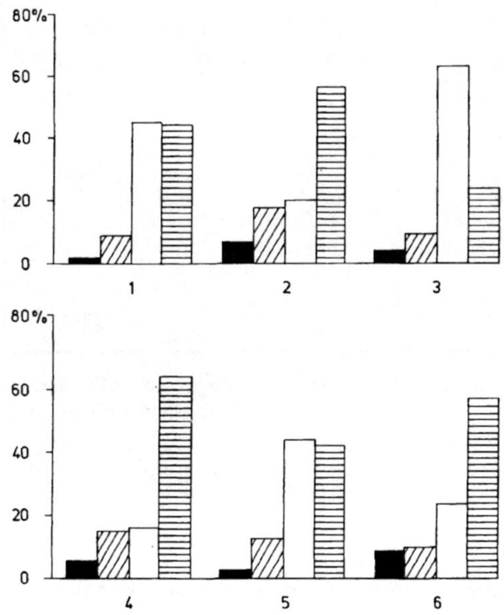

Fig. 3. Palynological spectra (total composition) of the gastrointestinal contents of the Yuribei mammoth and mammoth-embedding deposits

1 — stomach; 2 — mid-intestinal portion; 3 — colon; 4 — rectum; 5 — section 1, sample 15; 6 — section 1, sample 12, plant detritus under mammoth. Symbols as Fig. 2

[1]) Diagram 1 is based on data by L. A. KUPRIYANOVA (TIKHOMIROV and KUPRIYANOVA 1954), diagram 7 is based on results of palynological analyses of the sediments containing the mammoth remains (UKRAINTSEVA et al. 1981).

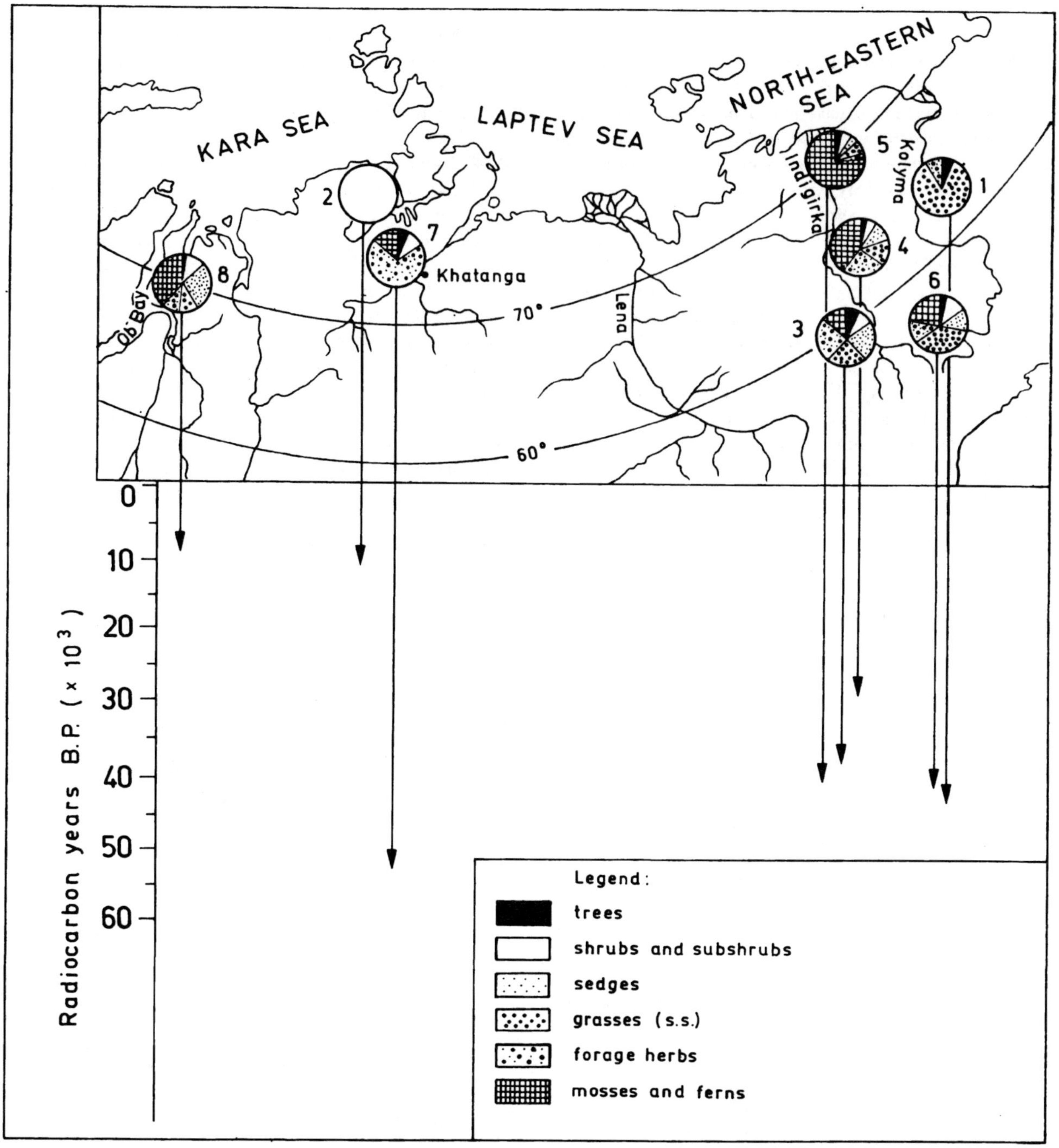

Fig. 4. Percentage of the main plant groups from Pollen and spores analyses of the gastrointestinal contents of herbivorous animals perished during the Upper Pleistocene and Holocene in different regions of Siberia

1 — mammoth, Beresovka River, 1900; 2 — mammoth, NW Taimyr, 1948; 3 — horse, Selerikhan River, 1968; 4 — bison, Indigirka River, 1971; 5 — mammoth, Chandrin River, 1971; 6 — mammoth („Dima") Kirgilyach River, 1977; 7 — mammoth, Large Forested Rassokha River, 1978; 8 — mammoth, Yuribei River, 1979

Abundance of moss remains, i.e. leaf and branch fragments as well as spores in the forage mass of the Chandrian mammoth (Fig. 4,5), bison (Fig. 4,4), and the Yuribian mammoth (Fig. 4,8) shows that the above mentioned plants although of little value forage (KOKSHINA 1961), were nonetheless ingested because being widespread in the given regions. The gastrointestinal tract of the Chandrian mammoth shows abundant more remains of 15—16 ecological related species (SOLONEVICH et al. 1977). Most frequent among them were the remains of *Polytrichum communis* HEDV., *Polytrichum strictum* SM., *Aulocomnium turgidum* WAHLENB.) SCHWAEGR., *Sphagnum angustifolium* C. JENS, *Sphagnum girgensohnii* RUSS., *Sphagnum* sp. ex sect. *Subsecunda*, *Sphagnum* sp. ex sect. *Palustria*. Remains of *Tomenthypnum nitens* (HEDV.) LOESKE and of some other five or six species, so far unidentified, have been found to be less abundant. Moss spores account for 76% (Fig. 4,5) of the total spore and pollen amount of the spectrum. The macro- and micro-remains (pollen and spores) in the tract contents of the animal indicate that sedges, cotton-grasses, low shrub and shrub branches and leaves (*Betula* sp., *Salix* ssp.), mosses and, occassionally, larch branches served as a source of its food. The quantities of larch needles, bark, remains of seed scales and pollen have been found to be very small. The animal was also grazing on different grasses and forbs whose soft epidermal tissues digest well, and their macro-remains are, therefore, only slightly, if at all, represented, while their pollen is present (*Dryas punctata*, *Valeriana capitata*, *Artemisia vulgaris*, *Pedicularis* sp., *Saxifraga* ssp., Caryophyllaceae).

It should be stressed that moss species composition as well as that of the main mass of macro-remains (sedges, cotton-grass), of pollen and spores bear witness of the swampy character of the area of the mammoth pastures, an area that was alternating with open light forests, and closed forests. The present-day vegetation of the Chandrian river basin is characterized by widespread bushy tundras, moss hummocky, marshy tundras as well as by sedge-hypnum swamps.

The data on the vegetation patterns of the region of the other animal caracasses are shown in table 1 suggesting that the excavated animals (individuals), at least the majority of them, existed when the environmental conditions were more favourable than (Berosovsky mammoth, Selerikhan horse, Chandrian mammoth, etc.), or very similar to the present conditions in the regions they died (Yuribian mammoth). As was determined by radiocarbon analyses (HEINZ and GARUTT 1964; ARSLANOV and CHERNOV 1977; ARSLANOV et al. 1980, 1981; SHILO and LOZHKIN 1981), at least eight of the nine animals in question perished during different relative warm periods of the Upper Pleistocene and Holocene (table 1). It should be mentioned here too that these are specific finds and it is due to them that we gat an admirable opportunity to have an idea of the environmental conditions of the Mammoth Faunistic Complex representatives.

However, a proper understanding of these conditions in a wider sense (i.e. considering the representatives of the Mammoth Faunistic Complex as species) can be gained if one views them on the background of global climatic changes in

Table 1. Data on the vegetation of the region of the animal caracasses

Fossil animal	^{14}C-dated time of perish	Pattern of reconstructed vegetation	Modern vegetation pattern of the find district
Mammoth Large Forested Rassokha River, 1978	>53170 yrs.	grass-forb and sedge-grass communities[9])	bush and low bush tundras, polygonal mires
Mammoth Terekhtikh River, 1971	44540 ± 1870 yrs.	tundras, forest tundras (?)[10])	larch taiga
Mammoth Beresovka River, 1900	44000 ± 3500 yrs.	larch taiga, occassionally with *Betula platyphylla*, *Betula pendula*, *Alnus hirsuta*, *Pinus sibirica*. Meadows, dry meadows, steppe like meadows[1])	larch taiga
Mammoth ("Dima") Kirgilyach River, 1977	41000 ± 1100 yrs. 39590 ± 870 yrs.	hypnum and cotton-grass swamps, bush and low bush tundras, larch taiga, occassionally with *Betula platyphylla*, grass-forb communities[6]) mountain (dry) tundras[7]), alpine bush and low bush tundras, larch forests along river valleys, mountain steppe like communities[8])	larch taiga, *Betula platyphylla* only on south slopes. Bushy tundras with *Betula exilis*
Mammoth Chandrin River, 1977	40350 ± 880 yrs.	larch taiga, bush and low bush tundras, sedge and hypnum bogs[4])	bushy tundras, heavily paludified
Mammoth East Taimyr, 1948	11450 ± 450 yrs.	bushy tundras forest tundras (?)[2])	cotton-grass tundras, sedge-grass tundras, polygonal tundras, "spotty" tundras
Mammoth Yribei River, 1979	9730 ± 100 yrs.	flood plaine meadows, shrub moss meadows, sedge-cotton-grass tundras with *Betula exilis*[9])	bushy tundras with *Betula exilis*, *Salix* ssp., moss tundras, polygonal bogs
Horse Selerikhan River, 1968	38590 ± 1120 yrs.	larch taiga with spruce and woody birch, grass-forbs meadows in the valleys and on the slopes[3])	larch taiga with *Betula platyphylla* on the south slopes, flood plaine meadows, mesoxerophytic conditions
Bison Indigirka River, 1971	29500 ± 1000 yrs.	larch taiga, heavily paludified, grass-forb and sedge-grass communities[5])	larch taiga

[1]) SUKACZEV 1914; TIKHOMIROV and KUPRIYANOVA 1954; KUPRIYANOVA 1957
[2]) TIKHOMIROV 1950; ZAKLINSKAYA 1954
[3]) TIKHOMIROV and KULTINA 1973; EGEROVA 1977; SOLONEVICH et al. 1977; UKRAINTSEVA (KUL'TINA) 1977
[4]) SOLONEVICH et al. 1978
[5]) UKRAINTSEVA et al. 1978
[6]) BELAYA and KISLEROVA 1978
[7]) ABRAMOV and ABRAMOVA 1981
[8]) UKRAINTSEVA 1981
[9]) UKRAINTSEVA 1982
[10]) MEDVEDEV and FORENOVA 1977

the Upper Pleistocene and Holocene (Fig. 5). The Mammoth epoch, i.e. the life-time of mammoth as a species, as well as of the other representatives of the Mammoth Faunistic Complex, is characterized by a cyclic, fluctuating pattern of the environment, i.e. alternation of cold (glacial) and warm (interglacial) periods. By this is following that the mammoth and the other representatives of the complex having adapted to cold (glacial) epochs, had to exist under environmental conditions not suitable to their morpho-functional pretensions. Afforestation and paludification resulting from the warm and interglacial periods dramatically diminished the pasture area. The forage quality deteriorated: dominating among the plants were wet swamp habitat species, i.e. sedges, cotton-grasses, grasses, mosses which are less valuable than plants of dry habitats and meadow grasland with respect to protein, albumin, lipid and mineral contents (LARIN 1958), while green mosses, although abundant, do not appear to be a suitable kind of forage (KOSHKINA 1961). Sedges and grasses whose remains are found to be dominant in the gastrointestinal contents of the fossil animals in question, are practically of the same forage value. However, wet habitat sedges are significantly inferior to desert and mountain sedges both in their nutrional proporties and in the mineral content, particularly calcium content (LARIN 1958). Reduced quality of the forage during the warm Pleistocene and Holocene intervals diminished birthrate and increased mortality in the warm

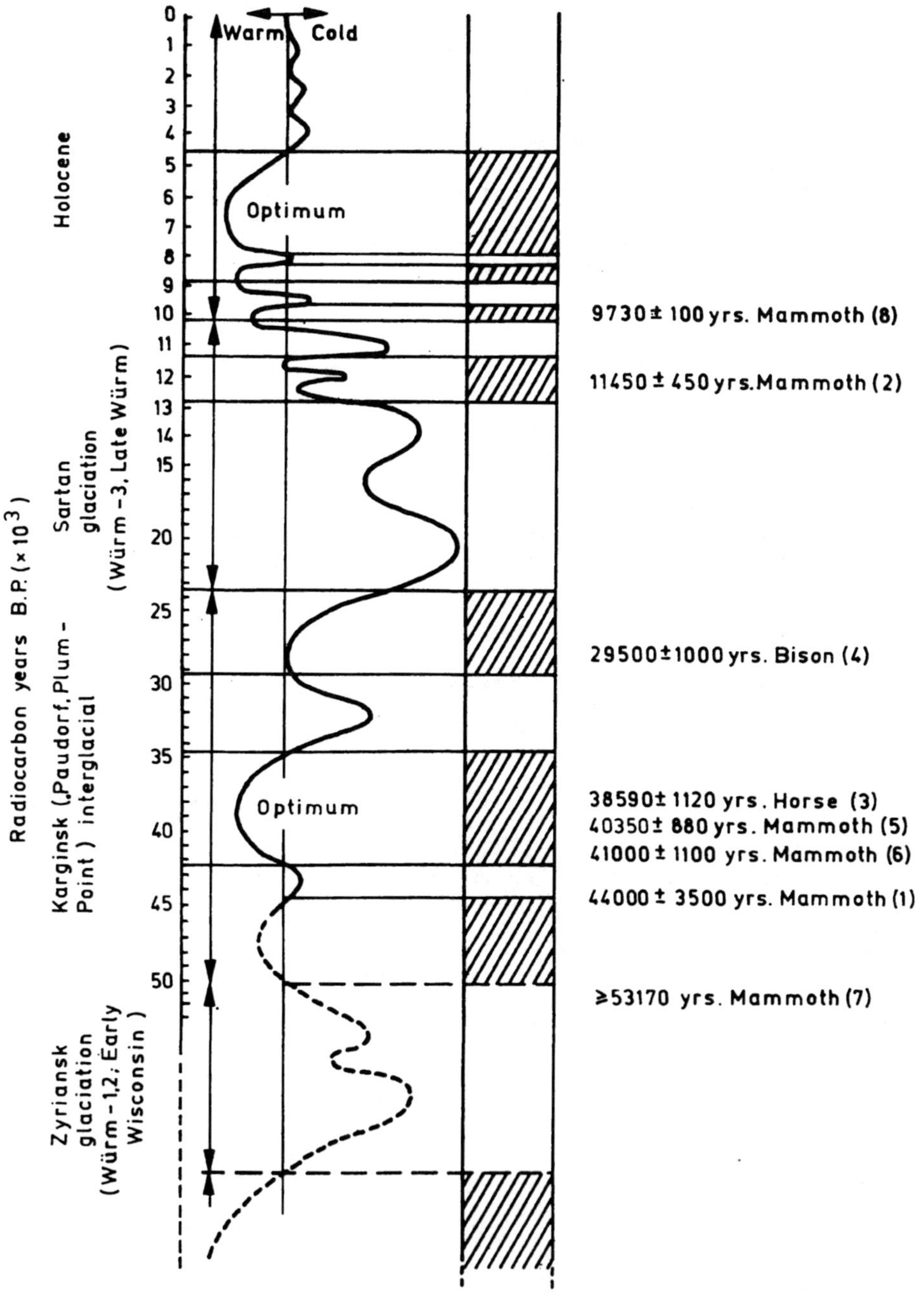

Fig. 5. Scheme showing the periods of the mass perish of large herbivorous animals of Siberia on the background of global climatic changes

periods, which invariably resulted in the decrease of the animals' total abundance. Genetically, such species as mammoth and wooly rhinoceros proved to be inadequately adapted to the warm epochs. The cyclic, fluctuating nature of the environment reduced the animals incapable of developing a species common strategy of adaption, since environmental fluctuations were sharp enough, while alternation of cold and warm rhythms was too rapid in terms of the geological time to allow a readjustment of the correlation mechanisms in the animal organisms. As a result, species of more narrow specialization, i.e. mammoth, wooly rhinoceros, appear to be extinct first, since their genetic possibilities of adaption to sharp changes in the environmental conditions proved to be most limited. Thus, mammoth extinction, as well as extinction of some of their "trabants", was due to set of causes which, in turn, resulted from the Pleistocene and Holocene fluctuating environment, alternation of cold and warm rhythms. The animals, as can be concluded, proved incapable of adapting to the drastic changes in the ecological conditions.

Summary

The majority of fossil animals discovered in Siberia is coming — according to C^{14}-dating — from warm-time deposits of the past 53000—10000 years. The essential part of plants, determined by results of research of the gastrointestinal contents of the fossil animals, give evidence that in these periods in Siberia forests, swamp communities of different types and swamps have been widely distributed. In dependence of this fact the favourable feeding-places have been reduced to a great extent. The types of the feeding-plants have been changed fundamentally too: sedge-grasses, cotton-grasses, grasses, and mosses became dominant. In the contents of basic food-elements (protein, albumin, and fat) as well as of minerals (calcium, calium, phosphor and others) these plants were distinctly poorer than those of dry places and meadows with various grass-communities essential for the large herbivorous animals.

These new palaeobotanical date allow conclusions as to the natural environment of the mammoth and its "sputniks" giving the possibility to solve the problem of Pleistocene extinction on a palaeobiological basis. The relative sudden extinction of the mammoth is undaubtedly explainable by a complex of causes. Our opinion, therefore, is that these causes should bee seen as a function of physical structure and the sum of natural processes in the Pleistocene, the cyclic alternation of the natural environment and the interaction of optimal cold periods and infavourable warm periods to cold-climate adapted animals. So the cold-climate adapted animals existed in the late Pleistocene mainly under natural conditions not suitable to their morpho-functional adaptions (altogether 2/3 of the time).

Zusammenfassung

Die Mehrzahl der in Sibirien entdeckten fossilen Tierkadaver stammt nach C^{14}-Daten aus Warmzeit-Perioden der letzten 53000 bis 10000 Jahre. Der Pflanzenbestand aus dem Magen-Darmtrakt der genannten Tierkadaver zeigt, daß in diesen Perioden in Sibirien vorherrschend Wälder, Sumpfgesellschaften verschiedenen Typus und Sümpfe weit verbreitet waren. Als Folge dieser Umstellung verschlechterten sich die für die Großsäuger günstigen Lebensbedingungen deutlich. Grundlegend veränderte sich auch das Nahrungsangebot: es herrschten Pflanzen feuchter und versumpfter Standorte vor — Riedgräser, Wollgräser, Gräser und Moose. In der Zusammensetzung der Grundnahrungselemente (Protein, Eiweiße, Fette) und im Mineralbestand (Calcium, Kalium, Phosphor und andere) waren diese Pflanzen bedeutend ärmer als diejenigen trockener Standorte und Wiesen mit vielfältigem Grasbestand, die für die herbivoren Großsäuger lebenswichtig waren.

Diese neuen paläobotanischen Daten erlauben nun, die Lebensbedingungen des Mammuts und der anderen assoziierten Großsäuger zu rekonstruieren und auf diese Weise an die Lösung des Problems des Aussterbens einiger Großsäuger im Pleistozän auf paläobotanischer Grundlage heranzugehen. So erscheint das verhältnismäßig schnelle Aussterben des Mammuts infolge eines Komplexes von Ursachen verständlich. Wir nehmen demzufolge an, daß diese Gründe durch den Verlauf der Naturprozesse im Pleistozän bedingt sind, durch den zyklisch schwankenden Charakter der Umwelt, durch die Wechselwirkungen der Dauer der für diese Tiere optimalen Kaltzeiten und ungünstigeren Warmzeiten. Diese kälteliebenden Tiere existierten also zumeist unter Bedingungen eines ihrer morphofunktionalen Anpassung nicht entsprechenden Milieus (insgesamt 2/3 dieser Zeit).

Резюме

Большинство из обнаруженных в Сибири ископаемых животных погибло, согласно данных C^{14} метода, в периоды потеплений в последние 53 — 10 тысяч лет. Состав растений, установленный в результате изучения содержимого их желудочнокишечных трактов, свидетельствует, что в эти периоды на территории Сибири широкое распространение получали леса, различного типа заболоченные сообщества, болота. Площади кормовых угодий в связи с этим резко сокращались. Кардинально менялся и состав кормов: в них преобладали растения увлажненных и заболоченных местообитаний — осоки, пушицы, злаки, мхи, которые по содержанию основных питательных элементов (протеин, белки, жиры) и минеральному составу (кальций, калий, фосфор и др.) значительно беднее растений сухих местообитаний и лугового разнотравья, что было жизненно важно для крупных растениеядных животных.

Эти новые палеоботанические данные позволили составить представление об условиях обитания мамонта и его „спутников" и, таким образом, подойти к решению проблемы причин вымирания некоторых животных в плейстоцене на палеобиологической основе. То что относительно быстрое вымирание мамонта было вызвано комплексом причин не вызывает сомнений. Однако, эти причины, мы полагаем, были обусловлены структурой и ходом природного процесса в плейстоцене, циклическим, колебательным характером природной среды, соотношением длительности холодных, оптимальных для них, и теплых, пессимальных, ритмов; тем, что эти холодовыносливые животные большую часть времени (в сумме 2/3) существовали в условиях среды не соответствующих их морфофункциональным приспособлениям.

References

Abramov, I. I.; Abramova, A. L., 1981: Mosses from the deposits of mammoth child. — In: Magadan mammoth child. — Leningrad: Nauka. — P. 247—253, in Russian.

Arslanov, Kh. A.; Chernov, S. B., 1977: Absolute age of the Selerikhan horse. — In: Anthropogene Fauna and Flora of the Siberian North-East. — Leningrad: Nauka. — P. 76—78, in Russian.

Arslanov, Kh. A.; Lyadov, V. V.; Tertichnaya, T. V., 1981: On the absolute age of the mammoth child. — In: The Magadan mamoth child. — Leningrad: Nauka. — P. 50—51, in Russian.

Arslanov, Kh. A.; Vershchagin, N. K.; Lyadov, V. V.; Ukraintseva, V. V., 1980: On the dating of the Khargin interglacial periods and Siberian landscape reconstruction as resulting from the investigating of mammoth corpses and their "satellites". — In: Geochronology of Quaternary period. — Moskva: Nauka. — P. 208—213, in Russian.

Belaya, B. V.; Kislerova, J. B., 1978: Palynological data obtained from the mammoth child find of the summer of 1977. — In: Materials on geology and mineral resources of the North-East of the USSR. — Magadan **24**. — P. 250—251, in Russian.

BELORUSOVA, Y. M.; LOVELIUS, N. V.; UKRAINTSEVA, V. V., 1977: Palaeogeography of Late Pleistocene and Holocene in the region of Selerikhan horse find. — In: Anthropogene Fauna and Flora of the Siberian North-East. — Leningrad: Nauka. — P. 265 to 276, in Russian.

EGEROVA, T. V., 1977: Carpological analysis of plant remains of Selerikhan fossil horse. — In: Anthropogene Fauna and Flora of the Siberian North-East. — Leningrad: Nauka. — P. 218 to 221, in Russian.

FLEROV, C. C., 1965: On the origin of Canadian fauna in connection with the history of Bering Land-Bridge. — In: Quaternary period and its history. — Moskva: Nauka. — P. 121—128, in Russian.

GERZ, O. F., 1902: Reports of the head of the Beresovka mammoth unearthing expedition of the Imperial Academy of Sciences. — In: Izvest. Imperial Acad. Sci. — Moskva **16** (4). — P. 137—174, in Russian.

HEINZ, A. E.; GARUTT, V. E., 1964: Absolute age of the mammoth and wooly rhinoceros fossil remains from Siberian permafrost as established by radiocarbon data. — In: Dokl. Akad. Nauk SSSR. — P. 1367—1370, in Russian.

KIND, N. V., 1973: Chronology of the Late Anthropogene according to radiometric data. — In: Stratigraphy, Palaeontology. — Moskva **4**. — P. 5—49, in Russian.

KOSHKINA, T. V., 1961: New data on the nutration of the Norwegian lemming (*Lemmus lemmus*). — Byull. Mosk. obshch. izpyt. prir., otd. Biol. — Moskva **66** (6). — P. 15—32, in Russian.

KUPRIYANOVA, L. A., 1957: Pollen analysis of plant remains from Beresovska mammoth stomach (on the problem of vegetation pattern for Beresovska mammoth epoch). — Sb. pamyati AFRIKANA NIKOLAEVICHA KRISHTOFOVICHA. — Moskva, Leningrad, IV. — P. 334—358, in Russian.

LARIN, I. V., 1958: Food value of sedges. — In: Dokl. VASKHNIL. — Moskva **8**. — P. 15—22, in Russian.

SHILO, N. A.; LOZHKIN, A. V., 1981: Radiocarbon dating of the mammoth child. — In: Magadan mammoth child. — Leningrad: Nauka. — P. 48—49, in Russian.

SOLONEVICH, N. G.; TIKHOMIROV, B. A.; UKRAINTSEVA, V. V., 1977: Preliminary results of the investigation into the remains from Chandrin mammoth gastrointestinal content (Yakutia), — In: Anthropogene Fauna and Flora of the Siberian North-East. — Moskva: Nauka. — P. 203—221, in Russian.

SUKACZEV, V. N., 1914: Research into the plant remains from the food content of the Beresovka mammoth find. — In: Scientific results of the expedition, fitted by Academy of Sciences for unearthing the mammoth on the river Berezovka in 1901. — St. Petersburg. — P. 1—17, in Russian.

TIKHOMIROV, B. A., 1958: On North Siberian environmental conditions and vegetation of mammoth epoch. — In: Problems of the North, 1. — Leningrad: Nauka. — P. 168—188, in Russian.

TIKHOMIROV, B. A.; KUL'TINA, V. A. (UKRAINTSEVA), 1973: Pollen and spores analysis from the Selerikhan fossil horse stomach. — In: Dokl. Akad. Nauk SSSR. — Moskva **209** (6). — P. 1464 to 1466, in Russian.

TIKHOMIROV, B. A.; KUPRIYANOVA, L. A., 1954: Pollen analysis of plant remains from the forage content of the Beresovska mammoth. — In: Dokl. Akad. Nauk USSR. — Moskva **95** (6). — P. 1313—1315, in Russian.

UKRAINTSEVA (KUL'TINA) V. V., 1977: Flora and vegetation reconstruction for Selerikhan fossil horse life-time based on palynological data. — In: Anthropogene Fauna and Flora of the Siberian North-East. — Leningrad: Nauka. — P. 223—253, in Russian.

—, 1979: Vegetation of warm intervals of Late Pleistocene and the extinction of some large herbivorous mammals. — In: Bot. Ž. — Moskva **64** (3). — P. 318—330, in Russian.

—, 1981: Environment and the causes of the death of the mammoth child. — In: Magadan mammoth child. — Leningrad: Nauka. — P. 254—261, in Russian.

—, 1982: Environment and the causes of mammoth death in the middle-reaches of the Yuribei River basin (Gidan Peninsula). — In: Yuribean Mammoth. — Moskva: Nauka. — P. 19—29, in Russian.

UKRAINTSEVA, V. V.; ARSLANOV, KH. A.; BELORUSOVA, Y. M.; BOCH, M. S., 1981: Vegetation and natural conditions of the Great Forested Rassokha River basin in Upper Pleistocene (in connection with mammoth find). — In: Bot. Ž. — Moskva **66** (10). — P. 1444—1453, in Russian.

UKRAINTSEVA, V. V.; FLEROV, C. C.; SOLONEVICH, N. G., 1978: Botanical analysis of plant remains from gastrointestinal tract of the Yakutian bison find. — In: Bot. Ž. — Moskva **63** (7). — P. 1001—1004, in Russian.

UKRAINTSEVA, V. V.; KOZHEVNIKOV, YU. P., 1979: The vegetation of Kirgiljach mammoth find region (the upper part of Kolyma River basin). — In: Bot. Ž. — Moskva **64** (8). — P. 1091—1098, in Russian.

UKRAINTSEVA, V. V.; KOZHEVNIKOV, YU. P., 1981: The vegetation cover in the area of the Taimyr's mammoth find (South-Western Taimyr, Great Forested Rassokha). — In: Bot. Ž. — Moskva **66** (7). — P. 978—992, in Russian.

VELICHKO, A. A., 1973: The natural process in the Pleistocene. — Moskva: Nauka. — 256 pp., in Russian.

—, 1981: To the problem of the succession and principal structure of main climatic Pleistocene rhythms. — In: Problems of Pleistocene palaeogeography in glacial and periglacial regions. — Moskva: Nauka. — P. 220—245, in Russian.

VERESHCHAGIN, N. K.; BARISHNIKOV, G. F., 1980: Palaeoecology of late mammoth fauna in the arctic zone of Eurasia. — In: Byull Mosk. obshch. izpyt. prir., otd. Biol. — Moskva (2). — P. 5—19, in Russian.

ZAKLINSKAYA, E. D., 1954: On the problem of the vegetation cover during Taimyr mammoth life-time. — In: Dokl. Akad. Nauk SSSR. — Moskva **98** (3). — P. 171—174, in Russian.

Quartärpaläontologie 6 239—242 Berlin 1986

Elks in the Pliocene of the USSR

By Inesa Anatolevna Vislobokova, Moscow

With one Plate

Contents

Abstractum

Data on finds of elk remains from the territory of the U.S.S.R. in Upper Pliocene deposits are given. An analysis of the systematic position of *Pseudalces mirandus* Flerov, 1962 from the Ciscaucasus Upper Pliocene is provided.

In der vorliegenden Arbeit werden Daten über Elch-Funde aus oberpliozänen Ablagerungen der UdSSR gegeben (entsprechend des offiziellen Geologie-Schemas der UdSSR). Dazu wird weiter eine Analyse der systematischen Stellung von *Pseudalces mirandus* Flerov, 1962 aus dem Oberpliozän des Vorkaukasus vorgelegt.

Приводятся данные о находках остатков лосей на территории СССР в отложениях верхнего плиоцена. Дается анализ систематического положения *Pseudalces mirandus* Flerov, 1962 из верхнего плиоцена Предкавказья.

1. Introduction

Problems of the systematics and pre-Pleistocene evolution of elks are being solved by current research based on European data, principally from France and Britain, materials from the USSR being practically left out of account. There are at present in the USSR quite a number of finds from the Upper Pliocene deposits (according to the USSR's officially recognized scheme) which permit us to form a more adequate idea of the history of elks.

2. Comments on the Systematics and Phylogeny of Elks

There is no universal opinion at present about the systematics of fossil elks. Simpson (1945) placed in the tribe Alcini the genera *Alces* Gray, 1821 from the Pleistocene of Eurasia, to which the present-day elk belongs, and *Cervalces* Scott, 1885 from the Pleistocene of North America. He included this tribe in the subfamily Odocoileinae Pocock, 1923. Flerov (1952, 1962) placed elks in the sub-family Cervinae Baird, 1857. Azzaroli (1952, 1953) brought all known fossil Eurasian elks into the genus *Libralces* with four species: *L. gallicus* Azzaroli, 1952 from the Upper Villafranchian of France and Britain, *L. minor* Azzaroli, 1953 from the Upper Villafranchian of Britain, *L. reynoldsi* Azzaroli, 1953 from Britain's Lower Pleistocene[1]), and *L. latifrons* (Johnson, 1874) from the Pleistocene of Eurasia. Later (Azzaroli 1979) he reduced *L. minor* to synonymy with *L. gallicus*, and *L. reynoldsi* to synonymy with *L. latifrons*. At present, Azzaroli (1981) is in favour of placing *Libralces* as a subgenus of the genus *Cervalces* on the basis of the similarities in the structure of their skulls. Viret (1961) included in the tribe Alcini all three genera: *Alces*, *Cervalces*, and *Libralces*. Heintz and Poplin (1981) bring all Eurasian elks into the genus *Alces* in which they include four species: *A. gallicus* (Azzaroli, 1952), *A. carnutorum* (Laugel, 1862), *A. latifrons* Johnson, 1874, and *A. alces* L.

The history of elks is usually described as a chronocline in which evolution is associated with the shortening of the beam and the development of the palmation. It is the morphology of antlers that attracts the majority of researchers of this group, who often neglect the craniological data. Attempts are made to use the known finds to elaborate a streamlined tine sequence of different types from the early ones with a long thin beam (*L. gallicus*) to the late ones with a shortened massive beam (*A. alces*). The details of antler morphology are made the basis of generic systematics. With such an approach errors are unavoidable, since the tendencies for the shortening of the beam and the development of the palmation may develop simultaneously in different deers; thus, for example, they are observed in megacerines.

An analysis of the literature, as well as the investigation of both fossil and recent osteological material, indicate that the degree of difference between *Libralces* and *Alces* corresponds to that between genera of the same tribe (e.g. *Cervus* and *Axis*) and is greater than the degree of difference between species of the same genus. Thus, the living species *C. nippon* and *C. elaphus* differ in details of colouring and antler structure. Quite significant are the following differences in the cranial

[1]) According to the British scheme, the species of "*L. reynoldsi*" and *L. latifrons* are early Middle Pleistocene (approximately Cromerian).

structure between *Libralces* and *Alces*: 1. The form of basicranium: in *Libralces* it is wide and long, while in *Alces* it is narrow and short; 2. The form of the forehead: in *Libralces* the forehead is relatively little concave, while in *Alces* it has a conspicuously saddle-like shape; 3. the length of the facial and rostral parts of the skull: in *Libralces* they are shorter than in *Alces*; 4. The length of the superior processes of the premaxillary bones: in *Libralces* they reach the nasal bones, while in *Alces* they do not contact the nasal bones; 5. The details of the basioccipital structure: in *Libralces* it is more primitive; 6. Degree of development of the palaeomeryx fold: in *Libralces* it is expressed on M_1 and occasionally also on M_2, whereas in *Alces* this fold is absent; 7. The orientation of antlers: in *Libralces* they are directed not only laterally but also slightly posteriorly, while in *Alces* they lie practically in a straight line. These differences, in my opinion, do not allow the referrat of all elks from Eurasia to the genus *Alces*, as was proposed by Heintz and Poplin (1981). Furthermore, I cannot agree with the suggestion of Azzaroli that the genera *Libralces* and *Cervalces* should be combined. It will be necessary to obtain further comparative data on the osteology of these genera, with most attention being given to investigations of the crania.

3. The Finds of Pliocene Elks in the USSR

Some elk remains have been found in the USSR in deposits more ancient than 0.7 mill. yrs.

In the European part of the USSR the remains of *Libralces* sp. were found in the vicinity of Rostov-on-Don in the Liventsov quarry (Baigusheva 1971). The locality belongs to the so-called Khaprov complex; the age of which is determined as Middle Villafranchian. We may regard Saint Vallier as its analogue in western Europe. From there, according to Baigusheva, originate a fragment of palmation, No. L-931, RSU, and a beam fragment of a cast antler, No. L-1221 RSU. The length of the beam from the burr to the palmation is 270 mm, the base diameters being 45 mm and 38 mm.

In Tadzhikistan unique elk remains, initially determined as Alcini gen. indet., were found at the Navrukho locality belonging to the Kuruksai complex, similar in age to the Khaprov one (Vislobokova 1980). They are represented by a fragment of antler broken off at the base of the palmation, upper premolars P^2 and P^4, a mandible with D_2-D_4, and a metatarsus (Plate I, Fig. 2). Antler measurements (mm): length of the beam from the burr to the beginning of the palmation — 300, horizontal and vertical diameters of the beam base — 53,5 and 45,0. Metatarsus measurements, No. 3848/219-69, GIN, (mm): length — 365,5, proximal epiphysis width — 47, distal epiphysis width — 63,5, diameter — 41,5. With respect to the dimensions and massiveness of the beam and the dimensions and proportions of the metatarsus, the Tadzhik elk is similar to *L. gallicus*.

In Western Siberia, in the Pavlodar area of the Irtysh river near Podpusk village, in deposits correlated with the Upper Villafranchian, there was found a fragment of elk antler broken off at the base of the palmation, No. 608/6, GIN. Its measurements (mm): length from the burr to the beginning of the palmation — 230, horizontal and vertical diameters of the beam base — 59,5 and 61,5. With respect to the length and massiveness of the beam it is close to *L. latifrons*.

The oldest elk from the USSR appears to be *Pseudalces mirandus* Flerov, 1962 from the Upper Pliocene of the Kosyakinsk quarry in Ciscaucasus (Flerov 1962; Flerov and Shevyreva 1963). *Pseudalces mirandus* was identified from an incomplete skull of a male (Plate I, Fig. 3). Its full description was given by Flerov and Shevyreva (1963). The Kosyakinsk fauna belongs to the Skotselsk complex which is correlated by Soviet researchers with the Montopoli fauna.

4. Comments on the Systematic Position of Pseudalces mirandus Flerov, 1962

Flerov and Shevyreva pointed out the following diagnostic features: a conspicuous elongation of the facial part of the skull, the narrowness and great length of the lacrimal-maxillary process, a conspicuous constriction of the frontals behind the orbits, and an "echelon" arrangement of the molars inclined outwardly. The pedicles in this deer are short and tend posteriolaterally and slightly superiorly. Flerov and Shevyreva presumed that *Pseudalces* belonged to a blind branch and was not an ancestor of the genus *Alces*. They indicate that in general traits it was similar to the genus *Eucladoceros*, widespread in the Eurasian Villafranchian, but the observed differences permit them to speak only of common ancestory. Following this remark, Heintz and Poplin (1981), in their article devoted to the description of *A. carnutorum* from the Pleistocene of Saint-Prest, France, as well as the problems of the systematics and evolution of elks, placed *Pseudalces* in the group *Eucladoceros*, excluding it from elks.

However, Flerov (1962) did not call the deer from the Kosyakinsk quarry "*Pseudalces*" by accident. This deer shows an indisputable affinity to elks because of certain morphological details of the cranium: an insignificant width of the post-antler part, which is not more than 2/3 of the postorbital width; the saddle-like form of the forehead; an insignificant length of the frontals and its thickening in the area of tuber frontale; direction and dorsoventral flattening of the pedicle; an elongated form of os lacrimale exceeding in length the antero-posterior orbital diameter; a narrow and long processus lacrimalis ossis zygomatici; and a great height of the choanae, as expressed in the elevation of the anterior ends of the frontal bones. In addition, some other features make it similar to elks: an "echelon" arrangement of the molars, their brachyodonty and conspicuous constriction from the base to the masticatory surface, and the large dimensions of P^4. In certain features it differs from the genera *Libralces* and *Alces*. These are a practically complete absence of a preorbital fossa on the lacrimal bone, and a slight conspicuousness of the tuber frontale which are presumably evidence of its primitiveness. At the same time, *Pseudalces* turns out to be more progressive compared to *Libralces* and *Alces* in such features as the elongation of the facial part, which is indirectly deduced from the conspicuous anterior displacement of the tooth row. The anterior orbital margin is removed from the posterior margin of the tooth row not less than half the length of M^3 (in projection on the longitudinal cranial axis).

Thus, it should be aknowledged that *Pseudalces mirandus* is the oldest (Early Villafranchian) representative of the elks in Eurasia. The details of its morphology do not allow it to be considered an ancestor of later genera.

5. Conclusions

Little is known about the early history of elks. Information on Pliocene elks from the USSR contributes to its understanding. It indicates that the time of origin of the alcini is more ancient that was believed. It is becoming clear that they originated not later than the Early Villafranchian. In the Middle Villafranchian the range of elks was already quite extensive — they were in existence in the south of European USSR and in Tadzhikistan. In the Late Villafranchian their remains have been recorded in Western Europe and Western Siberia. In the Early Pleistocene the range of elks covered the entire temperate zone of the Palaearctic.

The development of the alcines proceeded along several lines. This was presumably the result of a divergence which appears to have taken place not later than the Middle Pliocene. In the Late Pliocene these lines were represented by the genera *Libralces* and *Pseudalces*. The question of the phylogenetic relationships of *Libralces*, *Cervalces* and *Alces* calls for additional research, based on detailed comparative investigation of the skull.

Abbreviations:

PIN — Palaeontological Institute, USSR Academy of Sciences, Moscow
GIN — Geological Institute, USSR Academy of Sciences, Moscow
RSU — Rostov State University.

Summary

There exist in the USSR some elk finds from Upper Pliocene deposits (according to the USSR's officially recognized scheme). The oldest ones are remains of *Pseudalces mirandus* from the Kosyakin quarry in Ciscaucasus. The enclosing deposits are correlated by Soviet researchers with the Lower Villafranchian of Europe. Antler fragments of *Libralces* sp. are known from the Liventsov quarry near Rostov-on-Don, in deposits correlated with the Middle Villafranchian of Europe. Beam fragments, teeth, and a metatarsus of an elk similar to *L. gallicus* were found in deposits of a similár age in Tadzhikistan at the locality of Navrukho. In Western Siberia finds of an elk have been made in the Pavlodar area of the Irtysh river near Podpusk village, in deposits correlated with the Upper Villafranchian. The genus *Pseudalces* represents one of the lines of development of elk tribe Alcini.

Zusammenfassung

Aus der UdSSR liegen nur wenige Elchfunde aus oberpliozänen Ablagerungen (entsprechend dem offiziellen Geologie-Schema der UdSSR) vor. Die ältesten sind Reste von *Pseudalces mirandus* aus dem Steinbruch Kosjakin, Vorkaukasus. Die entsprechenden Fundhorizonte werden von den sowjetischen Wissenschaftlern mit den Ablagerungen des unteren Villafranchium Westeuropas korreliert.

Geweih-Fragmente von *Libralces* sp. sind aus dem Steinbruch Liventsov bei Rostov am Don bekannt, aus Ablagerungen, die dem Mittel-Villafranchium Westeuropas entsprechen. Stangen-Fragmente, Zähne und das Metapodium eines Elches, der *Libralces gallicus* entsprechen dürfte, stammen aus Ablagerungen etwa gleichen Alters von Navrucho, Tadžikistan.

In Westsibirien konnten fossile Elch-Funde im Raum Pavlodar am Irtyš, nahe der Siedlung Podpusk geborgen werden. Die Ablagerungen werden in das obere Villafranchium gestellt.

Das Genus *Pseudalces* vertritt eine der Entwicklungslinien der Elche des Tribus Alcini.

В СССР имеется ряд находок остатков лосей из отложений верхнего плиоцена (по официально принятой в СССР схеме). Наиболее древними являются остатки *Pseudalces mirandus* из Косякинского карьера Предкавказья. Вмещающие отложения советскими исследователями коррелируются с нижним виллафранком Европы. Фрагменты рогов *Libralces* sp. известны из Ливенцовского карьера близ Ростова-на-Дону из отложений, коррелируемых со средним виллафранком Европы. Фрагмент штанг, зубы, плюсневая кость лося, близкого к *L. gallicus* встречены в близких по возрасту отложениях в Таджикистане в местонахождении Наврухо. В Западной Сибири остатки лося встречены в Павлодарском Прииртышье у села Подпуск в отложениях, коррелируемых с верхним виллафранком.

Род *Pseudalces* представляет одно из направлений развития лосей.

References

AZZAROLI, A., 1952: L'alce di Sénèze. — In: Palaeontogr. Ital. — Pisa 47. — P. 133—141.

—, 1953: The deer of the Weybourne Crag and Forest Bed of Norfolk. — In: Bull. Brit. Mus. (Nat. Hist.), Geol. — London 2 (1). — P. 1—69.

—, 1979: On the occurrence of the cervid genus *Libralces* in the Pleistocene of Italy. — In: Palaeontogr. Ital. — Pisa 71. — P. 48—54.

—, 1981: On the Quaternary and recent cervid genera *Alces*, *Cervalces*, *Libralces*. — In: Bull. Soc. Palaeont. Ital. — Pisa 20. — P. 147—154.

BAIGUSHEVA, V. S., 1971: Iskopaemaja teriofauna Livenčovskogo kar'era (severo-vostočnoe Priazov'e) (Fossil theriofauna from the Liventsov quarry (north-eastern Azov area). — In: Tr. zool. Inst. — Leningrad 79. — P. 35—42.

FLEROV, C. C., 1962: Semnejstvo Cervidae. In: Osnovy paleontologii mlekopitajuščie (The family Cervidae. In: Principles of Palaeontology. Mammals). — Moskva. — P. 337—340.

FLEROV, C. C.; SHEVYREVA, N. S., 1963: *Pseudalces* — pliocenovyj olen'iz Predkavkaz'ja (*Pseudalces* — the Pliocene deer from the Cis-caucasus). — In: Paleont. Ž. — Moskva (3). — P. 120—123.

HEINTZ, E.; POPLIN, F., 1981: *Alces carnutorum* (LAUGEL, 1862) du Pleistocène de Saint-Prest (France). Systématique et évolution des Alciñes (Cervidae, Mammalia). — In: Quartärpaläontologie. — Berlin **4**. — P. 105—122.

SIMPSON, G., 1945: The principles of classification and a classification of Mammals. — In: Amer. Mus. Nat. Hist., Bull. — New York 85. — 350 Pp.

VIRET, J., 1961: Artiodactyla. — In: Traité de Paleontologie, Tome 6, 1. — Paris. — P. 887—1021.

VISLOBOKOVA, I. A., 1980: Oleni iz mestonachoždenija Kuruksaj. In: Granica neogena i četvertičnoj sistemy (Deer from the Kuruksai locality. In: The boundary of the Neogene and the Quaternary system). — Moskva. — P. 256—258.

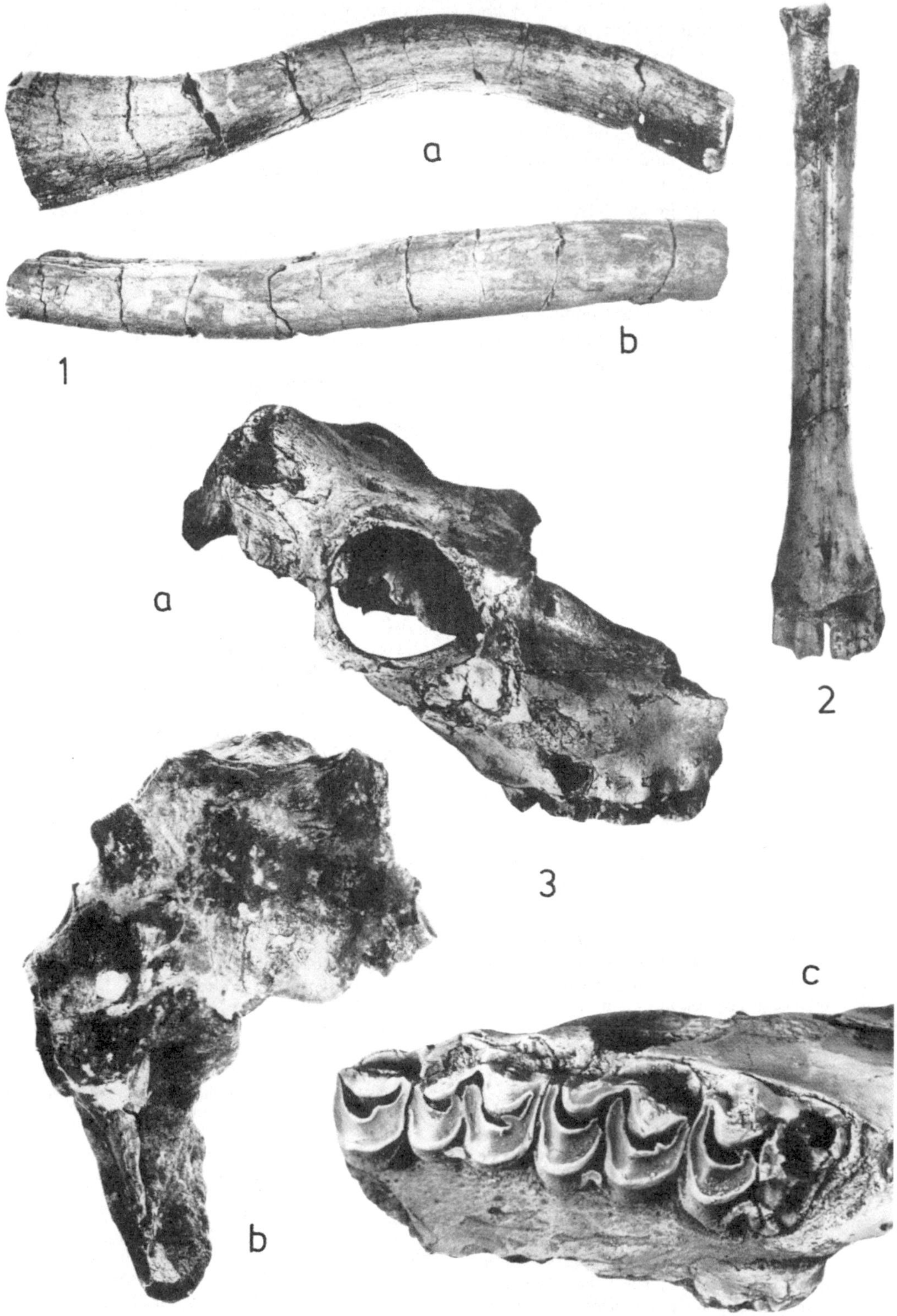

PLATE I

Fig. 1: *Libralces* cf. *gallicus* AZZAROLI, 1952, antler, spec. No. 3120-600, PIN. Tadzhikistan, Navrukho, Upper Pliocene, about 1/4 nat. size
a — ventral view; b — dorsal view

Fig. 2: *Libralces* cf. *gallicus* AZZAROLI, 1952, metatarsus, spec. No. 3848/219-69, GIN, frontal view. Tadzhikistan, Navrukho, Upper Pliocene, about 1/4 nat. size

Fig. 3: *Pseudalces mirandus* FLEROV, 1962, holotype, skull No. 255, PIN. Cis-Caucasus, Kosyakin quarry, Upper Pliocene
a — lateral view; b — view from above; about 1/4 nat. size; c — teeth, ventral view, about nat. size

Anschriften der Autoren

Prof. Dr. N. I. Burčak-Abramovich, Akademie der Wissenschaften der Grusinischen SSR, Paläobiologisches Institut, Potochnaja 4, Tbilissi, UdSSR

Dr. E. L. Dmitrieva, Akademie der Wissenschaften der UdSSR, Paläontologisches Institut, Profsojuznaja 113, Moskau 117321, UdSSR

Dr. R. A. Dubrovo, Akademie der Wissenschaften der UdSSR, Paläontologisches Institut, Profsojuznaja 113, Moskau 117321, UdSSR

Dr. M. A. Erbaeva, Akademie der Wissenschaften der UdSSR, Burjatische Zweigstelle der Sibirischen Abteilung, Geologisches Institut, Fabrichnaja 6, 670042 Ulan-Ude 42, UdSSR

Dr. I. V. Foronova, Akademie der Wissenschaften der UdSSR, Sibirische Abteilung, Institut für Geologie und Geophysik, 630090 Novosibirsk, UdSSR

Dr. A. Forsten, Universität Helsinki, Zoologisches Institut, P. Rautatiekatu 13, 00100 Helsinki 10, Finnland

Dr. P. L. Gibbard, Sub-Department of Quaternary Research, Botany School, Downing Street, Cambridge CB 2 3EA, Großbritannien

Prof. Dr. E. W. Guenther, Lehenhof, 7801 Ehrenkirchen bei Freiburg im Breisgau, BRD

Dr. W.-D. Heinrich, Museum für Naturkunde der Humboldt-Universität zu Berlin, Paläontologisches Museum, Invalidenstraße 43, 1040 Berlin, DDR

Prof. Dr. H. Hemmer, Universität Mainz, Institut für Zoologie, Saarstraße 21, 6500 Mainz, BRD

Dr. Dr. H. D. Kahlke, Institut für Quartärpaläontologie Weimar, Steubenstraße 19a, 5300 Weimar, DDR

Dipl.-Geol. R.-D. Kahlke, Institut für Quartärpaläontologie Weimar, Steubenstraße 19a, 5300 Weimar, DDR

Dr. B. S. Kožamkulova, Akademie der Wissenschaften der Kasachischen SSR, Zoologisches Institut, Paläobiologisches Laboratorium, Akademiestadt, 480032 Alma-Ata, UdSSR

Prof. Dr. B. Kurtén, Universität Helsinki, Institut für Geologie, Abteilung für Geologie und Mineralogie, Snellmaninkatu 5, 00170 Helsinki 17, Finnland

Prof. Dr. M. Malez, Jugoslavenska Akademija Znanosti i Umjetnosti, Radna zajednica istraživački centra, Zavod za paleontologija kvartara, ul. A. Kovacića 5, Zagreb YU 4100, SFR Jugoslavien

Prof. Dr. R. Musil, Universität Brno, Lehrstuhl für Geologie und Paläontologie, Kotlarská 2, 61137 Brno 2, ČSSR

Dr. T. Nötzold, Robert-Koch-Straße 18, 1144 Berlin, DDR

Dr. C. Radulescu, Institutul de Speologie, Str. Mihail Moxa 9, 78109 București 12, SR Romania

Dr. V. Ju. Rešetov, Akademie der Wissenschaften der UdSSR, Paläontologisches Institut, Profsojuznaja 113, Moskau 117321, UdSSR

Dr. P. Samson, Institutul de Speologie, Str. Mihail Moxa 9, 78109 București 12, SR Romania

Dr. A. V. Sher, Akademie der Wissenschaften der UdSSR, Institut für Evolutionäre Morphologie und Ökologie der Tiere, Lenin-Prospekt 33, Moskau 117071, UdSSR

Akademiker Prof. V. E. Sokolov, Akademie der Wissenschaften der UdSSR, Institut für Evolutionäre Morphologie und Ökologie der Tiere, Lenin-Prospekt 33, Moskau 117071, UdSSR

Dr. A. J. Stuart, University of Cambridge, Department of Zoology, Downing Street, Cambridge CB3 3EJ, Großbritannien

Dr. E. B. Sumina, Akademie der Wissenschaften der UdSSR, Institut für Evolutionäre Morphologie und Ökologie der Tiere, Lenin-Prospekt 33, Moskau 117071, UdSSR

Akademiker Prof. Dr. L. P. Tatarinov, Akademie der Wissenschaften der UdSSR, Paläontologisches Institut, Profsujuznaja 113, Moskau 117321, UdSSR

Dr. F. A. Tleuberdina, Akademie der Wissenschaften der Kasachischen SSR, Zoologisches Institut, Paläobiologisches Laboratorium, Akademiestadt, 480032 Alma-Ata, UdSSR

Prof. Dr. H. Tobien, Universität Mainz, Institut für Geowissenschaften, Saarstraße 21, 6500 Mainz, BRD

Dr. B. A. Trofimov, Akademie der Wissenschaften der UdSSR, Paläontologisches Institut, Profsojuznaja 113, Moskau 117321, UdSSR

Dr. V. V. Ukraintseva, Akademie der Wissenschaften der UdSSR, Botanisches Institut, Prof. Popov-Straße 2, Leningrad 197022, UdSSR

Dr. Ju. M. Vasilev, Akademie der Wissenschaften der UdSSR, Geologisches Institut, Pyževskij pereulok 7, 109017 Moskau, UdSSR

Dr. I. A. Vislobokova, Akademie der Wissenschaften der UdSSR, Paläontologisches Institut, Profsojuznaja 113, Moskau 117321, UdSSR

FSC
www.fsc.org
MIX
Papier | Fördert gute Waldnutzung
FSC® C083411

Zeitfracht Medien GmbH
Ferdinand-Jühlke-Straße 7
99095 Erfurt, Deutschland
produktsicherheit@kolibri360.de